CW00923934

The Weights and Measures of England

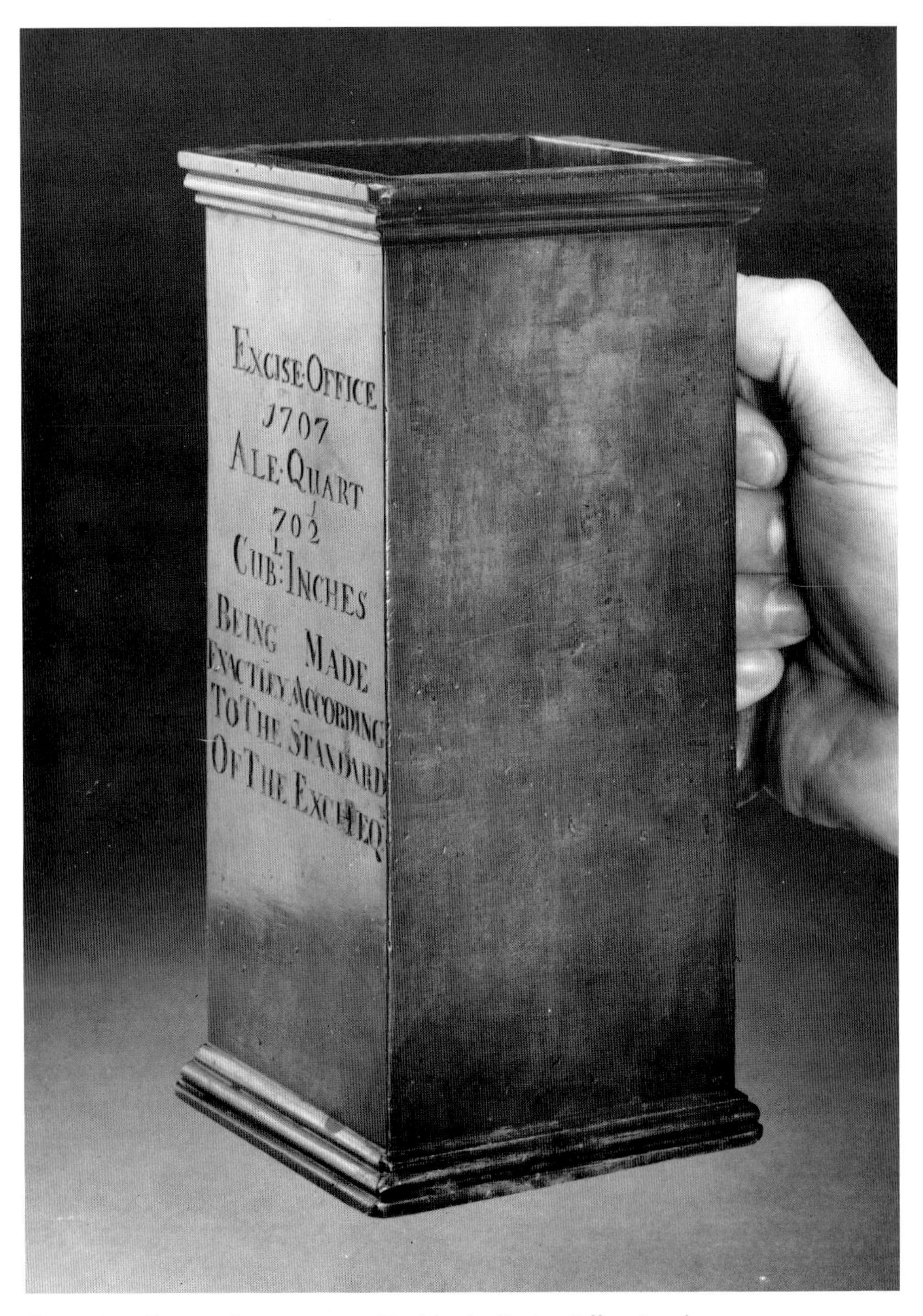

Frontispiece: Bronze ale quart, 1707. Used in the Excise Office, London.

Science Museum

The Weights and Measures of England

R.D. Connor

London: Her Majesty's Stationery Office

First published 1987

ISBN 0 11 290435 1

Printed in the United Kingdom for Her Majesty's Stationery Office
Dd. 738913 7/87 C18 46825

Assessment by weighing or measuring is a prime necessity of life in a human society. It is essential to the making and exchange of goods, to the erection of buildings and to the devices of transport. There is no commerce or industry of human beings but depends on it, no dealing in property but is defined by it. Without it there can be no civilization and no society but the primordial. It is the first essential tool of material creation and private and public economy are its dependants.

J. A. O'KEEFE
The Law of Weights and Measures
2nd Edition 1978
Introduction I

Contents

(Notes and references appear at the end of each chapter)

List of Illustrations xii
Abbreviations xiv
Reference to Statutes xv
The Kings and Queens of England xvii
Foreword xix
Acknowledgements xx
Introduction xxiii

I **From Early Beginnings to Roman Times** **1**
Body Measures *1*
Seeds as Weights *2*
The Early Coinage *5*
Greek Measures of Length *8*
Greek Capacity Measures *10*
Roman Measures *10*
The Roman Foot *10*
Roman Capacity Measures *12*
Roman Weight and Money *13*
British Currency Bars *16*
The Celtic Pound *18*
The First British Coinage *19*
Roman Coins in Britain *20*
Scales, Steelyards, and Bismars *21*

II **The Natural and Drusian Foot** **27**
The Natural Foot *27*
The Drusian Foot and the Pes Manualis *28*
The Shaftment *29*
The Historical Setting *31*

III **The Saxon Gyrd, the Rod, and the Acre** **35**
The Rod and the Acre, the Furlong and the Acre's Breadth *36*
The Gyrd and the Rod *39*
The Length of the Rod *43*
The Woodland Perch *45*
The Building Perch *46*

IV **The Hide and the Knight's Fee** **54**
The Hide *54*
The Sulung *56*

The Size of the Hide 57
The Knight's Fee 61

V **The Mile and the League** **68**
The Mile of the Saxons 68
The Length of the Mile 69
The Old English Mile 70
The League, English and Gallic 74

VI **The Shorter Linear Units** **79**
The Short Units 79
The Establishment of the Present Yard 83
The Foot of St. Paul's 85
The Yard and the Inch for Cloth 87
The Assize of Measures, 1196 90
The Measure of Cloth 91
The Ell 94

VII **The Early Currency** **100**
The Beginning of the English Coinage 100
The Shilling and the Sceatta 101
The Shillings of Mercia and Wessex 104
The Pound 106
The Introduction of the Penny and the Tower Pound 107
Procedures at the Mint 110
Manipulation of the Currency 113

VIII **The Origins of the Units of Commercial Weight** **117**
Troy and Apothecaries' Weight 117
Troy Weight in Saxon England 119
Troy Weight at the Exchequer 121
Troy and Tower Weight at the Mint: The 'Tractatus de Ponderibus et Mensuris' 123
The Libra Mercatoria 125
Haberty-poie Goods and Avoirdupois Weight 127
The Weighing of Wool 130
The Origin of Avoirdupois Weight 131
The Domestic Scene 132
Weighing in the Thirteenth and Fourteenth Centuries 133
The Abolition of the Auncel 138
Wool Weights 141

IX **Measures of Capacity** **149**
The Gallon, Bushel, and Quarter 151
The Measures of Henry VII 155
The Measures of Elizabeth I 159
The Ale and Wine Gallon 162

The Winchester Bushel 164

X **Particular Measures of Capacity mentioned by Statute** 170
The Tun of Wine 170
The Barrel 172
The Wirksworth Dish 177
Water Measure and Heaped Measure 178
Coal Measures 180
The Imputed Bushel 183
Apothecaries' Measure 185
The Reputed Quart 187
Capacity Measures of the United States of America 188

XI **The Assize of Bread and Ale** 193
The Assize of Bread 194
Troy Weight for Bread 197
Payments and Penalties 197
The Baker's Dozen 198
Enforcing the Assize 199
Fourteenth-century London 201
Setting the Assize 202
The Assize of c1500 202
The Flaw in the Assizes 203
The Assizes of 1600 and 1684 205
The Assize of 1710 207
The Assize of 1757 209
The Baker's Allowance, Eighteenth Century 210
Events at the Turn of the Century 211
The Nineteenth Century 214
The Twentieth Century 216
The Assize of Ale 220
The Ale-Conners 222
The Price of Ale 223
The Thurdendel 224
The Size of the Barrels 225
The Levying of Duty on Ale and Beer 226
The End of the Assize 226

XII **The Physical Standards** 232
(a) The Establishment of English Standards to Henry VII 232
Measures 232
Weights 233
Yards 234
Merchant Taylors' Yard 234
The Mercers and Drapers Company 236
(b) The Issues of Henry VII 237
The Winchester Yard 238

Henry VII's Exchequer Yard 239
Measures of Henry VII 239
The Weights of Henry VII 239
(c) The Issues of Elizabeth I 240
Weights 241
Measures of Capacity 243
(d) Progress from 1588 to 1760 243
Yards 243
The Carysfort Committee 246
(e) Events 1760 to 1834 249
General Roy's 42-inch Scale 249
The Shuckburgh Scale 250
The Committee Report of 1814 251
The Commission Reports of 1819, 1820, and 1821 253
The 1821 Report of the Select Committee 254
The Weights and Measures Act, 1824 255
Construction and Adjustment of the Principal Standards 257
(f) The Standards Restored 261
The Report of 1841 262
The Production of the Yard 264
The Production of the Pound 267
The Weights and Measures Act, 1878 271
Present Location of the Standards 272

XIII **The Onset of Metrication: The Nineteenth Century** **279**

XIV **Britain Goes Metric: The Twentieth Century** **289**
The Hodgson Report, 1950 289
The Weights and Measures Act, 1963 296
The Report of the Halsbury Committee, 1963 298
The Move to Metrication 301
The Metrication Board 302
The Weights and Measures Act, 1985 306
A Personal Comment 307

Appendix A *(a) Assisa Panis et Cervisie* **312**
(b) Judicium Pillorie 315
(c) Statutum de Pistoribus 317
(d) Tractatus de Ponderibus et Mensuris 320
(e) Statutum de Admensuratione Terre 322

Appendix B Regulation and Enforcement **323**
The Leet Courts 324
The Clerk of the Market 325
The Livery Companies 329
The Coopers 329

The Plumbers 329
The Founders 330
The Fruiterers 331
The Inspectorate 331
The Survival of the Eight-pound Stone 335
The Modern Period 336
The National Weights and Measures Laboratory (NWML) 337
The Institute of Trading Standards Administration 338

Appendix C A Brief Account of the Development of the Metric System **344**
The Metre 346
The Optical Metre 351
The Kilogram 352
The Litre 354

Appendix D Tables of Pre-metric British Measures **358**
Metric Equivalents 359

Glossary of Unit Terms 361
Bibliography 373
Index 388

List of Illustrations

Frontispiece: Bronze ale quart, 1707. Used in the Excise Office, London.
1 Leonardo da Vinci's Vitruvian Man.
2 Measuring land in Egypt c1400 BC.
3 Replicas of Egyptian royal cubits 2500 BC — 1st century AD.
4 The length of the barley corn.
5 Egyptian stone weights c2900–2000 BC.
6 Bronze 'ox-hide' ingot c1200 BC.
7 The Arundel relief, Ashmolean Museum, Oxford (c450 BC).
8 Roman footrules, second century AD.
9 Roman *denarii*.
10 Iron currency bars from the Winchester area.
11 Silver 'axe-head' ingot c400 AD.
12 Roman scales and weights AD. 79 (Pompeii)
13 Roman-Etruscan steelyard.
14 (a) Weighing scene, Sweden, sixteenth century, showing a bismar in use.
(b) Norwegian bismar, sixteenth century.
15 Malaysian bismar, twentieth century.
16 Steelyard counterpoise weights.
17 Surveyor's rule showing inscription '3 barley corns = 1 inch', 1574. (detail)
18 Diagram of acre and rood.
19 Tarrage map, Manor of Godbegot, Winchester, 1416.
20 Manor of Godbegot, 1980.
21 Determination of the German land rod, 1531.
22 Field system of village of Whitehill, Oxfordshire, 1605.
23 Histogram of the length of William Worcestre's itinerary mile.
24 Public bronze plaque of linear measure,
25 A *sceatta* and a penny.
26 Weight variation of silver penny c960 — 1090 AD.
27 Lead weight, 8 ounce Troy, eighth or ninth century.
28 (a) Winchester City weights of Edward III.
(b) Detail of 56 pound weight.
29 Edward I, 6¼ pound Avoirdupois clove.
30 Market scene, fifteenth century. From a window in Tournai Cathedral, Belgium.
31 Seven-pound wool weight, Henry VII.
32 Twenty-eight pound wool tod, James I.
33 Fourteen-pound wool weights, George I and III.
34 Exchequer bushel and gallon, Henry VII 1497.
35 Exchequer bushel, gallon, quart, and pint, Elizabeth I, 1601-2.
36 Derby City's 1601 pint and its leather case.
37 Exchequer wine gallon, Queen Anne, 1707.
38 Standard quart and pint, William III, City of Winchester, 1700.
39 The Wirksworth Dish.
40 Exchequer standard coal bushels of George II, 1730.

41 Drawings from *Liber Albus:*
(a) A baker at work.
(b) A baker being drawn on a hurdle through the streets of London with the offending loaf about his neck.
(c) The pillory.
42 A bread weight, quartern loaf, 4 lb 5 oz 8 dr Avoirdupois, 1801.
43 Churchyard wall, Great Wishford, Wiltshire showing the price of bread since 1800.
44 A steelyard for weighing 4 and 2 lb loaves.
45 The silver yard of the Merchant Taylors' Company and officers of the Company.
46 (a) Winchester's bronze yard of Henry VII.
(b) Detail of 'h' end.
(c) Detail of 'E' end.
47 (a) Exchequer standard yard, Henry VII.
(b) Henry VII at the assize of weights and measures, 1496.
48 Avoirdupois weights 112 lb — 2 lb Henry VII, 1497.
49 Exchequer standard Avoirdupois weights, Elizabeth I: First Series, 1558.
50 Standard Avoirdupois weights of Elizabeth I:
Third Series 1588. Bell-shaped 56 lb — 1 lb.
Disk-shaped 8 lb — 2 drams.
51 Standard Avoirdupois (disk) and Troy weights (cup):
Elizabeth I Third Series 1588.
Disk, 8 lb — 2 drams Avoirdupois.
Cup, 256 oz — 1 oz and 10, 5, 2, and 1 dwt Troy.
52 Conical standard measures, quart, gallon, pint, George IV, 1824.
53 Imperial standard bushel, 1824.
54 Imperial standard yard (1845) and pound (1844).
55 (a) Butter scales, nineteenth century.
(b) Egyptian scales of similar design, c2500 BC.
56 Tresca's winged section for metre bar.
57 Standard kilogram.
58 French *'cadil'*, 1793-4.

Sources of Illustrations.

Author: 1, 4, 7, 15, 18, 19, 20, 23, 25, 26, 43.
The Avery Museum: 44
The British Library: 21
Trustees of the British Museum: 6, 9, 11
Derby Museums and Art Gallery: 36
Merchant Taylors' Company: 45
Museum of London: 8, 42
National Library of Scotland: 14(a), 22, 41
Trustees of the Science Museum: Frontispiece, 2, 3, 5, 12, 13, 14(b), 16, 17, 24, 29, 30, 31, 32, 33, 34, 35, 37, 39, 40, 47(a)(b), 48, 49, 50, 51, 52, 53, 54, 55, 56, 57, 58
The Streeter Collection, Yale University: 27
Winchester City Museums: 10, 28(a)(b), 38, 46(a)(b)(c)

Abbreviations

Throughout the references to Anglo-Saxon charters, as a check to authenticity, the corresponding listing in SAWYER, P. H. *Anglo-Saxon Charters* London: Royal Historical Society 1968 is given; e.g. BIRCH, Walter de Gray *Cartularium Saxonicum* Vol. 1 p. 108 No. 74 of AD 688-90 (SAWYER No. 252).

The frequently quoted *Statutes of the Realm* (11 vols) London: HMSO 1810-28 are abbreviated to *Statutes*, e.g. Assisa de Ponderibus et Mensuris *Statutes* Vol. 1 p. 204.

Cambridge University Press and Oxford University Press appear as C.U.P. and O.U.P. respectively while N.S. means New Series.

Antiq. J.	Antiquaries Journal
Brit. Numis. J.	British Numismatic Journal
Econ. Hist. Rev.	Economic History Review
E.E.T.S.	Early English Text Society
Eng. Hist. Rev.	English Historical Review
Eng. Hist. Soc.	English Historical Society
Geog. J.	Geographical Journal
Numis. Chron.	Numismatic Chronicle
Phil. Mag.	Philosophical Magazine
Phil. Trans. Roy. Soc.	Philosophical Transactions of the Royal Society of London
Proc. Prehist. Soc.	Proceedings of the Prehistoric Society
Proc. Roy. Soc.	Proceedings of the Royal Society, London
Proc. Roy. Soc. Edin.	Proceedings of the Royal Society of Edinburgh
Proc. Soc. Antiq. London	Proceedings of the Royal Society of Antiquaries of London
Roy. Hist. Soc.	Royal Historical Society
Trans. Essex Arch. Soc.	Transactions of the Essex Archaeological Society

Reference to Statutes

A large number of Statutes are cited in this book. To facilitate reference for users, the citation is normally given as in the *Chronological Table of the Statutes* (C.T.S.) published by HMSO on behalf of the Statutory Publications Office, where this differs from previous listings. For old Acts, the most important part of the citation is the regnal year and chapter number – e.g. *19 Geo III, c.3* (an Act on Malt Duties of 1770). In the actual documents, the chapter number may be printed in Roman numerals and the regnal year in Latin, but the citation is given in all cases as above. Variations in numbering and dating between the different sets of Statutes exist, and are normally footnoted and rehearsed in the Table of Variances given in the *Chronological Table.*

It is useful to remember that from 1797 (38 Geo III), Acts are divided into *Public and General,* the Acts of each regnal year being numbered in Arabic figures, and *Local and Personal,* numbered in Roman figures. Regnal years are not quoted for Acts later than 1962, when the citation method was changed to that of calendar years.

To assist the reader, the location of a Statute in the particular set or series by the author may be given in brackets: readers should be aware that other series may serve them every bit as well.

Modern Statutes have both a long title and a short title, the former stating in some detail the purpose of the Act, the latter encapsulating in a few words its general subject matter. This has been the case only since the mid-nineteenth century: the Short Titles Act 1896 (59 & 60 Vict, c 14) gave short titles to 2076 earlier Acts which previously had only cumbersome long titles. By no means every previous Act was so dealt with. In this book, the actual or so-allocated short title is used where available; if not, an ad-hoc short title is used in the body of the text. This may, of course, not appear in all listings.

Some Acts of the eighteenth century and before covered more than one subject. For instance, the Statute of 1729, 3 Geo II c. 29, is noted in the Chronological Table merely as 'Price of Bread etc' but in fact covers the price and weight of bread, relief of debtors, frauds committed by bankrupt persons, Papists' Will registration, poor-law settlement certificates, relief of Protestant lessees, and removal of migrant paupers. These are known as 'omnibus Acts' and can cause considerable confusion. Readers checking certain of the Statutes cited in this book are therefore advised to search further, using the regnal year and chapter number given, even if the description given in the *Chronological Table* does not at first appear apposite.

Parliamentary Papers

All references to nineteenth-and twentieth-century papers are to the House of

Commons set of Parliamentary Papers divided by Session. They are given in the form *Parl. Papers 1950-51 (Cmd 8178) vii, p. 162,* which means the paper will be found in the sessional set for 1950-51, in Volume vii, at page 162 onwards. The number in brackets, given when available, refers to the original paper number, in case a non-standard unbound set is used. Readers will most often find the set available in libraries is the standard one, or its microform edition issued by Chadwyck-Healey Ltd.

The Kings and Queens of England

West Saxon Kings

Egbert	802 – 839
Ethelwulf	839 – 857
Ethelbald	857 – 860
Aethelbert	860 – 866
Aethelred I	866 – 871
Alfred (the Great)	871 – 899
Edward (the Elder)	899 – 924

Kings of England

Aethelstan	924 – 940
Edmund	940 – 946
Edred	946 – 955
Edwig	955 – 959
Edgar	959 – 975
Edward	975 – 978
Aethelred II (the Redeless; the Unready)	978 – 1013 1014 – 1016
Edmund (Ironside)	1016
Edward (the Confessor)	1042 – 1066
Harold II	1066

Danish Kings

Swein	1013 – 1014
Cnut	1016 – 1035
Harold I (Harefoot)	1035 – 1039
Harthacnut	1039 – 1042

Norman Kings

William I (the Conqueror)	1066 – 1087
William II (Rufus)	1087 – 1100
Henry I	1100 – 1135
Stephen	1135 – 1154

Plantagenet Kings

Henry II	1154 – 1189
Richard I (the Lionheart)	1189 – 1199
John	1199 – 1216
Henry III	1216 – 1272
Edward I	1272 – 1307
Edward II	1307 – 1327
Edward III	1327 – 1377
Richard II	1377 – 1399

House of Lancaster

Henry IV	1399 – 1413
Henry V	1413 – 1422
Henry VI	1422 – 1461
Henry VI restored briefly	1470 – 1471

House of York

Edward IV	1461 – 1470
	1471 – 1483
Edward V	1483
Richard III	1483 – 1485

The Tudors

Henry VII	1485 – 1509
Henry VIII	1509 – 1547
Edward VI	1547 – 1553
Mary	1553 – 1558
Elizabeth I	1558 – 1603

The Stuarts

James I (VI of Scotland)	1603 – 1625
Charles I	1625 – 1649

The Commonwealth

(Cromwell)	1649 – 1660

The Stuarts Restored

Charles II	1660 – 1685
James II	1685 – 1688
William III & Mary	1689 – 1694
William III alone	1694 – 1702
Anne	1702 – 1714

House of Hanover

George I	1714 – 1727
George II	1727 – 1760
George III	1760 – 1820
George IV	1820 – 1830
William IV	1830 – 1837
Victoria	1837 – 1901
Edward VII	1901 – 1910

House of Windsor

George V	1910 – 1936
Edward VIII	1936
George VI	1936 – 1952
Elizabeth II	1952 –

Foreword

To many people weights and measures may seem to be a dry and impersonal subject, concerned only with numerical relationships, but, as Professor Connor makes clear in this very readable book, this is far from the truth. In fact English metrology is a very personal subject as its origins are intimately connected with the human body and its activities. For example, the smaller units of length, such as the inch, the foot, and the fathom, are related to the size of various parts of the body, and some of the larger units are a legacy of agricultural practices which have long ceased. The perch was used to lay out the strips of land in the open field system and the furlong was the length of the furrow which a team of oxen could conveniently plough between rests. In general, each unit came into use because it was the most convenient for a particular purpose and this led inevitably to a great multiplicity of units. There were, for example, different volume measures for different commodities, such as the corn bushel and the coal bushel, the wine gallon and the ale gallon. When we take into consideration the many regional variations, we are left with a bewildering array of units through which Professor Connor skilfully guides us.

In the past there has been a regrettable tendency for some historical metrologists to ascribe a common origin to units widely different in time and place of use, merely on the basis of a similarity in size. This is what Professor Grierson in his Stenton Lecture on English Linear Measures has aptly described as 'mathematical romanticism and diffusionism run mad'. I am pleased to say that it does not feature in this book.

The person who is interested in the weights and measures of England is now well served, by this book, by the enlarged edition of Professor Zupko's *Dictionary of Weights and Measures for the British Isles,* and by Maurice Stevenson's work on the marks on English weights and measures in the revised *Weights and Measures,* by J. T. Graham. We can also look forward to the appearance, some time in the future, of Professor Connor's history of the weights and measures of Scotland which will be the second volume in his projected history of the metrology of the British Isles.

I am very pleased that Professor Connor has been able to make such good use of our unrivalled collection of English weights and measures and it is fitting that this book should be published by the Science Museum.

DENYS VAUGHAN,
Curator of Weighing and Measuring,
Science Museum, London.

Acknowledgements

Thanks are due to many people, agencies, and institutions for their contributions to this work. As sources of documentary material, the libraries of the University of Manitoba, the National Library of Scotland, the Library of the Science Museum, London, and that of the Trading Standards Administration (T.S.A.) (University of Sussex) have been pre-eminent, while for artifacts and information relating thereto, the Science Museum, the British Museum, the Museum of London, and the Winchester City Museums possess by far the greatest number of the standards referred to in this book and assistance and information has always been readily obtainable from all of them. In addition the significant collections at York, Chester, Derby, Eastbourne, Canterbury, Norwich, Oxford, Cambridge, Verulamium, and St. Albans must be recognized — a list which, though extensive, is far from exhaustive.

Thanks are also due to the University of Manitoba for the grant of two sabbatical leaves to undertake this work and to the Research Board of that University for the provision of a summer assistant and a grant for typing. I am indebted to the Social Sciences and Humanities Research Council of Canada for a research grant 1979-80 in support of this work.

Especial mention must be made of the help willingly given by Mr George Turnbull, former Director of Trading Standards, Highland Region, in the early stages of this work and to Dr Derek Keene, Miss Elizabeth Lewis, Dr Denys Vaughan, Mr Maurice Stevenson, and Professor Philip Grierson. Dr Keene, formerly with the Winchester Research Unit, was of great assistance in the interpretation of Anglo-Saxon data from Winchester. Miss Lewis, Curator of the Winchester City Museums, and Dr Vaughan of the Science Museum have been continuing sources of information over the years, especially with reference to matters relating to their museums' very considerable collections of standards. Dr Vaughan has read and offered valuable suggestions on the final draft of the text which have been especially useful in the light of his detailed knowledge of the subject in general and of the historical standards in his care in particular. Mr Stevenson, Hon. Librarian, Trading Standards Association, and Professor Grierson of Gonville & Caius College, Cambridge, made available reprints of their writings and gave full access to their files of notes relating to the subject. They have each read two drafts of the entire book and have saved me from many errors. The extent of my indebtedness to them is reflected in the frequency of reference to their published work. I am especially indebted to Professor Grierson for sharing his vast knowledge of the early and mediaeval period with me; Mr Stevenson has been most helpful for the post-medieval period, but, nonetheless, any remaining sins of omission or commission are entirely the responsibility of the undersigned.

Dr Chris Pond, Head of the Public Information Office, House of Commons, has also read the complete text. He has collated the references to the Statutes with the *Chronological Tables of the Statutes* (C.T.S), 1981, London, published by HMSO for Statutory Publications Office, and has advised as to the best method of referring to reports of Parliamentary committees and commissions of the nineteenth- and twentieth- centuries. I am grateful to him for sharing with me his profound knowledge of Parliamentary matters.

The index was prepared by Mrs Judy Batchelor. I offer her sincere thanks for bringing her scholarship and careful assessment to this important part of the book.

Thanks must go to Mrs N. Laxdal for demonstrating yet again her expertise in transforming what often was an illegible manuscript into a beautifully typed document for the press, and to Ms Pippa Richardson and Mr John Mason, who, as editors, have been unfailing guides and supports throughout the entire process from typescript to printed page. Permission to quote is gratefully acknowledged from the following publishers.

Cambridge University Press:	Robertson, A. J. (ed.) *Laws of the Kings of England from Edmund to Henry I,* Cambridge 1925. Attenborough; F. L. (ed.) *The Laws of the Earliest English Kings,* Cambridge 1922.
Clarenden Press, Oxford:	Downer, L. J. (ed. and tr.) *Leges Henrici Primi,* Oxford 1972. Oschinsky, Dorothea (ed.) *Walter of Henley and other Treatises,* Oxford 1971.
Daily Telegraph:	News item, 9 June 1980, p.2.
Essex Archaeological Society:	Nichols, J. F. (ed.) The Extent of Lawling in the Custody of Essex AD 1310. *Trans. Essex Arch. Soc.* 1933, Vol XX, pt. 2, Colchester.
Jarrold Colour Publications:	Hudson, Rev. William and Tingley J. C., *The Records of the City of Norwich,* Norwich and London, Jarrold and Son 1906.
Thomas Nelson & Son Ltd. and Oxford University Press	Johnson, Charles (ed. tr.), *The De Moneta of Nicholas Oresme and English Mint Documents,* London and New York, Nelson 1956.

Routledge and Kegan Paul:	Harris, M. D. (ed.), *Coventry Leet Book,* London. K. Paul, Trench, Trübner and Co. for E.E.T.S. 1907-13.
Royal Historical Society	Gibbs, Marion (ed.) *Early Charters of the Cathedral Church of St. Paul,* London, Camden Third Series Vol. LVIII, London 1939. Hall, Hubert and Nicholas, Freda J. *Select Tracts and Table Books Relating to English Weights and Measures (1100-1742),* Camden Miscellany Vol. XV, London 1929.
Selden Society:	Richardson H. G. and Sayles G. O. (eds. and trs.), *Fleta;* Pub. 72, London, Selden Society 1955.

To all I am very grateful.

R.D. CONNOR,
University of Manitoba,
Winnipeg, Canada.

Introduction

The purpose of this book is to endeavour to bring together the substance of some of the highly diffuse literature on the subject of the weights and measures of the land we now call England. The title would suggest a beginning date in about the mid-tenth century, when the unification of England had been completed, but to commence so late would leave many origins unexplained. The influence of the Romans on our metrology was considerable and the Romans had borrowed much from the Greeks, so it is with them that the story should commence. Nor shall we be overly concerned with the name of the land whose metrology is to be discussed. We will call it England. To speak of Britain would be misleading because the weights and measures of Scotland, Wales, and Ireland will not be considered here except when a brief reference would illustrate an aspect of those of England.

Much of what has been written on this subject is not in a form accessible to the general reader, being scattered throughout the chronicles, statutes, cartularies, journals, conference proceedings, and books in a variety of languages. There is no work recognized as standard on the history of English weights and measures. It is hoped that this book, the outcome of some thirty years of study of the subject, may be a step towards the realization of that end sometime in the future.

When considering origins, the study inevitably leads into law and trade and the sciences of anthropology, economics, numismatics, and archaeology, for during the period in question (over 2,000 years) the exchanging of gifts, ceremonial and otherwise, gave way to general barter which in turn yielded to trade. The need for a medium of exchange emerged, as did the need for a money or token of account. From the beginning money and weight have been inextricably bound up. It is not an accident that the unit of weight and the unit of currency in England are both called the pound.

Sources for a work such as this are of three kinds: material from legislative acts, charters, and other historical documents; surviving physical standards; and inferences drawn from sizes of buildings, the measures in use in other countries, etc. The three kinds are not of the same degree of acceptability. Inferences, unsupported by other information, can be misleading if not actually erroneous. Surviving standards, it should be remembered, were not constructed to the same degree of accuracy as their twentieth-century counterparts and it has long been recognized that not everything written down is necessarily correct. Ambiguities and inconsistencies even among the statutes have been a constant source of difficulty. All sources require verification and validation before acceptance. The plan of this book is to make the verified documents speak from themselves, but the written record can reveal only so much. It is invaluable when it interrelates an unknown measure with one the magnitude of which can be determined today. When the interrelations are missing or inconsistent, help can often come from the discovery of coin hoards, ancient weights, artifacts, and the like recovered from the earth by the plough or by excavation. A case in point is the late Roman pound. The written record says that the *solidus* of Constantine was minted in the early fourth century AD at 72 to the pound.[1] Coin hoards of *solidi* discovered in near-mint condition when weighed have enabled the weight of this pound to be determined and accepted as 327.45 grams,[2] although to claim five significant figures is quite unwarranted.[3]

Legislation on weights and measures in England runs for more than ten centuries to the present. Saxon laws, for example those of the fifth and sixth group of Aethelred (c.1000–1008): 'deceitful deeds and hateful injustices shall be strictly avoided, namely untrue weights and false measures and lying testimonies' (V Aethelred cap 24), and 'weights and measures shall be corrected with all diligence and an end put to unjust practices' (VI Aethelred, cap 32.2),[4] repeated the biblical injunctions against false weights and measures, the great to buy with and the small sell by, and proclaimed that throughout the land there should be equity in trade for all — injunctions and proclamations that were steadfastly ignored until the nineteenth century when an inspectorate was set up with real power to enforce the law.

Nor should it be forgotten that the history of weights and measures is an integral part of the history of the nation. It is far from being a footnote to the 'real' history of the doings of kings and statesmen, of battles and conquests. It is a part of the true record of a people, reflecting as it does their response to changing circumstances as they went about their daily lives throughout the centuries. If ever a subject could be called 'inter-disciplinary', this is it.

The available monographs on weights and measures fall into one of three categories: those giving practically no bibliography or references for the authority of their statements; those giving extensive bibliographies and few references, leaving the student to decide which of a great many works should be consulted in the hope of finding the further information needed; and those which give detailed references. In this book the third course is followed, placing the authorities at the end of each chapter so as not to interrupt unduly the flow of the narrative, yet enabling the primary or a good reliable secondary source to be found if required. In this way, too, the indebtedness of the writer to earlier authors will be made plain, while at the same time, it is hoped, preventing him from engaging in those flights of fancy which have so bedevilled the work of some whose writings have appeared in print.

Two of the less damaging pitfalls could perhaps be mentioned. A common error is to suppose that the meaning of a word today is the same as it had thirteen or more centuries ago. The Anglo-Saxon words *gyrd* (yard) and *scilling* (shilling), to give but two, indicate that this is not so, as will be seen later. Secondly, when a book (or a museum card!) gives the value of some early measure in terms of our present units to five or six places of decimals or eight significant figures one can be reasonably sure the writer has a calculator giving that number of digits. One of the worst examples of this must be the relation between the cubic metre and the cube of the French foot given in the late eighteenth century, namely: 'Le metre cube vaut en pieds cubes 29.202689827820139912 exactement'.[5] No standard ever achieved anything approaching that kind of accuracy. Early measures were essentially approximate and there was no need for them to be otherwise. Did it matter whether in a basket of corn there were one or two dozen grains more or less?

A more serious error occurs when an investigator encounters two measures which are approximately equal and proclaims their identity, claiming the one being derived from the other regardless of the fact that in distance they may have been continents apart and a millenium or two or three may separate them in time. Relationships like this must be proven by historical connection, not simply by measurements which show approximate agreement. Standards of one country have influenced those of another, especially those of its neighbours, but only when historical links are known to exist by the written record or by extant specimens of known provenance can one feel on safe ground.

Systems of metrology prior to those of the present day had regard more to the division of an article or unit than to ease of calculation. The former took place all the time whereas computation was the province of the few. It is therefore not surprising to find yards divided in binary fashion to the *nail* of $2\frac{1}{4}$ inches, the gallon similarly divided via the *pottle* ($\frac{1}{2}$ gallon) and quart ($\frac{1}{4}$ gallon) to the pint ($\frac{1}{8}$ gallon) and down to the quarter gill and sets of weights, such as those of Edward III, ranging from 56, 28, 14 to 7 lb with a 91 lb weight for the special purpose of weighing wool (it being one quarter of the sack of 364 lb, the half-sack being 182 lb). There are no pre-metrication decimal systems in England. In referring to old units, therefore, while metric equivalents will be given, the units will be discussed in their original context of aliquot parts. To do otherwise would be to obscure completely the interrelationships that existed and which are of importance. Thus the *mark* will be referred to as 8 Troy ounces and the Troy pound as 12 such ounces. To ignore this and to speak only of 248.9 g and 373.3 g respectively might not indicate immediately to everyone that of these two weights, the lighter is two-thirds of the heavier.

As we shall see, the earliest weights were seeds but coins too were used as is evidenced by the words 'penny' (coin) and 'pennyweight' (a unit of Troy weight). This indicates a careful control of the coinage, for were coins to be issued with variable weight, the heavy would soon disappear, to be melted down for their bullion content, leaving only the light in circulation.

A feature of English metrology is the vast amount of legislation which surrounds it. From early centuries kings and parliaments alike have sought to regulate and stabilize the weights and measures of the realm. The frequency of these laws indicates that they were largely ineffectual, for quite apart from deliberate fraud, the difficulty in making an accurate measure from a pattern was formidable. Even with accurate measures and weights there have always been the few who for small gain were prepared to risk jail or worse for giving light measure. As a small boy I can recall being very daring and peeping in the door of the local police station. There, in the main hall, stood the weighing machine, the arbiter of all complaints of light weight. And in those days what coal merchant dared to omit to carry his scales on his lorry lest he be challenged by an inspector to show that his bags weighed 112 lb?

These days are now gone and for a long time we have enjoyed fair trading on the widest of fronts, yet this happy state of affairs was not reached without effort. As we trace the history of our metrology from the beginning we shall have ample evidence of that effort which ensured that the exchange of goods was equitable, with the consumer relying ultimately on kingly support of his claim for justice in the market-place.

NOTES AND REFERENCES: INTRODUCTION

1. Isidore Hispalensis Episcopi, *Etymologiarum sive originum libri XX,* LINDSAY, W. M. (ed.) Vol. 2 Book XVI, XXV. 14. Oxford: Clarendon Press 1911.
 'Solidum apud Latinos alio nomine sextula dicitur, quod his sex uncia compleatur'.
 (The Roman *solidus* is called a *sextula,* six of which make an ounce.)
 At 12 oz to the Roman lb the *solidus* would weigh $\frac{1}{72}$ lb.
 (Isidore, Bishop of Seville, 560-636).
2. GRIERSON, Philip. Presidential Address, Weight and Coinage, *Numis. Chron.,* (1964) Vol.IV, pp. iii-xvii. London. See p. xiii.
3. GRIERSON, Philip. Presidential Address, Coin Wear and the Frequency Table. *Numis. Chron.,* (1963) Vol.III, pp. i-xvi. London.
4. ROBERTSON, A. J. *Laws of the Kings of England from Edmund to Henry I,* pp. 87, 101. Cambridge: C.U.P. 1925.
5. Anon. *Instruction sur Les Mesures* Poitiers: La Commission Temporaire des Poids et Mesures republicaines, Michel-Vincent, An. II (1793-4). See last (unnumbered) page.

CHAPTER I

From Early Beginnings to Roman Times

When primitive man needed to measure he used what he had. For linear measure there were the parts of the body: the cubit, the foot, the palm, the hand, the thumb, and the finger are all mentioned in the written record, some continuing in use to this day. If he had pottery then the measure was his bowl. He had no weights, but by picking up two objects, one in each hand, he could tell which was the heavier. He judged weight by thus poising the object in his hand and surprisingly the same practice was still going on in trade in the eighteenth century. The first edition of the *Encyclopaedia Britannica* (1771)[1] tells us that 'meat was sold by the hand without scales' and Derby Museum has on display a 4lb weight from a set made for, and inscribed, 'Henry Flint, Mayor, Derby; Meymott and Porter Fecit 1787'. The card reads:

> In May 1787 Henry Flint inserted an advertisement in the Derby Mercury to the effect that butcher's meat was being weighed by the hand and the 'lower class of inhabitants' had suffered impositions and he had set up in the Anti [sic]-Chamber of the Town Hall a pair of scales and standard weights where at any time persons could have their purchases weighed by Edward Broughton, Town Crier, without fee.

It had not been the custom to weigh meat. The Statute 24 Henry VIII c 3 of 1532 decreed that beef, pork, mutton, or veal should be sold by lawful weight, 'Haber-de-pois' (Avoirdupois), because of the very high price of victuals. This was ignored so the Act 25 Henry VIII c 1 of 1533 provided means for enforcement. However 27 Henry VIII c 9, 1535 suspended both Acts from 12 April 1536 to 24 April 1540, because 'the dearth of victuals had ceased'. The next year the suspended Acts were repealed upon petition of the butchers who claimed that, if compelled to sell flesh by weight, 'they should be utterly undone for ever'. Thus the custom of estimating weight without weighing in the eighteenth century reflects a much earlier practice.

Body Measures

The introduction of a system of weights and measures, however primitive by today's standards, marks the emergence of people into the civilized state. Without such a system there is no building, and protection from the elements must be sought in caves, under trees, or in bushes. But early in man's development the utility of parts of the body for building measurements became evident. The *cubit* was the length of the builder's forearm from elbow to finger tip, the *foot* was the length of his foot, the *palm* was the breadth of the four fingers. If the thumb was included the *hand* measure was obtained. The *thumb* and *finger* measures were, as one would expect, the breadths of these two appendages. Much could be achieved with these simple units, especially when some approximate interrelations were discovered and additions were made to the list of usable measures.

The *fathom* (the distance tip to tip of the fingers with the arms outstretched) was nearly one's height. One's step was roughly half a fathom, the cubit roughly a quarter, a foot roughly a sixth, the *span* (thumb tip to little finger with fingers outstretched) was roughly one-eighth. The hand was one third of a foot, the thumb (inch) one-twelfth, and the foot equivalent to four palms or sixteen finger-breadths (digits). Little wonder that Vitruvius, one of the most distinguished Roman writers on architecture of the first century BC, remarks 'Further, it was from the members of the body that they [the ancients] derived the fundamental ideas of the measures which are obviously necessary in all works, as the finger, the palm, foot and cubit'[2], and he describes the perfection of the parts of the human frame which much later was drawn up by Leonardo da Vinci (see Fig. I) as Vitruvian Man.[3]

By inference, these measures would have been the first to develop, for they would serve for the building of the family hut when construction passed from the primitive lean-to of branches, leaves, straw, and mud. Greater measures of length would only become necessary when the nomadic life, with flocks and herds continually on the move to fresh pastures, gave way to fixed settlement and to the cultivation of the land. Then one man's plot of ground must be set off from another's. Herodotus (fifth century BC) stated that the science of geometry, in its original meaning of 'measurement of the earth', originated in Egypt, having as its purpose the redetermination of the boundaries of the fields following the annual inundation by the Nile which washed away the markers.[4] But to do this, or to build a structure greater than that achievable by one man, required the establishment of an agreed standard, for one man's foot or cubit was not the same as another's. As soon as building became more sophisticated or one piece of land had to be measured off from another then standards were needed and the archaeological record shows their presence in very early times, certainly earlier than 2000 BC. Figure 2 shows an Egyptian wall painting (*c.*1400 BC) of workmen measuring with a knotted line rather like today's surveyor's chain. Figure 3 shows some early Egyptian measuring rulers dating from 2500 BC.

Seeds as Weights

Seeds provided the earliest units of weight. The terms *grain* and *carat* reveal their humble origin, the Paris grain of Charlemagne's time being the wheat grain while the English Troy grain was the barley corn. The carat was the seed of the carob plant *(Ceratonia siliqua)*, sometimes called St. John's Bread. Cereal grains were used as weights in northern Europe where the Mediterranean carob was all but unknown. We note parenthetically that much later the thumb or inch was taken to correspond roughly with the length of three barley corns (see Ch. VI) and this is repeated endlessly in mediaeval documents. Even earlier this century, with all the changes that had taken place in the development of cereal grains over the centuries, the same still held true. Figure 4 shows a 15-inch ruler and 45 randomly chosen English barley corns from the 1936 crop placed end to end.[5] The agreement in length is to within one half of one per cent.

The first scales were for the weighing of gold and silver, pearls, and precious stones, with the earliest known examples of balances dating from the second millennium BC. Small seeds served well for weighing gold, for few transactions would mean the exchange of more than a modest amount of the metal. Later, little weights were made to go with the scales and, when balances moved into the market-place for the weighing of common goods, larger weights corresponding in function to those of present-day trading were needed. The earliest artificial weights were of stone as museum collections show (Fig. 5). They had the merit of being non-corrosible unlike the later metallic weights, and so, if unchipped, tend to yield a more faithful record of the weights of the time. The Arabic glass weights of the seventh and eighth centuries AD are pre-eminent in this respect, while a moneyer's balance of AD 350 was sensitive to 30 mg.[6]

Some seeds would appear to be variable in weight from year to year, while others show little deviation from their original weight over long periods. In the Avery Museum, Birmingham are exhibited five vials of Indian rati seeds. Their contents and average seed weights are as follows: 53 seeds (96 mg); 100 seeds (97 mg); 96 seeds (100 mg); 100 seeds (101 mg); 78 seeds (102.5 mg). This represents a considerable spread. In London in 1873 a sample of 640 wheat grains increased in weight on exposure to the air by 1 per cent in a 24-hour period, no doubt due to the absorption of atmospheric moisture. Drying the seeds by means of heat led to a 6 per cent loss in weight. On the other hand the mean weight of 90 carob seeds bought two decades ago in the Cairo market is 195.7 mg.[7] The Arabic glass weights already mentioned were on the carat (carob) scale and they show the average carat to be 194.7 mg. The constancy here is very good. It is likely that the most stable seeds with respect of weight are those which have little or no moisture exchange with their surroundings.

By chance there exists a simple mathematical relationship between these basic units of weight which has been known since Roman times. The fourth century AD Greek *Tabulae Oribasianae* tell us that the *'keration'* seed weighs four wheat grains; a Latin text, *Tabula Codices Bernensis*, says the *'siliqua'* weighs 3 grains of barley while the text *Prisciani libro de Figuris Numerorum* says the *siliquae* and the *keratia* were the same[8], as indeed we know from many texts. We therefore have it that 3 grains of barley are equal in weight to 4 wheat grains, and both equal the weight of the carob seed.

Ridgeway found this still to be true in the late nineteenth century: 'In September 1887 I placed on opposite scales of a balance 32 wheat grains from the midst of the ear and 24 grains of barley grown in the same field at Fen Ditton near Cambridge and repeated the experiment thrice. Each time they balanced so evenly that a $\frac{1}{2}$ grain turned the scale'.[9]

But such an experiment cannot be repeated with the same certainty today. The addition of chemical fertilizers to the land, with advances in plant breeding and genetics, have changed the relative weights of the grains of wheat and barley. Whereas formerly wheat and barley grains weighed one quarter and one third of the carob seed (194.7 mg) respectively, or 48.6 and 64.8 mg, we nowadays find wheat and barley grains approximately equal at a weight of 37 mg in world markets.[10] From 243 samples of English wheat from the 1979 crop at 15 per cent moisture content, we sometimes maintain the old weight of 48 mg but the average is only 45.0.[11] The moisture content plays a significant role in the weight of a cereal grain. The old two-row barley (ears with two rows of grain) has through selective breeding given way to four- and six-row barley with the result that the more numerous the grains are in the ear the lighter they will be, other things remaining equal. The average of 195 samples of English barley from the 1979 crop at 15 per cent moisture content was 43.0 mg.[11]

From the Science Museum, London, stock of wheat and barley seed (1936 crop), four random samples of 100 barley corns and three samples of 100 wheat grains were weighed without further treatment on a precision direct-reading Oertling balance. The average weights found were 42.389 mg for barley and 37.733 mg for wheat. When large grains were deliberately selected the weights found were 57.644 and 43.254 mg respectively.

For average grains, the wheat/barley ratio is therefore 0.89, while for larger grains it is 0.75. This latter is the old, traditional value and as early English legislation[12] specified grains as weights 'from the midst of the ear' — that is, the largest grains — we here find in the grain sample the same ratio although the actual weights of the large individual grains of wheat and barley have both fallen by 12.4 per cent. Some of this may be due to the grain drying out over fifty years.

That these grain weights are not just of passing interest is seen in the fact that the Troy grain on which the pre-metric weights of England were based, at least in theory, was the grain of barley, while the early statutes define the weight of the English silver penny as that of 32 grains of wheat. The individual weights and the ratio of wheat grain to barley corn is the theoretical, if not the practical, basis of our metrology, the more so as three barley corns placed end to end were supposed to generate an inch and on average seem still to do so with fair precision.

The Early Coinage

As weight and coinage are so closely related, a little should be said as to the origins of the coinage, the establishment of which meant the existence of a weight standard, and one which had to be accurately known and to which reference could be made by the several moneyers. The worth of a coin until recent years was the value of the bullion it contained. To vary significantly from the standard either way was to invite problems. If the weight was exceeded, the State was simply presenting bullion to the receiver who would promptly melt down the coin for sale at his own profit. To issue light coin would lead to its rejection and a demand for payment by weight rather than by *tale* (coins counted and taken at face value). The smallness of the spread in the weights of the extant coins of a given issue in good condition is evidence that the standards were adhered to, particularly in respect to gold and silver, and this is true of all countries, ancient and modern, whose coinage is of stated weight and composition.

Originally metal of value was exchanged by weight; 'and Abraham weighed to Ephron the silver which he had named in the audience of the Sons of Heth, four hundred shekels of silver, current money with the merchant' (Gen. XXIII. 16). If, later, the shekel were to be a coin and a weight at the same time, then they must be of equal weight. Aristotle (fourth century BC) says that in men's dealings using iron, silver, and the like as articles of exchange 'the value was at first measured simply by size and weight, but in process of time they put a stamp on it to save the trouble of weighing and to mark the value'.[13] We have been reminded[14] that in the Homeric poems (eighth century BC but describing events some centuries earlier) only gold was weighed, silver being valued according to size, and there was no coinage then. The standard of value all over the Homeric world was the ox, equivalent in worth to the talent of gold, a small weight of 8.5 to 8.75 g.

The prizes for the foot race at the games following the funeral of Patroclus at Troy were: first prize a mixing bowl of chased silver; second, a cow; third, half a *talent* of gold.[15] The values of the three prizes were probably in the ratio of 4:2:1 with the silver bowl valued by size at 2 *talents* and the cow at its customary worth of 1 *talent*.

This *talent*, the so-called Euboic-Attic standard of later times, was the basis on which Philip II of Macedon (359-336 BC), the father of Alexander the Great, coined his gold *stater* of about 8.75 g (135 grains).[16] This *stater* was destined to become the prototype for the Germanic and British coins of Roman times.

Some sixty years ago archaeologists advanced the suggestion that, if a talent were to be the equivalent of an ox, the weight of copper of equal value would be considerable. Copper ingots were known which dated from early times, shaped like ox-hides, and it was supposed that this represented the value of an ox in copper.* While this representational view has not been demolished, considerable evidence has accumulated that these 'ox-hide' ingots do not represent the copper equivalent of an ox;

1. They are too variable in weight (Bass[17] records 34 ingots of average weight about 20 kg, about half that already given, from a sunken ship *c.*1200 BC).
2. The earliest ingots have no legs, but if the ox-hide representation were to be valid we would expect them to be clearly visible, only later being reduced to symbolic legs.
3. Ingots of about the same size and weight are known but in different metals. What would a lead or tin ingot represent?

So it is suggested that the legs are merely handles, devices for holding the ingot when carried on the shoulder. That they were carried as cargo in a sunken ship indicates that they were tradable objects, not coins or currency but objects of value, not unlike the spits of Pheidon or the currency bars of a later time in England (see pp.7 and 16).

We may therefore reluctantly have to give up the earlier concept that these ox-hide ingots represented in base metal the worth of an ox (and hence a *talent* of gold) but for commercial weighings something much larger than the 8.5 gram gold weight would be needed. Naturally these weights would be fabricated in base metal.

In Greece at later times, the word '*talent*' came to mean the heavier unit of base metal. Two weights systems developed, the Solonic (Attic-Euboic) for the mintage and the Aeginetan for commercial purposes. Each heavy talent was divided into 60 *minas* and each *mina* further sub-divided into 100 *drachmae*. One drachma could be further divided into six *obols*.[18] The unit weights of the two systems were somewhat variable. Typically the *mina* on the Attic-Euboic system was 425-437 g; that of the Aeginetan, 587-630 g.[19] The corresponding *talents* were 25.5-26.2 kg and 35.2-37.8 kg respectively.

Although the ox or cow was the money of account it was not the normal medium of exchange, being too large a unit for everyday business. Other articles were used prior to the introduction of the coinage, but long after the advent of minted money we come across the ox being used not only as money of account but also as the medium for payment. Thus in the Laws of Alfred (871-99) in Saxon England, we read in Chapter 18.1, at a time when the coinage was well established:

*The bronze ingot shown in Figure 6 is in the British Museum. It weighs 81 lb 10 oz (37.02 kg) and is approximately 70 cm long, 40 cm wide, and 5 cm thick. It was found as part of a hoard in Eukomi, Cyprus and dates from about 1200 BC.

> If a young woman who is betrothed commits fornication she shall pay compensation to the amount of 60 shillings to the surety [of the marriage] if she is a commoner. This sum shall be paid in livestock, cattle being the property tendered and no slave shall be given in such payment.[20]

In the ninth century payment had to be made for a sinful crime in the ancient money of account, but only by women. There was no similar act requiring payment on the hoof by men.

It is understandable that cattle would not normally be the medium of exchange when we reflect on the very considerable value of an ox. If a small debt was to be settled, a quarter or an eighth of an ox was not a transferable amount. Meat did not have the value of the live animal. Another element was needed for small change.

According to Xenophanes of Colophon (*c.*550 BC)[21], the Lydians of Asia Minor were the first to coin money, and this is repeated by Herodotus[22] (*c.*450 BC), but here Herodotus is referring to the gold and silver coinage of a century later, which is attributed to Croesus (*c.*564 BC). These very first coins, so-called, were bean-shaped pieces of electrum (a naturally occurring alloy of gold and silver) formed by dropping molten metal on an anvil and striking the upper face with three punches. Dating from the reign of Gyges, King of Lydia (685-652 BC), they bore no other inscription. There was no indication as to their weight or by whose authority they were made. Prior to the institution of the coinage, long iron spits *(obols)* were the medium of exchange (Greek, *obelos* = a spit). Six spits made a *drax* or *drachma* (Greek, *drakhme* = handful). Greek tradition asserts that Pheidon of Argos was the first in Greece itself to coin silver, as opposed to electrum, and the coins were made on the island of Aegina.[23] The introduction of silver coins of equivalent worth to the *obol* and *drachma,* and called by the same names, displaced the iron spits and Pheidon is said to have hung up in the temple of Hera at Argos bundles of these iron spits, the former currency, as a votive offering. The spits were first discovered in Italy in 1760 [24,25] and in Greece in 1894 during the excavations of the Argos temple, but it is unlikely that Pheidon coined silver as he flourished about 750 BC, while modern scholarship dates the first Greek silver coins at the last decades of the sixth century BC. Later discoveries in the present century have shown the spits of the Argive Heraeum to have been wasted by much corrosion and to be fragmentary. They are therefore shorter and lighter than they were originally. The silver *obol* weighed about 1 gram; the silver *drachma* about 6 grams. The corresponding iron *obol* has been shown to weigh something over 2 kg with the *drachma* correspondingly greater.[26]

The situation has been summed up by Kraay in 1976[27] as follows:

> A reasonable interpretation of the ancient tradition would be to suppose that

> Pheidon was remembered for having been the first to determine the weight of silver which would be accepted in his kingdom as the equivalent of the handful (drachma) of six spits. The silver drachma of Pheidon's day would have been simply a weight of silver of which standard samples may have been available in temples for purposes of verification.

Pheidon may not have coined silver but it is not impossible or even improbable that a weight standard attributable to him was the basis of the first coinage, as tradition enshrined in a statement of Isidore of Seville (AD 560–636) tells us specifically, in speaking of weights, that 'Phidon of Argos was the first who established the system in Greece, and although there may have been others, more ancient, he was a greater expert in this art.'[28]

The old spits were the first iron currency bars, a concept to be encountered again in England in the second century BC. The Greeks also instituted the use of engraved dies for the striking of coins, inscribing them with the name of the issuing authority, a practice that has continued ever since.

We require to proceed a little further into the units of Greece and Rome in order to appreciate the subsequent development of the subject, for much that was Greek was taken over by the Romans whose influence, in turn, was immense in all areas occupied by them. The Roman pound was the prototype of many European pounds and its subdivisions were frequently copied. The Roman names for measures were commonly used in the Middle Ages even if the volumes were different. The Roman thousand paces, *mille passum*, gave us the concept of the mile, their word *uncia* (twelfth) is the origin of our 'inch' ($\frac{1}{12}$ foot) and 'ounce' ($\frac{1}{12}$ Troy lb). The word *'uncia'* is etymologically connected with the Latin *'unguis'*, a finger nail, and came to mean one-twelfth, for the Roman foot, divided into sixteen fingers (digits) after the fashion of the Greeks, was roughly equivalent in length to the breadth of twelve thumbs. When an inch became established as one-twelfth of a foot, the breadth of the thumb at the root of the nail was frequently used as an equivalent body measure in England as elsewhere. Other examples of the links between the classical systems and those of England will be given as they are encountered. We shall therefore spend a little time with these ancient systems before returning to subsequent developments in England.

Greek Measures of Length

The *fathom* was the distance from finger tip to finger tip with arms outstretched and the noted Greek historian Herodotus* tells us that the fathom was four cubits or six feet in length. We should not suppose that any agreed standards of length were employed throughout the Greek city states. Indeed this is an area in which we are almost completely ignorant from lack of

*Herodotus Bk II, S149. (Ref. 4 vol. 1 p.186).

extant artifacts. Plutarch[29] tells us that the width of the step or platform on which the Parthenon of Pericles at Athens stood was 100 Olympic feet. Its detailed measurement in the nineteenth century yielded a value of 12.14 English inches (308 mm) for this foot. We shall accept this in what follows, realizing that this can hardly be an accurate measure of the foot of Olympia. Nor can it be said that the foot of Athens is well known today. Whereas Roman (and Egyptian) sculptured rulers and measuring rods are relatively common there would appear to be only one surviving representation of Greek lengths, namely the Arundel relief now in the Ashmolean Museum at Oxford.[30] Dating from about 450 BC, its likely origin is on the island of Samos. The relief shows the bust of a man with arms outstretched from which the fathom and cubit can be determined (Figure 7. The missing right-hand end was reconstructed as a mirror image of the left). Above the arm a representation of a human foot is engraved. The cubit is one-fourth of the fathom as usual but the foot is one-seventh rather than the customary one-sixth. The cubit is 1.75 and not 1.5 times the length of the foot.

The actual lengths are:

Fathom	207 cm
Cubit	51.75 cm
Foot	29.6 cm

Herodotus tells us the Samian cubit is the same as the Egyptian.[31] If this is to be interpreted as the royal Egyptian cubit, then its length is known, namely 52.3 cm. The fathom taken from the arms on the relief is a little short, but if the base of the frame of the sculpture is taken (209 cm), the cubit becomes very nearly 52.3 cm: moreover the other dimensions of the stone slab are consistent with the view that a cubit of 52.3 cm was used in its construction. The most recent interpretation of this slab is that it is not a standard of measure at all but is a symbolic representation of measures. It has been suggested[30] that the relief was once located over a doorway, perhaps the doorway of a weights and measures office of long ago. There was therefore no need to make the portrayed lengths of high accuracy. The Athenian foot *may* have been 29.6 cm long. Formal proof is lacking. It has been given in the past as 30.83 and 32.7 cm. While in all probability we still lack an authentic artifact giving the true lengths of Greek measures, there is evidence of a foot of 29.6 cm being used elsewhere in Greece. The length of the Samian foot is not known. It may have been two-thirds of the cubit of 52.3 cm. We observe in passing that 29.6 cm is the currently accepted length of the Roman foot (see p.10).

The *stade* (the origin of our word 'stadium') was 6 *pletheren* or 100 fathoms *(orgyias)* long and the fathom was 6 Greek feet. The *plethron* was therefore 100 Greek (Olympic) feet and the *stade* 600, (185.0 m, 607 English feet), eight of which made a *milion* rather like our mile of eight furlongs though shorter,

1479.5 m as compared to 1609.3 m or 4854 feet vis à vis 5280 feet. A tradition preserved by Isidore[33] is that the original *stadium* or *stade* was paced out by Hercules using his own feet and was the normal distance that could be run with all speed without getting out of breath.

Greek Capacity Measures

While there is some doubt as to the absolute sizes of Greek capacity measures it appears from early writers that the subdivisions were, for the most part, those adopted by the Romans. (see p.12.)

Roman Measures

The Romans took the same track length for the *stade* as the Greeks had used but divided it into 125 paces or 625 Roman feet.[33] In England in mediaeval times the *stade* was to be much confused with the furlong of 660 feet, as we shall see.

The Roman Foot

The actual length of the Roman foot has been the subject of some discussion. As we have seen, its length would be $\frac{600}{625}$ or $\frac{24}{25}$ of the Olympic foot of 308 mm, that is 296 mm, yet it has been recorded anywhere in the range 290 to 300 mm throughout its history. Greaves[34] from an examination of Roman monuments concluded in 1639 that the Roman foot measured 11.60 English inches (294.6 mm). The average of ten determinations of the foot by Raper[35] in 1760, also from the dimensions of monuments, was 11.63 inches (295.3 mm). Hussey[36] in 1834 concluded the most likely length to be 11.65 inches (296 mm). This latter figure has become the 'accepted' value.[37]

The average length of nine Roman folding footrules found in southern England and measured by the writer was 292.9 mm. The spread was 290 to 294.5 mm. Five of these nine have been dated as belonging to the twenty-year interval AD 100-120, having been found in the old bed of the Walbrook, a tributary of the Thames in the City of London. The site of the stream has long since been built over. The finding of so many in a restricted area might suggest that the rulers were deliberately thrown into the stream, perhaps as a votive offering. All are short of 296 mm, but a Roman brass ruler recovered during the contractor's excavation at Winchester Castle Yard in the 1960s measured 295.5 mm.[38]

All the nine metal rulers examined are centrally hinged and are made of brass (orichalcum) or bronze.*† The upper edge is divided into sixteenths

*Orichalcum is an alloy of about four parts of copper to one part of zinc. Bronze is a copper-tin alloy.

†They may be seen in the Museum of London and in the British Museum.

after the Greek fashion, the lower into four palms, and one side is divided into twelfths. The subdivisions are very crude, each 'twelfth' and 'sixteenth' being noticeably different in length from the next. The quarters are likewise irregular. This is no new observation. Greaves[34] writing in 1647 quotes the great French scholar Peiresc (1580-1637) as follows:

> I cannot sufficiently wonder at the inequality which I have found in the divisions by digits, and inches, of the ancient Romane feet; which seem to me to have been made for fashion's sake and *dicis causa* (as lamps that are found in tombs incapable of oile) more to express the mystery, and profession of those that were to use them, then [sic] for to regulate the measure of anything besides them.

Thus irregularities in the subdivisions are to be found not only in the remains of Roman Britain but also in rules from the territories of classical Rome. The subdivisions had no need of accuracy in Roman times as has been stated previously. The digit and thumb's breadth were imprecise to begin with. Present opinion favours the view that these rulers were working tools. Some of these interesting foot rules are shown in Figure 8.

Eight of Raper's ten measurements[35] relate to early Roman buildings and from these he decided that the length of the foot used in their construction was 295.7 mm. The two remaining measurements were from the Baths of Diocletian and the Arch of Septimius Severus. Both gave a foot of 293.9 mm. He concluded that prior to Titus (AD 79-81) the foot equalled or exceeded 295.7 mm but in the reigns of Severus (AD 192-211) and Diocletian (AD 284-305) it had fallen below 294 mm.

The evidence of the rulers found in southern England points to a foot 292-293 mm in length, well short of the 296 mm (11.65 inches) length currently accepted. Flinders Petrie[39] states that the mean length of twelve footrules to which he had access (not any of those already mentioned) was 11.616 ± 0.008 inches or 295.0 ± 0.2 mm and continues: 'In Britain, the Romans used a rather longer form of about 11.68 inches (296.7 mm)'. We would observe that the Roman foot was never maintained or known to an accuracy of better than one part in a thousand, and that the evidence given above does not support the widespread use of a Roman foot in Britain in excess of 296 mm throughout the entire period of the Roman occupation of England.

The standards of Rome were kept in the Temple of Juno Moneta at the Capitol within which was also the mint. Because of this, the measure of the foot was later distinguished by Julius Hyginus (60 BC-AD 10) by the term '*pes monetalis*'.[40]

Two and a half Roman feet made a step *(gradus),* 5 feet a pace, 10 feet a *decempeda* or perch (though sometimes their perches were of 12 feet), 125 paces made a *stade,* and 1000 paces a *millarium* from which our word 'mile' is derived. What appears to be a *decempeda* below the equestrian figures of Castor and Pollux is shown on the reverse of an early silver Roman coin, a *denarius* dated 209-208 BC.[41] The mile of 5000 Roman feet was 4854 of our present feet (1479.5 m), somewhat shorter than the mile of today of 5280 feet (1609.3 m). For the measurement of land, the unit above that of the perch was the *actus* of 120 feet, the square of which, doubled, gave the *jugerum*, 200 of which (almost exactly 125 of our present acres) corresponds closely to the Anglo-Saxon *hide* which will be discussed later.

Roman Capacity Measures

For capacity there was the *congius* and its sixth, the *sextarius*. The *amphora* contained eight *congii*. In 1781 Norris[42] decided that the *congius* would be 207.597*in^3 (3.4l) in capacity and the *amphora* 1660.776*in^3 (27.2l). There appears to be a basis for something like these figures, for of six extant *congii*, three are close to 3.48l.[43] The *sextarius* would then be about 567ml. (This is very close indeed to our present pint measure of 568ml.) The *congius* was supposed to be the volume of ten Roman pounds of water or wine and the copy of a *congius* of Vespasian of AD 75 in the British Museum is so marked. Moreover, the early Roman writers asserted the *amphora* of eighty Roman pounds to be equal in volume to the cube of a Roman foot, hence generating the equivalent name '*quadrantal*'.† Greaves[44] caused such a cube to be made but found it to be filled with only about seven and a half measures of the *congius* of Vespasian, this vessel having been made available to him 'by special favour' during his stay in Rome in the early part of the seventeenth century. Greaves rejected the equivalence of the *amphora* and the cubic Roman foot on the basis of his measurement, declaring the correspondence approximate only. Hussey[45] records a measurement of the volume of this *congius* in 1824 made by weighing the volume of water which it held. The volume so determined was 206.12 cubic inches (3.38l). The corresponding *amphora* would thus have a capacity of 27.04l while the volume of eighty Roman pounds of water each of 327.5 g (see p.xxiv) would be 26.2l. Petrie concludes the best value for the *amphora* to be about 1650 cubic inches or 27l.[46] The volume of a cube of side 296 mm (that is, a Roman foot) would be somewhat less, namely 25.9l. Greaves, then, was not far out in his measurements, for $7\frac{1}{2}$ *congii* of 3.38l gives a volume of 25.4l, very close to the cube of his determination of the Roman foot of 294.6 mm, namely 25.57l.

The correspondence in volume of the *amphora* and the cube of the Roman foot is approximate, but of the values for the capacity given above the spread is only from 27.2 to 25.9l or some 5 per cent. The difference, though

*The precision given is quite absurd.

†An erroneous name to start with, for the amphora was to be the cube and not the square of the Roman foot.

noticeable, can be attributed to the difficulty of making a capacity measure to a high degree of accuracy and perhaps also from the practice of having small measures somewhat larger than the appropriate fraction of the larger standards. For example, the Exchequer pint of Elizabeth I measures 34.46 cubic inches, eight of which would generate a gallon of 275.68 cubic inches, yet her standard Exchequer gallon vessel closest in volume to that of Henry VII had a measured volume of 268.97 cubic inches, a variation of $2\frac{1}{2}$ per cent. This is quite comparable to that which we see in these Roman measures some fifteen centuries earlier. It would appear that the reason for making the small aliquot parts of a measure oversized is to allow for spillage which might well occur if a large measure were to be filled by repeated application of the contents of a smaller.[47] This being so, it is not at all improbable that an identity between the amphora and the cubic foot was intended, and the equivalence has been continuously proclaimed down to the present.[48]

The subdivisions of the *amphora* have been given in detail by Spelman[49], namely:

> The quadrantal which is now called the *amphore* contains two pots (urnas), three corn measures (modios), six half corn measures (semodios), eight congios, forty-eight sextarios, ninety-six heminas, one hundred and ninety-two quartarios and five hundred and seventy six cyathos.

Whereas the work of scholars from the seventeenth to the nineteenth century appears consistent in indicating a single Roman *amphora, modius,* and *congius,* the picture in the twentieth century is not so simple. At least five different volume measures were called the '*modius*', namely those of 8.62l, 10.53l, 11.64-11.85l, 12.93l, and 15.8l.[50] Greaves's $7\frac{1}{2}$ *congii* would yield a *modius* of 8.47l using Hussey's figure for the volume of the *congius*. Presumably the *congius* of Vespasian belongs to the first family of measures.

Roman Weight and Money

As for weight and the coinage, the first metal of exchange at Rome dates from the ninth century BC. Termed '*aes rude*' by Pliny, it consisted of lumps of unstamped copper or bronze, crudely cast rather than struck owing to the size and weight. The pound weight was the *libra* subdivided into 12 ounces, and the unit of exchange, the *as,* was the pound of bronze or copper, and was termed the '*libral as*'. Its weight was the measure of its value and it was bulky and heavy. Livy (59 BC-AD 17) says[51] that in the fourth century BC money was moved around in wagons, while a Lex Tarpeia of 451 BC stated the equivalents of an ox and a sheep to be 100 and 10 *asses* respectively[52], while another dated three years earlier calculated fines in terms of cattle.[53]

The raw chunks of metal gave place to cast ingots stamped to indicate their weight, and therefore their value, which we call *aes signatum,* created for distribution as booty after a victory. They frequently bore the impression of an ox, sheep, or pig. Whereas previously the metal had to be weighed out at each transaction, this was now no longer needed, but the *aes signatum* were not coins but ingots of tradeable value. Then about 280 BC, the *aes grave* appeared in the form of circular or rectangular bronze coins weighing one pound each. Both the *aes signatum* and the *aes grave* continued till 242 BC.

The early coinage in southern Italy of silver *didrachms*[54] each of six scruples *(scripula)* and of weight 6.75-6.8 g can be dated no earlier than the onset of the Pyrrhic War, 280 BC. This Republican coinage bearing the inscription ROMANO lasted till the end of the war, 275 BC, and was followed by a new issue at the end of the First Punic War (264-241 BC) and a final issue was introduced about 220 BC.

During the sixty years from 280 to 220 BC, the weight standard of the *as,* nominally a pound of 327 g (see p.xxiv), underwent some downward changes as is seen from extant examples. The following table illustrates this for Roman issues.

Date	280-276 BC	275-270	269-266	265-242	241-235	230-226	225-217
Weight of *as*	c322 g	331 g	265 g	270 g	272 g	266 g	268 g

Then, to help meet the economic pressures occasioned by the Second Punic War (218-210 BC), the *as* was devalued to half a pound (163.5 g) in 217 BC, to a quarter pound (81.75 g) three years later, and to one-sixth of a pound (2 oz) in 211 BC. Then, in that year or a little before, Rome minted its first *denarius,* a silver coin destined to play a major role in European metrology.* Originally 96 per cent pure silver and weighing 4 scruples, ($\frac{1}{6}$ Roman oz or 4.55 g) they were issued at 72 to the Roman pound. Their history is long and chequered. As the name *denarius* implies, this coin was to be worth 10 *asses* and the symbol 'X' appears on the left of the winged head of Roma on the obverse of many early issues (see Fig. 9).

By 207 BC the weight of the *as* was slipping from the 2 oz mark and in 141 BC was set at 1 oz. In this year the *denarius* was set at 16 *asses,* a value at which it remained to the end of the coinage of *denarii,* although the army could still get a *denarius* for 10 *asses.*[57] The middle photograph of Figure 9 shows the *denarius* displaying its new value, with XVI behind the head of Roma. A little later, as the third *denarius* of Figure 9 shows, the XVI gave way to the symbol *, seen to the right of the bust.

The decline of the *as* had not ended. The imminence of the Social War (90-89 BC) necessitated a further reduction to $\frac{1}{2}$ oz in 91 BC. The *as* had long

*The dating of the minting of the *denarius* at Rome as 268 BC has been shown to be an error by the Yale excavations at Morgantina, Sicily where *denarii* appear together with datable coins from Syracuse.[55] The error originally, would seem to be traceable to Pliny.[56,57]

been no more than a token of declared value, having no relation to the intrinsic worth of the metal, but it now had the merit of being much more portable. In this year the *denarius* itself was reduced in weight to save silver, being minted at the rate of 84 to the pound (3.86 g each).

If we ignore the early *staters,* which were based on those of Greece and Macedon, gold enters the Roman currency lists with the Empire. In addition to gold, Augustus also introduced brass coins. The value scale was as follows:[58]

1 *aureus* (gold)	=	25 *denarii* (silver)
1 *denarius*	=	4 *sestertii* (orichalcum)
1 *sestertius*	=	2 *dupondii* (orichalcum)
1 *dupondius*	=	2 *asses* (copper)
1 *as*	=	2 *semisses* (halves) or 4 *quadrantes* (quarters) all in copper

The table shows the *denarius* as 16 *asses,* its later value.

Augustus minted 40 *aurei* to the pound of gold bullion. He also made 84 *denarii* to the pound of silver, a reduction from its weight in 211 BC when it was $\frac{1}{6}$ oz or $\frac{1}{72}$ lb. Claudius (AD 41–54) reduced the gold to 42 to the pound and Nero (AD 54-68) made both the gold and silver coins lighter, minting them at 45 and 96 to the pound respectively.

Thus far, though the weight of the *denarius* had been reduced, the quality of the silver had been maintained. Now, imperceptibly, the silver content began a fearful decline. For more than two and a half centuries its worth had been faithfully represented by the value of the silver it contained. From Nero to the end of the third century AD it, or its double, the *antoninianus,* became more and more debased until the latter became no more than 'a mere argentiferous bronze'[59] with frequently no more than 2 per cent silver content.

Denarii were reintroduced by Diocletian at a rate of 96 to the pound, that is, each being $\frac{1}{8}$ oz, but gold was reduced in weight to 60 to the pound (5 to the ounce) by Diocletian in AD 294 and to 72 to the pound (6 to the ounce) by Constantine in AD 309. Some *denarii* are marked XCVI (96); some *solidi* carry the legend LXXII (72). This gold *solidus* of Constantine and its eventual third, the *tremissis,* were destined to play a major role in the weight standards of mediaeval England. Actual *solidi* in good condition weigh about 4.48 g. If $1\frac{1}{2}$ per cent is allowed for loss by wear the weight standard of the *solidus* would be close to 4.55 g (70 grains Troy). This enables the weight of the later Roman pound to be deduced as 327.6 g. The presently accepted value is 327.453 g[60], a figure of exaggerated precision for the allowance for coin wear is by no means as definite as one might wish, nor do other means of computation arrive at precisely this figure. Although this value of a Roman pound was arrived at well over a century ago a recent careful analysis of the evidence shows it 'cannot be far wrong'.[61]

The weight of the *solidus* of Constantine was defined as 24 carats *(siliquae* or *keration)* and the *tremissis* weighed 1.5 g, 8 carats, or closely 24 grains Troy, a figure we shall frequently encounter in England from Saxon times onwards.

The subdivisions of the *as* as a monetary pound weight parallel those of the commercial pound, the *libra*.

As: *Semis* ($\frac{1}{2}$), *triens* ($\frac{1}{3}$), *quadrans* ($\frac{1}{4}$), *sextans* ($\frac{1}{6}$), *sesuncia* ($\frac{1}{8}$), *uncia* ($\frac{1}{12}$), *semuncia* ($\frac{1}{24}$), *sicilicus* ($\frac{1}{48}$). *sextula* ($\frac{1}{72}$), *scripulum* ($\frac{1}{288}$).

Libra: *Uncia* ($\frac{1}{12}$), *sicilicus* ($\frac{1}{48}$), *drachma* ($\frac{1}{96}$), *scripulum* ($\frac{1}{288}$), *obolus* ($\frac{1}{576}$), *siliqua* ($\frac{1}{1728}$).

From this the various interrelations may be obtained.

British Currency Bars

England was to experience, at first indirectly and later directly, the influence of these matters which have just been discussed. Beginning in the late second century BC and continuing into the first century AD, invasion and settlement brought hitherto undreamed of development and the application of strange ways to the indigenous population. Caesar recorded (though the text is not without its difficulties) the fiscal state of Britain following his exploratory invasion of 54 BC: 'For money they used either bronze or gold coins or iron ingots of fixed weight' (Ut untur aut aere, aut nummo aureo, aut taleis ferreis ad certum pondus examinatis pro nummo). The text has been repeatedly studied in order to determine the meaning to be assigned to the words *'taleis ferreis'*. Some translate the Latin as 'iron cuttings' and for long the use of iron as here described remained a mystery. Then in 1824 a hoard of iron rods was discovered at Meon Hill, Warwickshire, to be followed by other finds in 1856 at Malvern and in 1857 and 1860 in Gloucestershire. Since then more than 1100 of these rods have been unearthed in Hampshire, Somerset, Dorset, Wiltshire, Gloucestershire, Warwickshire, and Worcestershire. Twenty-one were recovered from a hoard at Danbury Hill, Hampshire, as recently as 1969. Allen[63] has categorized these bars into four groups of types with the first resembling unfinished swords being 'legitimately described as currency bars'. These bars are 2 to 3 feet long, $\frac{3}{4}$ to $1\frac{1}{2}$ inches broad, $\frac{1}{8}$ inch thick with a 3 to 4 inch hilt. In England these sword blades have been found only in the south and west. Their weight, if these are the iron ingots to which Caesar referred, is something of a problem but some regard them as belonging to a series of weights of a binary sequence: 4, 2, 1, $\frac{1}{2}$, and $\frac{1}{4}$ units. They date from early first-century BC pre-Belgic Britain. It has been suggested [63,64] that they may have been related to the lighter Celtic pound of 309 g referred to later on the ground that the 'corrected weights' of twenty specimens is $20\frac{1}{2}$ oz (581 g) which is 6 per cent less than 2 Celtic pounds, while of others, three of weights

302, 305, and 202 g are taken as being on a 309 g standard, 23 others are on a 619 g standard, and two of weight 1218 and 1150 g are to be considered as belonging to a standard of 1238.8 g, the standards being in the ratio 1:2:4.

If weight alone was to be the criterion, why go to the trouble of shaping the metal bars? Why not keep them in the form of cast ingots? There is no easy answer to these questions. If shape means something perhaps the length does also, so it was thought worthwhile to spend a little time to see if the weight per unit length yielded any light on the problem, it being granted at the outset that it would be difficult to see what possible advantage could be gained by having two parameters, weight and length, to determine value.

Seven currency bars (1-7) in the British Museum had their weights and lengths already recorded, so these were used together with four in the Winchester City Museum which the writer was allowed to weigh and measure, making eleven in all. The first seven have a varied provenance, coming from as far apart as Bourton-on-the-Water, Gloucestershire; Hamden Hill, Somerset; and Maidenhead, Berkshire. The remaining four (8-11) are from the Winchester area and are shown in Figure 10.

	Length (cm)	*Weight (g)*	*Weight per cm*	*Normalized* weight/cm*	*Possible scale*
1	81.9	357.4	4.4	28	30
2	69.9	601.1	8.6	55	60
3	78.7	1216.3	15.5	100	100
4	74.9	438.0	5.8	37	40
5	54.0	306.5	5.7	37	40
6	57.2	202.5	3.5	23	20
7	79.4	495.5	6.2	40	40
8	80.5	519	6.4	41	40
9	87.9	618	7.0	45	40
10	83.3	505	6.0	39	40
11	82.0	550	6.7	43	40

*Largest = 100

It was noticed that bars numbered 4, 5, 7, 8, 9, 10, and 11 were grouped around the scale number 40. The actual average of the seven is 40.29. These seven might each be called '40 units'. Thereafter numbers 1, 2, 3, and 6 would be 30, 60, 100, and 20 respectively. The average deviation from the proposed scale is 6.4 per cent. Considering weight alone, the average deviation from the weight standards of 309, 618, or 1236 g, choosing the nearest, is 13.5 per cent. It would be of interest to examine a large number of the currency bars in the same way but the data are not readily available.

Whatever their function, and whatever measure was used to determine their worth, there seems to be little doubt from the evidence of the hoards that they were regarded as items of value and were treated as such.

Allen[63] concludes 'nor can we assume the bars were used as coins, but were regarded as objects worth hoarding for their value and if we can apply Caesar's words to them [we may conclude] that they were a substitute for money in a context where barter must have been the normal means of exchange'.

The identification of these bars with Caesar's iron ingots has been contested by others who claim that the bars are not more than that which they appear to be, namely unfinished swords,[64] although if so, no reason is given why many of the hoards should contain between 100 and 200 specimens; in addition their parallel to Pheidon's spits is striking. Whatever their use in the system of exchange, they would pass into oblivion with the arrival of the first coins. Their date (mid to late Iron Age) fits well with known subsequent developments.

The Celtic Pound

Two Celtic pounds have been identified[65] as belonging to the pre-Roman and Roman periods, having weights of 309 g and 638 g respectively. Reference has already been made to the former. Both are represented in extant weights found in England and on the Continent. They were in use in pre-Roman and Roman England.

The table below summarizes some of the available data.

Location and Description[66,67]	*Weight*	*Presumed Standard*	*Weight Loss %*
Cranbourne Chase, Dorset, ingot	63.4g	$\frac{1}{10}$ of 638 g lb	0.6
Colchester, No. 1	38.5	$\frac{1}{8}$ of 309 g lb	0.3
Colchester, No. 2	126.8	$\frac{1}{5}$ of 638 g lb	0.6
Glastonbury, tin weight	127.2	$\frac{1}{5}$ of 638 g lb	0.3
Seven Sisters, Neath, Glamorgan, bronze weight	204.3	$\frac{2}{3}$ of 309 lb	0.8
Seven Sisters, Neath, Glamorgan, bronze weight	309.1	309 g (1 lb)	–
Melandra Castle, No. 1	306.9	309 (1 lb)	0.7
Melandra Castle, No. 2	229.1	231.9 ($\frac{3}{4}$ lb)	1.2
Melandra Castle, No. 5	111.8	115.9 ($\frac{3}{8}$ lb)	3.5
Melandra Castle, No. 8	76.6	77.3 ($\frac{1}{4}$ lb)	0.9
Melandra Castle, No. 20	19.3	19.4 ($\frac{1}{16}$ lb)	0.5
Melandra Castle, No. 25	12.2	12.9 ($\frac{1}{24}$ lb)	5.4
Melandra Castle, No. 28	9.5	9.7 ($\frac{1}{32}$ lb)	2.1

All are slightly less than the presumed standard but the smallness of the deviation is really quite remarkable. The weights from Melandra Castle (between York and Chester) are quite numerous. Only a representative selection has been given in the table.

The 18 g weight found at Mount Caburn and now in the Barbican Museum at Lewes, Sussex, has been thought to be $\frac{1}{16}$ of the lighter Celtic pound[68], but that would make it 8 per cent light. It is similar in shape to Melandra weight #20 which is significantly heavier. It is more likely to be $\frac{2}{3}$ of a Roman ounce — that is, 18.18 g on the standard and therefore only 1 per cent light.[69]

The First British Coinage

Around 150 BC several tribes of the Belgae moved into northern Gaul and there began a mintage of gold in the form of rather poor replicas of the mid-fourth-century BC gold *stater* of Philip of Macedon which, it will be remembered, weighed some 135 Troy grains or 8.75 g. The *philippus* served as the prototype for the Gallic mints. The new coins were cast at first, but later were struck, and were some 10-15 per cent lighter than the coin being imitated — that is, they weighed about 120 grains (7.8 g).[70]

The Belgae were not left in peace for long. They came under increasing pressure from the Germanic tribes to the east and from the Romans to the south and west, and eventually this pressure led to a series of emigrations of these Celtic peoples from northern Gaul to Britain, settling first in Kent and Essex, later in regions north and south of the Thames, and eventually in the south-west. Six distinct waves of invaders have been identified as coming to the shores of England in the last decades of the second century BC, bringing their coins and the concept of a coinage of gold with them. The British coinage began, then, as a result of immigration and with imported coin. Locally produced money came later. Most of the early *staters* were on the 120 grain standard; by 100 BC the weight standard was 100 grains or less.

Piggott[71] writes that 'coinage was introduced into Britain with the first Belgic colonization at the end of the second century BC and was then widely used in a variety of forms in Belgic and para-Belgic territory, side by side with the West of England use of iron ingot bars of standard weights in the form of roughed-out swords'.

By about 90-80 BC coins carrying inscriptions were being produced by many if not all the Belgic and mixed tribes of southern and eastern England. The first local coinage was that of the Durotriges of Dorset, Wiltshire, and Somerset. Gold, bronze, and silver were minted in succession and the coinage of the Durotriges lasted into the first century AD. Kent was minting gold *staters* early in the first century BC on a standard of about 93 grains (6 g).

The bronze which Caesar reported was probably more correctly termed speculum, an alloy of tin and copper, rich in the former. This was an early British coinage not made by invading Belgae and was in use before and after the Roman conquest. This so-called 'Tin Currency' closely resembles the '*potin*' coins of Gaul. The weight was about 1.5 g, that is, about $\frac{1}{4}$ of the *stater*. All silver coins appear to be subsequent to Caesar's visit as they are not mentioned at all in his writings. The coinage of silver, then, occurred first in England, for only gold coins appear to have been brought over by the Belgae, nor is there evidence of a silver coinage among these peoples in northern Europe at this time. But, with Caesar's suppression of the mintage of gold in Gaul after 52 BC, further importations of gold coin were unlikely. The date for the commencement of the coinage of silver cannot be far from that of the birth of Christ.

Major mintages were produced by all the main tribes with the weight standard of the *stater* being originally about 7.8 g, that is, the same as that in northern Gaul, but later falling to and settling at 5-5.5 g as the evidence of the tribal coin hoards attest. There is a considerable spread in the weights of individual coins due to wear and to the initial difficulty of procuring blanks all of substantially the same weight, but to achieve that which was accomplished implies the existence of an accepted, well-recognized weight standard.

Roman Coins in Britain

The Belgo-British coinage was of short duration. The importation of Roman coins following Claudius' expedition of conquest and colonization in AD 43 led at first to its imitation by local mints, especially for silver and bronze, but by the end of the century the local currency had been completely displaced by the Roman with the latter in use throughout England. Currency from the mints at Rome and Lugdunum (Lyons) entered the country but the latter mint closed in AD 77 or 78, after which Rome was effectively the sole supplier. A small part of the bronze coinage of *asses (aes)* of Antoninus Pius in and after AD 155 was minted in England,[72] the coins showing the defeated figure of Britannia on the obverse, doubtless to remind the inhabitants who was master following the suppression of the revolt of the Brigantes tribe by the Roman governor, C. Julius Verus, in the same year. (A happier Britannia was present on the pennies of the twentieth century but, following the issue of decimal coins, is now to be seen solely on the reverse of the present 50 pence piece of cupro-nickel issued first in 1969.) Only after AD 287 did local mints for a short time supply a major part of the currency needs of the land.

The London mint of the Romans was at the site of the Tower of London built by Gundulf Bishop of Rochester in 1078. While digging the foundations there for new offices for the Board of Ordinance in 1777 a silver ingot was unearthed which is now in the British Museum[73] (Fig. 11). It was some 10 cm long, 1 cm thick, and about 4.3 cm across at the narrow part. It had been hammered out at both ends to resemble a double-edged axe. Stamped '*Ex Offic(ina) Honorii*', it clearly belongs to the time of the Emperor Honorius whose reign (395-423) covered the period of the Roman withdrawal from

Britain. Its weight was given as 4992 Troy grains (323.5 g), that is, 4 g short of the Roman pound of 327.5 g. On reweighing some two centuries later[74] it appeared to be 301 g, far short of the original 323.5 g. In all probability a portion had flaked off. Also found at the same site were three gold *solidi*, each $\frac{1}{6}$ oz Roman.[72]

The British Museum purchased another axe-head ingot in 1970 (B.M. Reg. No. P 1970 7-2, 1) weighing 319.47 g stamped '*Ex Off CVRMISSI*'. Analysis has shown the high degree of purity of the silver, 95.2 ± 1%, copper 4.1%, gold 0.81%, lead 1.22%, and iron 0.1%.

These one pound silver 'axe' ingots are not rare. Two more are in the Canterbury Museum weighing 316 and 314.5 g and this museum also has a silver ingot of 320 g in the pre-beaten-out state. Painter[74] has listed 44 of these Roman ingots of the 'axe-head' type with full details, giving photographs of many.

Scales, Steelyards, and Bismars

There is little doubt that Roman weights and measures were the official standards of occupied Britain. Large numbers of Roman weights are to be found in museums all over England. Three devices were employed for weighing; scales, steelyards, and bismars. An example of a set of Roman scales and weights is shown in Figure 12 and a Roman steelyard is seen in Figure 13. What is thought to be the earliest woodcut showing a bismar is that from the 1555 book of Olaus Magnus,[75] Archbishop of Uppsala, Sweden, shown in Figure 14 together with an actual Norwegian bismar of the sixteenth century which it greatly resembles. Scales of the equal-arm type had been used in antiquity as we have seen. Bismars and steelyards were later developments, with the former known from Greek times to the present century in Europe in spite of many prohibitions. The steelyard was a Roman invention and was originally called the *romana* or the *statera*. It is to be found throughout the extent of the Roman Empire. The bismar and steelyard were unequal-arm devices, the former highly inaccurate with a fixed counterpoise attached to the beam and one or more sets of pins or nails stuck into the shaft to denote the relevant weight on one or more scales. The fulcrum or balance point could be varied until equilibrium was reached with the load to be weighed, the position of the fulcrum being determined by a loop of string or a thong or, more commonly for light commodities, by simply lifting the beam with a finger or the edge of the hand and noting the nearest pin. The inaccuracy of such a device will be clear at once. If the rod of the device is 100 cm long and the counterpoise is 1 kg in weight a balance will result with a 1 kg load with the hand at the mid-point of the rod. With a load of 9 kg the balance point is 90 cm from the counterpoise. With a 10 kg load the balance point is at 90.9 cm from the counterpoise. When the pivot point is only fuzzily seen, probably no better than to the nearest centimetre, especially in the turmoil of the market-place, the liability of error, whether innocent or fraudulent, is great. It would be hard to distinguish 90 from 90.9 cm yet the error would be 10 per cent of the weight. The steelyard was much more precise. Unlike the wooden bismar, it was usually constructed of metal, although one from Orkney of oak of the eighteenth century is to be seen in the Avery Museum, Birmingham, and the wooden one shown in Figure 15 was purchased in 1976 in the market at Johore Bahru, Malaysia, where such devices are still in service for purposes of trade but, being of slight construction, are only used for light goods such as meat, chicken, fish, etc. Rows of small brass pegs sunk into the rod give the weight in kati.

In the process of using the steelyard the object to be weighed was suspended from the end hook, the beam being suspended by the other hook, and the counterpoise was moved along the arm until the beam was horizontal. The position of the counterpoise relative to a scale engraved or notched on the long arm of the beam gave the weight reading. The great popularity of the bismar and steelyard with traders stemmed from the fact that there were no loose weights at all. The steelyard had the counterpoise which could only be removed deliberately from the beam. These counterpoises were often attractively carved or cast in metal showing the heads or busts of individuals, human or divine, or coats of arms and on occasion approached the level of serious works of art (Fig. 16). The term 'steelyard' is of later date than the device itself and is the word in use today for this weighing instrument. The name originated with the steel traders of the Hanse who in 1250 settled at a centre on the Thames near Cosen Lane at which place these weighing devices were much in evidence. Many privileges were given to these traders by Henry III in 1259. The Treaty of Utrecht of 1474[76] reflected these privileges:[76] 'certain houses and mansions shall be appropriated . . . to them and their successors for ever . . . that is to say a certain court situated in London called the Staelhoeff or Stylgerd [Steelyard]. . . .' But less than a century later Edward VI withdrew them. Elizabeth went so far as to expel the traders in 1597. The site of the Steelyard is now occupied by Cannon Street Railway Station. Later the word *auncel* appeared, being used somewhat indiscriminately for a bismar or a form of steelyard (see Chapter VIII).

Roman footrules found in London and elsewhere have already been referred to (Fig. 8). Buildings are known to have been constructed in England using this foot as a module and it continued in use long after the Roman Empire had disintegrated. And not only in England. In Normandy, to give but one example, the great abbey at Cluny was built in successive stages. The chapel (Cluny A), dating from the tenth century, was on a module of $1\frac{1}{2}$ Roman feet and right up to the end of the eleventh century, and indeed beyond, the additions to the abbey were made using the Roman foot.[77]

Roman influence endured wherever it had been felt. England was a part of the empire for 400 years, so it is not surprising that Roman methods, ideas, and values remained. England acquired Roman weights and measures at a time when these had largely stabilized themselves and the turmoil and change of the Republic had passed away. From the Christian era onwards uniformity was expected throughout the Roman world. Dio Cassius records that Roman measures were to be used everywhere. Pretextatus, Prefect of Rome, Praetorian Prefect of Italy, Pontiff of the Sun and of Vesta, to give him his full title, in AD 367 from Rome ordered the same weights to be employed everywhere. This did not happen, of course. Celtic weights are found together with Roman. For official purposes the standards of Rome were used throughout the period of the occupation but the people continued to employ the local measures to which they were accustomed. Nor did the Roman influence die with the withdrawal of AD 410, for Roman measures continued in use.

NOTES AND REFERENCES: CHAPTER 1

1. *Encyclopaedia Britannica,* 1st ed., Vol.1, p.508, 'Auncel Weight', Edinburgh: A Bell & C Macfarquhar 1771.
2. Vitruvius, *The Ten Books on Architecture* Book III, Ch. 1, S. 5, p.73. MORGAN, Morris Hickey (tr.), New York: Dover Publications Inc. 1960.
3. Leonardo da Vinci. The most accessible is probably *The Notebooks of Leonardo da Vinci,* RICHTER, J. P. (tr.) New York: Dover Publications Inc. (1970). See Plate XVIII with text on p.182. The drawing was printed in the Venice edition of 1511.
4. MACAULAY, G. C. (ed. & tr.). *The History of Herodotus* Book II, Ch. 109, (Vol.1, p.163) London: Macmillan & Co. 1904.
5. The grain was obtained by courtesy of the Science Museum, London.
6. COPE, L. H. Roman Imperial Silver Coinage Alloy Standards *Numismatic Chronicle,* (1967) Vol.VII, p.107, London.
7. GRIERSON, Philip. Presidential Address, Weight and Coinage. *Numismatic Chronicle* (1964) Vol.IV, p.xvi. London.
8. HULTSCH, Friedrich Otto. *Metrologicorum Scriptorum Reliquiae* Vol.I, p.248, No.70; Vol.II, p.128, No.133 and p.84, No.118. Stuttgart: Teubner 1971 (Reprint of edn of 1864-66).
9. RIDGEWAY, W. *The Origin of Metallic Currency & Weight Standards* p.182, Cambridge: C.U.P. 1892.
10. KENT, N. L. *Technology of Cereals* p.21. Oxford: Pergamon Press 1975.
11. Home Grown Cereals Authority, The Quality of the 1979 Wheat and Barley Harvests, Supplement to *Weekly Bulletin* Vol.14, Issue No.20, 17 Dec. 1979.
12. *Assisa de Ponderibus et Mensuris: Statutes* Vol.1, p.204.

13. ARISTOTLE, Politics Book I, Ch.9, 1257a, 35, JOWETT, Benjamin (tr.) Oxford: Clarendon Press, 1905.
14. RIDGEWAY, W. Loc. cit. p.117.
15. HOMER. *Iliad* XXIII, 740 p.432. RIEU, E.V. (ed. & tr.), Harmondsworth: Penguin Books, 1969.
16. A Guide to the Principal Coins of the Greeks, based on the work of B. V. Head, London: British Museum 1959.
17. BASS, G. Cape Gelidonya: A bronze age shipwreck, *Trans. Am. Philos. Soc.* (1967) N.S. Vol.LVII, Pt.8. Philadelphia.
18. See for example:
SKINNER, F. K. *Weights and Measures* London: Science Museum & HMSO 1967, also HULTSCH, Friedrich Otto, Loc. cit. Vol.1 p.228 No.57.
19. SMITH, C. (ed.) *A Guide to the Exhibition illustrating Greek and Roman Life*, p.146 et seq. London: British Museum, Dept. of Greek and Roman Antiquities, 1908.
20. ATTENBOROUGH, F.L. (ed.) *The Laws of the Earliest English Kings*, p.73. Cambridge: C.U.P. 1922.
21. POLLUX, Julius *Onomasticon* IX, 83 Frankfurt: C. Marnium 1608.
22. MACAULAY, G. C. Loc. cit. Book I, 94, (Vol.1, p.49).
23. See for example:
STRABO, *Geographica* JONES, H. L. (ed.) Book VIII 376, London and Cambridge, Mass.: Loeb Classical Library Vol.IX, p.181, Harvard University Press and William Heinemann 1927. 'Ephorus says that in Aegina silver was first struck by Pheidon.'
24. WALDSTEIN, C. *The Argive Heraeum* Vol.1, p.63. Boston and New York: Archaeological Institute of America; School of Classical Studies at Athens 1902.
25. SELTMAN, Charles Theodore. *Greek Coins* 2nd ed. p.35 et seq. London: Methuen 1955.
26. COURBIN, Paul, Valeur comparée du fer et de l'Argent lors de l'introduction du monnayage. *Annales: Economies, Societés, Civilisations* (1959) Vol.XIV, pp.209-33.
27. KRAAY, Colin M. *Archaic and Classical Greek Coins* p.314 Berkeley: University of California Press 1976.
28. ISIDORE Hispalensis Episcopi, *Etymologiarum sive originum libri XX*, LINDSAY, W. M. (ed.) Bk. XVI XXV. 3 Oxford: Clarendon Press 1911.
29. PLUTARCH, *The Lives of the Noble Greeks and Romans*, NORTH, Thomas Lord (tr.) Vol.II (Pericles) p.21. Stratford on Avon, Shakespeare Head Press: Basil Blackwell (Oxford) 1928.
30. For what follows on this topic see:
FERNIE, Eric. The Greek Metrological Relief in Oxford. *Antiquaries Journal* (1981) Vol.LXI, pp.255-263.
31. MACAULAY, G.C. Loc. cit. Book II 168 (Vol.1, p.195).
32. IBID. Book II 149 (Vol.1, p.186).
33. ISIDORE Loc. cit. Bk. XV XVI. 3.
34. GREAVES, J. *A Discourse on the Romane foot and Denarius* pp.22, 40-1 and 22-3. London: William Lee 1647.
35. RAPER, Matthew. An Inquiry into the Measure of the Roman foot. *Phil. Trans. Roy. Soc.* (1760) Vol.LI (2), p.774. London.
36. HUSSEY, Robert. *Ancient Weights and Money* p.230 Oxford: J. H. Parker 1836.

37. ZUPKO, Ronald Edward. *British Weights and Measures* p.6 Madison, Wisconsin: University of Wisconsin Press 1977.
38. Private Communication, Winchester Research Unit 1973.
39. PETRIE, W. Flinders *Encyclopaedia Britannica* 11th ed. 'Weights and Measures' Vol.XXVIII, p.483. Cambridge: C.U.P. 1911.
40. HULTSCH, Friedrich Otto. Loc. cit. Vol.II, pp.60-1; HYGINUS, Julius, De Condicionibus Agrorum.
41. CRAWFORD, M. H. *Roman Republican Coinage* p.174, also plate 15 No. 10 78/1 Cambridge: C.U.P. 1974.
42. NORRIS, H. On Greaves's Weights and Measures. *Archaeologia* 1781 (Vol. dated 1782) Vol.VI, p.221. London and Oxford.
43. SKINNER, F. G. Loc. cit. p.68.
44. GREAVES, J. Loc. cit. pp.34-5, 38.
45. HUSSEY, Robert. Loc. cit. pp.126-7.
46. PETRIE, W. Flinders. Loc. cit. p.485.
47. STEVENSON, Maurice. The Size of liquid measures in the 17th and 18th Centuries *Libra* (1964) No.3, p.17. Eastbourne, also private communication 1 Oct. 1980.
48. See for example:
    (a) YOUNG, W. *A Treatise on Weights and Money,* p.204 London 1836.
    (b) HUSSEY, Robert. Loc. cit. p.204.
    (c) RENFREW, Paul. *Report on the 50th National Conference on Weights and Measures.* National Bureau of Standards Misc. Publication 272, Washington, D.C.: U.S. Government Printing Office 1965.
49. SPELMAN, Henry. *Glossarium Archaiologicum,* DUGDALE, Sir W. (ed.) p.29 'Amphora' London: Alicia Warren 1664.
50. JONES, R. P. Duncan. Length Units in Roman Town Planning: The Pes Monetalis and the Pes Drusianus *Britannia* (1980) Vol.XI, pp.127-33. London. See Note 32 on his p.133.
51. LIVIUS, Titus, *Histoire Romaine.* BAYET, Jean (ed.), BAILLET, Gaston (tr.) Book IV p.98. Paris: Belles Lettres 1946.
52. RIDGEWAY, W. Loc. cit. p.370.
53. GRIERSON, Philip. *The Origins of Money* pp. 19 & 28 (Note 86) (The Creighton Lecture in History 1970) London: The Athlone Press, University of London 1977.
54. For much of what follows respecting the Republican Coinage see: CRAWFORD, M. H. Loc. cit.
55. GRIERSON, Philip. *Bibliographie Numismatique* 2nd ed. p.72 Brussels: Cercle d'Etudes Numismatiques, 1979.
56. CRAWFORD, M. H. Loc. cit. Vol. I, p.36.
57. PLINY. *Natural History,* RACKHAM, H. (ed.) XXXIII 45, Vol.IX, pp.36-9. Cambridge, Mass. & London: Harvard University Press and William Heinemann 1952.
58. JOSSET, Christopher Robert. *Money in Great Britain and Ireland* p.22. Newton Abbot: David and Charles 1971.
59. COPE, L. H. Roman Imperial Silver Coinage Alloy Standards: *Numis. Chron.* (1967) Vol. VII, p.107. London.

60. HULTSCH, Friedrich Otto. *Griechische und Romische Metrologie* p.119. Berlin: Wiedmann 1862.
61. GRIERSON, Philip. Presidential Address. Weight and Coinage. *Numis. Chron.* (1964) Vol.4, p.xiv. London.
62. CAESAR, C. Julius. *Commentarii De Bello Gallico* HOLMES, T. Rice (ed.) Bk.V, Ch.12, pp.185-6. Oxford: Clarendon Press 1914.
63. ALLEN, Derek. Iron Currency Bars in Britain. *Proc. Prehistoric Soc.* (1967) Vol.XXXIII, p.307. London.
64. SMITH, R. A. Currency Bars and Weights. *Proc. Soc. Antiq. London* 1903-5 (Pub. 1905) Vol.XX, pp.179-195. London.
65. See SCHWARTZ, G. T. Gallo-Romische Gewichte in Aventicum, *Schweizer Munzblatte* (1964), 13/14, pp.150-7.
66. SPRATLING, Mansel. Iron Age Settlement of Gussage All Saints, Part II, *Antiquity* (1973) Vol.XLVII, p.117. Cambridge.
67. CONWAY, R. S. *Melandra Castle 1905* Manchester: Manchester University Press 1906.
68. CURWEN, Eliot & E. Cecil. Excavations in the Caburn, near Lewes. *Sussex Archaeological Collections* (1916) Vol.LXVIII, pp.1-56. Lewes.
69. The author is indebted to Miss Fiona Marsden, Curator, Barbican Museum, Lewes for permission to weigh and examine this weight.
70. For a detailed description of the Belgic coinage see:
    (a) MACK, R. P. *The Coinage of Ancient Britain* 2nd ed. London: Spink & Son; B. A. Seaby 1964.
    (b) ALLEN, Derek. *An Introduction to Celtic Coins.* London: British Museum Publications Ltd 1978.
    (c) ALLEN, Derek. *Belgic Dynasties of Britain and Their Coins.* London: Society of Antiquaries 1944.
71. PIGGOTT, Stuart. *Ancient Europe* p.252. Chicago & Edinburgh: Aldine Publishing Co. & Edinburgh University Press 1970.
72. TODD, Malcolm. Romano-British Mintages of Antoninus Pius *Numis. Chron.* (1966) Vol.VI, p.147. London.
73. MILES, J. Observations on Some Antiquities found in the Tower of London in the Year 1777 *Archaeologia* (1778) Vol.V, p.291. London and Oxford (Vol. dated 1779).
74. PAINTER, K. S. A late Roman silver Ingot from Kent. *Antiquaries Journal* (1972) Vol.LII, p.84. Oxford. The ingot of Honorius is No. 9 on Painter's list.
75. MAGNUS, Olaus. *Historia de Gentibus Septentrionalibus* Vol. IV, p.468. Rome: Joannes Mario de Viottis in aedibus Brigittae 1555.
76. MEYERS, A. R. (ed.). *English Historical Documents 1327-1485*. Vol. IV, p.1046. London: Eyre & Spottiswoode 1969.
77. CONANT, Kenneth J. Mediaeval Academy (of America) Excavations at Cluny IX. Systematic dimensions of the buildings. *Speculum* (1963) Vol.XXXVIII, p.1. Cambridge, Mass.

CHAPTER II

The Natural and Drusian Foot

The Natural Foot

We would expect that the length of the human foot would play a large role, especially in the rural areas and in early times. The term appears in Latin texts as *pes naturalis,* that is, natural foot. It is a man's unshod foot so we would expect its length to be somewhat variable according to the district and its people, and we do find it given by metrologists in the range 246 to 252 mm (9.7 to 9.9 in). It is frequently taken as 252 mm, or $\frac{3}{4}$ of a foot of 18 digits, as we shall see. The human foot is one of the primitive body measures whose origins are lost in antiquity, but it appears repeatedly in the written record over the centuries.

In the second appendix to the sixth group of the Laws of Aethelstan[1] dating from the tenth century, in which the trial by ordeal is described, it is laid down that the nine feet the hot iron is to be carried 'shall be measured by the foot of him that goes to the trial'. Likewise, in the *Dean of Lismore's Book*[2] — a collection of Gaelic poems gathered in the first half of the sixteenth century but composed long before — we read of the contrived death of Diarmad O'Duine when Finn asks him to measure the slain boar. The translation runs, in part:

> 'Measure, Diarmad, the boar from the snout,
> And tell how many feet's the brute in length'. . .
> Along the back he measures now the boar . . .
> 'Measure it the other way against the hair,
> And measure, Diarmad, carefully the boar'. . .
> He went, the errand grievous was and sad
> And measured for them once again the boar.
> Th'envenomed pointed bristle sharply pierced
> The sole of him, the bravest in the field,
> Then fell and lay upon the grassy plain . . .

Recently, Grierson[3] has suggested the natural foot as the key to a mystery relating to Old St. Paul's Cathedral. The length inside was given as 690 feet in 1313 while a measurement in 1657 gave the length as 560 feet.[4] The cathedral had not been foreshortened in the interval, so either a gross error had been made in one or other measurement or the feet used were not equal. The seventeenth-century measurement would, of course, be in statutory feet (our present measure). If the fourteenth-century measure was made in natural, that is human, feet, the agreement is excellent, for 690 feet each 9.9 inches is 569¼ feet. The agreement is sufficiently good for the hypothesis advanced to be correct.

This foot, or something like it, appears in the Venedotian Code of north Wales of the tenth century where it states (Book II, Section XVII)[5] '5. And that measure Dyvnwal measured by a barley corn; 3 lengths of a barley corn in the inch: 3 inches in a palm breadth: 3 palm breadths in a foot . . .' This would give a 9-inch foot, the so-called Welsh foot, but the natural foot is defined quite precisely in the entry for 26 August 1395 in the *York Memorandum Book*[6] which, after giving the usual entries of 3 barley corns to the inch, 12 inches to the foot, etc., proceeds to state: 'and 3 inches make a palm and three palms and three [barley] corns make a foot . . .' (et tres pollices faciunt palmam; et tres palme et tres grana faciunt pedem . . .). Here we have the usual entry relating to a twelve-inch foot followed by an entry relating to a customary foot of 3 palms and 3 barley corns, that is a foot of about 10 inches.

The Drusian Foot and the Pes Manualis

An interesting local measure which in time had far from local significance is the foot used by the Tungri, a tribe which inhabited the regions of the lower Rhine north of Liège. Its length was recorded a few years before the birth of Christ by the Roman general Nero Claudius Drusus and is given as 'two digits longer than the Roman foot', that is 1⅛ Roman feet.[7] The Roman foot of 16 digits measured 296 mm. This 18-digit foot would measure 333 mm (13.11 inches) and is known to metrologists by the alternative names of the Drusian foot or the Great Northern foot. This is only one of several feet known to have been in use in various parts of the world at various times, all of which measure 330–335 mm. (Flinders Petrie[8] lists no fewer than ten, but the temptation to think of them all as one and the same must be resisted.) The Drusian foot cannot relate to any average length of the human foot as it is too long. Rather it is related to a measurement generated by hand *(ad manus)*. It is the *pes manualis* of Latin documents (see Ch.III p.45), and the *pied manuel* of French documents.[9] W. H. Prior, who wrote the earliest scholarly article of this century on the meaning of metrological terms, could make nothing of this term[10] but the word is described in an early mediaeval text, thus:[11]

> 'For it must be known that feet have two meanings according to the customs of the ancients. One, that which is the natural foot: the other, improperly taking the name of foot measured off with the hands, for the manual foot exceeds the natural foot to the extent of the thumb extended lengthwise'. (Illud enim sciendum est, quod pes dupliciter secundum morem antiquorum pronuntiatur: uno modo eo quod sit naturaliter pes, alio modo quod usurpative per manus metiatur tantum enim praecellit pes manualis pedem naturalem, quantum pollex in longitudinem protendi potest).

A natural foot of nearly 10 inches plus a thumb's length of about 3 inches generates a measure close to the accepted value of 13.1 inches. The meaning has been further explained by Prell[12] and by Grierson.[13] If a stick or rod is grasped by two hands with the thumbs extended and touching, a *pes manualis* is the distance between the extremities of the hands, and the stick can be measured by moving hand over hand along its length. Such measures are, by their nature, approximate, but in course of time they become standardized. This, then, was the foot of the Tungri, the Drusian foot, which continued in use for many centuries. It was brought to England, as we shall see, by the Saxon invaders following the withdrawal of the Roman armies from Britain early in the fifth century AD.[14] This foot was used extensively on the continent and in England as a building unit. Flinders Petrie[15], a very distinguished archaeologist of the late nineteenth century, believed that this foot was the most common English building unit in the Middle Ages, but in view of the lack of extant measures this is unlikely.

The Shaftment

The half-*pes manualis,* called a *shaftment* (fist with thumb out-stretched), is also noticed in northern Europe and measuring sticks based on it are extant. The word appears in the Anglo-Saxon document 'Pax', dated AD 910-1060 by Liebermann,[16] which defines the bounds of the king's peace from his residence as: 'That is 3 miles, 3 furlongs, 3* acre's breadths, 9 feet, 9 shaftments and 9 barley corns' (ðaet is III mila, III furlang, III* aecera braeðe, IX fota, IX *scaeftemunde,* IX berecorna).

The version given in the compendium of the old laws known as the *Leges Henrici Primi* (Laws of Henry I) in cap. 16.1 reads:[17] 'that is to say 3 miles, 3 furlongs, the breadth of 3 acres, 9 feet, the breadth of 9 hands and of 9 grains of barley' (hoc est iii miliaria et iii quarentine et iii acre latitudine et ix pedes et ix palme et ix grana ordei).

But a hand's breadth was not a shaftment. This is one of the several errors of commission and omission present in the various versions of the extent of the king's peace. The version in 'Quadripartitus'[18] leaves out shaftments and hands altogether, going straight from nine feet to nine barley corns, not that the real distance throughout which the king's peace was to extend was based on a formula of such exactitude, down to the nearest $\frac{1}{3}$ inch or barley corn. How far then did the king's peace extend? The editor[19] of the 'Laws of Henry I' agrees that most of the manuscripts give the number of acres breadths as IX but he prefers the reading of III as given in 'Pax' and in the 'Quadripartitus'. It will be readily conceded that in rapid copying the number IX frequently was written as III and vice versa. If we assume the formula should actually read: '3

*Text says IX, that is 9. See below and p.30.

miles, 3 furlongs, 9 acres breadths, 3 *perches,* 9 feet, 9 shaftments, 9 *hand's breadths* and 9 barley corns', retaining the reading IX acres breadths and restoring the 3 perches which had been omitted and the 9 hand's breadths which appear in some versions but not in others, then the formula is very close indeed to 3½ miles, which in all probability was the distance intended in this poetic play on threes and nines.[20]

The word 'shaftment' was still in use in the later Middle Ages for the common objects of daily life for in the *Extent of Lawling* of 1310 we read of the service to be given by a customary tenant, Joice atte Shameles, in part payment for his rent and the provision made for his food and drink during his period of service in the fields:[21]

> And be it known that at the ale-boon he ought to have food twice a day, that is to say, for dinner a wheaten loaf, milk and cheese and in the evening he shall have a wheaten loaf, ale, pottage, a platter of meat and a dish of bread and milk with cheese. *And he shall sit drinking with his fellows so long as five candles of Tallow each a shaftment in length are burning together.*★

Again in the Coventry Leet Book of 1474 for wood sizes we find:[22]

> And his ffagott of wodde of an ob [½ d] shall be iii schaftmond and a halfe about and a yerde of length. Ane his ffagott of 1 d schale be VII schaftmond about, kepyng the same length.

The Drusian foot and the shaftment were therefore of north European origin and known since the beginning of the Christian era. Tables have come down to us in which this foot, *pes manualis,* and *pes ad manus* is mentioned, while in the Middle Ages (1218) it takes the form *'ad pedem palme'*.[23] Two tables, the *Mensurarum Tabula Balbi* and the *Tabula codicis Gudiani,*[24] state that the Roman land rod, the *decempeda,* has 10 feet, each of 16 digits, while the *pertica* (perch) has 12 feet each of 18 digits, that is 12 Drusian feet.† Unfortunately the many texts do not have the merit of consistency for we also frequently find a *pertica* of 10 feet.[27] Isidore[28] speaks of perches sometimes 10 feet, sometimes with 2 additional feet; some of 15, some of 17 feet. Some were of local or provincial importance only, others were more widely used. Different perches were used depending on the richness of the land. Marsh or woodland was measured by a larger perch than worked, arable land, for a bigger area of poor land was needed to generate the worth of the products of a smaller area of rich land.

★The burning of a candle was frequently used as a measure of time. A further example of this is given later (Ch.VI p.79).

†Du Cange[25] quotes an Italian text of the eighth century containing the words *'pedes manualis'*. A number of wooden and iron length measures based on the Drusian foot are extant in France.[26]

The Historical Setting

The natural and the Drusian foot have played a significant role in English metrology. We shall have further need of them later when considering developments after the Roman withdrawal from England. Alaric the Visigoth sacked Rome in AD 410. Troops were recalled from all boundaries of the Empire. An appeal to Rome for help from defenceless England was answered by the Emperor Honorius telling the population to arm and defend itself against raiders and incursions from those parts of the island that had not submitted to Rome, but not to expect help from the central authorities.

The *Anglo-Saxon Chronicle* for the year 418 states: 'In this year the Romans collected all their treasures that were in Britain and hid some in the earth, that no man might afterwards find them; and conveyed some with them to Gaul'. The Roman withdrawal spelt disaster. Urban life rapidly declined. There are no datable finds, for example, in Winchester, the *Venta Belgarum* of the Romans, between AD 410 and 650. Lankhills cemetery outside the walls ceased to be used in about 425. We are told[29] of large desolate areas within the city walls from the seventh to the ninth centuries, with no sign left that this was a township. Only in the last decade of the ninth century do we find a semblance of urban life returning to this once important centre. Even in the Poll Tax return of 1377 Winchester's population is shown to be only 2,700, ranking twenty-ninth in the kingdom, and it was not much greater (population 3,800) at the time of the tarrage of 1416.[30]

The Britons, left defenceless by Rome, fell easy prey to the incursions of the Picts from Scotland and the Scots from Ireland. Gildas,[31] the first British historian, recounts the appeal of Vortigern to the Saxons for help, and Bede[32] tells of the arrival of Hengist and Horsa in 449. Having once invited the visitors as protectors it was hard to restrict the inflow of their peoples. The situation became delicate, then critical, as the helpers assumed the role of masters. Further incursions of 477 led to major battles between the Saxons and the Britons in 485 and 491, as the *Anglo-Saxon Chronicle* reports, with the Britons of what was to be Sussex virtually exterminated. Bede says that the invaders were drawn from three very powerful sectors of Germanic people, the Saxons, the Angles, and the Jutes. It is now believed that the main Saxon concentration was in Holstein, that of the Angles in Jutland, and that of the Jutes to the east of the lower Rhine.[33]

Cerdic, destined to be the founder of the West Saxon dynasty in England, arrived according to tradition in 495 and by 501 had conquered that which was subsequently called Wessex. The Britons, however, won a signal victory at Mons Badonicus which led to peace for a generation in which there was no further encroachment on British-held territory. This check was so severe as to cause some of the Germanic settlers to leave Britain to seek fresh fields. The first phase of the conquest of Britain was over. The greater part of southern England had been occupied, but with the defeat at Mons Badonicus a reverse migration to the continent took place during the first half of the sixth century. Saxon fortunes rose again after their victory at Bedcan Ford in 571 by which they regained lost land, and slowly the conquest of England recommenced with such success that by about 615 the Angles and Jutes had reached the Irish Channel, with the Saxons in all of southern England. The continental peoples had made themselves masters of what is now modern England. The Britons were either slain or driven into Wales and the north.

Seven kingdoms emerged, to become known as the Heptarchy: Essex, Wessex, and Sussex were all predominately Saxon; Kent was peopled predominantly by the Jutes; the Angles were to be found in East Anglia, Mercia, and Northumbria. The kingdoms vied with each other for supremacy. Kent predominated, with Aethelbert as king at the time of Augustine's arrival in 597, to be succeeded by Northumbria under Edwin, and then Mercia, which reached its greatest peak under Offa in the closing years of the eighth century. Finally Wessex, in the person of Egbert, conquered Mercia in 829. He was recognized as overlord of Northumbria and absorbed Kent, Sussex, and Essex into the kingdom of the West Saxons. Cornwall was occupied in 838 and, with the submission of all the kings, Egbert became the first true King of All England. He died the next year.

Internal strife is one thing, external attack is another, and even before the forging of England was complete new invaders appeared off the coasts. At first the raiders were from Norway. An attack was made at Portland and in 793 and 794 Lindesfarne and Jarrow were raided and plundered, as the *Anglo-Saxon Chronicle* tells us. The main Norse settlements were not in England but in the extreme north of Scotland and in the Scottish islands. The Vikings who changed the face of England were the peoples from a little further south; the lands bordering the Western Baltic, Denmark, and parts of Frisia and Saxony, known to the Franks collectively as the 'Kingdom of the Danes'. Raids preceded the major invasion at Sheppey in 835. There were more than twelve separate attacks on England during the next thirty years[34], by which time there was a single but major Danish army in England which moved around the country for nine years in a manner rather reminiscent of Hannibal in Italy. Finally the land was divided between Alfred and Guthrum in 886 following the invasion of that year which left only Wessex as a unit. Guthrum received north-east England — the 'Danelaw' (see Ch.IV p.54) — but this did not bring peace. A vast armada of 300 ships brought a Danish army to Kent in 892. The times were desperate indeed. Alfred died in 899 but his son Edward carried on the struggle and embarked on the reduction of the Danelaw. The re-conquest was completed by Edward's son Aethelstan. England was again a unit and it has remained so ever since. But the last had not been heard of the Danes. They were to become kings of England for thirty years in the first half of the eleventh century but on the death of Harthacnut the throne reverted to the Saxon line in 1042.

It is against this historic outline that we must measure the contributions of the Anglo-Saxons to English metrology. While much that was Roman was retained, the Saxons, Angles, and Jutes in their turn brought much with them.

NOTES AND REFERENCES: CHAPTER 2

1. ATTENBOROUGH, F. L. (ed.). *The Laws of the Earliest English Kings,* p.171. Cambridge: C.U.P. 1922.
2. MCLAUGHLAN, Revd. Thomas (ed. & tr.) *Dean of Lismore's Book.* Edinburgh: Edmonson & Douglas 1862.
3. GRIERSON, Philip. *English Linear Measures — The Stenton Lecture* 1971 p.23. Reading: University of Reading 1972.
4. COOK, G. H. *Old St. Paul's Cathedral.* p.105. London: Phoenix House 1955.
5. OWEN, A. (ed.) *Ancient Laws and Institutes of Wales* London: Record Commission 1841.
6. SELLERS, M. (ed.) *York Memorandum Book.* Part 1, p.142. Durham: Surtees Society 1912.
7. HULTSCH, Friedrich Otto. *Metrologicorum Scriptorum Reliquiae* Vol.II, pp.34, 61. Stuttgart: Teubner 1971 (Reprint).
   HUSSEY, Robert. *Ancient Weights and Money* p.239. Oxford: J. H. Parker 1836.
   GRIERSON, Philip. Loc. cit. p.35.
   SKINNER, F. G. The English Yard and Pound Weight. *Bulletin of the British Society for the History of Science* (1952) Vol.1, No.7, pp.179-87. London.
8. PETRIE, W. Flinders. *Inductive Metrology* (Synoptic Table) London: Hargrove Saunders 1877.
9. MACHABEY, A. *La Mètrologie dans les Musées de Province.* p.47 Paris: *Revue de Mètrologie Pratique et Légale, et Centre National de la Recherche Scientifique* 1962.
10. PRIOR, W. H. Notes on the Weights and Measures of Mediaeval England. *Bulletin du Cange* (1924) Vol.1, p.144. Paris.
11. HULTSCH, Friedrich Otto. Loc. cit. Vol.II, pp.137-8.
    HALL, Hubert & NICHOLAS, Frieda J. (eds.) *Select Tracts and Table Books Relating to English Weights and Measures* (1100-1742). Camden Miscellany Vol.XV, p.4. London: Camden Society 1929.
12. PRELL, Heinrich. *Bemerkungen zur Geschichte der Englischen Langenmass-System:* Berichte uber die Verhandlungen der Sachsischen Akademie der Wissenschaften zu Leipzig; Mathematische-naturwissenschaftliche Klasse, Band 104, Heft 4 Berlin 1962. See note 21, p.40.
13. GRIERSON, Philip. Loc. cit. p.35 Appendix.
14. This view is consistent, at least, with the apparent lack of use of the Drusian foot by the Romans in Britain. See: JONES, R. P. Duncan. Length Units in Roman

Town Planning: The Pes Monetalis and the Pes Drusianus. *Britannia* (1980) Vol. XI, pp.127-133. London.
15. PETRIE, W. Flinders. Loc. cit. p.107.
16. LIEBERMANN, F. *Die Gesetze der Angelsachsen* Vol. I, p.390. Halle a.S.: M. Niemeyer 1903-16.
17. DOWNER, L. J. (ed. & tr.) *Leges Henrici Primi* p.121. Oxford: Clarendon Press 1972.
    LIEBERMANN, F. Loc. cit. p.559.
18. LIEBERMANN, F. Loc. cit. p.391.
19. DOWNER, L. J. Loc. cit. p.334.
20. GRIERSON, Philip. Loc. cit. p.28.
21. NICHOLS, J. F. (ed.) The Extent of Lawling in the Custody of Essex AD 1310, *Trans. Essex Arch. Soc.* (1933) Vol.XX, pt.2, p.185. Colchester.
22. HARRIS, M. D. (ed.) *Coventry Leet Book* p.399. London: E.E.T.S. 1907-13.
23. *Patent Roll. Henry III 1216-25 AD.* pp.162-3. London: H.M.S.O. 1901.
24. HULTSCH, Friedrich Otto. Loc. cit. Vol.II, p.125, No.129 and p.129, No.134.
25. DU CANGE, Charles du Fresne. *Glossarium Mediae et Infirmae Latinitatis* Vol.VI, p.291. Niort: L. Favre 1883-7.
26. MACHABEY, A. Loc. cit. pp.108-9.
27. HULTSCH, Friedrich Otto. Loc. cit. Vol.II, p.107, No.123.
28. Ibid. Vol.II, p.136, No.137.
29. BIDDLE, Martin. Winchester, the Development of an Early Capital; *Vor - und Frühformen der europäischen Stadt im Mittelalter. Symposium in Reinhausen bei Gottingen 18-24 April 1972,* Teil 1. pp. 229-261. Gottingen.
30. ATKINSON, T. *Elizabethan Winchester* pp.30-1. London: Faber & Faber 1963.
31. GILES, J. A. (tr.) *The Works of Gildas and Nennius* p.23 London: James Bohn 1841.
32. BEDE. *A History of the English Church and People.* PRICE, Leo Sherley (tr.) Book I. Ch.15. Harmondsworth: Penguin Books 1972.
33. STENTON, Sir F. M. *Anglo-Saxon England* 3rd ed. p.11. Oxford: Clarendon Press 1971.
34. Ibid. p.243.

CHAPTER III

The Saxon Gyrd, the Rod, and the Acre

12 inches	=	1 foot
3 feet	=	1 yard
$5\frac{1}{2}$ yards	=	1 rod, pole, or perch
40 rods	=	1 furlong
8 furlongs	=	1 mile

Thus ran the table of length which many of us had to learn by heart as school-children. Sometimes there was an additional entry in the table at the beginning; '3 barley corns = 1 inch'. By the time of this writer's school-days it had been eliminated, but previous generations knew it well. Indeed, it is to be found inscribed on a surveyor's rule made by Humfrey Cole in 1574 (see Fig. 17). There seems to have been an overpowering desire throughout the centuries to have the measures relating to something natural rather than to something artificial. In English metrology we have not only the three barley corns already mentioned, but the weight table began '24 grains (barley) = 1 pennyweight'. We have seen already the role played by parts of the body as units of length in the early days. The carob seed was the basis of the ancient systems of weight and the ox was the unit of value. In the last decade of the eighteenth century, for the metric system, the metre was intended to be one ten-millionth part of the Earth's meridian, pole to equator through Paris, and from the metre was derived the unit of mass and hence of volume.

But in England, lengths were not in fact derived by lining up barley-corns any more than weights were establishing by weighing wheat or barley grains. Length measures were established by reference to other standards of length and the weight standard when lost was regenerated by going back to an earlier extant weight, massive and in good condition, to increase the accuracy. Never in the nation's history was recourse made to cereal grains, no matter what the first entry in the tables or statute books said. The metric system no longer rests on a measurement of the earth. In every-day practical terms recourse is made to a metal bar. The ox gave way to the coin and cereal grains to metal weights and though man begins with nature he is not long in turning to the artificial in preference to the natural. Only in the twentieth century have we seen the scheme beginning to go full circle back to natural units with the wavelength, or, very recently, the speed of light defining the unit of length, the metre.

If we now look again at the table of length we might agree that the relation of inches to feet and feet to yards is not unreasonable, nor is that relating rods to furlongs to miles, but the entry '$5\frac{1}{2}$ yards = 1 rod' strikes a discordant note, for who in his right mind would, from choice, establish any table of relationships involving fractional parts? Why not 5, or 6, or 10 yards to the rod? Whenever fractional parts are encountered, be it in the table of length or of weight, one can be sure there is a story to tell and it can also be taken for granted that the table was not set up *de novo* but that two or more systems were being fused together to meet the needs of the times.

We shall begin by considering the rod and furlong and the units of area which they define, the acre, the rood, and the hide, leaving the mile (a relative newcomer) and the league to be considered separately later on.

To begin this way is to recognize the importance of land in the minds of the people and in the development of the nation. Remove the Anglo-Saxon land charters and the record of land transfers and the taxes derived from them and the records — the *Cartularium Saxonicum,* Thorpe's *Diplomatarium,* and the *Codex Diplomaticus* — become slim volumes indeed, and the situation is no different in later times. The Rolls of the Middle Ages — Pipe, Patent, Charter, Fine, Chancery, Liberate, Calendar, and Close — would likewise shrink to an unrecognizable size. The Red Book of the Exchequer would disappear altogether. As we study the metrology of England we have to remember that wealth meant land and vice versa. The king and the nobles all held land. It was their first asset. Therefore nothing should happen to change the way in which it was measured.

The Rod and the Acre, the Furlong and the Acre's Breadth

The pivot of the table of length is the *rod.* It generates not only the *furlong* as a unit of length but also the *acre,* a unit of area used to this day to define land holdings. The words 'acre' and 'furlong' are Saxon in origin: *'aecer'* meaning a field or sown land, from the Old Saxon *'akkar'* which itself is derived from the Latin *'ager'* = a field, with no notion of a definite area; and *'furlang'* from *'furh'* = a furrow. Indeed the acre's length was the furlong or furrow's length of 40 rods.

In a land grant[1] by King Aethelred I to the Earl Aelfstan of about 870 we read in part; 'and six acres of meadow by the Frome . . .' (and sex made eres be Frume . . .) and also in the earlier grant of AD 732 by Ethelbert II King of Kent to the Abbot Dunn[2] 'also, I have given to it 100 acres of the same estate . . .'. Here we see the word 'acre' being used without need of definition or explanation. The word was well understood even then.

The concept of it being a rectangle 40 rods long and 4 rods wide (see Fig. 18) is clearly implied in the Burghal Hidage of the late ninth century as will be discussed later below (pp.39-43). The *Chronicle of Battle Abbey*[3] (twelfth century) is quite precise: 'An acre is forty perches in length and four in width'. The late thirteenth- early fourteenth-century *Registrum Vulgariter Nuncupatum — The Record of Caernarvon*[4] is equally clear: '40 perches in length and 4 in breadth make an acre of land'. The *rood* was the strip of land, 1 furlong by 1 rod, that is a quarter of an acre, and the equivalence of the *rood* and the quarter acre is reflected in the charter of AD 963 in which land is gifted by Bishop Aethelwold to Medeshamstede, Peterborough:[5] '14 quarter acres and 3 roods of seed (as tithes)' (XIIII giorde sed, III roda sed). The terms *'giorde'* and *'roda'* have the same meaning.[6]

Occasionally a rood is called a *yard* of land, the Latin equivalent being *virga* or *virgata,* that is *virgate* in English. Sometimes the word *acre* was used as a unit of length meaning, as we shall see, the acre's breadth of four rods and likewise the word *'roda'* could mean a length of one rod, or the breadth of the rood, as is shown in the charter leasing lands from Werfrith, Bishop of Worcester to Aethelred, Earl of Mercia and his wife Aethelflaed (eldest daughter of King Alfred) dated AD 904.[7] '28 rods long . . . 24 rods broad' (XXVIII roda lang . . . XXIIII roda brad). The length of the furrow, the furlong, was the distance a team of eight oxen could pull the plough without having the animals puffing and panting or requiring a rest. The furrows were a foot or so wide[8], giving 66 furlongs to the acre, the ploughing of which was thought to be a day's work, or rather a morning's work,[9] for in the afternoon the oxen went to pasture. Thus the acre corresponds to the German *morgen* as the land ploughable between dawn and noon or early afternoon. We read from Aelfric's *Dialogue*[10] of about AD 1000: 'having yoked my oxen and fastened my share and coulter [the vertical cutter placed in front of the plough share] I am bound to plough every day a full acre or more.' In later times oxen were supplemented or replaced by horses so that we read in 1310 in the Custody of Essex:[11] 'Each plough team should consist of 6 horses and 4 oxen and the plough team is ordinarily able to plough one acre of land in a day'.

If the furrow was a foot wide, doubtless measured if measured at all by the shod foot of the ploughman, then man and beast would walk some $8\frac{1}{4}$ miles (66 furlongs) while ploughing an acre. The ox team might plough at a rate of two or so miles per hour so that, with short rests at the end of each furrow when the plough was turned round, the work could be accomplished in about five hours which is about the daylight span till midday at the ploughing season.

This calculation is very similar to that given in Walter of Henley's *Husbandry* of *c.*1286.[12]

> First know that a quartentine [acre] ought to have 4 roddes in breadth and 40 roddes in length and the Kings rodde [or perch] is sixteene foote and a halfe and then hathe the acre in breadth 66 foote. When you have gone up and down 33 times with a furrowe of a fote broade, then is an acre ploughed.

A little later in his text Walter says that, having gone 72 furlongs, the beasts will have gone 6 leagues*, and he continues: 'Now shall the steere or oxe be verie poore whiche cannot goe faire and softely three leagues of waye and retourne again from morning tylle noone'. Clearly there would be variations from one part of the country to another, depending if on nothing else on the nature of the soil and on the weather, but the acre was originally a work unit in Saxon and mediaeval times, notionally a rectangle, 40 rods by 4, and only later thought of as an acre of 160 square rods.† The acre's breadth of 4 rods (66 feet, 20.12 m) itself became a measure of length in its own right and has been so used throughout the centuries, for example in Saxon times as a measure of the defensive walls of the *burh* (see page 39 et seq.); as the length of the surveyor's chain introduced in the seventeenth century by Edmund Gunter for the decimalization of the acre; as the distance apart of the wickets on a cricket pitch; and as the road allowance in parts of the Commonwealth overseas, being the width of the strip of land acquired by a municipality for the construction of a road and its shoulders. We find the acre's breadth appearing in charters of the Saxon, Norman, and Plantagenet periods as a unit of length. For example, Charter No. 542 of Dale Abbey dated as being prior to 1231 reads in part:[12] 'in width two acres' (et latitudinem duarum acrarum terre . . .).

We tend to think of town planning as something very modern, but new lines of research have revealed that towns were in fact planned in the modern sense in Roman and mediaeval times and that the unit used in laying out the terrain for building was on a module which itself in mediaeval times was a multiple of the acre's breadth of four poles. Preliminary studies have shown Salisbury, Winchester, London, Bury St. Edmunds, Colchester, Chichester, Wareham, and Wallingford to have been laid out on modules of 16 or 20 poles.[14] For instance, New Salisbury, dating from 1219, had streets running from north to south on a 16-pole unit while east to west the module was 20 poles. Winchester was laid out on a 16-pole unit. Part of London (Upper Thames Street) was subdivided on a 24-pole basis and the Bury St. Edmunds extension of the latter half of the eleventh century was on a grid of 32 poles subdivided into 20 and 12 poles. Similar results have emerged for the other towns and cities mentioned.

While there are obviously inaccuracies and variations which might lead one to doubt the correctness of some of these results, these doubts are dispelled when repeated modules of the same length within a stated error limit appear, for the probability that these repeated lengths are random is vanishingly small. For example the main street frontage at Winchester on the north side yields seven 16-pole units (total 112) each accurate to within 1.75 poles which is no great precision but the probability that this sequence of seven lengths is

*This shows the thirteenth-century English league to be $1\frac{1}{2}$ miles. See Ch.V. p.75.

†As for example in the Statute 24 Henry VIII c 5 s.5 of 1532/3 'And it is further ordered that the acres shall be acompted after the rate of VIII^{XX} (i.e. $8 \times 20 = 160$) perches for the acre . . .'.

random, occurring purely by chance, is only 1 in 50,000.[14] It is highly likely therefore that the interpretation given is correct and that the module used in laying out new towns or extending old ones was a multiple of four poles (the acre's breadth) from the ninth to the thirteenth century.

The Gyrd and the Rod

It remains to demonstrate that the rod or *gyrd* was Saxon in origin, was in use in England prior to the Conquest of 1066, and was of the same length as that of twentieth-century England, that is, 16½ feet (5.029 m), and in so doing the magnitudes of the acre and furlong will be shown as the same in Saxon times as at the present through the relationships given earlier. To do this, recourse will be made to two pieces of evidence from the many that could be given, both of which refer to the city of Winchester, the old Saxon capital. The royal residence shifted to London only towards the latter part of the reign of Edward the Confessor (1042-66) with the construction of his church and palace at Westminster.* In Saxon documents the word *gyrd* clearly means a length, but how long was it? We have translated it as *rod.*

The first piece of evidence for the length of the *gyrd* is drawn from the Burghal Hidage,[18] the second from the Manor of Godbegot[19] in the city of Winchester. These two are chosen because of their historical significance and authenticity.

The Burghal Hidage.[17] This Anglo-Saxon document gives a list of the Saxon *burhs* or fortified locations in the reign of Edward the Elder (901-24) in which the local inhabitants could find shelter from marauding armies and from which they might defend themselves and safeguard their few possessions. Though described in the early years of the tenth century, probably between 911 and 919,[20] the *burhs* of Wessex to which we shall refer were in being in 892. The street plan of Winchester itself within the walls was laid out before 904.[21] The fortifications described belong therefore to the late ninth century and are in all likelihood to be associated with Alfred the Great (871-99).[22]

To each town named in the document a number of hides of land is attributed. (The *hide* will be discussed in the next chapter. For the present it may be thought of as 120 acres.) The text then says:

For the maintenance and defence of an acre's breadth of wall [of the *burh*], 16 hides

*It has been suggested[15] that London was beginning to take precedence over Winchester in the previous century because London is named first in three of five extant manuscripts of Edgar's will[16] for uniform weights and measures: 'And there shall be one system of measurement and one standard of weights as is in use in London and Winchester'. However the other two manuscripts omit London altogether and speak only of Winchester. One of these is considered to be the original, with 'London and' a later interpolation. There is also further evidence displaying the vitality of Winchester in the tenth century, for the treasury was located at Winchester not London from the days of Cnut till the 1180s.[17]

are required. If every hide is represented by 1 man, then every *gyrd* of wall can be manned by 4 men. Then for the maintenance of 20 *gyrds* of wall 80 hides are required and for a furlong, 160 hides by the same reckoning as I have stated above.

This tells us unequivocally that the acre's breadth is 4 *gyrds* and the furlong is 40 *gyrds* in length. Hence the *gyrd* played the same role to the Saxon acre and furlong as the rod did to the modern acre and furlong. What then was the length of the *gyrd*? The word appears frequently in Saxon charters from the eighth century[23] (if not earlier) to the eleventh.[24] To endeavour to answer, we now refer to the entry in the Hidage for Winchester: '. . . and to Winchester, 2400 hides'. At one man per hide and 4 men per *gyrd*, the defences would have been 600 *gyrds* long. Of the first defences, which were probably of wood, nothing remains. The perimeter of the later walls is given as 603.3 rods (3034 m),[25] sufficiently close as to warrant the identification of the *gyrd* with the modern rod of 16.5 feet (5.03 m) with the corresponding identity of the Saxon and modern acres and furlongs.

However two objections to this deduction could be made. The first is to the calculations, which do not seem to work for the neighbouring town of Southampton which is attributed only 150 *hides* and so only 150 men for the defence of its walls. Southampton was not *that* insignificant in those days (it had two moneyers in 930 to Winchester's six according to the second group of the Laws of Aethelstan[26] Cap 14, §2), so this casts doubt on the validity of the figures given in the Hidage, because a perimeter for this *burh* of only $206\frac{1}{4}$ yards is unlikely, being far too short. The second objection is that the defences of Saxon Winchester followed closely the line of the Roman walls and to a very great extent used the Roman foundations.[27–29] The perimeter therefore could only come to 600 *gyrds* (rods) by chance because the Romans did not employ a measure approximating to $16\frac{1}{2}$ feet as one of their standards.

The problem of the shortness of the wall at Southampton has been solved recently after years of discussion.[30] There was a Roman fort of great strength three quarters of a mile north of the town site. Its walls were of stone ten feet thick, it had three bastions, two ditches, and an outer bank, and such was its durability that it lasted until Elizabethan times when it was largely demolished. It is therefore likely that the inhabitants, rather than build a new and probably weaker shelter, would seek refuge in the fort in times of trouble. The fort was protected by water (the river Itchen) on three sides. Attackers could penetrate only on the fourth, the landward side, so that this would be the only stretch of wall requiring to be manned apart perhaps from the occasional look-out on the river sides. Excavations at the fort *(Clausentum)* have shown the walls to be about 625 feet (37.9 rods) long. The 150 men at 4 men to the *gyrd* could man 37.5 *gyrds* of wall, again close enough to the measured figure of 37.9 rods as to restore confidence in the Hidage figures.

Nor is this an isolated example. The Hidage gives 500 *hides* to nearby Porchester. Again, standing Roman fortifications were most probably used with water defences requiring only the landward side to be manned. The 500 men could serve a wall 125 *gyrds* in length. The measured wall at Porchester is approximately 126.7 rods in length. Clearly, if there was a water barrier the Romans and the Saxons made good use of it with the latter only counting on having to defend the landward portion of the perimeter. The Burghal Hidage is therefore an accurate description of the various portions of the fortifications requiring manpower.[31]

The objection with respect to Winchester's walls is not so easily disposed of, other than on the basis of chance, for the Saxon walls were largely a rebuilding of the Roman, so to obtain confirmation of the evidence of the Burghal Hidage respecting Winchester we turn now to the Manor of Godbegot within that city.

The Manor of Godbegot (Goodbegot). A Saxon character of Aethelred II (978-1016) dated 1012[24] records the King's gift to his wife Aelgyfa (Emma of Normandy) of 'a certain estate on the north side of the city and near the market place'. The charter is in Latin except for the description of the bounds which are in Anglo-Saxon. A translation follows of the relevant parts of the charter as given by Goodman,[32] who in 1922 studied the history of the manor:

> (From the preamble)
> Wherefore, I, Aethelred, sole ruler of noble and wealthy Britain, to pleasure my lawful wife Aelgyfa have granted her a certain estate on the north side of the city and near the market place and in witness thereof I have given command to stamp this deed with my seal.
> On this estate stands a fair church erected in honour of St. Peter by a provost of the city named Aethelwine.
> (The bounds are given thus:)
> First, from the north-east corner of the church nine *gyrda* along the street out to marketplace; so up along the marketplace eight *gyrda* until it comes to Wistan's boundary; so along Wistan's boundary nine *gyrda* until it comes to the water pit; from the water pit ten *gyrda* along the street to the north-east corner of the church.

The boundaries of this estate remained unaltered for over nine centuries changing only in 1957 when a slice was taken off the north side during the widening of the present St. George Street.[33] The terrain of the Manor had been built and rebuilt upon several times in the course of these centuries but the frontages on the several sides did not change until 1957.

The old name of the present High Street was *Cyp* or Market Street. The church mentioned is that of St. Peter's in Macellis (meaning 'in the shambles'). It existed prior to 1012 but by the mid-sixteenth century it had fallen into a dilapidated condition. Its foundations were excavated during the road-widening process. As the manor can no longer be surveyed directly on the ground as in Goodman's day, recourse has to be made to the Tarrage Map of 1416* and to the 1873 Ordnance Survey 1/500 maps for Hampshire-Wiltshire, sheets XLI 13.13 and 13.14. In Figure 19 the length of the east side AB on the survey maps measures 146.6 feet instead of the 148.5 feet required by the charter (9 *gyrds*), if the *gyrd* and the rod are to be identical. However a re-examination[34] of the boundaries in the light of the Winton Domesday Surveys of *c*1110 and 1148 (the latter made at the instance of Bishop Henry de Blois) indicates that a slight modification can be made in Goodman's interpretation with resulting improvement in the agreement between the charter and this measurement. The twelfth-century Domesdays show that the roadway on the east side of the manor had been encroached upon sometime between 1012 and 1110. Stalls of a more-or-less permanent nature had been erected in the street. The survey of 1110 reports two, the 1148 survey six such stalls. It now appears that it is the front of these stalls which determines the location of the present frontage on the east side rather than the eleventh-century boundary of the manor. If a 16-foot allowance in depth is made for these unauthorized stalls, the High Street frontages of today would fit very precisely the dimensions of the charter (eight *gyrda,* with the *gyrd* equal to the rod) and the measured east side would come much closer to the length equivalent to nine *gyrds*. The other sides also fit very well.[34]

It is therefore not unreasonable to conclude that the length of the Saxon *gyrd* is that of the rod of the twentieth century and the acre, the rod, and the furlong have not changed in thirteen centuries and probably much longer. An error that persists to this day is to identify the *gyrd* with the yard of three feet. The word '*gyrd*' means rod and was applied to the length of the perch, later becoming applied to the three-foot yard measure introduced in the twelfth century.

As for Emma, after Aethelred's death in 1016, ending a marriage of 14 years, she married Cnut the Great in 1017. After Cnut's death in 1035, Harold I (Harefoot) ruled till 1039, to be succeeded by Harthacnut, the son of Cnut and Emma, who ruled till 1042, when Edward (the Confessor), son of Aethelred and Emma, ascended the throne. Unlikely as it sounds, Emma appears to have disliked the revival of the kingship in Aethelred's line, even through her own son by her first marriage, and would have preferred the throne to have by-passed him. During the first months of Edward's reign the evidence indicates that she supported the claim of Magnus of Norway to the English crown. This led Edward in the autumn of 1043 to go to Winchester

*The Tarrage Survey was a census of ground rents due to the King. The roll is dated 1416. Winchester College has a copy of about 1417. The Tarrage Map is reproduced by Goodman.[32]

and seize his mother's lands and property.[35]* Some must have been returned to her, for on her death in 1052 she willed the land and the house of Godbegot to the Prior and convent of St. Swithin's, a gift confirmed by Edward in the same year and reconfirmed by William Rufus, son of the Conqueror, in 1096.[36] The property, which is shown on a sixteenth-century map of the city, remained with the Church (apart from a short time in 1541 when Henry VIII took it into his own hand), until the last years of the nineteenth century when it was sold to private owners by the Ecclesiastical Commissioners.

Today a jeweller's shop in the High Street (Fig. 20) proclaims the site of the manor. The following dates are shown on a sign above the door: 1052, 1558, 1896, 1908, 1910, 1940, and 1971. The first is the date of Edward's confirmation of his mother's will; the second is the date of the construction of the building on the manor site, the frontages of which are those of the present structure; the third gives the date of the construction of an hotel on the western side following its passage into private hands; the fourth marks the extension of the hotel and the restoration of the frontage onto the High Street. The remaining dates record modifications to the fabric of the building.

The Length of the Rod

We must now ask how the rod came to acquire the length it did, that is, 198 English inches or 503 cm. It has been argued that fifteen Drusian feet make a rod† and that the Jutes, coming from those parts of northern Europe which had previously been occupied by the Tungri, would naturally bring with them the units to which they were accustomed.[38] But five centuries separate the Tungri and the Jutes, and it cannot be said with certainty that the Drusian foot was a current measure then with the Jutes. The coincidence of 15 Drusian feet to better than 1 per cent with the rod should be treated with caution. But having said this we must note the frequent use of *manupedes* (Drusian feet) in the declared lengths of the woodland perches (see p.45). So the possibility exists that the rod was originally 15 Drusian feet but another explanation can also be advanced.

The score (20) and its multiples make their appearance repeatedly in Germany and in Scandinavia. The Carolingian monetary system reckoned 20 shillings to the pound. The gold shilling of the Kentish Saxons was worth 20 *sceattas* and weighed 20 grains of barley. In the earliest English poem, 'Widsith', we read of a necklace worth 600 shillings, the poem itself being in a

*The *Anglo-Saxon Chronicle* (C) for 1043 describes this falling-out in a kinder though less realistic light: 'Soon in this same year the king had all the lands which his mother owned confiscated for his own use and took from her all she possessed, an undescribable number of things of gold and silver because she had been too tight-fisted with him'.

†The suggestion that the rod is based on the Drusian foot may have been advanced by Petrie by implication when he wrote 'The Belgic foot of the Tungri is the basis of the present land measures which we thus see are neither Roman nor British in origin but Belgic'.[37]

Scandinavian setting. With 20 so favourite a number it is not unreasonable to suppose that the land measure might be 20 feet long — but *natural* feet. This suggestion has been advanced by Grierson[39] in his 1971 Stenton lecture. Twenty natural feet measure 504 cm, again well within 1 per cent of the rod's length, in fact within 0.2 per cent. On balance this must be considered to be the more probable origin.

The rod's length is itself a compromise. To make it much longer would make it too unwieldy for field use; to make it much shorter would add greatly to the number of times the measure had to be laid down and picked up again when measuring a piece of land. The twenty natural feet would be a happy medium, neither too long nor too short.

But although the rod may be the standard measure, if it is a wooden pole it is still an awkward object. It is therefore not surprising to find a rope, presumably knotted at intervals, being used on occasion to facilitate the measurement of land.[40] As we have seen, just such a knotted rope was used in ancient Egypt (see Fig. 2). True, the length of the rod would vary from village to village and from settlement to settlement depending on whose hands or feet were used, but the variation would not cause many problems. These suggestions cannot be advanced as certainties but in the light of the evidence they appear not unlikely. Since the late twelfth century the standard perch of the king appears repeatedly in the extant documents; for example a charter of Richard I dated 11 November 1198[41] reads in part: 'one hundred acres by the king's perch' (Centum acris per perticam regis). But not all measurements were by the king's perch. Variants have existed ranging from 16 to 37½ feet.

As late as 1531 we are told by Jakob Kobel[42] how to generate a rod (in Germany) and we are shown a woodcut (Fig.21) of the method. Sixteen men are to be lined up, heel to toe, as they come from church while someone marks off, from the aggregate length of their feet, the length of the legal rod. The woodcut shows the men wearing shoes.

Whether shoes are to be worn or not is left unspecified in the Assize of Weights and Measures attributed to King David of Scotland (1124-54) though the text is of the early 1300s.[43] Here we read, giving the text in modern English: 'The rod of land shall contain 6 ells which make 18 feet of an average-sized man, neither big nor small. The land rod in towns shall contain 20 of these feet'. Here we see that not only is there likely to be local variation as to what constitutes a man of average size but the statute itself defines two rods of different lengths to be used according to location. Nor was the 20-foot rod confined to Scotland. It is recorded in 1280 at York and in 1329 in Lincoln, again for building within the cities.[44]

Walter of Henley declares, as we have seen, that the 'Kings rodde' has 'sixteene foote and a halfe', as does the *Extent of Lawling* of 1310;[45] but in the anonymous *Husbandry* (not that of Walter of Henley but of the same period) written between 1286 and 1300 we read in Chapter 58:[46] '. . . and all the land ought to be measured with a rod of sixteen foot . . .'. This *Husbandry* in Chapter 63 goes on to explain further:

> And acres are not all of the same size because they measure them in some acres by the perch of 18 feet and in some by the perch of 20 feet and in some by the perch of 22 feet and in some by the perch of 24 feet.

The twelfth-century lease of the Manor of Aldovesnasa in Essex, belonging to St. Paul's Cathedral, also states that the perch is of 16 feet, while that for Belchamp in the same county gives the perch as 16½ feet.[47] The often quoted Chronicle of Battle Abbey (1066-76)[48] also reads 'The perch is sixteen feet in length', although this may be a scribal error for sixteen and a half feet. The earlier but equally anonymous text 'Seneschauchy', dated 1260-76, gives a general direction to the Steward:[49]

> Cap. 4. On his first visit to the manors, the steward ought to arrange for all the demesne lands of each manor to be measured by lawful men. He ought to know by the perch of the country how many acres there are in each field.

The Cartulary of Tutbury Priory (Staffordshire)[50] shows perches differing in length by only six inches; for example, Charter No. 61 of about 1200 speaks of 12 acres measured by the perch of 18 feet while No. 274 of 1210-20 gives a measurement of land by the perch of 18½ feet. Even this kind of fine adjustment was apparently felt necessary.

The Woodland Perch

If a man by the sweat of his brown has to deliver rent in cash or kind to the lord of the manor and pay taxes as well as maintain his family he must have the wherewithal to do it. If his land is poor and his neighbour's land is rich, his neighbour will have a smaller acre. Yields were based on the notional acre. The rents expected per acre had to be comparable, but in no way could marshy land yield as much as good arable land nor could woodland be equated with meadow. Thus we find the larger perches and the broad acres where productivity was low with a balance being achieved between area and productivity. Woodland was usually allotted a larger perch (the largest encountered by this writer being 37½ feet)[51] as were also recent reclamations from the forest *(assarts)*. For example, the Charter Rolls for Richard I (1189) gives:[52] '614 acres of assarts measured by the perch containing a length 25½ feet *manupedes*' (614 acres de essartis mensurates per perticam continentem in longitudine 25 pedes et dimidian per *manupedes*) and in the Close Rolls of Henry III for 1229[53] we read of the use of the king's perch for arable land, '3 acres by the king's perch . . .' (III acras ad perticam regis . . .), but for forest land the perch of 24 or 25 feet *manupedes* which had been in use since the time of Henry II in the preceding century is to be used, the text reading:[54]

> which rod contains usually either 24 or 25 feet *manupedem* of the time of King Henry, grandfather of the King, Richard, uncle of the King and of King John, father of our Lord the King . . .' (quod pertica solet vel XXIV vel XXV pedes manupedum temporibus H. regis avi regis, R regis avunculi, et J regis patris domini regis . . .),

but in 1231 there is an entry:[55] 'one acre of woodland by the king's perch' (1 acras bosci ad perticam regis). The perch of 20 feet is also mentioned in the early fourteenth century; the Calendar of Patent Rolls for Edward II 1321-24 under the date 4 February 1324 refers to the perch of the New Forest in the words: '223½ acres by the forest perch of 20 feet'. The length of the woodland rod could be quite variable, even down to the length of the standard rod of the king. Not only did these woodland perches usually contain a larger number of feet but the feet themselves were often *manupedes* or Drusian feet.

Even at one locality two perches could be used for measuring different sorts of land. In the Cartulary of Dale Abbey, Derbyshire there is an entry of the mid-thirteenth century which tells of a grant of 24 acres of land:[56]

> . . . 12 of his demesne within the old hedges of the arable field measured by the perch of 18 feet and 12 acres outside the old hedges for assarting, measured by the perch of 20 feet, at a rental of a pair of gilt spurs of 6 pence.

In the *Surveyor's Dialogue* of 1607[57] in an imaginary conversation between the Bailiff and the Surveyor, the discussion proceeds as follows:

B. 'But woods are always measured with the pole of 18 foote.'

S. 'It is as the buyer and seller agreeth, for there is no such matter decreed by any statute neither is any bound of necessity.'

B. 'Why is it then in use?'

S. 'I take it because in woodlands (for they are they that are thus measured) for sale they have in many places sundry void places and galles wherein grow little or no wood or very thin. And to supply these defects the buyer claimeth this supply by measure.'

The Building Perch

The perch as a unit was not restricted to measuring land. Recent excavations have shown it (or a measure of equal size) to have been in widespread use as a unit of width for dwelling-houses. Instances of this have been particularly numerous in twelfth-century Winchester. Most houses were of timber though some were of stone. Frequently the width module was 5 metres, that is, one rod of 16½ feet, as excavations by the Winchester Research Unit have shown.[58] Further, the frontages of the street properties of Winchester in the first half of the twelfth century are predominantly in terms of rods of 16½ feet and half rods.

AVERAGE TENEMENT FRONTAGES — WINCHESTER AD 1148[58]

	English feet	*Rods*	*Deviations* (ft)*
High Street	25	1½	+0.25 (1.0%)
Snidelingestret	41	2½	−0.25 (0.6%)
Brudenestret (a)	42	2½	+0.75 (1.8%)
Brudenestret (b)	51	3	+1.5 (2.9%)
Scowrtenestret (b)	42	2½	+0.75 (1.8%)
Alwarnestret	51	3	+1.5 (2.9%)
Sildwortenestret	73	4½	−1.25 (1.7%)
Wunegrestret	48	3	−1.5 (3.1%)
Tannerestret	34	2	+1.0 (2.9%)
Outside West Gate	50	3	+0.5 (1.0%)
Outside East Gate	65	4	−1.0 (1.5%)

(a) shorter later mediaeval streets
(b) longer early mediaeval streets
* The difference between the measured number of feet and the presumed number of rods.

The precision with which these average tenement frontages agree with multiples of the rod of 16½ feet indicates that throughout Winchester the rod was a module closely adhered to. This is further exemplified by recent studies on the remains of mediaeval villages.[59] Plans of houses from Sussex, Cornwall, Northamptonshire, Yorkshire, and Dorset have been shown to be based on widths, external or internal, equivalent to 5 metres or 16½ feet. A recent study[60] of Ely Cathedral, the building of which began between 1081 and 1093, indicates a module of one third of a rod, that is 5½ feet.

It must be conceded that a width of 5 metres or some multiple thereof does not prove that it was the land rod of 16½ feet that was used. It could have been a unit of the same length, that is, of 15 Drusian feet or for that matter 20 natural feet. Flinders Petrie[61] was firmly of the belief that the Drusian foot of 13.11 inches was the common English building unit of the Middle Ages and a case can be made that this was the foot used in the construction of the Old Minster at Winchester in the seventh century,[62] a structure proposed by Cynegils but actually built by his son Kenwalh of Wessex about AD 648. This cathedral continued in existence, being repeatedly rebuilt or added to, until construction began on the present cathedral in 1079 on a site immediately adjacent to it. The day after the dedication of the present structure on 8 April 1093 the demolition of the Old Minster was begun with scarcely a stone being left. Excavations extending over a decade have revealed the plan of the Old Minster and the measure used in its construction has been deduced largely from the evidence of the robber trenches dug to quarry out its foundations.

Looking back over the centuries we find one rod, that of the Saxons, being preserved and protected by statute or legal notice to the present, while others, used locally for particular purposes, have disappeared, but they are not long gone.

The Second Report of the Commissioners on Weights and Measures, 1820, in the Appendix, after stating the rod, pole, or perch to be $5\frac{1}{2}$ yards or $16\frac{1}{2}$ feet, continues:[63] 'In many counties a perch of 8 yards is used for fencing. The forest pole is 7 yards; in Sherwood Forest 25 feet. A coppice pole is 6 yards'. The Report goes on to list seven perches ranging from $16\frac{1}{2}$ to 24 ft in length in eight counties with Lancashire having no fewer than six in different parts of the county.

We should not be too surprised at these variations. People then lived and died without venturing further than twenty miles from the house in which they had been born. A local unit sufficed. As late as Holinshed (Henry VIII and Elizabeth I) we find a declaration, in the dedicatory epistle to his *Chronicle*, as follows:

> Indeed, I must needs confesse that until now of late, except it were from the parish where I dwell unto your Honour in Kent, or out of London where I was borne unto Oxford and Cambridge where I have been brought up, I never travelled 40 miles forthright at one journey in all my life . . .

In the early part of the sixteenth century we are told that there was little or no communication between the various parts of England, with no dependable means of transportation from the capital. The produce of the land was consumed locally.[64] As people travelled afield the need for uniformity grew but seldom beyond the county boundaries even up to the nineteenth century. The coming of the railway changed everything. Travel was then safer and more rapid. The county boundary was no longer the end of the world. Uniformity became not only desirable but a necessity. We can date true uniformity from the accession of Victoria, no earlier. It has been said that taxation was a strong force in the direction of standard land measures. This can hardly be so unless we are prepared to believe taxation began with the mid-nineteenth century, a date few would accept!

Throughout this discussion the acre has been defined as a length times a breadth based on the linear measure of the rod. One acquired a 'feel' for an acre by looking at the land. The concept of square measure, and the idea that the acre was 160 square rods, was a late arrival and dates from the sixteenth century when more frequent use was made of actual surveys carried out with proper instruments. Relevant to this are some of the pages of Mowat's *Sixteen Old Maps*.[65] One of the earlier entries, drawn in 1605, shows the land round the village of Whitehill in Oxfordshire, one half of which belonged to Corpus

Christi College, Oxford, the other half to 'Edwarde Standerd, Yeoman'. A portion of the map is reproduced in Figure 22.

Here we see a survival of the early strip system in which land belonging to two or more is apportioned in a number of small strips with no two adjacent pieces going to the same owner. In this way all share equally the good and the bad, the richer land and the poorer. At Whitehill, with only two owners, alternate strips are all nominally of two acres. The college has had its lands surveyed and to each strip is given the actual area in acres, roods, and square poles; Mr Standerd has not put himself to this expense and in the map all his strips appear shown as 'II ac' (two acres). The survey of the college property shows how approximate these 'II ac' strips are. None is larger than the nominal two acres; all are smaller and some are only just over half an acre. One of the smaller strips is given as '0 acres, 2 roods, 13 sq. poles'.

Even up to this comparatively late date the acre is notional. It is found by looking, not measuring. Thus only in later years does the acre appear consistently as a fixed area to be defined by quasi-legal documents, and ultimately by statute[66] although continuing as a customary local measure despite legal enactments till a late date.[63]

We might note in passing that the last surviving example of the mediaeval open-field system is that at Laxton, near Newark-on-Trent, Nottinghamshire.[67] Here the South Field consists of 140 acres divided into 47 strips rather like those shown on the map of 1605. The West Field of 150 acres is divided into 51 strips while the Mill Field of nearly 200 acres is in 69 strips. Here the strips are rather larger than those of the map of Whitehill, being some three acres each. The whole estate covers some 2000 acres and is governed by a leet court. The Minister of Agriculture was the lord of the manor. In November 1979 the Government announced that the entire estate was to be sold, and fears were raised that this unique link with the past would disappear. Subsequently it was announced that a continuation of mediaeval farming practices would be a condition of sale. Happily the matter was resolved when it was announced in May 1981 that Laxton had been sold for one million pounds to the Crown Estate, thus preserving the estate and its traditional practices for a long time to come.[68]

NOTES AND REFERENCES: CHAPTER III

1. ROBERTSON, Agnes Jane (ed.) *Anglo Saxon Charters* 2nd ed. pp.22-5. Cambridge: C.U.P. 1956. (Sawyer No. 342.)
2. WHITELOCK, Dorothy. *English Historical Documents c.500-1042 AD* Vol.1, p.450

No.65, London: Eyre & Spottiswoode 1955. See p.451. (Sawyer No. 23.)

3. SEARLE, Eleanor (ed. & tr.) *Chronicle of Battle Abbey*. pp.50-1. Oxford: Clarendon Press 1980.
4. ELLIS, Sir Henry (ed.) *Registrum Vulgariter Nuncupatum — The Record of Caernarvon* p.242. London: Record Commission 1838.
5. ROBERTSON, Agnes Jane. Loc. cit. pp.72-5. No. XXXIX (Sawyer No. 1448.)
6. Ibid. p.329.
7. Ibid. pp.36-7. No. XIX (Sawyer No. 1280.)
8. MAITLAND, Frederic William. *Domesday Book and Beyond* p.377 et seq. Cambridge: C.U.P. 1897.
   (See also the quotation in the text from Walter of Henley's *Husbandry*, ref.12)
9. MAITLAND, Frederic William, Loc. cit. p.377 et seq.
10. THORPE, Benjamin. *Analecta Anglo-Saxonica* p.19. London: J. R. Smith 1868. KEMBLE, J. M. *The Saxons in England* pp.96-7. London: B. Quaritch 1876.
11. NICHOLS, J. F. (ed.) The Extent of Lawling in the Custody of Essex AD 1310. *Trans. Essex. Arch. Soc.* (1933) Vol.XX, pt.2 p.183. Colchester.
12. OSCHINSKY, Dorothea. *Walter of Henley and other Treatises* Ch.28. Oxford: Clarendon Press 1971.
13. SALTMAN, A. (ed.) *The Cartulary of Dale Abbey*. p.366. London: Historical Manuscripts Commission (joint publication II) 1967.
14. For the information given in these three paragraphs see: CRUMMY, Philip. The System of Measurement used in Town Planning from the Ninth to the Thirteenth Centuries. *British Archaeological Reports No. 72* (1979) No.8, pp.149-155. Oxford.
15. SHARPE, R. R. *London and the Kingdom* Vol.1, p.10. London: Longmans 1894-5. ROBERTSON, A. J. (ed.) *The Laws of the Kings of England from Edmund to Henry I* p.305. Cambridge: C.U.P. 1925.
16. ROBERTSON, A. J. (ed.) Loc. cit. p.29. (III Edgar 8.1).
17. BIDDLE, Martin (ed.) *Winchester in the Early Middle Ages*. Winchester Studies Vol.1, pp.556-7. Oxford: Clarendon Press 1976. See Addendum to Note 5 on p.461, also p.291.
18. ROBERTSON, Agnes Jane (ed.). Loc. cit. 2nd ed. pp. 246-9.
19. GOODMAN, Arthur Worthington. *The Manor of Goodbegot in the City of Winchester*. Winchester: Warren & Son 1923.
20. ROBERTSON, Agnes Jane (ed.). Loc. cit. 2nd ed. p.494.
21. BIDDLE, Martin (ed.). Loc. cit. p.273.
22. STENTON, Sir F. M. *Anglo-Saxon England* 3rd ed. p.265. Oxford: Clarendon Press 1971.
23. See for example:
    BIRCH, Walter de Gray. *Cartularium Saxonicum* Vol.I, p.293, No.207 (dated AD 772) London: Whiting 1885-93. (Sawyer No. 140.)
24. See for example:
    KEMBLE, J. M. *Codex Diplomaticus aevi Saxonici* Vol. III, pp.358-9. No. DCCXX (dated AD 1012) London: English Historical Society 1845. (Sawyer No. 925.)
25. BIDDLE, Martin (ed.) Loc. cit. p.272.
    The area within the walls was given very roughly as 860 × 780 yards, that is a perimeter of roughly 3280 yards (596.4 rods or 2999 m) in the *Victoria County*

History Hants. Vol.I, p.285, Westminster, London: Constable 1900. The more recent figure of 603.3 rods (3034 m) given in the text is believed to be considerably more accurate but both values lie very close to 600 rods.

26. ATTENBOROUGH, F. L. *The Laws of the Earliest English Kings* pp.134-5. Cambridge: C.U.P. 1922.
27. BIDDLE, Martin. Excavations at Winchester 1962-63, Second Interim Report. *Antiquaries Journal* (1964) Vol. XLIV, pp.188-219. Oxford.
28. BIDDLE, Martin. Winchester — The Development of an Early Capital; *Vor - und Frühformen der europäischen Stadt im Mittelalter. Symposium in Reinhausen bei Gottingen 18-24 April 1972* Teil 1. pp.229-261. Gottingen.
29. BIDDLE, Martin. Excavations at Winchester 1969. Eighth Interim Report *Antiquaries Journal* (1970) Vol.L., Pt.II, p.289.
30. HILL, David. The Burghal Hidage — Southampton. *Proc. Hampshire Field Club for the Year 1967* (1969) pp.59-61. Winchester.
31. For further information on the relation between the Saxon *burhs* and the Burghal Hidage, see:
    RADFORD, C. A. Ralegh. The Later Pre-Conquest Boroughs and their Defences. *Mediaeval Archaeology* (1970) Vol. XIV, pp.83-103.
32. GOODMAN, Arthur Worthington. Loc. cit. pp.6-7.
33. CUNLIFFE, Barry. *Winchester Excavations 1949-60* p.25 et seq. Winchester: City of Winchester Museum and Library Committee 1964.
34. BIDDLE, Martin (ed.) *Winchester in the Early Middle Ages* Winchester Studies Vol. I., p.37. Oxford: Clarendon Press 1976.
35. STENTON, Sir F. M. Loc. cit. pp.426-7.
36. ATKINSON, T. *Elizabethan Winchester* p.19. London: Faber & Faber 1963.
37. PETRIE, W. M. Flinders. *Encyclopaedia Britannica* 11th ed. 'Weights and Measures' Vol. XXVIII, p.481. Cambridge: C.U.P. 1911.
38. NICHOLSON, Edward. *Men and Measures* pp.86-7. London: Smith, Elder & Co. 1912.
39. GRIERSON, Philip. *English Linear Measures. The Stenton Lecture 1971* p.22. Reading: University of Reading 1972.
40. See for example the extract from Orderic Vitalis (first half twelfth century) given by FREEMAN, E. A. *The Reign of William Rufus* Vol.II, p.562 (Note U). Oxford: Clarendon Press 1882.
41. *Charter Rolls of Richard I 1198*.
    This charter is printed in full in *Calendar of Charter Rolls 1257-1300* Vol.II, p.335. London: HMSO 1906 when it was reconfirmed in 14 Edward I 1286.
42. KOBEL, Jakob. *Geometrei*, f 4V. Frankfurt am Meyn: Christian Egenolffs Erben 1535.
43. THOMSON, T. and INNES, C. (eds.) *Acts of the Parliament of Scotland* Vol.1, Appendix III, p.751 (inner pagination 387). London: Record Commission 1844.
44. PRIOR, W. H. Notes on the Weights and Measures of Mediaeval England. *Bulletin du Cange* (1924) Vol.I, p.144. Paris.
45. NICHOLS, J. F. Loc. cit. p.183.
46. OSCHINSKY, Dorothea. Loc. cit. Ch.58.
47. HALE, William Hale. *The Domesday of St. Paul's of the Year MCCXXII* Camden Society Publications Vol. LXIX, p.130 and 138. London: Camden Society 1858.

48. SEARLE, Eleanor (ed. & tr.) Loc. cit. pp.50–1.
49. OSCHINSKY, Dorothea. Loc. cit. Ch. 4.
50. SALTMAN, A. (ed.) *Cartulary of Tutbury Priory* pp.71 and 193. London: HMSO 1962.
51. FOWLER, W. On the Ancient Terms applicable to the Measurement of Land. *Royal Institute of Chartered Surveyors' Transactions* (1884) Vol.XVI, pp.275-316. London (See comment by NEALE, C. J. p.301).
52. *Charter Rolls Richard I 1189*
This charter is printed in full in *Calendar of Charter Rolls 1327-1341* Vol.IV, p.338. London: HMSO 1912, when it was reconfirmed in 9 Edward III (1335).
53. *Close Rolls Henry III 1227-1231* p.271 for AD 1229. London: HMSO 1902.
54. Ibid. pp.186-7 for AD 1229.
55. Ibid. p.522 for AD 1231.
56. SALTMAN, A. (ed.). *Cartulary of Dale Abbey*. p.18. London: Historical Monuments Commission (joint publication II) HMSO 1967.
57. NORDEN, J. *The Surveyor's Dialogue* Book IV, pp.180–1. London: H. Astley 1607.
58. BIDDLE, Martin (ed.) *Winchester in the Early Middle Ages*. Loc. cit. p.377.
59. BERESFORD, M. and HURST, J. G. (HURST, J. G. (ed.)). *Deserted Mediaeval Villages* pp.104, 106, 107, 108, 110 and 114-5. Woking: Lutterworth Press 1971.
60. FERNIE, Eric. Observations in the Norman Plan of Ely Cathedral. *British Archaeological Association Conference Transactions for the Year 1976*. DRAPER, Peter and COLDSTREAM, Nicola (eds.) Pt.II, Mediaeval Art and Architecture at Ely Cathedral 1979.
61. PETRIE, W. M. Flinders. *Inductive Metrology* p.107. London: Saunders 1877.
62. BIDDLE, Birthe Kjolbye-. Preliminary Report. *The Old Minster at Winchester in the 7th Century*. February 1974.
See also 1) BIDDLE, Birthe Kjolbye-. The 7th c. Minster Church at Winchester Interpreted. *The Anglo-Saxon Church,* Studies in History, Architecture & Archaeology in Honour of Harold Taylor, editors Butler, L.A.F. & Morris R.K. CBA Research Report 60, London: 1986.
2) BIDDLE, Martin & Birthe Kjolbye-. The Anglo-Saxon Minsters of Winchester, *Winchester Studies,* Vol.4.1, Oxford University Press (1984).
63. Second Report of the Commissioners on Weights and Measures 13 July 1820. *Parl. Papers* 1820 (HC 314) vii, p.473 (Appendix A).
64. PERCY, Thomas (ed.). *Regulations for the Establishment of the Household of H. A. Percy, 5th Earl of Northumberland 1512-25*. p.xviii. London: 1770.
65. MOWAT, John Lancaster Gough. *Sixteen Old Maps of properties in Oxfordshire (with one in Berkshire) in the possession of some of the Colleges of the University of Oxford, illustrating the open field system*. Oxford: Clarendon Press 1888.
MAITLAND, F. W. Loc. cit. p.381.
66. See for example:
    (a) Statute for Measuring Land (supposed to be 33 Edward I but probably earlier, i.e. thirteenth century) *Statutes* Vol. 1, p.206.
    (b) SHEPPARD, J. B. (ed.). *Certa Mensura Cantuariensis*. Second Report on Historical Manuscripts belonging to the Dean and Chapter of Canterbury, in the 8th Report of the Royal Commission on Historical Manuscripts. Appendix to Pt.1. pp.315-55 London 1881.

67. *Daily Telegraph.* Saturday, 10 May 1980.
68. (a) *The Times.* 20 November 1979, 4h.
    (b) Ibid. 4 February 1980, 13a.
    (c) Ibid. 1 May 1981, 4h.

CHAPTER IV

The Hide and the Knight's Fee

The Hide

A larger Saxon land unit, now obsolete, was the *hide*. It is mentioned in charters throughout the Saxon period commencing not later than the seventh century. There is a charter[1] dated 'before AD 675', describing a grant of lands by Frithewald to Chertsey Abbey, which speaks of 'fifteen hides of land at Egeham' (vifteen hide lond in Egeham). This unit, the hide, represented the land needed for the support of a family[2] throughout the year so may well antedate the acre. Being therefore a measure of productivity the hide should not be thought of as a fixed area. Like the acre the hide was a notional entity, only later, in about the thirteenth century, becoming standardized numerically as so many acres.

Moreover the hide was the unit by which service or taxation could be levied. We have already seen the hide as the unit for generating the defence of the Saxon *burhs* of the ninth century. Thus if a piece of land was taxed as a hide, it was regarded and recorded as a hide regardless of its superficial extent. This also created the idea of the annual value of a hide, either from the tax to be collected from the tenant-in-chief or from the rental to be obtained from the sub-tenant by the tenant-in-chief, who in turn, certainly in Norman times, held the land from the king under the feudal terms of service or cash payment. The hide was therefore a unit of assessment.

The hide was the ploughland of a team of eight oxen and to each hide there was nominally one plough. Henry of Huntingdon in his *Chronicle,* written shortly after 1135, defines the hide when speaking of the Domesday Survey as: 'acres sufficient for one plough for a year'. In the Danelaw* (late ninth century) the hide was known as the *'carucate'*,† and the eleventh-century Domesday Book uses the word *'caruca'* meaning a team of eight oxen. Throughout the kingdom in the seventh and eighth centuries the hide was to

*The Danelaw was the part of north-east England ceded by Alfred to Guthrum the Dane. It included Northumbria (most of modern Yorkshire), the districts of the Five Boroughs, (Leicester, Nottingham, Derby, Stamford, and Lincoln), with parts of the Midlands and East Anglia. The date of the treaty is uncertain but is about AD 886. The agreement as to the boundaries of the Danelaw is given in more picturesque form in the original[4] which translates: 'Up the Thames then up the Lea and along the Lea to its source, then in a straight line to Bedford and then up the Ouse to Watling Street'.

†Orderic Vitalis in his 'Ecclesiastical History of England and Normandy', composed in the first half of the twelfth century says: 'carucates which are called hides in England' (. . .carucatas quas Anglia hidas vocant . . .).

become known under a variety of names: *cassatura, cassatus, familiarum, manetium, mansatus,* and *mansura,* besides the term *carucate* just mentioned.[3]

Reasonably enough the eighth part of a hide was an *oxgang* or, in East Anglia, a *bovate*. Because it was ploughland, it almost always referred to arable land, excluding meadow, pasturage, woodlands, or swamps. An exception to this rule is found in the inquest of the lands of St. Paul's Cathedral of 1181, relating to the manor of Keneswrtha in Hertfordshire. There we are told:[5] 'Of 10 hides, 5 were in the lords demesne and so till now, in which 5 hides contain 20 virgates . . . Of the 20 demesne virgates three hundred (ccc) are in arable land and two hundred (cc) in woodland'.* This is an unusual declaration. Land other than arable land is only very rarely found included in the description of a hide. The statement gives, however, the *virgate* as a quarter of a hide. It had been previously met (Chapter III) as a quarter of an acre. Numerous authorities state that the quarter hide was the *virgate* or *yard-land.*[6]

Much has been made of the anomalous entry in the Chronicle of Battle Abbey,[7]: 'Eight virgates make a hide. A wist is four virgates'. This is not an error, for in totalling up the various parcels of land that made up the Abbey's *leuga,* the reckoning is at eight virgates to the hide. This quotation refers to the very early years of the twelfth century. A later comment of about 1125 says that the *leuga* of the abbey is now divided into *wists,* called elsewhere '*virgates*'. The implication is that the Abbey lands are now subdivided into virgates of the usual size, four of which are regarded as a hide.

The explanation[8] of this drop from eight to four *virgates* to the hide is that in the early days following the establishment of the Abbey on the battlefield at Hastings by William in the late eleventh century as a thank-offering for his victory, the lands of the Abbey were poorly developed, consisting mainly of forest together with the little that had been cleared. Such land could not yield for tenant or tenant-in-chief anything approaching that to be expected from well established, fully worked land. Thus eight virgates and not four were thought to be a reasonable hide. But once the land became productive the normal assessment was reverted to. Sussex in around 1100 was a far cry from rich farmland and years were needed to bring about the changed productivity.

Not only did the hidage of a parcel of land increase with development and improvement but occasionally in the records we come across substantial reductions. For example, the assessment of the manor of Tidwoldintun (Heybridge in Essex) belonging to St. Paul's Cathedral, London, declined from 8 to $7\frac{1}{2}$ and then to 3 hides. The first reduction is attributed to damage to woodlands by the inroad of the sea in the first half of the twelfth century, and a similar natural disaster is thought to be the cause of the subsequent major reduction to 3 hides at a point in time between 1181 and 1222, the dates of two

*Here five *hides* are said to equal ccccc acres. For the value of the *hundred* see pp.58-9.

inquests as to the value and extent of the lands of St. Paul's.[9] Similarly the manors of Chingford and Ardley fell in hidage from six to five and Drayton from ten to nine, but we read at the same time of the manor of Navestock rising from seven to eight hides in assessment. The inquest of Tuesday, 20 January 1181 already referred to established the hidage assessment of Tidwoldintun as $7\frac{1}{2}$ of which 4 were in the lord's demesne and $3\frac{1}{2}$ were in the hands of tenants but the 4 hides paid hidage and danegeld.[10]

That Battle Abbey was free of all taxation, and more especially was not in the jurisdiction of the Bishop of Chichester, rested on oral tradition which in the days of Henry II and on later occasions came under close review. Not surprisingly charters suddenly appeared in the mid-twelfth century to support the claims of the clergy. There is little doubt that the charters are forgeries and were seen to be so at the time. Thus if dues and gelds were now to be exacted from the revenues of the Abbey the normal modes of reckoning, 'four virgates are one hide', were to be applied. The developmental years had passed, and Battle Abbey was going to have to pay on the same basis as everyone else.

The Sulung

In Kent and only in Kent we find the land unit, the *sulung,* and its quarter, the *yoke* (again showing the conceptual link to the team of eight oxen). This is further clarified in the will of the Reeve Abba of AD 833-835[11] in which we read: 'he is to be given half a sulung . . . and with the land are to be given him 4 oxen, 2 cows, 50 sheep, and a horn' (him man saelle an half sulung . . . mon selle him to ðem londe IIII oxam, II cy, L scaepa, aenne horn . . .). Here the gift of half a sulung is accompanied by the oxen necessary to till it. By implication the sulung is the same as a hide but it had not always been so. We find it to be equal to two hides prior to about AD 830 as, for example, in a charter of AD 812 in which lands were exchanged between Coenwulf, King of Mercia, and Wilfred, Archbishop of Canterbury. The latter gave two pieces of land for Graveney near Faversham and Kasingburn in Kent.[12] 'This parcel of land is two hides i.e. a sulung, as the inhabitants of Graveney call it' (Hoc est terrae particula duarum manentium id est an sulung ubi ab incolis Grafoneah vocitatur).

Thus we see that the Kentish sulung is equal to two Mercian hides in the earlier part of the ninth century but thereafter the sulung is at par with the hide, at least nominally, for we read in the charter of Edgar granting land to Aelfstan, Bishop of Rochester, at Bromley, Kent, dated AD 975:[13] 'which is 10 hides called in Kent 10 sulungs' (hoc est decem mansas quod Cantigene decunt X sulunga . . .). Moreover the extent of the sulung did not change with the Conquest. The Chronicle of Battle Abbey in Sussex of the later part of the twelfth century records for the donation by the Conqueror of the manor of

Wye in Kent to the Abbey:[14] 'seven sulungs, that is hides' (septem swulingarum, id est hidarum . . .). Conceptually the hide and the sulung were a little different though no doubt they were each regarded as equivalent to the other. If the hide was the land necessary to support a family throughout the year, the sulung was the land a team of oxen could plough in a year.[15]

It is clear from the Domesday Book that a hide did not always represent the same area of land any more than did the sulung. One cannot but wonder whether the parcel of land mentioned in AD 812[12] is not another and much earlier instance of the beneficial hidation of undeveloped land as witnessed at Battle Abbey, with the basic tenement, here the sulung, shown as twice the normal size for rating purposes during developmental years.

The Size of the Hide

How big, then, was this unit of land, the hide or sulung? The acreages of an eight-ox plough most frequently encountered are 160, 120, and 40 principally in Kent, Essex, and Wessex respectively. The 120-acre hide was the unit of Mercia later in the ninth century and in course of time it became the predominant unit. In the tenth century (*c* 958) we read of a hide of 120 acres in the will of Aelfgar,[16] where thoughtfully this information is given in the last sentence of the document.

Again we read in a land grant[17] of the early twelfth century relating to the See of Kensington: 'two hides, 12 × 20 acres in extent and in addition a proportionate virgate' (duarum hidarum duodecies XX acris terra disterminata et insuper unius virgatae proportione). The sulung may, on occasion, have ranged as high as 200 but the 120 acres of the Mercian hide dominates although 160 is a number frequently encountered.

Fleta[18] in 1290 says 'If the lands are in three divisions then nine score acres make up a ploughland, inasmuch as sixty acres ought to be ploughed in the winter and sixty acres in the spring and sixty acres ploughed and left fallow in the summer. But in the case of lands lying in two divisions, eight score acres ought to be reckoned as the work of one plough, so that half is laid in fallow and the other half is sown in winter and in spring'.

The Domesday Book[19] referring to the lands of St. Martin states: 'four hundred acres and a half which make two solins and a half'. This gives the sulung 160.2 or near enough 160 acres. The quarter sulung was the *jugum* of 40 acres which also appears as the small hide of 40 acres in Wiltshire. Dorset[20] boasted a large hide of some 240 acres, in reality a double hide.

At Runwell in Essex in 1222 we are told that the manor contained 8 hides and the hide contained 120 acres but in ancient times it had contained 80 acres. The manor of Nastok had a hide of 140 acres in the same year while the hide of the manor of Drayton was only 64 acres in 1279.[21]

We must ask, then, what the hide or sulung was intended to be, even though fictitiously or imaginatively, as a land area. A clue is found in the *Dialogus de Scaccario*[22] of the twelfth century where we read that the *hide* was: 'at first a plan for a hundred acres' (primitina institutione ex centum acres constat). Hoveden[23] in 1198 mentions the *carucate* of a hundred acres. How many acres were in a *hundred?* Today the question sounds stupid for everyone knows that a hundred is 5 × 20, that is five score, but it was not always or universally thus.

The Teutonic (Saxon) hundred was six score or 120. The Old Norse hundred was 120. The Domesday Book (1086-7) takes the hundred as 120, as does the Domesday of St. Paul's of the year 1222,[24] ('cc sheep at 120 to the c') and Roger de Hoveden tells of the imposition of a levy of five shillings a hide by Richard I in 1198 and of Richard's agents fixing the unit area at one hundred acres. But here again the long hundred of 120 is meant, although some have interpreted Roger's narrative to mean that an attempt was made at this time to reduce the hide to 100 acres, thus generating more revenue, but the attempt failed.

Reverting to the 1181 inquest of the lands of St. Paul's Cathedral quoted earlier[5] we see that 5 hides are 20 *virgates* and total five hundred acres. Here again the hide is a hundred acres but the long hundred, 120, is meant and understood by the contemporary readers of the document.

Fleta[26] (1290) defines the *last* of herrings in the following explicit way: 'But a last of herrings consists of 10 thousands and each thousand of 10 hundreds and each hundred of six score [secies viginti]'. The eighteenth-century editor of the sixteenth-century record 'Regulations and Establishment of the household of H. A. Percy, 5th Earl of Northumberland 1512-25' comments:[27]

> It will be necessary to premise here that the ancient modes of compilation are retained in this book: according to which it is only in money that the hundred consists of five scores: in all other articles the enumerations were made by the old teutonic hundred of six score.

While it is a little exaggerated to say only money went by the hundred equal to 100, he was nearly right. For example, Fleta[26] a little further on says: 'But a hundred of canvas, cloth, and the like is reckoned as six score ells, while the hundred of iron is reckoned as five score'.

Even with respect to money the hundred was not always five score. We read in the will of King Eadred dated 951-955[28] 'I give to the archbishop two hundred mancuses of gold, reckoning the hundred at 120' (þaenne an ie ðam ercebiscop twa hund mancusa goldes, beo hundwelftigum). Here it is not clear whether the word *mancus* is a coin or a weight of gold. Earlier in the same will we read: 'gold to the amount of 2000 mancuses is to be taken and coined into mancuses' (þanne nime man twentig hund mancusa goldes and gemynetige to mancusan). The same word is here used in two senses, but with so great a coinage for one specific purpose it is regrettable that only five gold *mancus* coins are extant.[29] At the time of the Wantage Code[30] of Aethelred II, (*c* 1000), the hundred of silver consisted of 120 *orae*. (The *ora* was not a coin but a unit of calculation of weight).*

We are further told that at the end of the sixteenth century (so quite close in time to the writing of the 'Regulations') that[31] 'The hundred weight of gunpowder is but five skore poundes . . . But at the Kings beame at Cornwall yt (tin) is 120 pounds waight to the 100 . . .' Contemporaneously, we find in the Tables of Land Measure of the Rental of Henry, Earl of Northumberland[32] (latter part of the sixteenth century) that the knight's fee† was fixed at 640 acres at 5 score to the hundred or 540 acres at 6 score to the hundred [$(5 \times 120) + 40 = 640$, as before]. Here both values of the hundred are explicitly mentioned and worked out to yield the same number of acres, namely 640.

The hundred of 120 should not be thought of as pertaining exclusively to records many centuries old, for in 1820[33] no fewer than thirteen different items are recorded as being sold or measured by the hundred of 120 in various parts of the country, while the *hundred-weight* for 26 items was 100 lb (leaving aside any other variants including the common 112 lb cwt). It will not be too much to say that the term *hundred* sometimes meant 100 but very frequently it meant 120, particularly in the earlier centuries which are our present concern.

We may therefore think of the hide as a variable unit of land area not intended to be a fixed number of acres yet, in the course of time, doubtless under pressure of taxation, becoming crystallized at 'c acres', and most commonly encountered as 120 acres, but with wide variations.

Human nature being what it is, each district endeavoured to increase the number of notional hides within its boundaries as a matter of local pride. But when it was found that the hide formed a suitable base for taxation the number of hides in a district shrank markedly. For example, the Tribal Hidage,[34] the geld list of the Mercian kings of the first half of the seventh century, gave 240,000 hides to all England;‡ the Domesday Book found less than 70,000. Again the Tribal Hidage gave Sussex 7000 hides as did also Bede[35] in his eighth-century *Ecclesiastical History*. By the time of the Burghal Hidage,[36] dated *c*911–19, recording events of the ninth century, Sussex is down to 4350 while the Domesday Survey of 1086 gave it 3474.[37] Frequently we find the hidage of an estate in the Domesday Book reduced from that of the time of Edward the Confessor. For example, taking the lands of the

*See Chapter VIII

†See this chapter, p.61

‡Strictly there was no England at this time. Of the territories listed in the hidage about half have names which are not identifiable, so it is difficult to assess how much of what was later to be called 'England' was included.

Archbishop of Canterbury in Sillentone Hundred, Sussex: 'The Archbishop himself holds Loventone in demesne. In the time of King Edward it vouched for 18 hides. Now for 9 hides and a half' (p.16b col.1 of the original record).

Thus at least some of the reduction observed between the record of the Burghal Hidage and that of the Domesday Book occurred in the forty years preceding William's survey, and the instance quoted here is by no means an isolated example. But even these lower figures of the Domesday Survey give problems. Lennard[38] recalls that for 28 counties listed there were 71,785 plough teams. With one team to every hide of 120 acres, this means that there would have been some 8.6 million acres under the plough in 1086. In 1875 there were 10.5 million acres recorded as under wheat, barley, and oats, while in 1905 there were less than 8 million.[39] The official figure for 1914 for these 28 counties is 7.7 million. Lennard continues 'There are serious objections to the assumption that every Domesday plough implies the tillage of 120 acres'.[38] Maitland[40] long ago pointed out that to give Sussex 7000 hides created a problem, for this amounts to 840,000 acres under plough. In its entirety Sussex has only 940,000 acres.

But all is not lost. Returning to the Whitehill map of 1605 (see Ch.III pp.48-49), and taking 21 of the college strips at random, each nominally two acres, we find, instead of 42 acres, only 24 acres, 1 rood, and 34 square poles. The ratio of the statute to the nominal acre here is therefore about 0.58. If we were to take the total college acreage at Whitehill, including what are called in the map the West, East, Middle, and South Fields we find the estimated acreage to be 316, while the survey gave 199 acres, 2 roods, 18 square poles. The ratio of statute to nominal acre is 0.63. This is close to the answer obtained taking a few (say 21) strips which indicates that, though the attributed acreage may be notional, there was a high degree of consistency in what an acre was recognized as being.

This ratio of about 0.6 was still valid one hundred years later, for we read in the *Natural History of Oxfordshire* of 1705, speaking of the manuring of fields,[41] '. . . they lay about 12 loads upon a common field acre, i.e. about 20 upon a statute acre'. The ratio is about 12/20 or again 0.6.

If the acres of 1605 and 1705 were notional, those of 1086 would be no less so. The plough might match up to 120 acres but not necessarily 120 of our statute acres. The Domesday commissioners were not likely to embark on an on-the-spot survey, so the inflated figures would stand. The commissioners would do little to reduce the figures claimed, for the Conqueror had only two years before suddenly tripled the tax on the hide (from 2 to 6 shillings). An inflated hidage would ease William's fiscal needs.

The 'reduction' in the number of hides in the centuries between the Tribal Hidage and the Domesday Book of a factor of at least two is insufficient if we are to match a hide to 120 statute acres. The Whitehill map shows how a further reduction by a factor of almost two can be achieved, for 42 nominal acres of the early sixteenth century are just over 24 statute acres and the same is still true a century later, in 1705, as we have just seen. If this was so then, how much more so in 1086? If the attributed acres of the Domesday Book are roughly halved in number we will not be far wrong in terms of statute acres and the problems of eleventh-century England having more arable acres than there are *in toto* disappears. Support for this suggestion may be found in an earlier proposal,[42] namely that the hides of the Tribal Hidage were not hides at all but were *virgates* or quarter hides of 30 acres.

The hide was a useful unit. As we have seen it served to provide a man for the defence of the walls of the *burh*. It could be taxed, too, but not every hide, notional or otherwise, was taxed. The royal estates were free, as were also certain lands of the Church and the nobility. The Manor of Godbegot, already discussed, was to be 'untroubled by any yoke of earthly service so long as the torch of the Englishman's native land shall shine',[43] to quote the delightful words of the charter, though it lost most of the privileges and its right to its own courts at the Reformation, much to the gratification of the city.[44] Domesday Book shows the Bishop of Salisbury having sixteen carucates of land at Sherborne which had never been divided into hides, nor did they pay geld,[45] while for Havochesberie Hundred in Sussex we read: 'In Belingeham, the Earl has one hide in demesne. Queen Eddid held it. It has never paid geld.' In Cornwall the Domesday commissioners recorded[46] that in the time of Edward (the Confessor) the Bishop of Exeter held Iregel, geldable for two hides but containing twelve.

To proceed further into the land tenure and taxation of Saxon and mediaeval England would be a lengthy and complex undertaking beyond the scope of the present work. Suffice it to say that the acre is still with us though its future is in doubt, and it may shortly be replaced by the hectare of approximately 2.471 acres. The hide disappeared from use in the sixteenth century.

The annual value of a hide was £5. At 240 pence to the £1, an acre was therefore supposed to generate a rental in cash or kind of 10*d*. In Saxon times the landholding of a thegn was five hides, which had to provide one man for the army *(fyrd)* for the defence of the realm in time of need, with the thegn under personal obligation to the king. This is seen in a charter[47] of about AD 767 in which 30 hides are granted by Coenwulf of Mercia with the grantee having 5 hides to himself and the obligation of providing five men for the fyrd from the other 25.

The Knight's Fee

Following the Conquest the new aristocracy of England was essentially of either a military or an ecclesiastical nature. The barons, churchmen, and great abbeys held their lands from the king and in return were expected to give military service if called upon. This particular system of knight's service was introduced from Normandy but was well understood by the Saxons of earlier generations.[48] From a tenant-in-chief holding land from the king, a certain number of knights (depending on the size of the land holding but without exact relationship thereto) would be required in return each to give forty days and nights service a year. One knight was to be provided for each land unit known as a *knight's fee*. Each man had to be trained and properly equipped[49] and the nobles, as tenants-in-chief, were expected to attend the field in person. The forty days were to be without cost to the king. Service thereafter was at the expense of the royal coffers. Such a system had its drawbacks. The call to arms might come at a particularly inconvenient time with the attendant reluctance to serve. Poorly trained men, less than completely fit and supplied with indifferent arms, might be sent as substitutes. The host so gathered would lack cohesion and many problems would arise if the army was kept in active service for long periods or more or less continuously, as happened towards the end of the reign of Richard I. Wars did not proceed on a time quantum of forty days. Cash for the payment of mercenaries would be preferable to all concerned, so there arose the idea of *scutage,* a payment in lieu of service; so much money for each knight service due. Scutage, or shield money, existed as early as the reign of Henry I. The Latin word *scutagium* in the old records meant any payment levied or assessed on the basis of the knight's fees. Sometimes they were called *aids* and were not restricted to monetary levies for the upkeep of the army. For instance, the aid of £1 levied in 1194 on knight's fees for the raising of the huge ransom for Richard I (£100,000) was called 'scutagium ad redemptionem Regis'[50,51] but most commonly scutage was levied on land holdings to provide cash for a military expedition.[51] As with the hide, the knight's fee was a unit of assessment, notionally, and at times in practice, being the service due from the holding of a fixed acreage of land.

The knight's fee *(feodium militis)* was the land held of such a size or value as to warrant the provision of one knight. How large an area or land of what value made a knight's fee? There is no single answer. On being enfiefed a noble would receive lands declared as so many knight's fees. The number of fees could be, and frequently was, quite arbitrary, bearing no relation to the acreage or value of the land, or the number could be in some relation to its annual value or to its surface extent. Thus the fees at the time of William I ranged from 2 to 14 hides[52] and, on occasion, land of £20 annual rental was to provide one knight, and again on other occasions every 4 or sometimes 5 hides were to provide this unit of military service. The usual acreages recorded for a fee are most frequently 480 or 640, that is 4 or $5\frac{1}{3}$ hides of 120 acres.

A hide, with all its own vagaries, could be counted as anything from a half to a sixth of a fee even within the bounds of one county.[53] Of the six fees associated with the abbey at St. Albans, two were of 5½ hides and the others were of 6, 7, 7½, and 8½ hides, while in Dorset there is an instance of one of 2 hides and in Cambridgeshire one of 27 hides.[54] Madox[55] reports a fee of 2½ hides: 'and two knights hold the other 5 hides of the Bishop as 2 Fees'. Part of the difficulty in determining the acreage to be associated with a knight's fee arises from the fact that not all the lands of a baron might be held by 'service'. Thus a noble possessing many acres might be assessed only one or two knight's fees, that is, only a few of his hides might be held by this form of tenure. It would be quite wrong and would yield a highly inflated acreage if the total holding were simply divided by the number of fees assessed.

It has been mentioned in the literature more than once[56] that land of annual value of £20, that is 4 hides, would commonly make a suitable fee. Where this was the rule, the corresponding scutage would then have been £2 or 10 per cent but this was far from being universal. In the times of Henry I and Stephen scutage was one mark (13*s* 4*d*). Henry II and Richard I each levied scutage several times, most frequently at a rate of one pound rising in Edward I's reign to £2.[53]

Robertson[57] quotes an extract (without reference) from the Red Book of the Exchequer to the effect that: 'It is to be noted that when forty shillings are given as scutage from the great knight's fee, each virgate pays thirty pence, each half-virgate fifteen, each ferlingate sevenpence halfpenny and from each acre a *half penny*'.*†

This is an instance of the use of the word *virgate* to denote a quarter hide. We met it earlier as quarter acre. The payments given show there to be 16 virgates to the fee here — that is, the fee under discussion is one of 4 hides each of 120 acres, thus 480 acres in all. An example of the 640 acre fee has already been given (see p.59 this Chapter and Reference 32).

In the early seventeenth century the knight's fee for the Duchy of Lancaster[58] was given as 1920 acres, that is three times 640 acres, or four times 480 acres, but by the 'common accompt of England' the fee elsewhere was 5 hides of 96 acres, that is 480 acres. This area is the same as 4 hides of 120 acres.

Spelman's source[59] seems to be far out when he says the fee is 8 hides of 144 acres (1152 acres) but it is closer when later he twice gives 96 acres to a hide and 5 hides to a knight's fee which is the same 480 acres as given above. In Kent, Spelman tells us that two sulungs made the knight's fee. This would only fit if these sulungs were each two normal hides, a relationship which did prevail in Kent in the early ninth century as we have seen. The knight's fee, though of no fixed or uniform area, may be thought of as approximating in many instances

*Scribal error for *one penny*. Forty shillings are 2 pounds or 3 marks. The customary levy of *one* pound per fee (20 shillings or 240 *d* per 480 acres) meant a levy of ½*d* per acre, hence the error in the text.

†A page-by-page search of the Red Book failed to reveal to this writer the extract in question.

to a nominal 4 hides each of 120 or 160 acres.

The irresponsible levying of scutage at 26*s* 8*d* (2 marks) in times of peace in John's reign, culminating in the levy of 3 marks in 1214, did much to unite the barons against John and was a not inconsiderable cause leading to Magna Carta of the year following. The Crown was then prohibited from levying any scutage save 'by common counsel of our realm'.

The levying of one, two, or three marks or one or two pounds (three marks) was not entirely a whim of the monarch. Several entries in the Pipe Roll of Henry II tell of the pay of a knight being 8*d* a day. For example the entry under 1162 states[60] that seven knights received £84 18*s* 8*d* for a year's service. This works out at 8*d* a day each for a year of 364 days and other instances could be quoted. Thus one mark kept a knight in the field for 20 days and £1 for 30 days.

Others have correlated the scutage with the supposed annual value of £20 or with the nominal 480 or 640 acres of the fee. The former gives one mark as one thirtieth and £1 as one twentieth of the annual value, while for the latter Robertson's quotation shows the £2 levy as 1*d* per acre for a fee of 480 acres. If of the larger 640 acres the £2 levy is $\frac{3}{4}d$ per acre. The other levies are less convenient sums; one mark represents either $\frac{1}{3}d$ or $\frac{1}{4}d$ for the two acreages mentioned while £1 represents $\frac{1}{2}d$ or $\frac{3}{8}d$ respectively.

Richard I began to exempt the barons from personal service in the field on payment of a special 'fine'. These fines began to replace scutage. By the time of Edward I scutage was being levied only on the under-tenants, the lords having compounded with the king for an agreed sum. By the reign of Edward III scutage was obsolete, other forms of taxation being applied, but the idea of a knight's fee continued.

Long after the original basis of personal service in the field for lands held or its cash equivalent had passed into obsolescence we find the knight's fee being continued in use as the basis of money levies, 'voluntary' gifts to the Crown, fines, etc. Only with the abolition of knight's service by Statute (12 Car II c 24 of 1660), following the Restoration of the Monarchy, did these exactions and the basis for their computation disappear. But the king did not lose thereby, for in section 14 of the said Act we find a duty granted on 'beer; cider; perry; metheglin or mead; vinegar beer; aqua-vite; imported beer and on coffee, chocolate, sherbet and tea, made and sold'. No doubt these new imposts more than made up for the loss of revenues experienced with the abolition of the ancient fee and the income based thereon. The concept, albeit a rather loose one, had endured for six centuries.

NOTES AND REFERENCES: CHAPTER IV

1. BIRCH, Walter de Gray. *Cartularium Saxonicum* Vol.1, p.58 No.34. London: Whiting 1893. (Sawyer No. 1165).
2. Bede calls the hide '*familia*'.
3. See for example BIRCH. Loc. cit.
   (a) Vol.I, p.108, No.74 of 688-690 AD (*cassatorum*) (Sawyer No. 252)
   (b) Vol.I, p.156, No. 107 of 704 AD (*manetes*) (Sawyer No. 1164)
   (c) Vol.II, p.497, No.767 of 941 AD (*mansas*) (Sawyer No. 476)
   (d) Vol.I, p.210, No.143 of 725 AD (*manentium*) (Sawyer No. 251).
   (e) Vol.I, p.256, No.179 of 749 AD (*familiarium*) (Sawyer No. 258).
4. ATTENBOROUGH, F. L. *The Laws of the Earliest English Kings* pp.98-9. Cambridge: C.U.P. 1922.
5. HALE, William Hale. *The Domesday Book of St. Paul's of the Year MCCXXII* Camden Society Publications Vol. LXIX, p.140. London Camden Society 1858. (Inquisitio Maneriorum Capituli Ecclesiae S. Pauli 1181).
6. ZUPKO, Ronald Edward. *A Dictionary of English Weights and Measures*. p.177. Madison, Wisconsin: University of Wisconsin Press 1968.
7. SEARLE, Eleanor (ed. & tr.) *The Chronicle of Battle Abbey*, pp.50-1. Oxford: Clarendon Press 1980.
8. SEARLE, Eleanor. Hides, Virgates and Tenant Settlement at Battle Abbey *Economic History Review* (1963) 2nd ser., Vol.XVI, pp.290-300.
9. HALE, William Hale. Loc. cit. p.xiii and p.xii.
10. Ibid. p.142.
11. BIRCH, Walter de Gray. Loc. cit. Vol. 1., p.575, No. 412. (Sawyer No. 1482).
12. Ibid. Vol.I, p.476, No.341. (Sawyer No. 169)
13. Ibid. Vol. III, p.609-611, No. 1295. (Sawyer No. 671).
14. SEARLE, Eleanor (ed. & tr.). *The Chronicle of Battle Abbey* pp.76-7. Oxford: Clarendon Press 1980.
15. STENTON, Sir F. M. *Anglo-Saxon England* 3rd ed., p.281. Oxford: Clarendon Press 1971.
    See also;
    FORRESTER, Thomas (ed. & tr.) Henry of Huntington's *Chronicle* p.215. London: Bohn Antiquarian Library (Henry G. Bohn) 1853 where referring to the Domesday Survey,
    '. . . how many hides, that is, acres sufficient for one plough for a year'.
16. KEMBLE, J. M. *Codex Diplomaticus aevi Saxonici* Vol. VI, p.12, No.1222. London: English Historical Society 1848. (Sawyer No. 1483).
17. STEVENSON, J. (ed.) *Chronicon Monasterii de Abington* Vol.II, p.55. London: Record Commission, Rolls Series, Longman et al. 1858.
18. RICHARDSON, H. G. and SAYLES, G. O. (eds. & trs.) *Fleta* Book II, p.241. Selden Society Publications Vol. 72 of 1953. London: Selden Society 1955.
19. ROBERTSON, E. William *Historical Essays* p.94. Edinburgh: Edmonston and Douglas 1872.

20. EYTON, Rev. R. W. *A Key to Domesday* Dorset, p.3. London: Taylor & Co. 1887.
21. HALE, William Hale. Loc. cit. pp.69, 81 and lxxix.
22. JOHNSON, Charles (ed.) *Dialogus de Scaccario* Ch.XVII. London & New York: T. Nelson 1950
23. STUBBS, William (ed.) *Chronica Magistri Rogeri de Houedene* Vol.IV, p.47. London: Record Commission, Rolls Series. Longman et al. 1871.
24. HALE, William Hale. Loc. cit. p.13.
25. STUBBS, William. Loc. cit. Vol.IV, p.46.
26. RICHARDSON, H. G. & SAYLES, G. O. Loc. cit. Book II, Ch.12, pp.118-9.
27. PERCY, Thomas (ed.) *Regulations for the Establishment of the Household of H. A. Percy, 5th Earl of Northumberland 1512-25 p.xviii.* London: Privately printed 1770.
28. HARMER, Florence Elizabeth (ed. & tr.) Select English Historical Documents of the Ninth and Tenth Centuries pp.35, 65. Cambridge: C.U.P. 1916.
29. Ibid. p.121.
30. STENTON, Sir F. M. Loc. cit. p.510.
31. HALL, Hubert and NICHOLAS, Frieda J. *Select Tracts and Table Books Relating to English Weights and Measures* Camden Miscellany Vol.XV, pp.22, 24. London: Camden Society 1929.
32. SKAIFE, R. H. (ed.) KIRKBY, J. de *The Survey of the County of York* (Kirkby's Inquests) pp.442-4. Durham: Surtees Society Pub. Vol.49 1867.
33. Second Report of the Commissioners on Weights and Measures, 13 July 1820. *Parl. Papers* 1820 (HC314) vii, p.473. (Appendix A).
34. BIRCH, Walter de Gray. Loc. cit. Vol.1, pp.414-6, No.297.
35. BEDE. *A History of the English Church and People* PRICE, Leo Sherley (tr.) Book IV, Ch.13. p.227 Harmondsworth: Penguin Books 1972.
36. ROBERTSON, Agnes Jane. *Anglo Saxon Charters* 2nd ed. pp.246-9. Cambridge: C.U.P. 1956.
37. MAITLAND, Frederic William. Loc. cit. p.505.
38. LENNARD, Reginald Vivian. *Rural England 1086-1135*. Appendix 1. Oxford: Clarendon Press 1959.
39. TRUSLOVE, Roland and FREAM, William *Encyclopaedia Britannica* 11th ed. Vol.1, p.398 'Agriculture'. Cambridge: C.U.P. 1910.
40. MAITLAND, Frederic William. Loc. cit. pp.502-7.
41. PLOT, Robert. *The Natural History of Oxfordshire* 2nd ed. p.249. Oxford: L. Lichfield 1705.
42. MAITLAND, Frederic William. Loc. cit. p.508 et seq.; also p.511.
43. GOODMAN, Arthur Worthington. *The Manor of Godbegot in the City of Winchester* p.3. Winchester: Warren & Son. 1923.
44. ATKINSON, T. *Elizabethan Winchester* p.21. London: Faber & Faber 1963.
45. *Domesday Book - Sussex*. p.186 col.2 of original record 1086.
46. EYTON, Rev. R. W. Loc. cit. p.5.
47. BIRCH, William de Gray. Loc. cit. Vol.1, p.284, No.201. (Sawyer No.106).
48. FREEMAN, E. A. *The Norman Conquest* Vol.V, p.866. Oxford: Clarendon Press 1876.

    HERLIHY, D. *History of Feudalism* p.200. London & Basingstoke: Macmillan 1970.

49. STUBBS, William (ed.) *Gesta Regis Henrici Secundi Benedicte Abbatis,* The Assize of Arms Vol.1, p.278 (Clause 1). London: Record Commission, Rolls Series. Longman et al. 1867.
50. HALL, Hubert (ed.) *The Red Book of the Exchequer* Vol.1, pp.9, 79. Vol.2, p.747. New York: Kraus Reprint 1965.
51. MADOX, Thomas. *History and Antiquities of the Exchequer* p.431. London: John Matthews 1711.
52. MCKECHNIE, William Sharp. *Magna Carta* pp.234-5. Glasgow: James Maclehose & Sons 1905.
53. POLLOCK, Sir Frederick and MAITLAND, Frederic William. *The History of English Law* Vol.1, pp.230 et seq. Cambridge: C.U.P. 1895. See also Ref. 50, Vol.2 p. clii et seq.
54. POOLE, Austin Lane. *From Domesday Book to Magna Carta* 2nd ed. p.15. Oxford: Clarendon Press 1955.
55. MADOX, Thomas. Loc. cit. p.401.
56. For example:
STENTON, Sir F. M. Loc. cit. p.637.
MAITLAND, Frederic William. *The Constitutional History of England* p.25. Cambridge: C.U.P. 1926.
57. ROBERTSON, E. William. Loc. cit. p.97.
58. NORDEN, J. *The Surveyor's Dialogue* p.58. London: H. Astley 1607.
59. SPELMAN, Sir H. *Glossarium Archaiologicum*. DUGDALE, Sir W. (ed.) pp.87, 218, 292, 530. London: Alicia Warren 1664.
60. *Pipe Roll 8 Henry II* Vol.V, p.53, AD 1162. London: Pipe Roll Society 1885.

CHAPTER V

The Mile and the League

The Mile of the Saxons

Our word 'mile' is derived from the Latin *milliarius,* sometimes spelt *miliare,* meaning a thousand, or alternatively *milliarium* (often spelt *miliarium*), a milestone, meaning a distance of a thousand paces *(mille passuum)* each of five Roman feet. The Roman mile (1479.55 m) of 5000 Roman feet was divided into 8 *stadia,* each containing therefore 125 paces or 625 Roman feet.[1]

The Saxon word was *mil* (plural *mila*). The *Anglo-Saxon Chronicle* begins: 'The island of Britain is 800 miles long'* (Brytene igland is ehta hund mila lang), but when a Saxon document is written in Latin the bounds are almost invariably given in Anglo-Saxon or Early English depending on the date. Where Roman terminology is used in a Latin text, for example 'duo stadium longo', the document is frequently a late forgery.[2]

If it is correct (Ch.III p.44) that the Saxons used the natural foot of about 9.9 of our present inches, we might expect them to use 5000 of them to the *mil,* remembering the Roman influence. Five thousand natural feet would be 1257.3 m, that is 0.85 Roman or 0.78 English statute miles.

Grierson[3] has compared the Anglo-Saxon text of King Alfred's *Orosius* with the Latin text. In the former, the length of the walls of Babylon is given as $70\frac{1}{7}$ miles, presumably Saxon miles; in the latter the length is given as 60 miles. A conversion has clearly been made, presumably into Roman miles. The ratio is 0.86, close enough to that already calculated, 0.85. Some further support, though of poor precision, may also be obtained from the above quotation from the *Anglo-Saxon Chronicle*. Britain extends from 58° 39′ N to 49° 57′ N which, to within one half of one per cent, is a meridian distance of 600 statute miles. This, according to the *Chronicle,* is 800 (Saxon) miles. The ratio is 0.75. Remembering the rough nature of the figure 800,† this Saxon/statute ratio is sufficiently close to the value of 0.78 already given as to lend some credence to the suggestion that one of the measures of the Saxons in England was a mile of 5000 natural feet, and hence a distance somewhat shorter than the statute mile of today.

If the Saxons did use a mile of 5000 natural feet (4125 English feet or 1257.3 m), we have already observed another mile in use at this time. In the

*This is the same length as was given by the sixth-century British historian Gildas and by Nennius and Bede in the eighth century.
†Indeed, how this figure was arrived at is not known.

text designated 'Pax' of *c*1000, already discussed when considering the meaning of *shaftment* (Chapter II, p.29-30), we cannot escape the strong implication of an 8-furlong mile. The extent of the king's peace is to be 3 miles, 3 furlongs, 9 acres breadths, (3 perches), 9 feet, 9 shaftments, (9 handsbreadths) and 9 barley corns,★ — a strange mixture of units if ever there was one — but this is only slightly different from the round figure of 3½ miles, each of 8 furlongs, and each furlong some 40 perches in length.[4] The suggestion of an 8-furlong mile is strong, particularly when the deeply entrenched value of the length of the furlong is considered, a length which has endured to the present day.

The Length of the Mile

Roman influence is to be seen in the Saxon documents, and this influence continued well beyond the Conquest. In the fourteenth-century document 'Certa Mensura' we read:[5] '8 stades make 1 miliarium' (octo stadia unum miliare faciunt) and later:

> and the king's common perch contains 16½ feet. Forty rods are contained in a furlong. Seven and a half furlongs, 3 perches and half a foot make a miliarium' (et regina virga communis continet sexdecim pedes et dimidium. Item XL virgate continent unam quarentenam. Item septem quarentene et dimidia tres virgate dimidia pes continent unum miliare).

The first quotation is the Roman measure given above. The second proclaims the standard measure of the perch and furlong. The composition of the *miliarium* is then given as 7½ furlongs (4950 feet) plus 3 perches and half a foot (50 feet) for a total of 5000 English feet. The scribe knows that the *stade* is less than the furlong but he errs in making 5000 English feet the *miliarium* instead of 5000 Roman feet.

Confusion tended to be heaped upon confusion. From Saxon times to the end of the Tudor era there were two familiar concepts of the mile, namely that of 8 furlongs and that of 5000 feet, be they natural, Roman, or English and the confusion was enhanced by the simultaneous use of the words *stade* and *furlong* as representing the same length, which they were not; the stade was 184.9 m while the furlong was 201.2 m.

The matter had not been resolved by the opening of the sixteenth century. In the first edition (1502) of *Arnold's Chronicle* we read:

> XVI fote and a half makith a perch as is a boue [above] said, that is V yardis and half, VIC foote by fife score to the C [hundred] makith a furlog, [sic] that is XXXVIII perchis sauf [save] II fote, VIII furlong make an English myle, that is VM [5000] foote and so III C and III perchis [303 perches] also an English myle.

★The entries in parentheses are missing in the original.

This needs a little interpretation. After the formal definition of a perch in the present-day terms of $5\frac{1}{2}$ yards or $16\frac{1}{2}$ feet, we are told of a furlong of 600 feet which is instantly corrected to 38 perches less 2 feet, that is 625 feet (still not a furlong), and 8 of these *furlongs* make an English mile (of 5000 English feet). The equivalent measure given, namely 303 perches, is only half a foot short of 5000 feet. Arnold's mile is 8 furlongs totalling 5000 feet, but his furlongs are *stades* of 625 English feet.

We must wait till 1593 for any statutory definition of the mile. In that year it was laid down (35 Elizabeth I. c 6 1592/3) that the mile is 8 furlongs, the furlong is 40 rods, and the rod is $16\frac{1}{2}$ feet. This statute, entitled 'An Acte againste newe Buyldinges', prohibited new construction within three miles of the gates of the City of London and was in all probability intended for local use only, but from this time on we have a mile mentioned by statute which was adopted gradually throughout the kingdom, the adoption being nearly complete by the end of the eighteenth century, but only becoming of universal application through the all-encompassing Act of 1824 (5 George IV c 74).

It is thought that the crystallization of the mile as 8 furlongs, each of 40 rods, was a direct outcome of the confusion of the furlong with the *stade* and the mile with 1000 paces or 5000 feet. Elizabeth's advisors were well acquainted with the documented entries showing the English mile to be greater than the Roman, something over 7.5 furlongs certainly, whereas the Roman was less than 7.4. With the rod and furlong definitely established at $16\frac{1}{2}$ feet and 40 rods respectively, to proclaim the mile as 8 furlongs was relatively easy. It was, after all, only an itinerary measure, a matter of convenience. Adjustment from a past practice which rested on custom, tradition, or simple uncertainty had no fiscal overtone or implication. No rent or ferm was computed using miles. Land was not measured by this unit. It was safe and acceptable to call a halt to the confusion of centuries.

The mile of Elizabeth is the one in use today, though soon to be replaced by the kilometre, (1km = 0.6214 miles). We may note in passing that the first enactment of a statute mile for Scotland is the Act of 1685, James VII (and II of the United Kingdom) Parl. 1. Cap. 44 entitled 'An Act for a Standard of Miles', which lays down the mile as 1760 yards, as in England.

The Old English Mile

The standardization in the days of Elizabeth, even if only of a local character initially, was necessary not only to sort out *milaria* from miles and *stades* from furlongs, great though that confusion may have been. The map-makers and the compilers of early itinerary books (the forerunners of the guide-books of the present day), tended to use larger units which metrologists have frequently lumped together under the name of the *old English mile*. This is not to be confused with the Saxon or Roman miles, for, unlike these, the old English mile exceeded the statute mile of 8 furlongs quite significantly.

The fourteenth-century Gough Map in the Bodleian Library, Oxford, has the 'miles' conveniently marked on the individual roads. These 'miles' each average 1.3 statute miles.[7] The unit William Worcestre[8] used to describe his itineraries of the four years 1477-80 was somewhat greater. An analysis of 237 distances given shows the 'mile' used to be very nearly 1.5 statute miles, as his editor has in fact observed (see Fig.23 for an analysis of 161 of his distances). The Latin text begins using the term *mila* but soon changes to *miliaria*. If he were truly using Roman miles as his unit the value of 1.5 would drop to 1.38 statute miles but this point cannot be taken as proven.

In 1544 a table listing nine inter-city distances appeared[9] which showed the average length of the 'mile' used was again 1.3 statute miles, omitting one short distance in which the accuracy would be poor.

In 1588 William Smith[10] published his work *The Particular Description of England* giving these inter-city distances together with additional entries and alternative routes. Smith's unit amounts to very nearly 1.25 statute miles, and the itineraries of William Harrison[11] in his *Description of Britain* of the same period closely resemble those of 1544 and 1588, although there are additions, particularly those giving inter-city distances in Scotland. Harrison's unit is very nearly 1.25 statute miles also.

John Evelyn in his diary[12] records that his birth took place in 1620 at Wotton, saying that Wotton is 12 miles from Kingston (on Thames), 6 miles from Guildford, 3 miles from Dorking, and 20 miles from London. The modern distances would be 17, 8, 3, and 26 statute miles respectively giving a mean of 1.26 statute miles for his 'mile'.

These are all examples of the unit known by the romantic title of the *old English mile*. It was discussed by de Morgan[13] in the *Penny Cyclopaedia* (1833-43) and some fifty years later by Flinders Petrie[14] who concluded from an examination of ten itineraries or maps ranging in age from the Bodleian map to those of Gibson of 1695* that without exception the unit used exceeded the statute mile in length, the range being 1.255 to 1.32, and the old English mile had been 1.3 statute miles from the end of the fifteenth century to the end of the seventeenth century but that it had been 1.2 statute miles in the fourteenth century. There is little to support this contention. He further pointed out that the 1695 maps of Gibson carried no fewer than three scales which corresponded to miles of about 1.29, 1.17, and 1.04 statute miles. These are closely in the ratio of 10/9/8 and he suggested that the three miles shown corresponded to two customary miles of 10 and 9 furlongs respectively and one mile of 8 furlongs, this being the statutory mile.

Further, Plot in his *Natural History of Oxfordshire*[15] of 1677 states:

* Gibson's edition of Camden, Ref 14, pp.257, 262, and 264.

> As for the scale of miles, there being three sorts in Oxfordshire, the greater, the lesser, and the middle mile, as almost everywhere else it is contrived according to the middle sort of these, for this I conceive may be called most properly the Oxfordshire mile . . .
> I have found it to contain for the most part $9\frac{1}{4}$ furlongs.

In all probability this middle mile corresponds to the 9-furlong mile, while the greater was the old English mile and the lesser the statute mile.

The 9-furlong mile would appear to have had some, though no great, currency, for the map of England produced by William and Johan Blaeu in the *Theatrum Orbis Terrarum* of 1635 is provided with a scale of 30 miles called on the map 'Milliaria Anglica' (English miles). Checking this against some of the itineraries of Harrison and measuring the Blaeu map over considerable distances to improve the accuracy, it is clear that, when related to modern miles, the Blaeu mile is 1.12 statute miles. A 9-furlong mile would be 1.125 statute miles, near enough to suggest an identity.

Close[7] continued the analysis of the old maps and guide books, concluding that the unit involved was indeed 1.3 statute miles. Karslake's suggestion[16] of an 11-furlong mile may be based on an error, for his single definitive distance of Dover to Canterbury, given in the fourteenth century by Higden as 12 old English miles, later shown to be $15\frac{1}{4}$ statute miles,* indicates the former to be 10.17 furlongs, not 11. The analysis of Grundy[17] gave for the sixteenth-century itineraries of Leyland and those of Ogilby of the seventeenth century a value of 1.25 statute miles for the unit used, that is 10 furlongs, although Ogilby himself says that the number of modern miles is one-third more than in the old reckoning.[18] There is, after all, only a 6 per cent different between $\frac{5}{4}$ and $\frac{4}{3}$.

From whichever source we draw our information (with the possible exception of William Worcestre) we find the unit employed to be not far from 1.3 statute miles. Distances were never given other than roughly. There was neither the necessity nor the capability of doing otherwise. It is moreover difficult to convert a scale on an old map into modern units with any high degree of precision. One can do a little better with an itinerary which gives in figures the milage between two towns for this can be compared with the modern distance, but even here the modern thoroughfare may not coincide with that of several centuries earlier.

Some idea of the roughness of the estimates of milage in these early days can be obtained by taking an average of the ratio of a number of distances in old and modern miles and calculating the standard deviation. Doing this for 161 distances in William Worcestre's *Itinerary* for the years 1477-8 and using the

*Measured by the postmasters of the route in 1633. (HYDE, J. Wilson *The Early History of the Post Office* pp.53-4 London: Adam & Charles Black 1894).

data which fall under the curve the mean is 1.47 ± 0.32 statute miles. This comfortably embraces all the values given previously for the old English mile. The spread is more obvious from the accompanying histogram (Fig.23).

The old mile was not extinguished by Elizabeth's statute. Seventeenth-century milestones not uncommonly were inscribed with the distance in old English miles, although the principal roads served by the post office set up in the reign of James I* most commonly used statute miles in order to regularize the costs of delivery. Those with the old English mile had to be altered in the eighteenth century to conform to the then almost universal acceptance of the statute mile.

The origins of the old English mile are obscure, but this has not prevented several suggestions from being advanced, though none can be regarded as completely satisfactory or fully proven, but most have the merit of generating a unit of length of approximately the correct magnitude. Some of the more important suggestions are as follows;

(1) de Morgan[13] regards the old English mile as a survival in England of the *old Gallic league* of 1500 paces (see next section). Fifteen hundred paces, each of 5 Roman feet, would amount to some 2220 metres or 1.38 statute miles, while if the feet were to be thought of as English, 12-inch feet, then the distance would be 2286 metres or 1.42 statute miles. This view is strongly supported by Seebohm.[19]

(2) Petrie makes two suggestions; (a) that it is a distance of 1000 fathoms each of 6 Drusian feet, that is a distance of 2012 metres or 1.25 statute miles, and (b) that it is the old French mile, the *mille de Paris*. His suggestion is that the two are the same whereas the length of the *latter* is more correctly 1949 metres or 1.21 statute miles.[20] Grierson[21] quotes the work of Guérard to show instances of the French mile *(mille de Paris)* of the early mediaeval period to have been between 2000 and 2200 metres (1.24-1.37 statute miles).

(3) The evidence for the old English mile having been a distance of 10 furlongs (1.25 statute miles) has been given above.

(4) The statement of Ogilby that the number of statute miles to the number of old English miles is as 4 is to 3 has been referred to earlier. The editors of William Smith's *Particular Description of England*[18] drew attention to some observations on this point made over a century ago. The gist of the suggestion is this. By tradition an English league, as we shall see, was the distance one could walk in an hour. The league was equally regarded latterly as being a distance of three miles. Now if one could walk four (statute) miles or a little more on the average in an hour and this distance was to be *called* three miles

*The initial appointment by Elizabeth in 1590 was renewed in 1607 by James I to Lord Stanhope as 'Master of the Messengers and Runners commonly called the King's Posts', for the transport of official mail. Mail from the public was carried unofficially from about 1630 becoming a recognized service in 1633.

then each of these miles would be $1\frac{1}{3}$ statute miles or 2146 metres. The argument against this suggestion is that the league was three statute miles and not three old miles and did not attain this length until the fifteenth century, whereas the old mile was in use in the fourteenth century if the date of the Bodleian map is to be relied upon.

As for the suggestion that the old mile is nothing more than the Gallic league, we shall see that this unit began as 1500 paces, but by the twelfth century had increased to 2000 paces. With so early an increase, it is difficult to see how a unit current from the fourteenth to the seventeenth century would be based on the shorter length of 1500 paces.

Perhaps the most plausible explanation of the origin of the old mile is to be sought in France. The range of values for the old mile admittedly embraces the old Gallic league at the high end and 10 furlongs at the low end, with the *mille de Paris* in between, but this latter unit must at the present time be regarded as the most likely source of the old mile. It would most likely have been introduced by the Normans as a unit familiar to them. There is no evidence to show a unit of about this length in Saxon times, for then we had the short mile and the eight-furlong mile. Were its origin to be a unit of 10 furlongs, it would have been expected in the Saxon era. The old English mile only appears in the records well after the Conquest. The long gap in between the date of the Bodleian map and the itineraries of the fifteenth and sixteenth centuries makes it difficult to show the continuity of the existence of the unit. It would be of great interest if an earlier map or itinerary would turn up, together with additional information of the period 1350–1450.

The League, English and Gallic

The *league* was a measure of length in Roman Gaul.[22] The word appears in various forms, *leuva, leuca, leuga,* or *leweke,* but both word and unit are practically obsolete today. Saxon documents rarely mention the league[23]. It does appear, though this need not be taken to imply that either the word or the measure was in current or common use. For distance, the Saxons used the word *mil* (mile), but in land transfers it is rare to find mentioned any length exceeding the furlong.

The original Gallic league settled at 1.5 Roman miles *(miliaria)* for we read in a passage attributed to Isidore[24] (*c*600): 'with the Gauls a miliarius and a half make a league which has 1500 paces. Two leagues or 3 miliaria make a *rast* with the Germans' (miliarius et dimidius apud Gallos leuvam facit, habentem passus mille quingentos. Duae leuvae sive miliarii tres apud Germanos unam rastam efficiunt). This Germanic unit, the *rast,* did not become a unit in English metrology except when this measure of 3 *miliaria* became on occasion confused with the league itself.

The word '*league*' and the measure itself came into frequent use with the post-Conquest period and in all likelihood was a Norman importation. The suggestion that the league was brought to England with the Belgic invaders of the first century BC[25,17] lacks historical evidence but the possibility is not to be ruled out completely.

The Register of Battle Abbey[26] of the twelfth century gives the English league as 12 furlongs or 1½ English (statute) miles and Walter of Henley's *Husbandry* of 1286 says[27] 'for 40 perches in length make one coterie (furlong) and 12 coteries make one league' thus confirming the statement.* There was therefore initially some parallelism between the leagues of France and England, but in the later Middle Ages it was believed that the French league was 2000 *paces*.[23] The 2000-pace Gallic league is also mentioned in a fourteenth-century document[28] where we read: 'and 8 stades make an English mile; and 16 stades make a Gallic mile called in Gaul a league' (et 8 stadia faciunt miliare Anglicum; et 16 stadia faciunt miliare Gallicum, quod vocant Gallii unam leucam). This is confirmed by the entry for 26 August 1395 at Southall in the *York Memorandum Book*[29] which states that the length of the Gallic league was 16 *stades*. This is a longer league than that found by the Romans in Gaul (1500 paces) and this or its equivalent led to the later (fifteenth-century) interpretation that '2 English miles make a French league'.[22] This is repeated in *Arnold's Chronicle* of 1502: 'eight furlongs make an English mile and sixteen furlongs make a French league'.[6] There seems little doubt that in the century after the Conquest and for long thereafter, the French league was being taken as two *miliaria* or alternatively as two English miles.

Things were happening to the English league also. From the end of the thirteenth century or the beginning of the fourteenth it too began to lengthen beyond the previous 1½ miles. The fourteenth-century 'Certa Mensura' referred to earlier assures us that the league is two *miliaria* ('Item duo miliaria continent unum leucam secundus geometriam'[5]) and Bracton states that in the days of Henry III the league was two miles. These two-mile leagues in England probably arose as a result of a confusion between the Gallic and the true English leagues. But from the closing years of the fifteenth century onwards[30] we find the English league being equated to *three* English miles. This information comes from William Worcestre, where under the date 1478 he records the information given by his kinsman Robert Bracy, namely that 'each 'kennyng' contains 7 leagues or 21 miles'. The editor draws attention[32] to the fact that on occasion William's informants used eight furlong rather than old miles. This was one of these occasions. Again in Fabyan's *Chronicle* of 1494[33] we read of 'an hundreth legis whereof every lege conteyneth III Englysshe myles' while Johnson's *Dictionary* of 1755 gives: 'League. A

*It will not escape notice that a rectangle, 1 league of 12 furlongs by 1 furlong, is an area exactly the 120-acre Domesday hide.

measure of length containing three miles'.

In 1820 in the Appendix to the Second Report of the Commissioners on Weights and Measures, we find:[34] 'League, 3 miles. Nautical or Geographical league $\frac{1}{20}$ of a degree of latitude . . . the French league is $\frac{1}{25}$ of a degree'. The terrestrial league is thus given as 3 miles. The sea league is $\frac{1}{20}^\circ$. The degree was normally taken as 60 miles as Sir Isaac Newton's early calculations on gravity show. One-twentieth shows the distance as three miles. But following the work of Picard and of Lacaille and Cassini (1739-40) the degree was shown to be 69.1 English miles. One-twentieth gives 3.456 miles[22,35] for the nautical league and $\frac{1}{25}$ is 2.76 miles. This latter, the French league, is identical to 3000 Roman paces or 3 *miliaria*. It is the old *rast*.

One-third of the nautical league is the (British) nautical mile. It is therefore 1.152 statute miles or 6082 feet. The international nautical mile is 6076.103 feet. The United States of America formerly employed a nautical mile of 6080.2 feet but, since 1 July 1954, have used the international unit.[36]

Returning to Plot's Oxfordshire miles[15] for a moment, he tells of the mile which he had found to contain 'for the most part 9 furlongs and a quarter [see pp.69-70] of which about 60 answer to a degree . . . for reckoning 5280 feet (or 8 furlongs) to a mile as usually in England, no less than 69 will correspond to a degree; upon which account it is and no other, that of the middle Oxfordshire miles, each containing 9 Furlongs and a Quarter, about 60 will do it'. Indeed he is correct for 60 × 9½ furlongs = 69.4 statute miles. From this it is clear, too, that Plot is well aware of the true length of the statute mile.

The English and French leagues may be summarized as follows:

Date	*English league*	*French league*★
Prior to 1066	Rarely mentioned	1½ Roman miles (2.22 km)
12th Century	12 furlongs (1½ English miles) (2.4 km)	2000 paces (2 Roman miles)
13th Century	12 furlongs	2000 paces (2.93 km)
14th Century	2 miliaria or 2 miles (3.2 km)	2 miliaria
15th Century	3 miles (4.8 km)	2 miliaria
18th Century	3 miles	3 miliaria or 3 miles (4.4 km)
19th Century	3 miles	$\frac{1}{25}^\circ$ of arc or 3 miliaria
20th Century	——— virtually obsolete ———	

From the table we see that from the fourteenth century there has been a nominal parallelism between the two leagues.

We should pause for a moment to ask why it might be that the English league would lengthen in the course of time from 1½ to 2 and then to 3 miles. These distances are usually thought of as 8-furlong miles. But suppose in

★For variants see ZUPKO, Ronald Edward, *French Weights and Measures before the Revolution*. pp.95-6. Bloomington & London: Indiana University Press 1978.

measuring a league that the old English mile is used, initially perhaps by accident, later by custom; then $1\frac{1}{2}$ old miles would measure 2 statute miles very nearly (for example $1.3 \times 1.5 = 1.95$). Likewise, if the league was known to be 2 miles but old miles were used in its computation, then we are well on our way to 3 statute miles for $1.3 \times 2 = 2.6$.

It is not improbable that the old English mile becoming confused with the statute mile was responsible for the lengthening of the league.

NOTES AND REFERENCES: CHAPTER V

1. HULTSCH, Friedrich Otto. *Metrologicorum Scriptorum Reliquiae* Vol.II, p.58, l. 27. Stuttgart: Teubner 1864-6. Reprinted 1971.
2. E.g. BIRCH, Walter de Gray. *Cartularium Saxonicum* Vol.1, pp.33-4, No.22. London: Whiting 1893 (Sawyer No.68).
3. GRIERSON, Philip. *English Linear Measures, The Stenton Lecture 1971* p.29. Reading: The University of Reading 1972.
4. Ibid. pp.27-8.
5. SHEPPARD, J. B. (ed.) *Certa Mensura Cartuariensis.* Second Report on Historical Manuscripts belonging to the Dean and Chapter of Canterbury, in the 8th Report of the Royal Commission. Historical Manuscripts. Appendix Pt.1, pp.315-55. London: Royal Commission on Historical Manuscripts 1881.
6. ARNOLD, Richard. *The Customs of London, commonly called Arnold's Chronicle.* p.173. London: F.C. & J. Rivington 1811 (Reprint of 1st ed. *c.*1502).
7. CLOSE, Col. Sir Charles. The Old English Mile, *Geographical Journal* (1930) Vol.LXXVI, p.338. London.
8. HARVEY, John H. (ed.) *William Worcestre Itineraries* Oxford: Clarendon Press 1969.
9. ANON. *A Chronycle of Yeres.* London 1544 (Printed Wyllyam Myddylton), quoted by CLOSE, Col. Sir Charles. Loc. cit. p.339.
10. SMITH, William. *The Particular Description of England 1588.* WHEATLEY, Henry B. & ASHBEE, Edmund W. (eds.) London: Privately printed 1879.
11. HARRISON, William. *Description of Britain,* preliminary section to HOLINSHED, Raphael. *Chronicles of England, Scotland & Irelande.* London 1577. A.M.S. Reprint of Edition of 1807, 1965.
12. BRAY, William (ed.) *The Diary of John Evelyn.* Vol.I, p.3. Universal Classics Library, Washington and London: M. Walter Dunne 1901.
13. MORGAN, A. de. *The Penny Cyclopaedia* 'Mile' Vol.XV, pp.210-13. London 1839.
14. PETRIE, W. M. Flinders. The Old English Mile. *Proc. Roy. Soc. Edin.* (1882-4) Vol.XII, pp.254-66. Edinburgh.
15. PLOT, Robert *The Natural History of Oxfordshire* 2nd ed. Oxford: L. Lichfield 1705.
16. KARSLAKE, Lt. Col. J. B. P. Further Notes on the Old English Mile *Geographical*

Journal (1931) Vol.LXXVII, pp.358-60. London.

17. GRUNDY, G. B. The Old English Mile & Gallic League. *Geographical Journal* (1938) Vol.XCI, p.251. London.
18. SMITH, William. Loc. cit. p. x.
19. SEEBOHM, Frederic. *Customary Acres and their Historical Importance* pp.12, 79-93. New York: Longmans, Green 1914.
20. GRIERSON, Philip. Loc. cit. p.31.
21. Ibid. p.32.
22. MORGAN, A. de. *The Penny Cyclopaedia,* 'League' Vol.XIII, pp.375-6. London 1839.
23. PRELL, Heinrich. *Bemerkungen zur Geschichte der Englischen Langenmass-Systeme;* Berichte uber die verhandlungen der Sachsischen Akademie der Wissenschaften zu Leipzig; Akademie Verlag, Berlin 1962 p.50. Note 38.
24. HULTSCH, Friedrich Otto. Loc. cit. Vol.II, p.138.
25. KARSLAKE, Lt. Col. J. B. P. Silchester and its Relation to the pre-Roman Civilization of Gaul. *Proceedings of the Society of Antiquaries* (1920) Vol.XXXII, pp.185-201. (See p.198). London.
26. LARKING, L. B. (ed.) *The Domesday Book of Kent* p.184. London: J. Toovey 1869.
27. OSCHINSKY, Dorothea. *Walter of Henley and Other Treatises* Ch.29. Oxford: Clarendon Press 1971.
28. HALL, Hubert & NICHOLAS, Frieda J. *Select Tracts and Table Books relating to English Weights and Measures (1100-1742).* Camden Miscellany Vol.XV, p.7. London: The Camden Society 1929.
29. SELLERS, M. (ed.) *York Memorandum Book. Pt.1 (1376-1419)* p.142. Durham: The Surtees Society 1922.
30. ZUPKO, Ronald Edward. *A Dictionary of English Weights and Measures* p.99. Madison, Wisconsin: University of Wisconsin Press 1968.
31. HARVEY, John H. (ed.). Loc. cit. p.109.
32. Ibid. p.xxiv.
33. FABYAN, Robert. *The New Chronicles of England and France.* V.1. xxxv, 63. London: F. C. & J. Rivington 1911. (Reprinted from Pynson's Edition of 1514).
34. Second Report of the Commissioners on Weights and Measures 13 July 1820. *Parl. Papers* 1820 (HC 314) vii, p.473 (Appendix A).
35. KELLY, Patrick. *The Universal Cambist* Vol.I. London: Printed for the author 1821.
36. O'KEEFE, John Alfred. *The Law of Weights and Measures* London: Butterworth 1966.

CHAPTER VI

The Shorter Linear Units

We began by considering the rod and worked up to larger units of length and to units of area. It was shown that the rod was firmly established in Saxon times as were the land measures and longer linear measures of the acre's breadth and the furlong and there were two concepts of the mile. We now consider the units below the rod.

The Short Units

Something called an *inch (ynce)* was known to the Saxons. In the Laws of Aethelbert of Kent which have been dated AD 602–3 we read of the compensation to be paid the injured party for wounds inflicted.[1] '67.1, For a stab over an inch [long or deep?] - one shilling; two inches, two; over three, three shillings'. Beyond three inches we are told: '67. If a thigh is pierced right through, 6 shillings compensation shall be paid for each stab'. A comparable compensation is recalled from Saxon times in the so-called *Leges Henrici Primi* (Laws of Henry I) translated into Latin in the post-Norman period. In Section 93.3 we have:[2] 'If the wound under the hair is one inch long the compensation payable shall be five pence, that is one shilling'.* The inch is undefined but is not likely to be far from the traditional three barley corns mentioned earlier.

In Asser's *Life of Alfred,*[3] written in the ninth century, we are told of candles, or more correctly tapers from their length and weight, each 12 inches long and weighing 12 pennyweights. Their length is given as 'duodecim uncias pollices', literally '12 thumb inches'. The breadth of the thumb at the root of the nail has for long been a customary measure for an inch.

Asser states that six of these tapers burn for 24 hours so one lasts four hours. They therefore burn at a rate of three inches per hour. Each being similar to another we can assume an equal rate of burning for each taper.†

In the Assize of Weights and Measures attributed to King David I of Scotland (twelfth century) we read:

*For the number of pennies in a shilling in Anglo-Saxon England, see Ch.VII.

†We note in passing that this is the first indication of the division of a day into 24 equal parts as at present, such a division usually being attributed to the thirteenth century with the construction of mechanical clocks. Since Greek times the intervals between dawn and dusk and between dusk and dawn had been each allotted twelve hours, the duration of one of which would vary with the season. There were the long working hours of summer, each being one-twelfth of the daylight span. Greek and Roman water clocks were frequently fitted with rotatable drums marked with the 'hours' of the various months and identified by the appropriate sign of the Zodiac, and this practice continued into the Middle Ages.

> On the Ulna:
> The Ulna [ell] of King David ought to contain XXXVII inches measured by the thumb of three men, a large, a medium-sized, and a small man. And the thumb of the medium-sized man ought to stand or else by the length of three grains of good barley without tails. The thumb ought to be measured at the root of the nail.

This was also the measure of England, for the fourteenth-century 'Certa Mensura' states[5] that the measure of a thumb is its breadth at the root of the nail. Anomalously this document goes on to say four thumbs (instead of the usual three) make a palm and four palms a foot, but by 1474[6] we again find the inch as three barley corns: 'III barley corns take out of the middes of the Ere make a Inche and XII inches makith a foote and III fote makith a yarde'. Even more direction is given as to the barley corns at the beginning of the seventeenth century where we read[7] (transcribing into modern English):

> Three barley corns without tails set together in length make an inch, of which corns one should be taken off the middle ridge, one off the side of the ridge, and another off the furrow. Twelve inches make a foot of measure. Three feet and an inch make an ell. . . . This is the measure of Scotland.

In the same century, Spelman's *Glossary* under 'Pollex' (thumb) states:[8] 'The thumb in all measurements must be measured at the root of the nail and must stretch the length of three good grains of barley, without tails.' A century later in an Act of Queen Anne[9] (1711) we read that an inch was to be an inch 'instead of that commonly called a thumb's breadth'.

The equating of three barley corns to one inch was not a very satisfactory means of generating the latter unit. Early in the sixteenth century Sir Richard de Benese[10] pointed out the variations in length of the barley corns according to the richness or poorness of the land with resulting variations in the length of the inch. His strong recommendation was to use an artificer's rule which was made two feet in length and which contained 24 inches. This would give a reliable inch.

Another unit, the *ell (elne)* appears in the Laws of Aethelstan of the early tenth century:[11] 'and if the ordeal is by water he shall sink to a depth of 1½ ells on the rope' (II Aethelstan 23.1). Here the unfortunate accused is to be tied in a bundle, with his hands below his knees, and gently placed in water. If he sinks all is well: he is clearly innocent, for water being a holy element will only receive the guiltless. The guilty would be rejected and would float. Unfortunately we have no knowledge as to the length of the Saxon ell.

Reference has already been made to the description in these early tenth-century laws of the foot in the ordeal by the hot iron: 'Nine feet shall be measured by the feet of him that goes to the trial' (Appendix II to VI Aethelstan),[12] and consideration has already been given to the use by the

Saxons of the natural foot and the Drusian foot (See Ch.II).

In the Laws of Ine of Wessex dated 688-94 we learn that, like the Romans, the Saxons employed the *digit* (finger's-breadth) as a unit. The length of the Saxon digit is unknown but if we rely on their predeliction for the score there would be twenty to the foot. Whether the natural or the Drusian foot would be meant is a matter for conjecture. There are no extant Saxon footrules.

One of the relevant laws of Ine[13] which mentions the digit is delightful and worth quoting. It describes the return in kind, that is to say in pigs, to be expected if another man's pigs are allowed to feed in your mast pasture. Ine, 49.3;

> If pannage is paid in pigs every third pig shall be taken when the bacon is three fingers [*thry-fingrum*] thick, every fourth when the bacon is two fingers thick and every fifth when it is a thumb thick.

If pigs do well in your pasture you get not only heavier pigs but more of them from their owner as payment. This is a good example of payment by results and probably one of the earliest.

There is no evidence for the general use of a unit of three feet in Saxon times despite assertions to the contrary. These suggestions of a unit of three feet rest, for the most part, on a misinterpretation of the Saxon word *gyrd* by calling it a yard, an error to which allusion has already been made.

If something called an inch was to be three barley corns then some 36 barley corn or so would make a convenient measure of a foot. We have already seen (Ch.I p.3) how closely 36 barley corns of this century represent a foot of today's measure and there is evidence that something closely akin to the present twelve-inch foot (305 mm) was gradually emerging in England in late Saxon times. It has been suggested[14] from recent archaeological work on the Old Minster at Winchester that a foot of the modern length was employed in the tenth-century additions to the cathedral while the earlier fabric was built using the Drusian foot. The Great Hall at Wolvesey Palace in the same city, a twelfth-century structure, would also appear to have used a similar unit of twelve modern inches.

While documentary evidence is either weak or lacking, as are also artifacts other than walls or foundations, the following pattern emerges. Measuring in Saxon times fell into two categories: it was land that was to be measured or it was something else. Land already had its unit, the rod, from which the longer unit, the furlong, could be derived and the acre's breadth was just four rods. Other things, such as buildings or cloth, could be measured by any convenient unit and that unit could be completely independent of the rod in exactly the same way that cloth was sold in Britain in 1980 by the metre while road distances were still given in miles. As long as the two measuring systems remained separate and did not mix there was no need to fuse the two into one, or have a simple ratio of one to the other. Only if land was to be measured with greater precision than in units of rods would there be a need for smaller units tied into the rod.

Early Saxon charters did not always state the dimensions of the boundaries of the lands referred to, as we see from the charter of King Aethelwulf dated 22 April 854 which reads in part:[15] 'First to the north island eastwards; then to the great mead, then to Tydging mead; then to island dike; then up to Mealmehte lea; then to Scale's place . . .'. But from the beginning of the tenth century we find rods and feet occurring together in describing the charter lands, especially in towns, where more precision was needed than in the open country. For example in a charter of Edward the Elder of 901-909 we read[16] '. . . then due east 43 rods and 6 feet to the east street then due south 20 rods and 6 feet to the south street . . .'. How long would these feet be? They could equally well be the shod feet of the measurer or a unit derived from a measuring stick. They might even be our present twelve-inch feet. Clearly we are not going to get very satisfactory land transfers until we know the actual length of the foot being used in the measurement.

But it would be an error to suppose every land grant from the early tenth century onwards was given with such precision as that quoted above. The bounds of the Manor of Godbegot were given in the early eleventh century in whole rods without fractional parts or feet being mentioned, but this manor may be unusual in that its bounds do appear to be precisely whole numbers of rods, as we have already seen (see Ch.III). By the thirteenth century precision was common and the relation between feet and rods was given; for example in a land grant to Geoffrey de Lucy, Dean of St. Paul's, dated 1229-37 we read of:[17] '8 rods and 10 feet . . . by the rod of 16½ feet'.

Certainly by the late twelfth century the foot had been standardized at its present length and accommodated to the rod. Now there was no great range of values open for the foot. If the unit is to be called a foot it should bear some resemblance to the foot of a human even if it is quite undersized as in the case of the so-called natural foot of 9.9 inches or quite oversized as in the Drusian foot of 13.1 inches. The Roman foot of 11.65 inches was well known, especially on the continent.

One might have expected, as both the natural and the Drusian foot fitted closely into the rod, the former twenty times, the latter fifteen times, that in establishing a new table of length one or other would have been used. In fact neither was adopted. Even the Roman foot fitted almost exactly seventeen times into the rod, but seventeen is a very awkward number, having no factors suitable for ready subdivision. One can but speculate how the present twelve-inch foot came to be adopted. Perhaps other interests had to be served. Perhaps they were those of the cloth-makers, whose body units of the fathom and natural ell were obtained by holding out a piece of cloth with arms extended or by holding the material in the left hand at the left side of the chest and pulling till the right arm was fully extended. These were units of roughly six and four feet respectively which the rod could not accommodate.

One thing was certain. The official length of the rod could not change. The nation's wealth was in land and nothing was likely to be done to upset the measures relating thereto.

The Establishment of the Present Yard

It was not the foot but a longer measure, the *yard,* which was fitted into the rod, with the foot a mere subdivision, one-third thereof. The word for 'yard' as it appears in the Latin texts is *ulna,* but unfortunately the same Latin word was used for *ell,* the cloth measure.

William of Malmesbury (1095-*c*1143) tells of the origin of the yard as it is known today, alleging it to have been measured by the arm of Henry I. 'The measure of his own arm was applied to correct the false ell (yard) of the traders and enjoined on all throughout England' (Mercatorum falsam ulnam castigavit; brachii sui mensura adhibita, omnibusque per Angliam proposita).[18] Unfortunately this entry is not under any particular year but appears in William's summary of the life and times of Henry.

William's declaration respecting Henry's arm has been dismissed by some as nonsense, or at best as a myth, but there is nothing unique about this use of the dimensions of a monarch's body in creating the length of a standard. Apart from the present instance, perhaps the best known example is that of the French *pied du Roi* which purported to be a foot based on the length of the foot of Charlemagne. Moreover William is usually a reliable source and, in the present case, can be taken as a contemporary witness.

Whether our present yard was or was not the actual measure of Henry's arm is immaterial. The standard was so chosen that 5½ made the existing land rod and the royal attribution lent weight and authority to the measure. By tradition the yard's length was the distance from the tip of the royal nose to the tip of the finger with arm outstretched and the date usually accepted is the first few years of the reign of Henry I. But if the interpretation[14] of the tenth-century additions to the Old Minster at Winchester is correct, the twelve-inch foot of the present day may antedate the yard by over a century, in which case its ultimate origin is quite obscure.

The Statute for the Measuring of Land,[19] most probably of the thirteenth century, is quite explicit, giving 3 barley corns to the inch, 12 inches to the foot, 3 feet to the yard, and 5½ yards to the perch (see Appendix A (e)).

Surprisingly Fleta (*c*1290)[20] is silent on the yard although he goes into detail about many aspects of other measures. Britton, writing a year or two after Fleta, does better, the French text stating:[21] 'the yard of two cubits and two inches . . .' (le aune de deus coutes et deus pouz . . .). The cubit was the natural unit from elbow to finger-tip. Its usual equivalent was about 18 Roman inches (17.5 English inches, 44.5 cm), approximately half a yard, so here Britton is clearly referring to the yard and not to any other unit.

The Record of Caernarvon[22] of about the fourteenth century says: 'Three feet make a yard, five and a half yards make a perch . . .' (Tres pedes faciunt ulnam, quinque ulne et dimidia faciunt perticam . . .). A little later, in 1479, William Worcestre recorded the dimensions of the Church of the Friars of St. Francis at Norwich.[23]

> Note:
> But the foresaid width [length?] of the church from the west to the first doors is 35 yards or 105 feet by my measurement with a yard, 3 feet in length.

The foot from those days to the present was always subsidiary in law to the yard, the statutes declaring the foot to be one third of the yard of thirty-six inches. The yard had been made to fit the rod, albeit incongruously, with the anomalous figure of $5\frac{1}{2}$ to the rod, pole, or perch.

This then is the key to the abnormality of fractional parts in a table of length which was mentioned in Ch.III p.35-36. The table was not set up *de novo*. The lower part of the table was already set. There were, and would continue to be, 40 rods to the furlong and the acre was going to continue as 40 × 4 rods. The smaller units were to fit into the larger, not the other way round, and so from the twelfth century to the present, whenever a yard is mentioned, it always means the same thing, that is 3 feet, 36 inches, or 0.914 m by our present measure.

The yardstick was intended for general use but the Exchequer standard of Henry VII was subdivided in binary fashion as used by the cloth-makers into halves, quarters, eighths, sixteenths, and thirty-seconds, and one foot of the standard bar was subdivided into twelve inches. The sixteenth part of a yard is the *nail* of $2\frac{1}{4}$ inches. The word is derived from the fact that the finger, that is the nail, was $\frac{1}{16}$ of a Roman foot. Dr Johnson in his *Dictionary* of 1755 merely defines it as $2\frac{1}{4}$ inches. However, the term was in use centuries earlier, as we see from the Coventry Leet Book for 1474:[24] 'And hit [the yard] to be sysed and sealed and thereupon to be marked a True halfe yarde, quarter, halfe quarter, nayle, and halfe nayle', and from an earlier letter dated 20 September 1465 from John to Margaret Paston in which we read in part:[25] '. . . then buy me a quarter of a yard and the nail thereof for colours, though it be dearer than the other' (. . . thanne bye me a quarter and the *nayle* thereof for colers, thou it be dewer thanne the tother).* And again almost a century and a half later, about 1600, we have, in a commonplace book:[26] '. . . and in bredith one yard and a nayle, at the lest . . .''

*See also p.93 below, Statute 4 Edward IV cap. 1 1464/5.

Nor was the binary division of the yard restricted to cloth measure. We meet it in certain land charters, for example '9 yards and a quarter and half a quarter' as in a grant by William the Cook to the Church of St. Paul,[27] dated 1189-1211. Sometimes we meet a mixture of old body measures with the newer standards, for example[28] '7 yards less a full hand', which is contained in a land grant from one Nicholas to Herbert of Bulogne dated 1202-3. Sometimes land is measured only in feet, without mentioning poles or yards, for example '41½ feet wide and 120 feet long' as in a document of St. Paul's dated 1181-1202.[29] Another of the charters of St. Paul's Cathedral of about 1200 contains both yards and feet;[30] '. . . 10 yards by the king's measure, less one foot . . .'.

We therefore see that by the year 1200 the yard and the foot are commonly used in the record of land transfers, and moreover the yard is 'by the king's measure' as the last quotation shows. The king's measure as we shall see shortly took the form of an iron rod. This gave permanence to the standard and authority to the measure. Further it would be difficult to shorten an iron bar without it being noticeable.

The Foot of St. Paul's

The people of London were given early access to the new measure of the twelve-inch foot. Documents of the closing years of the twelfth century refer to a foot inscribed or sculptured on the base of a column in the church of St. Paul. A charter of St. Paul's dated prior to 1200, when referring to the manor of Lisson in Marylebone, London, states:[31] '57 feet by the foot of Algar which is sculptured on the base of a column in the church of St. Paul' (LVII pedes per pedem Algari qui insculpitur super basim columpnae in ecclesia Sancte Pauli).

This Algar[31] is thought to have been (but not the son of)[32] the first prebendary of Islington in the time of Henry I. Although here called 'Algar's foot' it is much more common to find it referred to as the 'foot of St. Paul's.[32] For example, in the Cartulary of St. Mary Clerkenwell we read of a grant of land from Aimer, the tailor, to his wife. The land is located in Chicken Lane, London. The document is dated 1216–20 and contains the words:[33] '60 and 12 feet of St. Pauls' (LX et XII pedes Sancti Pauli). Somewhat later the church of the Grey Friars of London which was started in 1306 by Queen Margaret, second wife of Edward I, and completed in 1327 is described thus:[34]

> 'Description of the length, breadth, and height of the aforesaid church. Firstly the church contains 300 feet in length by the foot of St. Paul's. Also in breadth 89 feet by the foot of St. Paul's. Also in height from the ground all the way to roof 64 feet by the foot of St. Paul's'
> (Descriptio longitudinis et latitudines ecclesiae et altitudines supradictae. In primus continet ecclesia in longitudine ccc pedum de pedibus Sancti Pauli. Item in latitudine IIII^{xx} IX pedem de pedibus Sancti Pauli. Item in altitudine a terra usque

ad tectum LXIIII pedem de pedibus Sancti Pauli).

Liber Albus, a compendium of London affairs compiled in 1419 but dealing with the century approximately 1280–1380, also mentions this foot: '. . . and that paviours [those who made pavements] shall take for the toise [a measure] $7\frac{1}{2}$ feet long and a foot of St. Paul [in breadth] — 2 pence'[35]

Fortunately we are able to ascertain that the foot of St. Paul's was none other than the prevailing standard, namely the twelve-inch foot of the present, from a line in a Cambridge deed of 3 June 1459 conveying land from Corpus Christi to Queens' College, where we find the words[36] '. . . $64\frac{1}{2}$ feet of the king's standard in length and in breadth $24\frac{3}{4}$ feet by the foot of St. Paul's'. It would be most unlikely to mix two foot measures in arriving at the length and breadth of the plot of ground, but to avoid any ambiguity the document itself a little later gives another length as: '$13\frac{1}{2}$ feet of the standard and of the foot of St. Paul's'. The words 'by the foot of St. Paul's' continued to appear in documents till the end of the fifteenth century but they appear progressively less frequently than such expressions as 'ped de Standardo Regio' (foot of the royal standard) or 'feet by the iron yard of King Richard, King John, or King Henry' or simply 'the king's iron yard', which came into use at the end of the twelfth century and thereafter appear with increasing frequency.

By mid-fifteenth century and continuing through the sixteenth and well into the seventeenth we occasionally find the measures to be used are those 'of the assize', for example in 1440 '. . . decem pedes assise . . .'; in 1531 '. . . on the easte parte 31 fote and 3 enches of assys . . .'; and in 1670 '. . . two hundred and seaventy foote of assize, or thereabouts in length . . .'.[37]

The St. Paul's foot certainly played its part in the story of English metrology but it passes in and out of history without notice by the early historians of the cathedral, the city of London, or of the monuments of the Church in England. Interestingly enough, Stowe gives the length of the Church of St. Paul as '240 Taylors' Yards', a reference to the silver yard of the Merchant Taylors' Company (to be referred to in Chapter XII), but the length of the Church is not 720 feet as his calculation would make it (see Ch.II p.27).

The engraved pillar carrying the measure of the St. Paul's foot has disappeared without trace. Publicly displayed standards of the length were few and far between until the nineteenth century when it became common for a City Corporation or Town Council to erect a device such as that in Figure 24 for the guidance and assistance of the people. Well known public measures are those in the window at the London Guildhall of 1, 2, and 3 feet and in the floor 100 and 66 feet (the old acre's breadth or 100 links of Gunter's chain) originally placed there by the Corporation in 1878 and replaced in 1954 following the damage to the Guildhall in Word War II. In 1973 a twenty-metre measure was set in the floor and a one-metre measure in a window adjoining that carrying the foot measures. Equally well known are the measures set into the north wall of Trafalgar Square, London (1876) and the 100 and 66 foot measures set into the steps of the National Gallery of Scotland, Edinburgh.

We have seen that the whole yard of Henry VII was divided in binary fashion down to the sixteenth part and only twelve inches are engraved on this standard. The implication is that the principal use was to be in the cloth trade.

The Yard and the Inch for Cloth

The yard measure was not without its idiosyncrasies. When selling cloth by yardage it was customary to place the thumb beyond the end of the measure and cut the cloth across just beyond the thumb. This gave an extra inch but ensured that, if the cut was a little oblique and not quite straight, there would be no loss to the buyer. This practice continued into the twentieth century. But the original practice became obscured, for by the end of the twelfth century it became customary in selling cloth for the buyer to take an extra inch for *each* yard, not an extra inch in the piece of cloth, and finally, but much later, the merchants of London took a full hand's breadth extra for each yard, as we read in the Statute Book for 1439;[38]

> . . . for where they were wont to measure cloth by the yard and the full inch, now they will measure by the yard and full hand which groweth to the increase of the buyer 2 yards of every cloth of 24 yards.

The 'hand' was a measure of four inches, so that by taking four inches per yard instead of the customary one inch the buyer acquired an extra three inches in every yard or two extra yards in a piece of cloth measuring 24 yards, as stated in the text. This statute prohibited the yard and the full hand, substituting the yard and the inch, so that the *dozen* of shrunk cloth would be 12 yards and 12 inches long and the length of *dry* (unshrunken) cloth 14 yards and 14 inches. The prohibition was quite ineffective as the 'yard and the full hand' continued in use until the reign of Elizabeth I.

The Act goes on to say that the line used by the *alnager* (cloth-inspector) for measuring cloth, and hence assessing the duty to be paid, was to be 12 yards and 12 inches long signed at every yard an inch and at the end of the half yard a half inch and the quarter yard a quarter inch.

In Scotland the cloth measure, the ell, was either 37 or 37.2 English inches. Both lengths were in use, the shorter being the older. The public ell of 1706 displayed to this day at Dunkeld was measured by the writer to be just $37\frac{1}{8}$ inches although locally it is said to be $37\frac{1}{4}$. Its subdivisions are irregular and do not form an easily recognizable pattern although no doubt intended to represent the binary subdivisions of the measure.* It has been said that the 37-inch Scottish ell is a manifestation of the acceptance of the yard and the inch with the extra inch incorporated directly into the measure for convenience. While this may be true, another possibility is that the Scottish ell is related to the *porter* (*bier* or *beer* in English), a device for carrying goods, in this case the yarn to be woven. Six hundred openings to carry the thread was a '30 porter' — that is to say a porter was twenty such openings. The full six hundred measured 37 inches. The single porter was 1.85 inches.[39]

The custom of adding an inch to each yard was so common in the thirteenth century that it spread to the measurement of land as well and we find that it often became necessary to state which measure was meant when using the word 'yard', that is whether it was to be the 'yard with the inch' or the 'yard without the inch', that is the bare 36 inches.

We find, for instance, in three of the Charters of the Monastery of St. Peter's, Gloucester,[40] references to so many 'yards of our lord the king with inches between' (virgas domini regis ulnaries cum pollice interposita). They leave little doubt that the land was to be measured 'to every yard an inch'. The distances involved range from 5 to 39 yards. While the shorter distances would present no problem, the longer would be quite burdensome unless measured by a line, not unlike the alnager's line, with an inch between each yard. The editor of the Cartulary assigns dates as follows to the three charters: No. 499 — 1139/48, No. 679 — 1149/79, No. 808 — 1257/8. It is unusual to find the yard and inch mentioned in land grants of the twelfth century for the iron yard of the king did not come into existence till the century's end† but it is quite common in the thirteenth. As stated earlier sometimes the inch was given, sometimes not. Frequently the fact that the inches were not to be given was stated quite emphatically and whereas no king's name is associated with the 'yard of our lord the king' in the charters mentioned above, other documents are not so reticent.

In a document of time of Henry III we read:[41] '. . . 12 yards by the iron yard of King Henry measured without inches', while a contemporary charter of St. Paul's Cathedral of *c*1240[42] reads: '. . . 29 yards of the iron yard of our Lord King Henry, son of King John, with inches'. Sometimes there is no reference at all to the inches, but a natural unit is intruded as in a charter dated 1195-1215 also from St. Paul's Cathedral:[43] '. . . 10½ yards less 1 palm by the iron yard of King John, King of England'.

Here we see examples of the yard (with or without inches) in regular use in the late twelfth and early thirteenth centuries,[44] but the charters tell us more. They speak of the king's *iron* yard so frequently that one is driven to conclude

*R. M. Wells (Notes & Queries, *Libra* VI 1967, p.21) gives the length as 37.2 inches with the subdivisions intended to be $\frac{1}{2}$, $\frac{1}{4}$, $\frac{1}{8}$, and $\frac{1}{16}$.

†It may be that charters 499 and 679 have been ascribed too early a date.

that a yard measure of iron was something of a novelty. We do not know the composition of the earlier yard measures but they must have existed in one form or another since 1100.

The earliest king's name to appear connected with an iron yard is that of Richard I.* One such entry from the Cartulary of St. Mary, Clerkenwell, dated soon after 10 February 1199, in which the Princess Isabel gives Henry Bacon land in the parish, reads, in part:[45] '. . . 45 yards and 2 feet by the iron yard of our Lord King Richard . . .'. The iron yards are attributable to Richard following the proclamation of his Assize of Measures,[46] of 20 November 1196,† and his subsequent issue of standard measures throughout the country. Although the names of various other kings, including John and Henry, are associated with these iron yards in the various texts, they are all the same measure.

In the Statute Book, listed as of uncertain date but probably of the thirteenth century, appears a document entitled 'Statutum de Admensuratione Terre' (Statute for the Measuring of Land). Inserted in the text (Ms Cott. Claudius DII fo 241 b) is the following:[19]

> It is ordained that three grains of barley dry and round do make an inch; twelve inches make a foot; three feet make a yard; five yards and a half make a perch and forty perches in length four in breadth make an acre.

In some printed versions this insert is shown as a separate article entitled 'Compositio Ulnarum et Perticarum'. At the end of the statute this is largely reiterated, though in different terms:

> And be it remembered, that the iron yard of our Lord the King, containeth three feet and no more. And a foot ought to contain twelve inches, by the right measure of this yard measured; to wit, the thirty-sixth Part of this yard rightly measured maketh an inch neither more nor less. And five yards and a half make one perch that is sixteen feet and a half, measured by the aforesaid iron yard of our Lord the King. [See Appendix A (e).]

It is clear that by the time this document was issued the inch, foot, yard, rod, and acre of the present day had been definitely established and this not long after Richard's Assize, for the iron yard of our Lord the King is much in evidence.

*It is not a coincidence that extant charters from Richard's reign onwards speak not only of the king's yard but also of the king's perch. See Chapter III p.44 and the corresponding reference 41.

†The year is still frequently but wrongly given as 1197. In Hoveden's *Chronicle* it appears among documents for 1197 but the rubric to the Assize is quite unambiguous as 'The Feast of St. Edmund at Winchester in the 8th year of the reign of Richard King of England', that is 20 November 1196. The matter is thoroughly discussed by Lady Stenton in her introduction to the Pipe Roll of 9th Richard I, pp.xxi-xxii.

The central part of the Statute for the Measuring of Land consists of a lengthy table giving the breadth of the acre when the length of the acre plot ranges from 10 to 80 rods at intervals of one rod. It begins: 'When an Acre of Land containeth Ten perches in length then it shall be in breadth Sixteen perches. When eleven in length, then Fourteen perches and a half, and one Foot and five Inches . . .' and so on.

Most of the breadths given are in error as a little calculation will show. When eleven in length the breadth should be 14½ perches and 9 inches, not 14½ perches, 1 foot, and 5 inches, but the errors were no doubt acceptable at the time and probably went largely unnoticed. The important thing in this statute is that the units of length are established, and established by authority.

The traditional date of the statute as given in the old legal compendia is 33 Edward I, that is 1304/5. For this there is no known authority and the document most probably belongs to an earlier period, much closer to Richard's Assize of Measures.[47]

The Assize of Measures, 1196

A proclamation by Richard, preceding his Assize, that all trading throughout the land shall be according to one weight and measure is recorded in *Chronica Johannis de Oxenedes* under the date of 1188, in the following words:[48] 'All just trade and selling throughout the kingdom shall be founded in one weight and measure' (Omnia commercia verum venalium per totum regnum constituta sunt unius ponderis et mesurae).

Bartholomaei de Cotton in his *Historia Anglicana* of the thirteenth century tells of a declaration of one weight and measure to be used throughout the land at the coronation of Richard in 1189 under the title 'De coronatione regis Ricardi' in virtually the identical Latin.[49] The same call is to be found recorded twice by Matthew Paris, once in his *Historia Anglorum,*[50] and again in his *Chronica Majora.*[51]

This was followed by the Assize issued at Westminster in 1196, the eighth year of Richard's reign, as mentioned above, in which a uniform measure, namely a good horse-load for dry goods, grain, etc., is laid down together with liquid measures for wine, ale, and the like 'according to the various liquors' and there are to be uniform weights 'according to the various merchandises'. The definitive version of the Assize is that given in full detail by the contemporary Roger de Hoveden in his *Chronica.*[46] It is given almost as fully in the *Historia Anglorum* of Matthew Paris,[50] the latter indicating that the Assize came from the king at the instance of Hubert, Archbishop of Canterbury and Justiciar of England (1193–8).

Roger de Hoveden's statement of the Assize contains the following: 'Throughout the realm there should be the same yard of the same size and it should be of iron'. There is no likelihood of error if we attribute the introduction of the iron yard measures to Richard I at the close of the twelfth century (1196), and with the yard the foot of twelve inches (0.305 m) came into general use.

The Assize of 1196 goes on to decree that all cloths are to be 'duabus ulnis infra lisuras'. This is usually rendered as '2 ells between the selvedges (lists)' but more correctly it is '2 yards' for Roger goes on to say 'the *ulna* is of iron' (et ulna sit ferrea). It was the *yard* measure that was of iron. The cloth ell is a later development. The Assize shows the importance of cloth. It is the only commodity for which a linear measure is specified and is on a par with bread, wine, and ale in importance in the life of the people.

The Measure of Cloth

The decreed breadth for cloth was too large for the trade. Richard's justices could not enforce it. Dealers ignored it, preferring to be fined rather than to submit to a measure at such variance with past practice. Cloth as wide as this would need to be woven on a double loom with two people sitting side by side to work the shuttle. Narrower cloth could be produced by one worker at a single loom. Some weavers could afford a broad and a narrow loom but not all. In some parts of the country broad and narrow weaving are kept separate, as distinct crafts, as at Winchester. Not until 1784, with the invention of the flying shuttle, could a single person weave broadcloth.

But this did not prevent John's Magna Carta of 15 June 1215 from repeating the breadth of cloth to be two yards, just as Richard had decreed, and this was repeated in the subsequent Magna Cartas of 1216, 1217, 1224/5, and 1251/2. There was considerable unrest among the weavers who preferred a narrower cloth, yet to offer such for sale was to invite its confiscation, or the imposition of a fine, or both. Nor was the matter helped by the irresolution of the statutes themselves.

The Statute of Northampton of 1328[52] gave a considerable concession to the weavers by narrowing the cloth:

> . . . the length of the cloth of Ray [striped cloth], by a line of 7 yards four times measured by the list and the breadth of every Ray cloth 6 quarters [of a yard] measured by the yard and of coloured cloths the length shall be measured by the back of a line of 6 yards and a half, four times measured, and the breadth 6 quarters and a half measured by the yard without soiling the cloths.

So Ray cloth was to be 28 yards long and coloured 26 yards. Ray was to be 6 quarters broad; coloured cloth $6\frac{1}{2}$. Six quarters are 54 inches; six and a half quarters are $58\frac{1}{2}$ inches. The cloth is to be measured by a line or cord for length but by a yardstick for breadth.

Nine years later, in 1337,[53] it was decreed that no man or woman should wear cloth other than that made in England, Ireland, Wales, or Scotland upon pain of forfeiture and continues: '. . . and that in the said lands of England, Ireland, Wales, and Scotland within the Kingspower, a man may make cloths as long or as short as a man will . . .'.

The new-found freedom did not last. By 1350[54] we are back to the provisions of the Statute of Northampton which were repeated in 1353.[55] The commonalty protested in 1373[56] that, whereas Ray cloths should be 28 yards long and 5 (sic) quarters in breadth, they are in fact barely 22 yards by 4 quarters and coloured cloths, which should be 26 yards and 6 (sic) quarters broad at least, scarcely reach 22 yards long and 4½ quarters in breadth. This led to the statute of the same year[57] reaffirming the 28 yards and 5 (or 6) quarters* for Ray and the 26 yards and 6 quarters (at least) for coloured, dimensions that were reaffirmed in 1388[58] with Ray cloth being unambiguously stated to be 5 quarters wide. Then in 1389 Richard II decreed that cloth shall be of such length and breadth as a man pleases, 'notwithstanding any Statute, made to the contrary'[59] and this is repeated in 1393/4 in the Statute 17 Richard II c 2. The weavers of the fourteenth century needed to be men of courage and resolution to sail unscathed through these choppy seas.

The situation was not improved with Henry IV's Act of 1405[60] which changed the breadth of Ray to 6 quarters from 5, as though the decrees of 1389/90 and 1393/4 had never existed. The outcry when cloths not according to these dimensions were seized was so great that this Act was repealed in 1407[61] and we are back to the familiar 28 and 5 for Ray and 26 and 6 or usually 6½ for coloured cloth. The waters were further muddied in 1409 when Henry IV made another ill-advised foray into the world of commerce by decreeing coloureds were to be 28 yards long.[62]

In an endeavour to create some semblance of order, the commonalty petitioned Henry VI in 1433,[63] reciting the Acts of 1393, 1405, and 1409, and pleading that the provisions contained therein should apply to whole woollen cloths called 'broadcloths' and 'broad dozens' and not to other cloths called 'streytes' which were to be made 14 yards long and a yard broad unwetted or else 12 yards wetted (with no mention as to breadth when wetted). This was agreed to by Statute the same year.[64]

But further perturbations were in store. By 1439 the length of a cloth was described by Statute as 24 yards[38] and this was repeated in 1464;[65]

> That every whole woollen cloth called Broad cloth . . . after a full watering, racking, straining or tenturing . . . shall hold and contain in length 24 yards [aulnes] and to every yard an inch containing the breadth of a man's thumb to be measured by the crest of the same cloth, and in breadth 2 yards or 7 quarters at least within the lists.

*Some versions have 5, some 6.

Here we have a total length of 24 yards and 2 feet for broadcloth but a breadth of anywhere from 63 to 72 inches, or 64¾ to 74 inches if the inch was to be given to the yard of breadth as well.

We are further told by this Act that the cloth called 'streits' is to be 12 yards and the inches long and in breadth one yard within the lists but the cloth called kersey (a coarse ribbed cloth) after a full watering etc. '. . . shall be in length 18 yards and the inches as is aforesaid and in breadth a yard and a nail or at least a yard within the lists'.

An Act of Richard III of 1483[66] followed by another of Edward VI in 1553 directed that the measure be by the yard and to every yard an inch, measured by a man's thumb.

As we might expect, all this interest in the length and breadth of the various types of cloths was not entirely altruistic. If a cloth met the requirements, the alnager sealed it and for a full piece received one half penny in 1353; a half cloth yielded him a farthing, but smaller cloths were measured free, and a royal duty was imposed amounting to from four to six pence for a full cloth and proportionately less for half cloths but if not of the correct size the piece was forfeit to the king.

The Act 27 Henry VIII c 12 1535/6 laid down that every broad-cloth should be seven quarters within the lists instead of the two yards specified in 1 Richard III c 8 1483 but 5/6 Edward VI c 6 (1551/2) specified the breadths and weights of a great number of cloths according to the place of manufacture. However, as the Act of 4 and 5 Philip and Mary c 5 (1557/8) shows, the merchants could not cope with the complexities of the 1551/2 Act. Unhappily that of 1557/8 was no better, for we find in 27 Elizabeth c 17 (1584/5) that the clothiers could not handle its provisions either. After stating it was difficult to make cloth seven quarters wide the merchants showed they were meeting the weight requirements of 4/5 Philip and Mary. The width requirement was accordingly reduced to 6½ quarters, and this was *made perpetual* by 35 Elizabeth c 7 s30, 1593 which also shows the 'yard and inch' measure to be still in force.

Moving into the next century the prescription of cloth by district continued as seen in 14 Charles II c 32 s VI (1662) for the West Riding of Yorkshire. Here every cloth was to be 30 to 36 yards long and 1½ yards (6 quarters) broad. So much for the declaration of perpetuity of 1592/3. Moreover every yard of cloth was to weigh 2¼ pounds of 16 ounces, that is Avoirdupois weight.

At the beginning of the eighteenth century (7 Anne c 13, 1708) every broadcloth made in the county of York was to be 'at least five and a half quarters by the standard yard wand in breadth.' A whole cloth was to be not more than 46 yards long, and in 1711 Queen Anne[67] decreed that every yard was to have an inch 'instead of that commonly called a thumb's breadth'.*

But enough was enough. The opening of the nineteenth century brought with it a total change in attitude to the cloth and wool trade. The Act 43 George III c 136 of 1803 suspended prosecutions under a variety of previous Acts till 1 July 1804 and the suspension was continued annually till 1809 when by the Act 49 George III c 109 most of the Acts which restricted cloth and woollen manufacture were abolished. No fewer than 32 Acts were repealed in whole, commencing with the Statute of Northampton (2 Edward III c 14 1328). Six further Acts were repealed in the portion dealing with the dimensions of cloth commencing with 27 Edward III c 4 of 1353. Thereafter we no longer hear of the lengths and breadths of cloths in the Statute Book.

In the First Report of the Commissioners on Weights and Measures of 1819 we have the somewhat inaccurate if nostalgic plea that 'whereas at the Conquest broadcloths were to be 2 yards between the lists there are now allowed in some counties only $1\frac{3}{8}$'.

The important measures were the breadths of the cloths and we have seen them variously as 4, 5, 6, $6\frac{1}{2}$, and 7 quarters. The 7-quarter width had but a transient life. The $6\frac{1}{2}$-quarter breadth lasted longer. But cloth widths of 36, 45, and 54 inches (4, 5, and 6 quarters) endured till the present century.

The Ell

Notwithstanding the repeated mention of seven and six quarters in the statutes, the weaver's measuring rod (ell) from the fourteenth century would appear to have been five quarters long. Pegolotti, writing of Lōndon in the early part of the century, states:[68] 'Canvas is sold by the hundred with 120 to the hundred and with 5 quarters to the ell'.

The Exchequer had no such standard until the days of Elizabeth, for on 12 May 1474 a Coventry merchant, J. Wyldgrys, presented to the Barons of the Exchequer two iron rods, one a yard measure, the other an ell, requesting that they be placed beside the Exchequer standard (singular) for verification.

> Whereupon on the same day the said yard and ell having been compared with the Exchequer yard, it appears to the Barons that the said yard brought by J. Wyldgrys is of standard length and that the ell is equal to the length of a yard and a quarter according to the standard yard of the exchequer.[69]

Elizabeth's standardization of weights and measures of 1588 completed the dominance of the 45-inch ell. A bronze standard bar was then made for the Exchequer, the first of which we have record, 0.6 inch in square section.

*'Every owner . . . shall have one table or board 12 feet long and 3 feet wide at least . . . with the length of a yard nailed or marked thereupon to which shall be added one inch more which shall be used instead of that commonly called a thumb's breadth so that the same length shall contain 37 inches' — 12 Anne c 16 s IV 1711.

Today it measures 45.04 inches long and is in the Science Museum, London.

Arnold's Chronicle of 1502[70] says that a Flemish ell is three quarters of an English yard and two Flemish ells make an English ell and a quarter of a yard, while five Flemish ells make three English ells. This useful piece of information, while making no use of feet or inches, gives us at once the English and Flemish ells in terms of the 36-inch English yard, the former being 45 inches, the latter 27. In case the arithmetic proved too much for his readers, Arnold had earlier stated '. . . and V quatirs of the yerde make an elle . . .', which removes any doubt.

That the weavers' measure was five quarters of a yard may tell us more than all the statutes just what the width of the woven material was likely to be. Is it not strange that the construction of the Exchequer standard of this length of 1588 should precede the 1592/3 legislation making the 6½ quarters 'perpetual'? Remarkably enough, although there was an Exchequer standard ell from the days of Elizabeth it had no formal status in law. As late as 1743 a paper read to the Royal Society contained the words:[71] 'The ell is not therein [the Acts of Parliament] particularly described though universally reputed equal to 1¼ yards or to 45 inches'. However the ell is noticed in the 1803 Statute relating to weavers in Scotland (43 George III c151 1803);[72] 'And be it further enacted that in all cases the length of the web for which the workman is to be paid shall be ascertained by the standard ell of 45 inches'. No doubt the length of the English ell had to be given in order to distinguish it from the Scottish ell of 37 inches. Had the Statute referred to England it is quite likely that the length of the ell would not have been given, as by common knowledge and custom it was 45 inches.

The Exchequer ell of Elizabeth was a primary standard (even if not legally defined), till 1824 when its useful life terminated and it ceased to be a legal standard by the Act of 5 George IV c74. Even here the ell is not mentioned by name, but the Act of 1824 reads that

> the 'Imperial Standard Yard' — shall be and is hereby declared to be the Unit or only Standard Measure of Extension, wherefrom or whereby all other Measures of Extension whatsoever, whether some be lineal, superficial, or solid, shall be derived, computed, and ascertained . . .,

with the further declaration that on and after 1 May 1825 all Contracts, Bargains, Sales, and Dealings shall be made according to the standards declared in the Act. This was the death knell of the ell, but from the days of Elizabeth to 1825 its length continued to be recorded. About 1600 it is given as 'contening 45 ynches in lenght' [sic].[73] In 1682 it is given as 'a yard and nine inches'.[74] In 1820 we read:[75] 'Ell – 5 quarters = 45 inches in England and legally in Scotland by the Act 43 George III 1803'.

The history of British metrology might have been simpler had Richard in his 1196 Assize promulgated a narrower measure for cloth. But he did not simply promulgate. He sent measures, iron yards, scales, and weights into all the counties of England in 1197 as we have seen.

Edgar kept his standards at London and Winchester but, to be used at all, replicas would be needed on the local scene. Crudely made, no doubt, but serviceable, the local standards would be used until something better came along as in 1196-7. Being then made of iron, the yardsticks would last and be a durable record of the true length, difficult to shorten in an unauthorized fashion. Unhappily none of Richard's standards has survived the centuries. We have no extant examples of his weights or capacity measures or 'iron ulnae of our Lord the King'. Since the beginning of the twelfth century, then, English linear and superficial measure has not changed. The yard, foot, and inch of today are therefore eight centuries old while the rod, furlong, and acre antedate these measures by several centuries.

NOTES AND REFERENCES: CHAPTER VI

1. ATTENBOROUGH, F. L. *The Laws of the Earliest English Kings* pp.12-13. Cambridge: C.U.P. 1922.
2. DOWNER, L. J. (ed. & tr.) *Leges Henrici Primi* p.293. Oxford: Clarendon Press 1972.
3. ASSER. *Life of King Alfred;* STEVENSON, William Henry (ed.) Ch.104, p.90. Oxford: Clarendon Press 1904, Reprinted 1959.
4. THOMSON, T. & INNES, C. (eds.) *Acts of the Parliament of Scotland* Vol.I, Appendix III, p.309 (p.673 of enumerated text). London: The Record Commission 1844.
5. SHEPPARD, J. B. (ed.) *Certa Mensura* Second Report on Historical Manuscripts belonging to the Dean and Chapter of Canterbury, in the eighth Report of the Royal Commission on Historical Manuscripts Appendix I, (Pt.1), pp.315-355. London: Commission on Historical Manuscripts 1881.
6. HARRIS, Mary Dormer. (ed.) *The Coventry Leet Book* Vol.I, p.396. London & New York: E.E.T.S.; K. Paul, Trench Trubner & Co. Ltd., H. Milford, O.U.P. 1907.
7. COWELL, J. *The Interpreter* Cambridge: John Legate 1607. (s.v. perch).
8. SPELMAN, Sir Henry. *Glossarium Archaiologicum,* DUGDALE, Sir W. (ed.) London: Alicia Warren 1664. (s.v. pollex)
9. 10 Anne 1711. c16 s4; *Statutes at Large*. Vol.IV, pp.509-510, 1786.
10. BENESE, Sir Richard de. *Boke of the Mesurying of Londe* Southwarke 1537; cited by PRIOR, W. H. Notes on the Weights and Measures of Mediaeval England. *Bulletin du Cange* (1924) Vol.I, pp.141-2. Paris.
11. ATTENBOROUGH, F. L. Loc. cit. pp.138-141.
12. Ibid. pp.170-1.

13. Ibid. pp.52-3.
14. (a) BIDDLE, Birthe Kjolbye-. The 7th c. Minster Church at Winchester Interpreted. *The Anglo-Saxon Church,* Studies in History, Architecture & Archaeology in Honour of Harold Taylor, editors Butler L. A. F. & Morris R. K. CBA Research Report 60, London: 1986.
(b) BIDDLE, Martin & Birthe Kjolbye-. The Anglo-Saxon Minsters of Winchester, *Winchester Studies,* Vol.4. 1, O.U.P. (forthcoming).
15. THORPE, Benjamin (ed.) *Diplomatarium Anglicum aevi Saxonici* p.109 London: Macmillan 1865 (Sawyer No.308).
16. Ibid. pp.156-7 (Sawyer No.1443).
17. GIBBS, Marion (ed.) *Early Charters of the Cathedral Church of St. Paul London* Camden 3rd Series Vol.LVIII pp.264-5. No.330 London: Camden Society 1939.
18. STUBBS, William (ed.) *Willelmi Malmesbiriensis Monarchi de Gestis Regum Anglorum* Vol.II, Book V. p.489 London: (Rolls series) Longman et al. 1887.
For translation see:
GILES, J. A. (ed.) *William of Malmsbury Chronicle* p.445 London: Henry G. Bohn 1847.
19. Statute for the Measuring of Land; *Statutes* Vol.I, pp.206-7.
20. RICHARDSON, H. G. & SAYLES, G. O. (eds. & trs.) *Fleta,* Selden Society Publication, Vol.72 of 1953, London: The Selden Society 1955.
21. NICHOLS, Francis Morgan (ed. & tr.) *Britton* Vol.I p.189 (Book 1, Ch.xxxi. Sect. 5 of text) Oxford; Clarendon Press 1865.
22. ELLIS, Sir Henry. *Registrum Vulgariter Nuncupatum — The Record of Caernarvon.* p.242. London: The Record Commission 1838.
23. HARVEY, John H. *William Worcestre Itineraries* pp.236-7 Oxford: Clarendon Press 1969.
24. HARRIS, Mary Dormer (ed.). Loc. cit. p.397.
25. DAVIS, Norman (ed.) *Paston Letters and Papers of the 15th Century* Vol.I, p.140. Oxford: Clarendon Press 1971.
26. HALL, Hubert & NICHOLAS, Frieda J. (eds.) *Select Tracts and Table Books Relating to English Weights and Measures (1100-1742)* Camden Miscellany Vol.XV, p.26. London: The Camden Society 1929.
27. GIBBS, Marion. Loc. cit. p.236, No.297.
28. Ibid. p.99, No.135.
29. Ibid. p.128-9, No.167.
30. Ibid. p.198, No.252.
31. Ibid. p.136-7, No.177.
32. NICHOLS, J. G. The Foot of Saint Pauls. *Gentleman's Magazine* (1852) N. S. Vol.XXXVIII, Pt.2, pp.276-7. London.
The information respecting the foot of St. Paul's is based on the above article and that by COOPER, C. H. The Foot of Saint Paul in the same journal (1852) N. S. Vol.XXXVIII, p.57. The matter is fully treated by GRIERSON, Philip *English Linear Measures, The Stenton Lecture 1971* pp.17-19. Reading: University of Reading 1972.
The general reader might first encounter this foot in NICHOLSON, Edward *Men and Measures* p.60. London: Smith Elder & Co. 1912, where it is mentioned without reference, or in SALZMAN, L. F. *English Trade in the Middle Ages* p.52.

Oxford: Clarendon Press 1931, where it is mentioned with references to the articles of Cooper and Nichols as given above.

33. HASSALL, W. O. (ed.) *Cartulary of St Mary Clerkenwell* p.205, No.314. London: Royal Historical Society 1949.
34. NICHOLS, J. C. (ed.) *Chronicle of the Grey Friars of London.* p.xiii, footnote 'a'. London: The Camden Society 1852.
35. RILEY, Henry Thomas (ed.) *Munimenta Gildhallae Londoniensis; Liber Albus, Liber Custumarum, Liber Horn.* Vol.I, p.728; Vol.III, p.282. London: (Rolls Series), Longman et al. 1859-62.
36. Ms BAKER xxx 27 b; Quoted by COOPER, C. H. Loc. cit. p.57.
37. Appendix to the *9th Report of the Historical Manuscripts Commission* Vol.IX, Pt.1, pp.11-12. London: Historical Manuscripts Commission 1883.
38. 18 Henry VI c 16 1439; *Statutes* Vol.II, p.312.
39. BURRELL, L. The Standards of Scotland. *The Monthly Review, The Journal of the Institute of Weights and Measures Administration* (March 1961) Vol.LXIX, pp.49-62 (see p.51).
40. HART, William Henry (ed.) *Historia et Cartularium Monasterie Sancti Petri Gloucestriae* Vol.II, p.15, No.449; p.153. No.679, and pp.236-7, No.808. London: (Rolls Series) Longman et al. 1865.
41. Anon. 'Original Documents' (Relating to Land adjacent to London in the Reign of Henry III) *Archaeological Journal* (1849) Vol.VI, pp.280-282 (see p.281).
42. GIBBS, Marion. Loc. cit. p.212, No.270.
43. Ibid. p.165, No.209.
44. Additional examples are to be found in:
GRIERSON, Philip. *English Linear Measures, The Stenton Lecture 1971* p.17 et seq. Reading: University of Reading 1972.
45. HASSALL, W. O. (ed.) Loc. cit. p.207, No.316.
46. STUBBS, William (ed.) *Chronica Magistri Rogeri de Houedene* Vol.IV, pp.33-4. London: (Rolls Series) Longman et al. 1868-71.
47. GRIERSON, Philip. Loc. cit. pp.13-14.
48. ELLIS, Sir Henry (ed.) *Chronica Johannis de Oxenedes* p.72. London: (Rolls Series) Longman et al. 1859.
49. LUARD, Henry Richards (ed.) *Bartholomaeus de Cotton, Historia Anglicana* p.83. London: (Rolls Series) Longman et al. 1859.
50. MADDEN, Sir Frederic (ed.) *Matthaei Parisiensis Historia Anglorum* Vol.II, p.10. London: (Rolls Series) Longman et al. 1866.
51. LUARD, Henry Richards (ed.) *Matthaei Parisiensis Chronica Majora* Vol.II, p.351. London: (Rolls Series) Longman et al. 1872-83.
52. 2 Edward III s 14 1328; *Statutes* Vol.II, p.260.
53. 11 Edward III s 2 1337; Ibid. Vol.II, p.280.
54. 25 Edward III Stat. 3 c 1 1350/1; Ibid. Vol.II, p.314.
55. 27 Edward III Stat. 1 c IV 1353; Ibid. Vol.II, p.330.
56. STRACHEY, J. (ed.) *Rotuli Parliamentorum* (47 Edward III No.3, AD 1373) London HMSO 1767-77.
57. 47 Edward III c 1 1373: *Statutes* Vol.II, p.395.
58. 12 Richard II c 14 1388; Ibid. Vol.III, p.60.
59. 13 Richard II Stat. 1 c 10 1389; Ibid. Vol.III, p.64.

60. 7 Henry IV c 10 1405/6; Ibid. Vol.III, p.154.
61. 9 Henry IV c 6 1407/8; Ibid. Vol.III, p.160.
62. 11 Henry IV c 6 1409; Ibid. Vol.III, p.163.
63. STRACHEY, J. (ed.) Loc. cit.; 11 Henry VI 1433.
64. 11 Henry VI c 9 1433; *Statutes* Vol.III, p.284.
65. 4 Edward IV c 1 1464/5; Ibid. Vol.III, p.403.
66. 1 Richard III c 8 1483; Ibid. Vol.III, p.484.
67. 12 Anne c 16 s 4 1711. London: HMSO 1786. *Statutes at Large* Vol.IV, pp.509-10.
68. PEGOLOTTI, Francesco Balducci. *La Practica della Mercatura* EVANS, Allan (ed.) p.255. Cambridge, Mass: Mediaeval Academy of America (Publication No.24) 1936.
69. HARRIS, Mary Dormer. Loc. cit. Vol.I, (Pts. I & II), p.394-5.
70. ARNOLD, Richard. *The Customs of London, commonly called Arnold's Chronicle* pp.204, 173. London: F. C. & J. Rivington 1811 (Reprint of 1st ed. of *c*1502).
71. ANON. An Account of the Proportions of the English and French Measures and Weights . . ., *Phil. Trans. Roy. Soc.* (1742/3) Vol.XLII, pp.541-556. London (see p.545).
72. 43 George III c151 s19 1803. London 1804; *Statutes at Large;* Vol.XV, p.1115.
73. HALL, Hubert and NICHOLAS, Frieda J. (eds.) Loc. cit. p.27.
74. Ibid. p.28.
75. Second Report of the Commissioners appointed by His Majesty to Consider the Subject of Weights and Measures 13th July 1820, *Parl. Papers* 1820 (HC314) vii pp.473-512 Appendix A, p.15.

CHAPTER VII

The Early Currency

The Beginning of the English Coinage

From the Roman withdrawal until the first half of the seventh century no coins were produced in England by Briton or Saxon.[1] As for the circulation of Roman or Romano-British coins, this had virtually ceased by AD 430. There are no Anglo-Saxon coins at all in the coin hoard of the Sutton Hoo, Suffolk, burial of AD 625-30.[2] All 37 of these gold coins are Frankish of various dates and this implies the absence of a local coinage.[3,4]

It is very hard to imagine life without coins, or with only the rare foreign import, especially in an environment which had once possessed a currency, yet all evidence shows that there was a period of more than two and a half centuries when virtually all commerce was carried out by the ancient methods of exchange, the giving of gifts and jewellery, and by barter. The earliest English poems (seventh century) speak of kings as the ring-givers or the bestowers of treasure but no money or coin is ever mentioned. Three centuries later, long after coins were common, this tradition is recalled to mind when the *Anglo-Saxon Chronicle* (A) for AD 937 states in poetic form 'King Aethelstan, lord of warriors, ring-giver of men . . .'.

Jewellery was the vehicle of exchange in the absence of money. For small change, pieces were cut from rings, necklaces, bracelets, etc. as the archaeological evidence shows and these served as pieces of value in a manner similar to coins.

A very different picture is given by the Crondall, Hampshire, hoard located in 1828 when 101 small gold coins were discovered.[5]* Of these 24 are Frankish (Merovingian) and 72 are of Anglo-Saxon origin. The evidence indicates that this hoard was buried about AD 640. Thus an Anglo-Saxon coinage in gold must have sprung up in the intervening ten to fifteen years which separated the burial of this hoard from that at Sutton Hoo. The general appearance of the Anglo-Saxon coins was Merovingian but the models were from available Roman coins.

It is generally agreed that this and subsequent Anglo-Saxon coinages were made in imitation of that of the Franks. Keary, who catalogued the Anglo-Saxon coins in the British Museum, writes:

*One piece was counterfeit[5] so this looks like the *wergeld* (see p. 104) of a common man which in the Laws of Aethelbert (AD 602-3) is given as: 'If one man slays another the ordinary wergeld to be paid as compensation shall be 100 shillings' (cap 20).

> We may take it as established that the whole class of anonymous gold and silver coins which constitute the earliest English coinage was derived from the coinage of the Franks under the Merovingian kings . . . it was from the Merovingian coins, in the first instance from gold, later on from the silver, that the earliest English coinage was derived.[6]

The first evidence that this is so arises from the similarity in appearance; the second is the identity of their weights. Of the 96 Frankish and Anglo-Saxon coins of the Crondall hoard, all in mint condition, the average weight was 19.8 Troy grains (1.28 g) while of the 72 coins known to be definitely Anglo-Saxon in origin the mean weight is 19.97 grains (1.29 g).[7] The steps leading up to this weight identity are as follows:

At first the Merovingian kings struck imitations of the Roman *solidus* and *tremissis* ($\frac{1}{3}$) on the Roman standard of 4.5 g and 1.5 g respectively, that is 24 and 8 *siliquae* (Ch. 1, p. 15–16). At the end of the sixth century the Frankish *solidus* became less used and the *tremissis* was reduced in the AD 570s from 8 to 7 *siliquae* in weight, that is to 1.3 g, and some of these coins (from the strongly Romanized parts of Gaul) carry the legend 'DE SELEQAS VII'.

At the same time a move away from reckoning in *siliquae* is in evidence in northern Europe. Seven *siliquae* are almost exactly the theoretical weight of twenty grains of barley, a grain destined to replace the Mediterranean *siliqua* (carob) in the north. From this time onwards less and less is heard of *siliquae* and more and more of grains, that is barley corns or the Troy grains of the present day.

As for the minting of the old *solidus*, this continued into the early part of the seventh century being reduced to between 20 and 21 *siliquae* (3.7–3.9 g) in keeping with the reduction in the *tremissis*, as the coins themselves show.[8,9] Thereafter the *tremissis* of twenty grains (1.3 g or 7 *siliquae*) held the field. Saxon and Frankish gold being of the same weight would be freely exchangeable and inter-changeable on both sides of the Channel.[10] It is not surprising then to find them hoarded together as at Crondall.

The Shilling and the Sceatta

The oldest document in Anglo-Saxon that has come down to us is the set of Laws of King Aethelbert of Kent ascribed to AD 602–3. They are largely concerned with the compensatory payments to be made for injury to or death of the individual according to rank. There are two units of payment mentioned, the *scilling* and the *sceatta*. The laws reveal the relation between the two.[11] In the laws concerning the loss of finger and toe nails we read:

> 55 For the nails of each [of the above mentioned fingers], 1 shilling [shall be paid as compensation].

> 71 For each of the other toes [a sum] equal to half that laid down for the corresponding finger shall be paid.
> 72§1 Ten Sceattas shall be paid as compensation for the loss of the other toe-nails [other than the big toe].

Therefore if ten *sceattas* is the compensation for a toe-nail, twenty *sceattas* is that for a finger-nail: hence 20 *sceattas* = 1 *shilling*[12].

'*Scilling*' is our word 'shilling'. It originally meant a small piece of precious metal cut from a ring, armlet, necklace, or other ornament and offered as payment in a time when there was no coin,[13] as mentioned at the beginning of this section. Though frequently employed in England and Scotland the practice was much more prevalent in Scandinavia. Value was determined by the bullion content only. The workmanship expended on the piece went for nought. With the advent of a coinage the practice of cutting up jewellery declined, though on occasion coins are found hoarded together with portions cut from a larger worked ornament, and the pieces are sometimes of the same weight as the coins and hence of equal value. With this standardization the name of the fragment is transferred to the gold coin. The twenty-grain gold coins such as those of Anglo-Saxon origin found at Crondall can thus be called shillings.

The oldest English poem known to us, 'Widsith', speaks of the *scillings:*[14] 'There me the Goth king . . . a ring gave in which six hundred was of beaten gold Treasure scored, in scillings reckoned', as do also the late seventh-century Laws of Ine of Wessex, the Laws of Alfred (ninth century), Edward the Elder (tenth century), and those of Aethelstan of the same century.

What, then, was a *sceatta?* If minted gold *scillings* weighed twenty grains and we are told unequivocally that there are twenty *sceattas* to the shilling in Aethelbert's laws of the first decade of the seventh century, may we not think of a *sceatta* as being equivalent to one Troy grain of gold?[15] It cannot have meant a coin, for at the time of Aethelbert's laws (AD 602–3) there was no coinage in England and moreover in the early documents it is used in the more general sense of 'money' or 'treasure'.[16] The suggestion that the word *sceatta* meant a grain of gold was first made some years ago[15] and more recent support for this view is afforded, as Stewart Lyon[17] has pointed out, by a gloss in the Lindisfarne Gospels, written about the year AD 700 in Latin. The gloss to the text was supplied interlineally in Anglo-Saxon in the tenth century. In Luke 15.8 where the Authorised Version of today speaks of 'ten pieces of silver', and the Latin 'ten dragmas', the Anglo-Saxon gloss reads 'fifty sceattas', making one *dragma* the equivalent of five *sceattas*. A *dragma (drachma)* is $\frac{1}{8}$ ounce Roman, so a *sceatta* is worth $\frac{1}{40}$ of a Roman ounce (of silver), that is approximately 10.5 Troy grains of silver. With a gold/silver value ratio in the range 10/1 to 12/1 in the seventh to tenth centuries we again find the *sceatta* to be worth closely one Troy grain of gold. Thus over a period of more than three centuries (seventh to tenth), we find the value of the *sceatta* confirmed.

The first English coinage, then, is represented by the gold pieces of the Crondall hoard, but the gold currency was not destined to last long, being displaced by the first silver coinage dating from about AD 680.★ A coinage in silver first appeared in northern parts of Europe in the second half of the seventh century AD with the coins themselves called the *denarius* or penny. (Abbreviation *d* = penny, as in pre-metric Britain, £. *s. d.*). England was quick to follow the universal change to the *denarius*. The Laws of Wihtred of Kent[18] (AD 695) speak of shillings and *sceattas* only; those of Ine of Wessex[19] (*c*690) speak of shillings and pennies, never *sceattas*. For example Ine 58 says 'The horn of an ox is worth 10 pence' (X paeniaga) so already before the eighth century the word 'penny' was in use in England. Because it was current on the continent in that century,[20] its occurrence in these laws does not appear to be a later interpolation.

Three centuries ago the lack of a coinage at the beginning of the seventh century was not realized and the term '*sceatta*' was taken by the early numismatists to mean the coins of this first silver currency which were therefore supposed to antedate the Laws of Aethelbert, with the term 'penny' being reserved for the larger thinner coin some 10 per cent heavier which was to dominate the entire currency picture from the last decade of the eighth century, that is a whole century after the appearance of these first coins. But there is documentary evidence to show that the word '*sceatta*' was adopted in the seventh and eight centuries to describe these first silver post-Aethelbert coins, as we shall see presently, with the terms 'penny' and '*sceatta*' being used interchangeably notwithstanding the greater weight of the flat penny.

The *sceatta*-coins were uninscribed, approximately circular discs, thick but of small (8 mm) diameter. The quality of the silver was good, being 90-95% pure, although later examples were frequently much debased. From the British Museum Catalogue we see their original weight was twenty grains (1.3 g) and this weight standard is confirmed by the coins of the Barth (Emden) hoard, all found in near-mint condition.[21] Being devoid of lettering these coins cannot be attributed to any particular ruler, moneyer, or location except on occasion where, for example, it can be shown that the same set of punches and dies were used to strike the blanks, but they carried a design on

★With the disappearance of the gold shilling there was no coin to represent this sum in Francia or in England. The shilling then became only a unit of account in a similar way to the term 'guinea'. The guinea as a coin was not minted after 1813 but as a money of account of value 21 shillings it continued to be used in professional circles most commonly as legal, dental, and medical fees until the introduction of the metric system. To send out an account for '100 guineas' looks better than '£100' and has the added merit of being 5 per cent more. Payment was made in current money, that is pounds and shillings. There was no silver shilling in England until 1504 when a *testoon* worth 12 pence was issued but its name soon changed to shilling. This coin continued until its recent replacement by the metric '5p' piece of cupro-nickel.

both sides and were current until displaced by the introduction of the thin 'penny'.

The weight standard fell from twenty grains to sixteen and there are some exceedingly light specimens of less than eight grains (0.5 g).[22] This is most probably due to wear and to the loss over time of part of the metal of the coin itself.

The Shillings of Mercia and Wessex

The *sceatta*-coins (*sceattas* of the numismatists) are mentioned in several important documents. In Mercia, we are told that the *wergeld** of a ceorl is 200 shillings (cc scill.), that of a thegn is 1200, and of the king 7200 shillings, which latter is stated to be 30,000 *sceattas* or 120 pounds.[23] Here the word '*sceatta*' must be taken as meaning the early silver coin and not the grain of gold, for 7200 shillings each worth twenty grains of gold would yield 144,000 grains not 30,000. The word appears then to have been used in the two senses already given, being applied to the silver coin on the displacement of the gold.† This rather parallels the use of the word *pence* to denote coins worth $\frac{1}{240}$ of a pound sterling prior to the mid-twentieth century and the same word (originally prefixed by the word 'new' which has now been dropped as superfluous) to denote $\frac{1}{100}$ of the same pound sterling following the recent decimalization of the currency.‡

With the silver coin the same weight as the displaced gold coin, the number of *sceattas* or pennies in the shilling gives the gold/silver value ratio. At the end of the eight century it was 12/1 on the continent and most likely the same in England, otherwise a considerable movement of gold and silver would have resulted with buyers exchanging silver for gold at the best price.

By Frankish reckoning the shilling was worth twelve pence for we read in Charlemagne's *Capitulare Saxonicum* of 28 October 797;[24] 'For silver, twelve pence make a shilling' (In Argento duodecim denarios solidum faciant), and

*The *wergeld* was the compensatory payment for the killing of an individual. It naturally varied with the station in life of the person killed and was levied on the killer personally or on the residents of his district.

†We read in the *Anglo-Saxon Chronicle* for AD 694, just after the introduction of the *sceatta*-coin, that the people of Kent paid to Ine, King of Wessex, 'Thirty thousand' (units unspecified) for their having burned to death Mul, brother of Caedualla, who had secured the kingdom of Wessex and whom Ine succeeded. Scholars are agreed that the units meant are *sceattas,* on the assumption that before the end of the seventh century these early coins are going by this name.

‡Here history is seen repeating itself, for just as the 'new' penny of the twentieth century was worth 2.4 of the old we see 30,000 *sceatta*-coins each weighing 20 grains amounting to 600,000 grains of silver while 7200 shillings each worth 20 grains of gold, or, with a gold silver ratio of 10/1, each worth 200 grains of silver, amount to 1,440,000 grains of silver. The ratio 1,440,000/600,000 is again 2.4. Of course the fact that the ratio (2.4) is the same is quite fortuitous and the gold/silver ratio did vary considerably.

two hundred and forty pence made the monetary pound as early as 755.[25]★ With the gold shilling and the silver penny of the same weight and a 12/1 gold/silver value ratio we would expect there to be twelve pence to the English shilling, too.

The later Norman reckoning can be obtained from a portion of the *Leges Henrici Primi*[26]† which records and explains earlier Saxon laws: Thus '76.6a . . . 5 mancuses which equal 12 shillings and 6 pence' (XII Solidos et VI denarios). With the *mancus* at thirty pence‡ the Norman shilling emerges as twelve pence also. But if twelve pence make a shilling what are we to make of the statement given earlier that in Mercia 7200 shillings are 30,000 sceattas or 120 pounds?

If 7200 shillings are to be 120 pounds each of 240 pence then each shilling is worth four pence not twelve. Likewise 7200 shillings each of four pence are 28,800 pence or *sceattas,* not 30,000. Clearly the figure 30,000 is to be taken as a rounded number. Further, in Wessex we are told of the shilling being five pence, as we see in *Leges Henrici Primi,* Chapter 70, entitled 'The customary law of Wessex', and even more clearly in Chapter 76, where we are given the value of the Wessex shilling:

> 76,4 A person is said to be a two-hundred man whose wergeld is two hundred shillings which equals four pounds.
> 76,4(a) A twelve-hundred man is a person of noble rank, that is, a thegn, whose wergeld is twelve hundred shillings which equal twenty-five pounds.§

Here we see again a certain rounding off to the nearest pound, for if 1200 shillings are 25 pounds then 200 shillings should be $4\frac{1}{6}$ pounds not 4. By Norman reckoning 25 pounds are 6000 pence (for there were 240 pence in the pound). For this to be worth 1200 shillings the shilling is five pence.

As for the Mercian shilling mentioned earlier we turn to the *Leis Willelme* (the so-called Laws of William I), written down in the twelfth century, which consists of three parts:[28]

1. Laws current in William's lifetime including many insertions of Anglo-Saxon law;

★It is important to note that the Roman *solidus* had always been equated to 40 *denarii* by Frankish law, but the monetary *solidus* was worth 12 *denarii* from AD 743 at the latest and this was repeated in 745 in a letter of 31 October from Pope Zacharias to St. Boniface (See GRIERSON, Philip, Ref.8, p.501, Footnote 2).

†As we shall be requiring further service from the *Leges* it may be well to explain more fully what these so-called Laws of Henry I really are. Although written in the reign of that monarch they are not a compilation of the laws issued by Henry. Rather they provide a summary of the then extant laws from the earliest Anglo-Saxon codes of which we have record (those of Aethelbert) right up to the time of the writing. The text enables comparison of values in the Saxon kingdoms with those of the Normans.[27]

‡See p.144 ref. 19.

§The 25-pound wergeld is also mentioned in II Aethelred, 5, of about AD 1000. See p.106.

2. a collection of Roman law;
3. a collection of Anglo-Saxon laws, principally those of the second group of Cnut's laws of the eleventh century.

From the frequent reference to Mercian law and custom it is thought likely that the document was written in Mercia. In Chapter 2, 2a we read:[29]

> and he who violates it [the king's peace] in Mercia or Wessex shall pay the king's fines which belong to the Sheriff: — 40 shillings in the province of Mercia and 50 shillings in the province of Wessex.*

The fines here are each intended to be 120 of the local shillings, for 120 Wessex shillings are 600 pence which at twelve to the shilling by Norman calculation make 50 shillings. In Mercia 120 shillings are 480 pence which at twelve to the shilling are 40 shillings.

Further evidence for this value of four pence for the Mercian shilling is to be found in the early charters. We will quote from one dated AD 857: '60 shillings in silver . . . one pound' (sexaginta solidorum argenti . . . uno libra . . .).[30] For the pound to have its normal value of 240 pence the shilling, as a unit of account reckoned in silver pennies, is here worth four pence. Other instances are known from the charters from the ninth to the eleventh century.

The reason for the four and five penny shillings instead of a twelve penny shilling cannot be established with absolute certainty but it is perhaps due to the serious and progressive debasement of the gold shilling of the mid-seventh century,[31] which made the gold coin worth substantially less than its proper value if minted from pure gold.

The Pound

The value of the pound† in those days is not given directly in the various groups of laws. In the late tenth century, the Laws[32] of Aethelred state the wergeld of an Englishman or a Dane to be 25 pounds. This is almost certainly intended to be the worth of a thegn, that is 1200 shillings. The reckoning of Wessex at this time was five pence to the shilling as we have seen. So 25 pounds amount to 6000 pence at 240 pence to the pound. It is generally agreed that this was also the value of the English monetary pound in earlier times. And the English monetary pound in Norman times was of 20 shillings for we find, again referring to the *Leges Henrici Primi*, C64 1 h,[33] 'In causes where the accused should undertake the threefold exculpation he shall submit to the

*In the same set of laws Section 8 we are told that the wergeld of a thegn in Mercia is 20 pounds and in Wessex 25 pounds. This latter at 240 pence to the pound and 5 pence to the shilling amounts to 1200 shillings. Twenty pounds at 4 pence to the shilling amount to 1200 shillings also.

†The pound was then a unit of account. In England there was no coin of this value until October 1489 when Henry VII issued his gold piece called a 'soveraign', worth twenty shillings.

ordeal of the iron of 3 pounds weight (that is the equivalent of 60 shillings)'. From this description of the trial by ordeal we see clearly that the pound weight used here is 'the equivalent of' 20 shillings, that is 240 pence. This number of pennies then generated not only the monetary pound of account but also a pound weight which was to be of great significance in English metrology, as we shall see.

Much later, a document of 1267, quoted by Stapleton,[34] speaks of ten pounds sterling paid in two portions, one at the feast of St. Michael, the other at Easter. Each payment was to be of 100 shillings. This value for the pound continued until the decimalization of the currency in the mid-twentieth century.

The Introduction of the Penny and the Tower Pound[35]

Sceattas were produced in abundance from the end of the seventh century to the third quarter of the eighth in England and Frisia. They were closely allied to the Merovingian silver *denier* so that in effect there was something of a common currency around the English Channel and parts of the North Sea, which may have stimulated trade among the countries concerned.

It should not be thought that this coin circulated throughout 'England', for there was no England as we know it today. Rather, circulation was confined to the southern and eastern parts of the country and even there for only some eighty years before being replaced by a totally different coin, the larger, heavier, but thinner *penny* which was to dominate the currency for the next five centuries. The first silver pennies are usually attributed to Offa of Mercia but were actually introduced by two lesser known kings of Kent, Heaberht and Ecgberht, about AD 775-80.[36] This change from the uninscribed rather dull *sceattas* to the penny bearing the name of king and moneyer was a turning point in English numismatics and metrology. (A photograph of a *sceatta* and a penny is given in Figure 25.)

As with previous mintages the English pennies were derived from those of the Franks. The sequence of events may be summarized as follows. After a series of weak rulers, the Merovingian line of the Frankish kingdom ended in 751 with the deposition of Childeric III and his removal to the monastery of St. Bertin. Pepin the Short, Mayor of the Palace, assumed the throne, ruling till his death in 768. A currency edict of 755 seems to have involved the generalization of an earlier Merovingian standard with the *denier* at 240 to the pound, though the actual coins were some 0.75 inches (19 mm) in diameter and thinner than those of the later Merovingians while of the same weight and value.

Charlemagne and Carloman, Pepin's sons, ruled jointly after 768 until Carloman's death in 771 after which Charlemagne was sole ruler. His first silver *denier* or penny of 768 continued the size, fabric, and rather coarse workmanship of that of his father Pepin. These continued to be issued at a weight of twenty grains Troy (1.3 g) (a figure much in evidence in previous chapters), till 793/4 when a general reform of the coinage and of weights and measures occurred, the weight of the *denier* being increased to *c* 1.70 g. This would have been about 26 (Troy) grains, a meaningless figure which is best explained on the assumption that the Troy grain (barleycorn) of 0.065 g had been replaced as a standard by the wheat grain of 0.053 g, the later *Paris* grain, for the new coin weighed exactly 32 of the latter. The Frankfort Capitulary[37] of 794 reflects these changes and speaks in Chapter 5 of the new pennies ('novi denarii'). Meanwhile in England, Offa of Mercia (ruled 757-96) was extending his kingdom, gaining control of Kent, and reducing the local rulers to vassals. His first coinage, initiated in the 770s, was of the same weight standard as the *sceatta* (twenty grains) and in fabric resembled the contemporary Frankish *denier,* though showing greater variety of design. Virtually all the coins were produced at Canterbury. One issue, with a finely designed royal bust, is sometimes treated as a separate class, but is of the same weight standard as the rest.

In 792 Offa instituted *his* currency reform, decidedly increasing the diameter and weight of his pennies. The condition of the few surviving specimens of the new coins makes it difficult to determine their weight standard with certainty but it was in excess of twenty grains and perhaps even as high as 24, like the later pennies of the ninth century. The bulk of Offa's reformed coins today lie in the range 18-20 grains. Most unfortunately, Offa's laws, though mentioned by Alfred in the preamble to his legal code of about 890, have not come down to us. From these lost laws we might have found how many pence were cut from the pound and how heavy the pound was in those days. We assume 240 pence were derived from the pound. Indications that this is so are twofold:

1. During the last eighty years of the Merovingians the kingdom of the Franks minted at the rate of 240 *denarii* to the pound and the connections with England were close.
2. Subsequent English coinage of pence, not very different in weight from those of Offa, at least in theory, were minted at that rate.

We shall follow the majority in assuming 240 pence to the pound. What then was Offa's monetary pound? The truth is that we really do not know with any degree of certainty. The most likely candidates are the Troy and Roman pounds. If the former the theoretical weight of Offa's penny would be 24 grains, if the latter, 21 grains. Seeking help from the ninth century, we find in the British Museum a lead weight in good condition which carries the impression of a coin die of the time of King Alfred (*c*880). The die is of his 'Type 5' coins which are rare. Extant specimens today weigh 17-21 grains

(1.1–1.36 g). The weight was found in St. Paul's churchyard, London and weighs 2515.31 grains (163 g). Presumably it was intended to be the weight of a certain round number of the pennies whose mark it bore. If it was intended to be the weight of ten shillings (120 pence) then the pound would have been somewhat in excess of 5030 grains when possible weight loss is allowed for. The Roman pound is 5050 grains. The suggestion that the Roman pound was indeed the monetary pound of Offa is not new, having been made at least as far back as 1905,[38] but this cannot be considered as other than conjecture at this time, especially as some are of the opinion that it was the Troy pound.

A myth has circulated for many years that from Offa's time the theoretical weight of a penny was 22.5 grains (1.46 g) and that all pennies of the next five centuries were minted at this weight. This would yield a pound of 5400 grains (350 g). It is true that such a weight is referred to in documents of the thirteenth and later centuries as the 'Tower pound' from its association with the Mint at the Tower of London (for a more detailed look at those documents see the next chapter p.123 et seq), but the Tower was built in the days of the Conquerer and did not exist in Saxon times.

It is equally true that such a weight was used at the Mint in later times. An ancient brass weight was found in 1842 in the old pyx* chamber in the cloister of Westminster Abbey. According to the official report it weighed 5391 grains,[39] clearly intended for 5400. By 1873 this weight had disappeared,[39] so its date cannot be ascertained, and no example of the Tower pound would appear to be extant, which is not surprising as they would not have been numerous, being used exclusively in the Tower and the few provincial mints still active in the later Middle Ages. Whereas Offa had perhaps no more than three mints (Canterbury, London, and Rochester) Edward the Confessor had over sixty, but by 1180 the number had been reduced to ten, and from the thirteenth century onwards the vast majority of English coins were struck in either London or Canterbury.

But the Tower pound does not go back to Offa any more than does the penny fixed at 22.5 grains. The coins show that the weight frequently changed with change in type and individual coins of a given issue are distributed about the standard weight, some heavier, some lighter. Figure 26 shows the variations in weight of late Saxon and Norman pennies from the currency reform of Edgar in *c*973. It is clear that the weight was deliberately varied, 22.5 grains being the standard only occasionally, and not till some years after the Conquest was the penny stabilized at this figure, at which it remained for the next two centuries.

Once the penny was fixed at the standard of 22.5 grains, the so-called Tower pound could and probably was mainly intended to speed operations at

*For the Trial of the Pyx see p.112.

the mints by removing the need to count. But it became a national standard, and continued so until 1527, when Tower weight was abolished by Henry VIII in favour of Troy weight. The Exchequer record reads:[40]

> Nowe yt is determyned by the Kingis Highnes and his said Councelle [that the aforesaid pounde Towre shall be no more used and occupied] but al maner of golde and sylver shall be wayed by the Pounde Troye which maketh 12 oz Troye which exceedith in the Pounde Towre in weight iii quarters of the oz.

The Troy pound mentioned here is of ancient origin (see next chapter). It was divided into twelve ounces each of twenty pennyweights (dwt) each of 24 grains. The ounce was therefore 480 grains (31.1 g), the pound 5760 grains (373.3 g). This was the standard 12 ounce pound of the Exchequer for many centuries. The 1527 statement shows that the Troy pound equalled the Tower pound increased by three quarters of a Troy oz, in other words the Troy pound is $5400 + 360 = 5760$ grains.

The origin of much confusion will now be evident. The weight of the later penny was 22.5 grains but the Troy pennyweight (dwt) was 24 of the same grains, that is the pennyweight was not the weight of a penny. We shall see in what follows the role of Troy weight at the Exchequer when payment by weight was demanded.

Procedures at the Mint

If we envisage an idealized operation of the mint the relationship in post-Norman days between the Troy and Tower pounds may become clearer. The mint might buy pure silver by the Troy pound of 5760 grains and pay for it by giving the same weight of minted coins in exchange, that is 256 silver pennies each of $22\frac{1}{2}$ grains.

To the pure silver, workmen would add sufficient base metal to make the silver alloy of the fineness of sterling, that is 92.5% pure*. This requires 467 grains of base metal for every Troy pound of pure silver and the resulting alloy would weigh 6227 grains for each Troy pound of pure silver processed. This would mint into 276.75 pennies† and as the bullion was paid for by 256 such pennies the profit is 20.75 pennies from which the workmen must be paid

*William de Turnemire's *Treatise on the New Money*[41] gives $18\frac{1}{2}$ dwt copper to the *standard* (Troy) pound. This makes the fineness 92.3%, but a document of the same approximate date, referring to the St. Edmundsbury Trial Plate,[42] gives the pound as containing 11 oz $2\frac{1}{4}$ dwt of fine silver. This is a fineness of 92.6%. We have taken the sterling standard of 92.5%. We have seen that the *sceatta*-coins were 90-95% pure silver and today sterling silver is still 92.5% pure. The *Treatise* also described the practice of the previous 30-40 years, stating that if anyone brought sterling silver to be coined, $14\frac{1}{2}$ *d* shall be retained, the king receiving 9*d* and the Master of the Mint $5\frac{1}{2}$*d*, and William remarks that whereas the king was content previously with 6*d* profit in the pound he now took 9*d*. If old money was brought in a further $1\frac{1}{2}$*d* was taken for a total of 16*d* (1247).

†From the eleventh and later centuries also called *sterlings*. See p.123.

and all running expenses met, not forgetting the king's 'right and lawful farm'.

This fits very well with the description of the costs and profits given in the indenture of William de Turnemire[43] dated 8 December 1279 in the 8th year of the reign of Edward I in which William (from Marseilles) was appointed Master of the Mint in England. For the coinage of pennies we read: 'And the king shall give him for every pound of sterlings seven pence' and towards the conclusion of the document: 'And there will remain to our lord and king a profit of twelve pence at least'. If William is to receive 7*d* and the king 12*d* from the minting of a 'pound of sterlings', that is 240 pence, they should receive 8.07 and 13.84 pence respectively from the minting of the 276.75 pennies given above. This amounts to 21.91 pence, close to but not quite that actually generated, for, as already shown, the profit was 20.75 pence. This probably explains why, within the year, William's share was reduced to 6*d* from 7*d* and the shear of the penny increased from 240 to 243 pence to the Tower pound (see below this page and p.113). These measures yielded the king 12.5 pence profit on every monetary pound of 240 pence which ensured his '12 pence at least'.

Of course, the Mint did not restrict its purchases to pure silver but received bullion in a variety of degrees of fineness at prices reflecting the silver content. In particular old coin was purchased by weight with a discount for alloys which proved to be weaker than that of sterling. In actual practice the procedure was not quite so straightforward.

The Duties of the Changer[44] at the Mint in 1248 stated that, if pure silver was bought in bulk, 6*d* was to be taken 'in each pound of the right and accustomed farm'. Hence for his Troy pound of silver the merchant received 250 pence. In our example, this would increase the profit to the rather large value of 26.75 pence on the minting of 276.75 pence or 23.2 pence profit on 240 pence and it has been suggested that the *seignorage* or royalty given in the indenture of William de Turnemire has been overestimated.[45]

William wrote a *Treatise on the New Money*, dated about 1280, in which it is clear that Edward had reduced the weight of the penny slightly from $22\frac{1}{2}$ to $22\frac{1}{4}$ grains giving 243 pennies from the Tower pound of alloyed silver. William says:[46]

> For example the shear of English money is 20*s* 3*d* to the pound as evenly as it can be cut but if the coins be found to be not more than a penny stronger or weaker than a pound, they will pass. By 'stronger' I mean if in the pound weight there are only found 20*s* 2*d* by tale. By 'weaker' I mean if in the pound weight there are 20*s* 4*d* by tale. Also in each pound one strong and one weak penny may pass uncounted, and two other pence stronger or weaker by one and a half grains than the just penny of 20*s* 3*d* to the pound sterling.

Any standardized coinage must submit to two tests of quality, the first as to weight of the individual coin and the second as to fineness of the alloy. These tests are the time-honoured Trials of the Pyx, for which the earliest known writ is dated 1282* but the first recorded trial took place on 11 March 1248[48] in the 32nd year of the reign of Henry III.

Such a trial is recorded in Mint documents of that date[48] under the title *The Assay of the New Money,* but the *Dialogus de Scaccario* (The Course of the Exchequer)[49] written about 1180 describes in some detail the testing of the money received by the Sheriff on behalf of the King and delivered by him to the Exchequer so there may well have been earlier trials. The procedure given in the *Dialogus*[50] may be summarized as follows. The coins received from the Sheriff are first thoroughly mixed and weighed out to balance an *Exchequer* (that is Troy) pound.† This would take 256 pennies of average standard weight. If more than six pence extra is needed to provide a balance, the coin is too light and cannot be accepted.[51] If six pence or less are required beyond the normal 256, the money is satisfactory. Once the money has been approved as to weight forty-four shillings in pence are taken and sealed up in a bag by the Sheriff paying in the money. Forty-four shillings at 12 pence to the shilling are equivalent to 528 pennies. To proceed with the assay, the 528 pennies are emptied out and mixed up again, and a pound (Troy) of pence weighed out and counted to verify that they are of lawful weight. For pennies of average weight they are 256 in number. The remaining pennies are returned to the bag except two which are given to the melter for his trouble. The Troy pound of pence is then fused in an ash *cupel* or small vessel in the furnace with materials to remove the alloy. The pure silver is made into an ingot which is then weighed against a Troy pound weight. Pennies from the same purse are added to the pan containing the ingot until a balance is achieved, whereupon it is recorded that the pound of pence had lost so many pence in the fire. If the Sheriff refuses to accept the result of the assay or if the melter admits that the testing failed, the bag is reopened, a second pound of pence taken, and the test repeated.

Thus far, if a second test is needed, $256+256+2=514$ pennies have been accounted for. The melter is not paid again for conducting the second test. Of the original 528 pennies only 14 remain on average which would only be called into service if both pounds were 6 pence light. The suggestion that the melter's fee was not two pence but two shillings[52] is not borne out by this accounting, which again shows clearly that the pound by which the pennies are weighed *at the Exchequer* is the Troy pound and not the Tower pound. The calculation does not accommodate a pound weighing 240 pennies.

*To mark the 700th anniversary of the issuing of this write Her Majesty Queen Elizabeth II attended the more recent Trial of the Pyx at Goldsmiths Hall in the City of London on 25 February 1982.[47] It was the first official visit of the Monarch to the Pyx since 1611.
†See Ch.VIII for the use of Troy weight at the Exchequer.

Manipulation of the Currency

The coinage can be manipulated in three ways: by reducing the silver content, by reducing the weight of the coins, or by both. All three have appeared in the history of the English coinage.[53] In 1280 the alloy was closely that of sterling silver 92.5% pure, that is of a Troy pound 11 oz 2 dwt was pure silver and 18 dwt was copper, and in this year the first weight reduction occurred. The Tower pound yielded not 240 but 243 pennies,[46] the weight of each being $22\frac{1}{4}$ grains. Things remained thus till 1344 when the pound yielded 266 pence. Two years later (20 Edward III) this was further increased to 270 and the weight of the penny was down to 20 grains. In 1395 the weight was 18 grains[54] and in the early fifteenth century we find the penny down to 15 grains. In 1464 a further reduction took place to 12 grains. *Arnold's Chronicle* of the early sixteenth century tells of 32 pence to the ounce which would yield a penny of 15 grains.[55]

As has already been stated, the Tower pound was abandoned at the Mint in 1527 in favour of the Troy pound and we find in 1543 that the penny was being coined from sterling silver at the rate of 576 to the Troy pound. The penny now weighed ten grains, less than half its original weight. Thereafter rapid debasement accompanied by a further weight reduction set in. By 1546 the alloy was one part silver to two of copper, and while in 1549 the standard had recovered to 50% silver no fewer than 864 pennies were being derived from the Troy pound of this 50-50 alloy. The penny then weighed a mere $6\frac{2}{3}$ grains. Elizabeth I fully restored the quality of the silver to the 11 oz 2 dwt fine standard while minting at 720 to the pound Troy (producing a penny of 8 grains) until 1601 when 744 to the pound were struck.

A document of 1590-1620 records these latter events:[56] 'The coyners in the Towre allowith but 24 grains to a peny sterlinge waight and the peny sterlinge valewid at 3*d* currant. One pence is 8 grayns and halfe pence 4 grayns in waight.' With the mint pound now 5760 grains and the penny at 8 grains, a pound yielded 720 coins, that is 3×240. A pennyweight (dwt) of sterling silver, the silver of the coinage, was worth 3 *d* in coin, for three pennies each weighing 8 grains are needed to produce the Troy pennyweight of 24 grains.

The general coinage of silver pennies terminated with the issue in 1672 of the copper $\frac{1}{2}$*d* and in 1797 of new, 1*d*, copper coins although silver continued to be struck as Maundy money. The silver in coins of larger denominations issued as need arose from the fifteenth century onwards maintained the purity of sterling (92.5%) until 1920 when the silver content was reduced to 50%. In 1947 silver disappeared from the coinage for the first time since the seventh century, being replaced by a copper (75%) – nickel (25%) alloy.

NOTES AND REFERENCES: CHAPTER VII

1. DOLLEY, R. H. M. (ed.) *Anglo Saxon Coins* London: Methuen 1961, Section I. KENT, J. P. C. *From Roman Britain to Saxon England,* p. 7.
2. MITFORD, R. Bruce-, *The Sutton Hoo Ship-Burial* Vol. I. London: British Museum Trustees 1975.
3. DOLLEY, R. H. M. (ed.), KENT, J. P. C. Loc. cit. p. 8.
4. GRIERSON, Philip. La Fonction Sociale de la Monnaie en Angleterre aux VII[e] - VIII[e] Siecles, *Settimane di Studio del Centro Italiano di Studi Sull'alto medioevo* (1961) p. 347 et seq.
5. GRIERSON, Philip. Loc. cit. p. 348.
6. KEARY, C. F. *A Catalogue of English Coins in the British Museum. Anglo-Saxon Series.* Vol. I. London: British Museum Trustees 1887; quoted in DOLLEY, R. H. M. (ed.) Loc. cit. Section II, p. 29.
   WHITTING, P. D. *The Byzantine Empire and the Coinage of the Anglo-Saxons.*
7. See for example, GRIERSON, Philip. Loc. cit. p. 351;
   SUTHERLAND, Carol Humphrey Vivian *Anglo-Saxon gold coinage in the light of the Crondall Hoard* Oxford: O.U.P. 1948.
8. BRAUNFELS, W. (ed.) *Karl der Grosse* Book I, p. 529. (GRIERSON, Philip *Money and Coinage under Charlemagne)* Dusseldorf: L. Schwann 1965.
9. LYON, Stewart Historical problems of Anglo-Saxon Coinage - (3) Denominations and Weights. *Brit. Numis. J.* (1969) Vol. XXXVI, pp. 204-222. See p. 209 et seq.
10. KEARY, C. F. Loc. cit. p. i. The entire Introduction can be read with profit.
11. ATTENBOROUGH, F. L. *The Laws of the Earliest English Kings* pp. 11, 13, 15. Cambridge: C.U.P. 1922.
12. These relations are further exemplified in the same laws sections 54 and 70 which allow 20 shillings for the loss of a thumb and 10 shillings for the loss of the big toe. In sect 54 § 1 the compensation for a thumbnail is 3 shillings while that of a big toenail (sect. 72), is 30 sceattas. Thus $1\frac{1}{2}$ shillings = 30 sceattas or one shilling is the equivalent of 20 sceattas, as before.
13. KEARY, C.F. Loc. cit. Vol. I viii.
14. MALONE, Kemp. (ed.) *Widsith* Lines 89-92. London: Methuen 1936.
15. In discussing the *scilling* and the *sceatta* I have followed the arguments of Philip Grierson, especially those of reference 4 (above) which are compelling.
16. RUDING, Rogers. *Annals of the Coinage of Great Britain and its Dependencies* 3rd ed. p. 108. London: J. Hearne 1840.
17. LYON, Stewart. Loc. cit. p. 217.
18. ATTENBOROUGH, F. L. Loc. cit. p. 25 et seq.
19. ATTENBOROUGH, F. L. Loc. cit. p. 37 et seq.
20. STENTON, Sir F. *Anglo Saxon England* 3rd ed. p. 222. Oxford: Clarendon Press 1971.
21. GRIERSON, Philip. Loc. cit. p. 354.
22. See for example MACK, R.P. *Sylloge of Coins of the British Isles* Vol. XX. London: British Academy; O.U.P. 1973.
23. CHADWICK, H. Munro. *Studies in Anglo Saxon Institutions* p. 13. Cambridge: C.U.P. 1905.

24. PERTZ, G. H. (ed.) *Monumenta Germaniae Historica,* Leges IV, p. 76. (Capitulare Saxonicum) Hannover 1835 Stuttgart Kraus Reprint 1965.
25. Ibid. Leges I p. 31.
Here we read of Charlemagne's father, Pepin, stating that no more than 22 shillings should be struck from the pound of bullion, with the moneyer receiving one shilling. This implies that the pound of minted silver would be issued at 20 shillings. As the shilling was a unit of account, consisting of 12 pence, and not a coin, we have 240 pennies to the pound.
26. DOWNER, L.J. (ed. & tr.) *Leges Henrici Primi* p. 241. Oxford: Clarendon Press 1972.
27. Ibid, Introduction.
28. ROBERTSON, A. J. (ed. & tr.) *The Laws of the Kings of England from Edmund to Henry I,* pp. 226-7. Cambridge: C.U.P. 1925.
29. Ibid, p. 253.
30. BIRCH, Walter de Gray. *Cartularium Saxonicum* Vol. II, p. 95, No. 492 (dated 18 April 857). London: Whiting 1893. (Sawyer No. 208)
31. GRIERSON, Philip. Ref. 3 p. 355.
32. ROBERTSON, A.J. Loc. cit. p. 59 (II Aethelred 5).
33. DOWNER, L.J. (ed. & tr.) Loc. cit. p. 205.
34. STAPLETON, T. (ed.) *De Antiquis Legibus Liber* p. xxvii. London: Camden Society 1846.
35. Valuable information on the material of this section is to be found in: DOLLEY, Reginald Hugh Michael (ed.) *Anglo-Saxon Coins* London: Methuen 1961.
DOLLEY, Reginald Hugh Michael (ed.) *The Norman Conquest and the English Coinage* London : Spink 1966.
LYON, C.S.S. Variations in Currency in Late Anglo-Saxon England, in CARSON, R.A.G. (ed.) *Mints, Dies and Currency* pp. 101-120. London: Methuen 1971
I am also indebted to Professor Philip Grierson for his suggestions on this section.
36. DOLLEY, Michael. *Anglo-Saxon Pennies* p. 14. London: British Museum 1970.
37. PERTZ, G.H. Loc. cit. Leges I, p. 72, Ch. 5. (Frankfort Capitulary of 794).
38. CHADWICK, H. Munro *Studies on Anglo-Saxon Institutions* p. 40. Cambridge: C.U.P. 1905.
39. CHISHOLM, H.W. *Seventh Annual Report of the Warden of the Standards* p. 16. London: HMSO 1873.
40. Ibid;
See also the Report from the Committee appointed to inquire into the Original Standards of Weights and Measures in this Kingdom (Lord Carysfort's Report) 26 May 1758 *Reports from Committees of the House of Commons* Vol. II (1737-65), pp. 411-51. See p. 420.
41. JOHNSON, Charles (ed. & tr.) *The De Moneta of Nicholas Oresme and English Mint Documents* p. 65. London & New York: Nelson, 1956.
42. Ibid, p. 87.
43. Ibid, p. 59 et seq. See especially pp. 60-1.
44. Ibid, p. 51.
45. Ibid, Introduction p. xxxiv, Note 1.
46. Ibid, p. 76.

47. *The Times,* London, 26 February, 1982.
48. JOHNSON, Charles. Loc. cit. p. 53; This is also given in ELLIS, Sir H. (ed.) *Johannes de Oxenedis Chronica* pp. 26-7 and p. 315 1859.
49. JOHNSON, Charles (ed. & tr.); Richard (Fitzneale) *Dialogus de Scaccario* London & New York: T. Nelson 1950.
50. Ibid, p. 11 and p. 36 et seq.
51. This is confirmed in the Statute Book a century later for we find (*Statutes* Vol. I, p. 219. London 1810) a document entitled 'Statuta de Moneta' listed as of uncertain date but attributed to 20 Edward I (1292). This date cannot be far out as variant readings are taken from the Patent Roll of 12 Edward I. The text reads in part: 'And if he find of the new money that the pound weigheth not XX shillings by the number of 4 pence then he shall have regard to the Tumbrel where the Default shall be'.
    Four pence in a Tower pound of 20 shillings is equivalent to 5.6 pence in the Troy pound, doubtless thought near enough 6 to pass. The Tumbrel is a coin balance. The money is to be weighed and if faulty, rendered useless by piercing.
52. RAMSAY, Sir James H. *A History of the Revenues of the Kings of England 1066-1399.* Vol. I, p. 23, Note 1. Oxford: Clarendon Press 1925.
53. For a description of the coinage see JOSSET, Christopher Robert. *Money in Great Britain and Ireland* Newton Abbot: David & Charles 1971.
54. ROUSE, Rowland; Correspondence. The Weight of an English Penny at Various Periods, *Gentlemans Magazine* (1797) Vol. LXVII, Pt. 1, p. 394 et seq.
55. ARNOLD, Richard. *The Customs of London, commonly called Arnold's Chronicle* p. 191. London: FC & J. Rivington 1811 (Reprint of 1st ed. of c.1502).
56. HALL, Hubert and NICHOLAS, Frieda J. (eds.) *Select Tracts and Table Books Relating to English Weights and Measures (1100 -1742)* Camden Miscellany Vol. XV, p. 22. London: Camden Society 1929.

CHAPTER VIII

The Origins of the Units of Commercial Weight

Troy and Apothecaries' Weight

The influence of Rome on European thought, manners, and custom can hardly be exaggerated. Not only did Roman ideas and concepts prevail during the centuries of Roman power but, once established, these ideas and concepts continued into the Middle Ages and indeed some are still with us today. Roman influence on weights and measures was no less than on other aspects of daily life and we see the form of the Roman subdivision of the *libra* reappearing throughout Europe in later centuries.

The Roman pound or *libra,* as we have already seen, was divided thus:

1 *libra* = 12 *unciae* (ounces) = 48 *sicilici* = 96 *drachmae* = 288 *scripula* = 576 *oboli* = 1728 *siliquae.*

Such a subdivision is to be found in the *Distributio*[1] of Lucius Volusius Maecianus*, tutor to Marcus Aurelius and Governor of Alexandria, written in the second century AD, and the same sort of subdivision was used for Roman capacity measures, with the *amphora* or *quadrantal* being subdivided into 576 *cyathi.* Similar duodecimal subdivisions reappear in not a few of the weights of European cities and states. To give but a few examples,[2] the *peso sottile* of Venice was divided into 1728 *carati;* the *libbra* of Rome for gold and medicines contained 288 *denari;* the old Parisian bullion weight had a sub-unit, the *mark,*† which was subdivided into 576 units called '*grains*', and the *mark* of Cologne contained the same number of *grains.* George Agricola[3] in the mid-sixteenth century used German weights with Roman names and his *mark* was divided into 2 × 576, that is 1152 *siliquae.* Furthermore there were twelve ounces in the pounds of Bologna, Florence, Naples, Parma, the Vatican, Turin, Valencia *(libra sutil),* and Venice.

Thus it should not be too surprising were we to find a pound of twelve ounces subdivided more or less in the Roman manner appearing in the earliest history of our weights. This pound,[4] the most ancient of the kingdom, is the *Troy pound* (373.242 g), subdivided as follows:

*Maecianus was murdered in Alexandria in AD 175.
†A mark was either half or two thirds of a pound, depending on the system in question. It was always 8 ounces.

> 1 Troy pound (lb) = 12 ounces (oz.) = 240 pennyweights (dwt)★ = 5760 Troy grains.

We have already seen the ancient connection between Troy grains and barley corns. Traditionally, the weight of a grain of barley is the same as that of a Troy grain.

This self-same pound has been used since early times for drugs under the name of *Apothecaries' weight,* the only difference being in some of its subdivisions, as shown here:

> 1 Apothecaries' (Troy) pound (lb.) = 12 ounces (℥) = 96 *drachms* (ʒ) = 288 scruples (℈) = 5760 grains.

Hence the pound, ounce, and grain in the two systems are identical. The *Antidotarium* of Nicholas of Salerno[5] of the twelfth century used Apothecaries' weight in the same manner as that given above, and Apothecaries' weight continued in England until formally abolished on 1 January 1971. Until then, medical prescriptions continued to be written in hieroglyphic† symbols for the various weights of drugs to be compounded, together with the Roman 'S' *(semis)* for 'half', for example ʒS meant half a drachm. That these symbols are of remote provenance can be seen from their use in documents dating from the early centuries of the Christian era such as the *Tabula codicis Bernensis* and the *Tabula codicis Mutinensis prioris.* Their continued use in the twentieth century stemmed from long tradition and served to remind the laity of the descent of modern pharmacy from classical times.

Belief in the antiquity of the Troy pound is to be found stated in the *Common Place Book* of John Colyn of London[7] dated 1517: 'Pleasythe hyt you to understonde that afore the Conqueste sondrye [sundry] kynges aprovyd and provyded how and by what maner of weyes they might beste and longeste endure; for a very trowthe, how far to make a lib. Troye . . .'. The fact that the pound, ounce, and grain of both the Troy and Apothecaries' system of weights are the same encourages one in the belief that these systems are in reality one and are of ancient origin. This much is almost universally granted, but there has been great discussion as to the origin of the name and the reason for the magnitude of the Troy pound itself. Chaney[8] has recorded some of the speculation as to the name itself. To some it reflects its place of origin as the city of Troyes in France; to others the name comes from *trone,* the word for a large beam scale used for weighing bulky goods; others prefer a derivation from the French word *trois* meaning three. Most scholars today

★In what follows the abbreviation *dwt* will be used to signify the Troy pennyweight of 24 grains (dwt = 1.56 g). The term *pennyweight* was abolished effective 31 January 1969 by the 1963 Weights and Measures Act.

†The term used in official papers for such symbols as ℥ , ʒ , ℈ etc.

accept the name as having been derived from the city of Troyes. The city stood at the cross-roads of transcontinental business traffic. The great fairs at Troyes, noted even in the days of Charlemagne, raised the city to pre-eminence in the twelfth and thirteenth centuries. The primary weights of England came therefore from those of this city. There have been other views, largely unsupported by evidence and not widely held; one was endorsed by no less a figure than Mr Davies Gilbert MP (and later a President of the Royal Society) before the Committee of the House of Lords under the chairmanship of the Earl of Harrowby in 1823. We cannot do better than to quote directly from the Minutes of Evidence.[9]

> The Troy weight appeared to us to be the ancient weight of this kingdom, having as we have reason to suppose, existed in the same state from the time of St. Edward the Confessor and there are reasons moreover to believe that the word 'Troy' has no reference to any town in France but rather to the monkish name given to London, of Troy Novant, founded on the legend of Brute. Troy weight, therefore, according to this etymology is in fact London weight. We were induced moreover to preserve the Troy weight because all the coinage has been uniformly regulated by it and all medical prescriptions or formulae are now and always have been estimated in Troy weight under a peculiar subdivision which the College of Physicians have expressed themselves anxious to preserve.

As to the magnitude of the weight itself, some claim it to be Arabic in origin,[10] others feel it to be derived from a pound of Constantine of 84 *solidi*,[11] while yet others consider the Troy pound to be $\frac{1}{125}$ of the Greater Talent of Alexandria.[12] Recently it has been considered to be a development of the pound of Bremen,[13] an even less likely suggestion than some of the others. Again the most reasonable suggestion to this writer's mind is that of Ridgeway[14] who proposes that Apothecaries' weight (and hence Troy weight) is a direct descendant of the Roman *denarius* which was originally minted at a weight of 4 *scruples* (that is $\frac{1}{6}$ Roman ounce, 4.55 g, or 70 grains Troy). This means the scruple had a weight of about 18 Troy grains. In the post-Neronian period the *denarius* fell in weight to 60 grains and this coin was frequently used in early medicine and pharmacy as a weight, but being light, was taken as 3 scruples, each of these new scruples therefore weighing 20 Troy grains. The ounce was kept at 24 scruples (420 grains) as before (Chapter 1, p.14) and the pound at 12 ounces. Thus the Apothecaries' pound contained then, and up to the twentieth century, 12 ounces or 288 scruples or 5760 grains Troy. Such an origin is by no means improbable and is far superior to the other suggestions given above.

Troy Weight in Saxon England

There is direct evidence of the use of Troy weight in Saxon England prior to the Conquest.

1. The weight shown in Figure 27 is said to date from the eighth or ninth century though the grounds for such a date are unknown. It weighs 247.4 g. Eight Troy ounces, that is $\frac{2}{3}$ of a Troy pound or a Troy *mark*, weighs 248.9 g. It is reasonable to suppose that the weight was intended to represent a Troy *mark*. The weight is described as being of lead 'surmounted by [a] bronze gilt disc engraved in a basket-weave pattern' and it is now located in the Streeter Collection at Yale University.[15]
2. A bronze weight found at Grove Ferry, Kent, is 37.3 g or 576 grains, exactly $\frac{1}{10}$ of a Troy pound. Other weights from early (sixth- or seventh-century) Kentish graves also seem to fit into such a pattern.[16]
3. A group of weights discovered at Silchester, Hampshire, was described in 1923 and it was said that one was only 1.7 grains (110 mg) heavier than a Troy pound, while two others were $\frac{2}{10}$ and $\frac{8}{10}$ of the same pound.[16] This evidence is less certain than one would desire, for the weights, said to be in the Reading Museum, are, in fact, not there and indeed the Silchester Collection at Reading is described as being of Roman weights.[17] It would be of great interest to find these missing weights and redetermine their magnitudes.
4. In the will of Aethelstan,[18] the eldest son of Aethelred II, who died in 1015, we find by comparing the price of three estates of supposed equal value that 5 'pounds of silver' are worth 50 *mancuses* of gold. A gold *mancus* was worth 30 pence[19] so a 'pound of silver' was 10 *mancuses* or 300 pence. But the monetary pound is worth only 240 pence. What then is the significance of this pound of 300 pence?[20]

We come across it again in William the Conqueror's Domesday Book of 1086. Repeatedly we read of payments from the royal estates both in King Edward's time and in William's day being calculated at 20 pence to the *ore*. Now the *ore* was a unit of account and of weight, of Scandinavian origin, and there were always 15 *ores* to the pound.[21] Traditionally there were 16 pence to the *ore*.[22] We read in Domesday Book of payments from estates owned by commoners at 16 pence to the *ore*, for example of the 26 *sokemen**, who have 4 *hides* and pay 8 pounds, 'which is 30 ores per hide'.[23] At 2 pounds a *hide* and 2 pounds being 30 *ores* we at once find the *ore* to be 16*d*, and many more examples could be quoted. Why were the royal revenues rated at 20 pence to the *ore*, or 300 pence to the pound, whereas others paid at the rate of 16 pence to the *ore*, or 240 pence to the pound?[24] No satisfactory explanation has yet been advanced.

An analysis of silver pennies[25] has shown that from 870 to 1136 the alloy was sometimes better than sterling (92.5% pure), sometimes poorer, but it would appear that a serious attempt was being made to keep the quality in the Saxon period at the sterling standard, and the Conqueror did likewise. We shall assume for our purposes that the alloy was of the quality of sterling. Thus

*A *sokeman* was one who held land by *socage*, that is by performing certain defined services but not knight's service for the sokeman was not of the nobility.

300 pennies would contain 6243.75 grains of pure silver if minted at the theoretical weight of the penny, 22.5 grains. The surviving coins of Aethelstan's day (issues of Aethelred II) show a scatter for the various types around a weight of about 22.5 grains,[26] but the mean weights of the first five types of the Confessor's coinage lie between 16.5 and 17.0 grains and four of the last five issues lie between 19.5 and 19.7 grains, separated by two heavier issues of 21.5 and 25.5 grains respectively.[26] In William's reign the coin types are distributed about a weight of 20.8 grains. If 2 per cent is allowed for wear,[27] we see that, apart from the coins of Aethelred, the majority of these issues are less than the nominal 22.5 grains and the mean weight of William's issues rises to a value slightly greater than 21 grains. Some support for this latter figure is given by Miss S. Harvey's observation[28] that, of the coins in the British Museum of the various types issued by William, 240 pennies would require 14-28 additional pennies to make up the pound weight (5400 grains). If the coins actually averaged 21 grains, 17 extra would be needed, which fits well into the range of weights given above.

We now observe that, if the average weight of the penny was precisely 20.76 grains, 300 would contain exactly one pound Troy of pure silver. Might a reasonable inference not be that, rather than weighing, blanching, or assaying, the Crown, wishing to secure a Troy pound of pure bullion for each pound owing, elected to accept 300 pence by tale as a good and reasonable approximation? From what we have seen above, something very close to the Troy pound of pure bullion would be received by the Crown from the royal estates for each pound paid, and there are supporting indications in the Domesday Book itself that payment at 20 pence to the *ore* or 300 to the pound was taken as equivalent to *blanched* coin (that is, coin melted and purified to determine the true bullion content), as Miss Harvey has shown,[29] for example for Hampshire 'pounds blanched of 20 to the ore' (Libras blancas de xx in ora), for Gloucestershire 'white [purified] pounds at 20 to the ore' (Libras de xx in ora et albas), (libras alborum nummorum de xx in ora).

Thus the suggestion that the 300 pence would generate a Troy pound of silver may indeed be the correct explanation for the puzzling entries in the record of the *ore* of twenty pence rather than the customary sixteen.

These indications surely proclaim the existence of the Troy pound in Saxon and Norman times much earlier than the frequently quoted date of post 1350, that is the latter part of the reign of Edward III. Such a date is too late by far.

Troy Weight at the Exchequer

The king's revenues were collected from the royal estates by the sheriff of the county or shire and he was responsible for meeting the terms of the various grants and indentures at the King's Exchequer, a Norman institution but replicating a similar Saxon court which existed before the Conquest of 1066.

The sum owing the Exchequer might be a simple payment made *numero* or *by tale,* that is, so many pounds at 240 pence to the pound with the requisite number of silver pennies being deposited on the appointed day. But the payment could be *ad scalam,* intended as an allowance for wear, in which current coin was accepted but at a very nominal surcharge of six pence in the pound, or again *blanched* in which the coins were melted, the silver purified, and an assay made, the payment being the weight of silver. Blanched payments were intended to make good deficiencies in weight and fineness. This payment by combustion could be real or nominal. If real, the assay was actually carried out; if nominal, a surcharge of twelve pence or one shilling in the pound was levied, a rent of one pound being discharged by a payment of 252 pennies. Payments to be made *ad pensum* were to be made by weighing. The legal tradition given by Madox in 1711[30] quotes a work dealing with the accounts of the sheriffs attributed to Sir Matthew Hale, Chief Justice of England from 1671 to 1676, to the effect that payment *ad pensum* at the Exchequer was payment in coin to the weight of a Troy pound, that is for coins of full weight (22.5 grains) this would require 256 coins; for light or clipped coin, more would be needed.

Payment was sometimes required in a variety of different modes of reckoning. For example in the Domesday Book of Kent[31] we read of payments 'to the king 24 pounds in pence of 20 to the ore but to the earl 30 pounds by tale' or again 'he himself renders from that manor 70 pounds weighed, and 111 shillings in pence of 20 pence to the ore and 7 pounds and 26 pence by tale'.

It is important to note that payment of the rents and farms* owing to the Crown was made *at the Exchequer.* The Exchequer has, since its establishment, been the repository of the national standard measures of length, capacity, and weight, a function which it lost only recently. Only two classes of weights were kept at the Exchequer, Troy weights and Avoirdupois weights, of which more will be said later. The Tower pound was not an Exchequer standard at all, being a poise used at the Mint to check the correctness of the weight of 240 newly minted pennies, in much the same way as modern banks had special weights of, for example, '100 shillings' whereby weighing could replace the

*When a royal estate was farmed out, the tenant-in-chief compounded with the king for a certain annual payment. He in turn levied a rent or *farm* on the undertenants who actually occupied and worked the land. Naturally this levy was set to yield a handsome profit. The sheriff of the county collected on behalf of the king making payment at the king's Exchequer and was responsible for the quality of the coin in which the farm was paid. The sheriff was held strictly to account. The office must have been profitable to make up for the nervous tension generated by the royal summons which opened thus after peremptory greetings:[32] 'See that, as you love yourself and all that you have, you be at the Exchequer at such a place on the Morrow of Michaelmas or on the morrow of Low Sunday and have with you whatsoever you owe of the old farm or of the new and in particular . . . etc'.

tedious occupation of counting. Madox and Hale are correct when stating that the weight at the Exchequer would be the Troy pound. Traditionally it has been the weight for bullion down to the present century and for centuries it has been the weight of the gold- and silversmith. Avoirdupois weight would not be so employed for that was the weight of the market-place for common commodities and bulky goods.

The Florentine merchant Pegolotti,[33] writing of London in the early fourteenth century, tells of two ways of weighing silver: by the *mark* of the Mint at the Tower which was the same as the *mark* of Cologne 'by which is sold and bought all manner of silver in plate or in bars or in coin or bullion in pieces'; and by the *mark* of the goldsmiths 'which is heavier and larger than the Tower mark by $5\frac{1}{3}$ sterlings, of sterling 20 the oz and 8 oz to the mark'. A *mark* was a monetary unit and a weight. For both in England it was $\frac{2}{3}$ of a pound: of the monetary pound it was therefore 160 pence or 13 shillings and 4 pence; of the unit of weight it was 8 ounces.

The first of Pegolotti's *marks* is quite straightforward. It is the *mark* of the Tower pound. The second definition is less clear. The *goldsmith's mark* was that of the Troy pound which is heavier than the *Tower mark* but not by $5\frac{1}{3}$ sterlings. The *Troy mark* exceeds the *Tower mark* by 240 grains or $10\frac{2}{3}$ sterlings each of $22\frac{1}{2}$ grains and when he speaks of 'sterling 20 to the ounce' he is confusing the weight of the penny with the Troy dwt. Pegolotti is usually a reliable witness. One can but conjecture how he came to be out by a factor of two in his definition. No juggling of coin weights or finenesses will bridge this considerable difference.

*Troy and Tower Weight at the Mint: The 'Tractatus de Ponderibus et Mensuris'**

In the old printed versions of the Statutes, tacked on to the Assize of Bread and Ale *(Assisa Panis et Cervisie)*, of uncertain date but usually attributed to 51 Henry III of 1266,[34] is a short paragraph sometimes designated as a separate item under the title 'Compositio Mensurarum' or 'Compositio Monete', reading:

> By the Consent of the whole Realm of England, the measure of our Lord the King was made, viz, that an English penny called a sterling, round and without any clipping, shall weigh XXXII wheat corns in the midst of the ear; and XX pence do make an ounce; and 12 ounces a pound; and VIII pounds do make a gallon of wine

*The Tractatus is strictly not a statute. It has been described as a history which in Elizabethan times was quoted 'as having Authority' (see the Report from the Committee appointed to inquire into the original Standards of Weights and Measures (The Carysfort Committee) 26 May 1758 p.420). It has also been taken as a private memorandum of extremely valuable information but not a piece of legislation (see GRIERSON, Philip. Stenton Lecture, University of Reading 1972, p.13). For full text of the Tractatus, see Appendix A (d).

and VIII gallons of wine do make a London bushel; which is the eighth part of a quarter.

This appears virtually word for word as the introductory part of a longer document on measures, entitled 'Assisa (or Tractatus) de Ponderibus et Mensuris' (Assize of Weights and Measures). This Assize, like that of Bread and Ale, is of uncertain date but is commonly attributed to 31 Edward I 1302/3. The text appears in various manuscripts* and is printed in Ruffhead[35] and in Statutes of the Realm[36] and there are some variations among the several sources. A version of 1290 is given by Fleta[37] in Chapter 12. An earlier source is the 'White Book' of Peterborough Abbey[38] of about 1253 and the passage quoted is in the Record of Caernarvon[39] of the thirteenth century under the title 'Complicio Monete et Mensura'. The evidence strongly suggests a mid-thirteenth century origin for the declaration as we have it today but it could well be earlier. The multitude of manscript copies and copied versions attests to its importance. We shall refer to this document as the 'Tractatus' and date it 1302/3 to avoid confusion with other Assizes, recognizing that it is most likely some fifty years earlier, and shall consider the paragraph already quoted before considering the rest of the Tractatus.

Looking then at the passage quoted, we find at once a declaration as to the weight of the ounce and the pound, a description of the units of capacity in terms of weight (to be discussed in Chapter IX) and a statement as to the weight of an English penny. The use of the word 'sterling' to mean an English silver penny is of some antiquity, dating from the eleventh century and in widespread use from the thirteenth. The word's derivation has been the subject of a special study.[40] Remembering that 32 wheat grains weigh as much as 24 barley grains (at least notionally and nearly so if one does not demand for the thirteenth and earlier centuries the precision of the twentieth), we find this to be a declaration of Troy weight, for 24 × 20 gives the number of barley grains or Troy grains in an ounce, that is 480, and at 12 oz to the pound we are back to the familiar 5760 Troy grains to the pound. In terms of wheat grains there would be 640 to the ounce and 7680 to the pound.

Notwithstanding this declaration there is no evidence that any large weight was ever generated or reconstituted from cereal grains but new weights were always established with reference to some older weight whose relationship to the standard in question had been determined previously and was known to be of a certain definite value.

The statement that an English silver penny weighs 32 wheat grains (24 Troy

*MS Reg. 9A II fo 170^{b}
MS Cotton, Galba E IV fo 28^{b} — 29
MS Cotton, Claudius D II fo 25^{a}
MS ADD. 6159 fo 146
among others

grains) is simply not true. On rare occasions, for brief periods, and in Saxon times only, did the penny ever attain that weight standard.* Rather the theroetical modulus of a penny was, from the evidence of the coins themselves, not 32 but 30 wheat grains or 22½ Troy grains.

With the weight of the penny coin being some 6 per cent lighter than the pennyweight of the weight scale, it is not difficult to identify the seat of the confusion, nor should we wonder at later errors which crept into the business affairs of the nation, not only in the market place but in the texts of the Statutes themselves. The coin itself was frequently used as a weight. The fact of its being some 6 per cent lighter than the pennyweight it represented would do little to discourage its use by merchants.

It can perhaps now be seen how the Compositio led to the consideration of two pounds; one the *Troy pound,* generated in theory, as the statement says, with 480 Troy grains to the ounce and 12 ounces to the pound, the other the *Tower pound,* the actual weight of 240 silver pennies, each of 22.5 Troy grains; the Troy pound of 5760 grains, the Tower pound of 5400 grains. Because the Tower pound is mentioned in thirteenth-century documents and is said to have been the Saxon money weight, it is usual to find this pound given pride of place as the original weight standard of the kingdom,[41] especially as several authorities have declared Troy weight to have been a creation of the mid-fourteenth century, and with the first mention of Troy weight in the Statute Book appearing to be that of 1414 (2 Henry IV c 4). This attribution of primacy must be an error, for the Tower pound was never a trade pound in England.

The Libra Mercatoria

Troy weight to this point has been spoken of as applicable to the weighing of pharmaceuticals, bullion, and coin exclusively. Indeed we find in the written documents that from the thirteenth century Troy weight was in fact exclusively used for these items, the only additional commodity to be so weighed being bread (see Ch.XI).

Towards the end of the Tractatus of '1302/3' we read:[42]

Item, it is to be known that the Pound of Pence, Spices, Confections as of

*The record of the coinage is that following the reform of Edgar in 973 there commenced a recall and reissue of the currency every six years, later becoming every three years. With each new issue the weight was deliberately varied. Heavy pennies became common, starting with Edgar's first 'short cross' series on a weight standard of 24 grains. This continued through the reign of Edward the Martyr and the first part of Aethelred II. Aethelred's 'Crux' coinage was on a standard of 25.5 grains and the 'long cross' (977-1003) series reached 27.5 grains. The pressure of Danegeld on the coinage drove the weight down below 18, but Cnut restored it to 18 grains. Edward the Confessor began at 18, returned it to 27 briefly, then down to 21.5 grains which the Conqueror continued till 1080, after which it rose to 22.5 at which it remained, theoretically at least, for the next two centuries. A diagram of the variation is given in Figure 26.

> Electuaries [pharmaceuticals], consisteth [in weight] of twenty shillings. But the pound of other things weigheth twenty-five shillings. But in Electuaries the ounce consisteth of twenty pence and the pound contains twelve ounces: but in other things the pound contains 15 ounces: but the ounce in either case is in weight twenty pence.

This extract from the Tractatus makes it clear that two pounds are being defined, both with an ounce of 20 pence. We must now ask, does the document mean an ounce of 20 pence, each $22\frac{1}{2}$ grains in weight, thus defining the Tower pound of 5400 grains, or does it mean an ounce of 20 pennyweights, thus defining the Troy pound of 5760 grains? Had the Tractatus only mentioned 'pounds of pence' we might have considered the Tower pound to be that which was meant, but when it mentions 'spices, confections, and electuaries' we see goods of trade and pharmaceuticals included and so we must conclude that it is the latter interpretation which is meant, that is the Troy ounce of 20 dwt and the pound of 5760 grains. The heavier pound for all common trade goods would be 25 per cent greater, 15 ounces instead of 12.

It cannot be claimed that this interpretation has been that of all previous writers on the subject. Some have argued that the first pound mentioned is that of the Tower and the heavier pound therefore weighs 6750 grains, being 15 ounces each of 450 grains,[43] but that interpretation cannot be maintained in the light of the use of Troy weight as the basis of pharmaceutical weights and the fact that the use of the Tower pound was restricted to the operations of the mint. Support for the interpretation favoured here has also appeared in the literature over the years.[44]

The heavier commercial pound is mentioned in the mid-thirteenth century in the 'White Book' of Peterborough Abbey,[45] thus:

> 'For electuaries and confections the pound is of 12 ounces; for all other things the pound is of 15 ounces. The ounce of all things is of 20 pence in weight'
> (En letuaris e confeciuns la liver est de XII uncis; en tutes autre chosis la li est de XV uncis. La unce de tote chosis est de XX deners en peys).

This commercial pound is mentioned by Fleta in 1290 under the name *libra mercatoria:*[46] 'and 15 ounces make a libra mercatoria' (et quindecim uncie faciunt libram mercatoriam). It is also termed the *London pound* in documents of the thirteenth and fourteenth centuries: for example 'Et uncia debit ponderare viginti denarios. Et quindecim uncie faciunt libram Londonie',[47] and again; 'The ounce must weigh 20 pence and 15 ounces make a London pound' (Uncia debet ponderare viginti denarios. Quindecim uncie faciunt libram Londonie).[48] We note in passing that the 15 ounce pound of 7200 grains is precisely 2 Tower marks each of 8 ounces of 450 grains. Fleta also tells us that the pound for wax, sugar, pepper, cumin, almonds, and wormwood is of the weight of 25 shillings. This partial listing of commodities is intended to show the wide range of goods weighed by the *libra mercatoria* and he is careful to add:[46]

> 'But a pound of gold, silver, electuaries and such like apothecaries' confections has the weight of 20 shillings sterling only'
> (Libra vero auri, argenti, electuariorum et huiusmodi apotecariorum confecte consistit solummodo ex pondere XX solidorum sterlingorum).

Haberty-poie Goods and Avoirdupois Weight

The *libra mercatoria* or London pound would therefore be $1\frac{1}{4}$ pounds Troy or 7200 grains (466.6 g) and remarkably enough no example of this weight would seem to have survived the centuries. This may be because two radical developments were to occur in a short space of time to change the mercantile unit of weight. The first was the change from 15 to 16 ounces. The second was its metamorphosis into the Avoirdupois pound so well known today. The Avoirdupois pound was of 16 ounces but only 7000 grains. The ounce (Avoirdupois) was therefore of 437.5 grains (28.35 g).

Of the first of these changes we find the raising of the pound from 15 to 16 ounces mentioned in fifteenth-century manuscripts, but the ounces are called Troy ounces. For example, in modern English we read:[49]

> Lying weight [weights that lie in the scale pan] is the third weight and this weight and 'Haber de Peyse' are the same. The pound of this weight contains 16 ounces Troy and by this weight all manner of merchandise is bought and sold, as tin, lead, iron, copper, steel, wax, wood, alum, madder, spices, corsys, laces, silks, thread, flax, hemp, ropes, tallow . . .

and in a Harleian manuscript[50] we have the following:

> . . . and twelve uncs maketh a pounde of Troye weight for silver, golde, breade, and measure. . . . The same tyme ordeined that XVI uncs of Troie maketh the Haberty poie a pound for to buy spice by . . .

Again in the Coventry Leet Book for 1474,[51] after being told that the pound for silver, gold, bread, and measure consists of 12 ounces, each of 20 sterlings, each weighing 32 wheat grains, we read: '. . . the seid XXXII graynes of whete take out of the myddes of the Ere makith a sterling penny and XX sterlings makith a ounce of haburdepeyse and XVI Ouncez makith a li . . .'. These declarations make the Troy and Avoirdupois ounces the same, and the belief that the Avoirdupois pound contained 16 Troy ounces was to cause all kinds of problems in later centuries, most notably in Elizabethan times, when it was only at the third attempt at providing Exchequer standards that the true

Avoirdupois weight was regained.

First, let us notice the use of the term 'Habur de Peyse' or 'Haberty poie' to describe a weight. In earlier usage this term and its variant spellings 'Haberdepois', 'Aver de pois', 'Averdepois', and 'Avoirdupois' meant, not the weight by which goods were bought and sold, but the goods themselves. In this context the word appears in a charter of 1303[52] and in the Statute Book for the early part of the fourteenth century,[53] and in the records of the City of London[54] of 27 July 1345 (19 Edward III) where we are told of certain men being elected 'per probos homines mester averii ponderis, piperariorum de Sopers Lane . . .' (by good men of the mystery [trade] of *averii ponderis,* pepperers of Sopers' Lane) but quickly the term became associated, not so much with the goods, as with the weights and the system of weights used in normal commercial transactions in the weighing of goods brought to market.

Some authorities[55] have pointed out that in the Middle Ages there was a 16 Troy ounce pound posing as an Avoirdupois pound and containing 7680 grains (497.7g). The documents speak of it, though they do not confirm this pound as being of 7680 grains. When we consider the derivation of the units of capacity it becomes clear (Ch.IX) that, although the documents speak of a pound of 16 ounces, it is a 7000 grain pound that is appropriate. It can therefore be supposed that the insertion of the word 'Troy' came from habit. The ounces in common use were Troy ounces of which the Troy pound contained 12 and the *libra mercatoria* 15. The true Avoirdupois pound did contain 16 ounces but they were not Troy ounces and Avoirdupois weights on the standard of 1 lb = 7000 grains were in use a century earlier than the date of the documents cited above.

Very few 16 Troy ounce pounds are extant but recently a weight in private hands came to this writer's attention (I am indebted to its owner, Dr J. E. Satchell of Kendal for much of what follows relating to this weight). It is a lead shield-shaped wool-weight carrying a single fleur-de-lys and shield-shaped frame in relief. Its weight is 3504 g. It was found on an abandoned farmstead about 10 miles north of Kendal, an important wool-exporting town (see Satchell, J. E., *Kendal on Tenterhooks,* Kendal: Frank Peters 1984). Work is in progress aimed at identifying the family whose arms are represented on the weight. One of several possibilities is the family of Edmund Langley, Duke of York, who in 1347 became Lord of the Manor of Wakefield. The fleur-de-lys is still represented on Wakefield's arms.

It is here suggested that from its shape this weight is a 7lb wool-weight, and if each pound weighs 16 Troy ounces (497.7 g) then 7 pounds should weigh 3483.9 g. The weight is therefore 20 g too heavy (0.6 per cent) and could have originated in the late fourteenth or fifteenth centuries. It could not have come into existence after 1588 when Elizabeth I succeeded in re-establishing true Avoirdupois weights in England. It is certain that true Avoirdupois weight standards date back, as we shall see, to prior to the mid-fourteenth century, coexisting, no doubt, with the *libra mercatoria* of 15 Troy ounces and the 16 Troy ounce pound.

The Tractatus of the first years of the fourteenth century or indeed of 50 years earlier speaks only of Troy weight and the mercantile pound of 15 ounces for general commerce, but a 6¼ pound Avoirdupois clove of Edward I (1272-1307) is extant (Fig.29) and a 56 pound Avoirdupois weight of Edward III's time (1327-77) was to be found in the Exchequer in the days of Elizabeth[56] and was used as the basis of her Avoirdupois standards (see Ch.XII). Surprisingly, Avoirdupois weight does not seem to have been recognized in law until 1532,[57] when meat was ordered sold by Avoirdupois weight, but Winchester has a unique set of weights of Edward III's reign of denomination 91, 56, 28, 14, 7 and 7 pounds Avoirdupois.[58] That these are Avoirdupois and not pounds of 16 Troy ounces is in no doubt for in 1927 they were weighed against standardized Avoirdupois weights with the result shown in the table below.

Winchester City Weights — Edward III

(1 lb Avoirdupois = 16 ounces) (1 oz Avoirdupois = 16 drams)

Nominal weight	*Actual weight*	*Percentage loss*
91 lb (6 shields)	90 lb 14 oz 6 dr	0.11
56 lb (4 shields)	55 lb 10 oz 2 dr	0.66
28 lb (4 shields)	27 lb 11 oz 4 dr	1.06
14 lb (3 shields)	13 lb 13 oz 13 dr	0.98
7 lb (3 shields)	6 lb 14 oz 12 dr	1.12
7 lb (3 shields)	6 lb 14 oz 9½ dr	1.26

The bodies of these weights are of bronze, the rings and staples are of iron. On the bottom is a piece of lead for sizing the weight. The 91 pound weight is a quarter sack for wool weighing.

It is clear that Edward's weights had been carefully sized to agree among themselves and with the standard. Each weight carries a number of embossed shields on which are the arms of Edward III. The Royal Arms are quartered with those of Old France, and as this device was only in use from 1340 to 1405 the date usually assigned to this set is 1357,[59] when Edward enacted[60] that certain weights were to be issued to all the sheriffs of England. They are shown in Figure 28, and are to be seen today in the City Museum, Winchester. (See Ch.XII p.233).

If the Tractatus is actually to be dated *c*1250 or 1260 and makes no mention of Avoirdupois weight, the inference is that Avoirdupois weights came into general use sometime in the interval 1250-1357, most likely close to 1280-1300, having in mind the clove of Edward I. (See Ch.XII p.233.) This

Avoirdupois weight is to be seen in the Science Museum, London. (See Fig.29).) What, then, occasioned the change in the *libra mercatoria* from 15 Troy ounces to the Avoirdupois pound of 16 ounces each of 437.5 grains rather than 480 grains?

To have a pound divided into sixteenths was more convenient than having one divided into fifteenths. Sixteen can be divided in binary fashion by 2, 4, and 8 whereas 15 has factors 3 and 5 only, neither of which is particularly convenient. The binary division of the yard has already been mentioned and successive division by two is a very natural and convenient procedure. Moreover a great number of European mercantile pounds were divided into 16 ounces.

The change to an ounce of 437.5 from one of 480 grains is not so easily seen, involving as it does the *libra mercatoria,* the method of weighing the principal export, wool, and England's trading partners of the thirteenth and fourteenth centuries. To see clearly how this change came about requires information on the wool trade and methods of weighing in the thirteenth century.

The Weighing of Wool

In the central portion of the Tractatus*, from which we are getting so much information, we are told that a *load* or *char* of lead, wool, tallow, and cheese is 12 *weighs (weys)* and the *weigh* is 14 *stone.* The load is therefore 168 stone and we are told that each stone is to be 12½ pounds. The sack of wool is to be 2 *weighs,* that is 28 stone or 350 pounds. We are also told that a load of lead is 30 *fotmals* each of 70 pounds and that a load is 175 stones each of 12 pounds. In both descriptions we are told the *load* is to be 2100 pounds and the pound is to be of 25 shillings, that is the *libra mercatoria,* but there are two stones, one of 12½ and the other of 12 pounds.† Fleta in 1290[46] gives the same information and a fourteenth-century entry in one of the registers of Durham Priory (Ms Addit. 24059, f 15)[61] reads in part:

> 'An ounce must weigh 20 pence. Fifteen ounces make a London pound. Twelve pounds and a half make a London stone. The sack of wool must weigh 28 stone . . . The load of London is 2100 lbs . . ., that is 6 sacks of wool'
> (Uncia debet ponderare viginti denarios. Quindecim uncie faciunt libram Londonie. Duodecim libre et dimidia faciunt petram Londonie. Saccus lane debet ponderare viginti octo petras . . ., duo milia et centum libras . . . sex sacci lane).

*See Appendix 1 (d) for full text.

†With Troy, Tower, Avoirdupois, and mercantile pounds and two stones for lead, not to mention the stone of 8 pounds for wax, sugar, pepper, cinnamon, nutmeg, and alum and the stone of 5 pounds for glass, also given in the Tractatus, one wonders whatever became of Article 35 of John's Magna Carta of 1215, confirmed by Henry III in 1216, 1217, 1218, 1225, and 1252 and by subsequent monarchs, that there should be one measure throughout the land 'and with weights it shall be as with measures'.

This confirms the sack of 350 pounds but also shows the pound of 15 Troy ounces still in use after the date already proposed for the introduction of Avoirdupois weight. However it should be noted that the 15-ounce pound differed by less than 3 per cent from the Avoirdupois pound, the former being the heavier. Where precision was not required the Avoirdupois pound could and frequently did replace the other. With time this substitution became complete, with the weights of all common goods going by the name Avoirdupois and having standards both in the Exchequer and in the cities.

The *sack* of wool, as we have seen, is 28 stones each of 12½ pounds, that is 350 pounds, and these pounds are *libra mercatoria* of 15 Troy ounces. The sack did not remain at this weight for long for there is an entry in the Records of the Borough of Leicester for 1281 as follows:[62]

> It is determined by the whole community that henceforth none may weigh wool by any 12 pound stone but it must weigh 13 pounds fully and if anyone be found hereinafter who has a 12 pound stone he shall pay the community of Leicester half a marc [6 shillings and 8 pence].

Thus the stones of 12 and 12½ pounds were disappearing from the wool trade as Fleta was writing.

Twenty-eight stones of 13 pounds is 364 mercantile pounds, but though the 13-pound stone generated the 364-pound sack it did not last long either, for 13 is such an awkward number. Thus we see in the Statute 14 Edward III Stat. 1 c 21 of 1340 that the sack is redefined, while keeping the weight at 364 pounds, as 26 stones each of 14 pounds. This relation was repeated in Statute 2 of the same year. We now have fewer stones and each is now 14 pounds, a figure which has lasted to the present for all Avoirdupois weighings. We shall consider why the wool sack shifted from 350 to 364 pounds a little later, but we are now in a position to see the probable origin of the Avoirdupois pound. The discussion which follows is that supported in general[63] by most authors who have written on the subject, but with some variations.

The Origin of Avoirdupois Weight

The mid-thirteenth century was a time of great expansion in the wool industry. England was affected as were also continental centres. Spanish wool dominated the market until the end of the century when English wool began to take over. In the fourteenth century the export duty on cloth was 2 per cent while raw wool paid 33 per cent and it is said that Edward III gave the Woolsack to the Lords that they might be reminded continually from whence came England's prosperity.[64] By 1421 the customs on wool amounted to 74 per cent of the entire custom revenue of the land.[65]

One of the European cities with which England conducted a large trade in wool was Florence. The Florentine *libbra* has been given as of 12 ounces, of estimated weight 342.6 g,[66] 339.5,[67] 340.4 or 339.84 g[68]★ with a 16-ounce *libbra* of 453.9 g.[67] It will be noted that this heavier pound differs by only one-third of a gram from the present equivalent of an Avoirdupois pound. The weight of the lighter pound will not be far from 340 g. The ounce would be therefore 28.33 g while that of the heavier pound would be 28.37 g. In all likelihood the ounce was intended to be the same for both pounds, say 28.35 g as the average. This is just the weight of the present Avoirdupois ounce. Hence the suggestion that the ounce of Florence was chosen as the unit of the new scale of weight intended for the wool trade but at once being adopted for all commonly traded goods, in the reign of Edward I.†

The standard weight of wool in Florence was 500 *libbrae*[69] of the lighter variety, that is 170 kg. This is almost exactly 364 mercantile pounds (169.82 kg). Hence a sack of wool of 364 mercantile pounds or near enough 364 Avoirdupois pounds would fit extremely well into the standard of Florence. But the weight of the sack of wool did not move from 350 to 364 pounds just to fit into the Florentine standards. More domestic considerations forced the change too and the happy coincidence of the two lines of development, both leading to 364 pounds, no doubt settled the desirability of the change.

The Domestic Scene

Wool had been an important item of trade since Saxon times. In the mid-tenth century, in the Laws of Edgar (III Edgar 8.2) we hear of a weight of wool, the *wey (weigh),* and its value. 'And a wey of wool shall be sold for 120*d*. And no one shall sell it at a cheaper rate'.[70] Two *weys,* or *weighs* as they came to be called, made a sack of wool, so the sack was worth 240*d* or one pound.

In the late eighth century we know from the correspondence between Charlemagne and Offa (besides other sources) of the international trade in English woollen cloth, for in 796 Charlemagne wrote [71] to Offa concerning the lengths (or widths) of English '*sagorum*'. It is uncertain whether this was a cloak, a blanket, or a length of cloth but it was certainly woven from wool.

The export of wool and cloth was a principal component of the English economy, but, as with other commodities, they became facets of the

★This seems to have been almost identical with a local variety of the Roman pound used in antiquity, since the British Museum has 8 Roman-Italian basalt weights (Cat. No. 1867.5-8.336 et seq) running from 3 to $\frac{1}{16}$ oz. From these weights we find an average value for the weight of the pound from which they were derived, that is 341 g giving an ounce of 28.4 g. The Grand Duchy of Tuscany when going metric at the beginning of the nineteenth century recorded its current pound as 339.54 g. See further, GUILHIERMOZ, P. Note sur les poids du Moyen Age, *Bibliothèque de l'Ecole de Chartres (1908) Vol.LXVII, p.413.*

†MISKIMIN, Harry A. Two Reforms of Charlemagne? Weights & Measures in the Middle Ages. Economic History Review (1967) Second Series XX, No.1, pp.35-52 (see p.43) asserts that the Avoirdupois pound was imported from France, but this cannot be sustained.

politico-economic game, suffering from underproduction then overproduction, prohibitions with respect to import and export, restrictions and regulations, with the failure of traditional markets thrown in for good measure, all in a manner totally recognizable to twentieth-century traders in the aftermath of two world wars.

Prior to the mid-fourteenth century the export of raw wool was much more important than that of woven cloth, but in the century that followed there was a drastic drop in wool exports, with cloth replacing wool as the principal English export, so much so that between 1350 and 1430 the total revenue from customs was halved.[65] Strangely enough, although the wool trade was depressed in the mid-fifteenth century it was established policy to restrict the export of raw wool by aliens, 'that sufficient plenty of these wools may continually abide and remain within the realm as may competently and reasonably serve for the occupation of clothmakers of England'.[72] Clearly the long-term economic advantages of exporting finished or semi-finished goods rather than raw materials were appreciated even then, and it could only further English cloth interests if the ban on wool exports meant a dearth of material with which the Flemish weavers could work.

Weighing in the Thirteenth and Fourteenth Centuries

Apart from gold, silver, and the like, which were always weighed by scales, the common business of the market-place, especially for small retail weighings, was conducted using *auncels* as weighing machines. This device has been stated to be a bismar or type of steelyard (Ch.I), but there has been some discussion as to the precise meaning of the term. Archbishop Chicheley in 1428[73] speaks of the 'Auncel, otherwise Scheft or Pounder'. Now a pounder was a bismar, but the Archbishop repeatedly refers to auncel *weight* and speaks of the 'aforesaid *weight* vulgarly called le Auncell, Scheft, or Pounder or *any of them*', which implies the possibility of more than one device. Cowell's *Interpreter* of 1607 states:[74]

> Awncell weight as I have been informed is a kind of weight with scales hanging or hookes fastened at each end of a staffe which a man lifteth up with his finger or hand and so discerneth the equality or difference between the weight and the thing weighed. . . .

He then states the 'awncell' to be forbidden for centuries: 'Yet a man of good credit once certified me that it is still used in London among butchers etc.'. Dr Johnson's *Dictionary* of 1755, which is largely based on literary quotations, ignores *auncel, bismar,* and *pounder* but the first edition of the *Encyclopaedia Britannica* of 1771 gives us:

> *Auncel weight,* an ancient kind of balance now out of use being prohibited by

several statutes on account of the many deceits practised by it. It consisted of scales hanging in hooks, fastened at each end of a beam which a man lifted up on his hand. In many parts of England auncel-weight signifies meat sold by the hand without scales.

Woodcuts in printed books show that by the early modern period the finger or hand had been replaced by a handle with a fixed pivot or fulcrum and the device resembled a steelyard[75] with a hook at one end for the goods, the pivot near this hook, and an arm or shaft extending beyond the fulcrum along which a counterpoise could be moved relative to an engraved or notched scale.

Both bismars and steelyards were open to grievous abuse by unscrupulous traders. The gross inaccuracy of the bismar will be obvious at once. The steelyard needed more deliberate action to be rendered fraudulent, for if properly made and engraved, with the correct counterpoise sized for the particular instrument, it can be highly accurate. Steelyards were sealed and certified correct for use in trade in England in the nineteenth and twentieth centuries. The Avery Co. catalogue for 1912 listed a large selection of highly accurate and sensitive steelyards for a variety of load ranges and for different purposes. What then was going on in the thirteenth and fourteenth centuries which caused these instruments to come under a cloud?

The answer is that it was just too easy in those days to modify the weight of the detachable counterpoise. Once the beam was made and the positions of the suspension hook and the hook for the goods determined, the scale was true only for the counterpoise which had been used in its calibration. Lighten the counterpoise by chipping out some of the metal and it would balance the goods at a point on the scale indicating a larger weight than that actually present — ideal for a fraudulent seller. If a heavier counterpoise was substituted for the proper one, the apparent weight would be given as less than that of the actual goods being weighed. This suited a fraudulent buyer. Further, it could be so positioned as to make the position of the counterpoise on the engraved scale difficult to be seen. An unscrupulous weigher could declare a quite different weight from that shown by the steelyard. It must therefore be concluded that in the Archbishop's mind the word '*auncel*' meant bismar and fraudulent steelyard alike if the text of his declaration is to be properly understood.

In the thirteenth century, then, small retail weighings were made by auncel, meaning bismar or steelyard. For wholesale transactions the king's great beam was used and all manner of heavy goods ('averia ponderis' or 'avoir du pois') was weighed on this great scale, of which the Guild of Pepperers (later to become the Grocers' Company of London) had charge. The king's lesser beam, a smaller balance, was used for the weighing by the pound of high-value commodities, silks, spices, etc.

These beam scales were equal-arm balances of the type familiar today. Weights were placed in one pan, goods in the other. Heavy goods were weighed by the hundred (hundredweight, cwt). In the late thirteenth and early fourteenth centuries there were no fewer than four different hundredweights but related to each other in pairs. The original two are to be found in the Tractatus. The first, not surprisingly, is a hundredweight of 100 pounds. It appears as the *seam* of glass, 20 stone of 5 pounds each. It is readily seen to be the equivalent of 8 stone of 12½ pounds each. The second is the hundredweight of 108 pounds, specified in the Tractatus for wax, sugar, pepper, cinnamon, nutmegs, and alum and subdivided we are told into 13½ stones each of 8 pounds. For the weighing of wool there was the *clove,* which was one sixteenth of a hundredweight. With two hundredweights, there were two cloves, that of 6¼ pounds for the 100-pound hundredweight and that of 6¾ pounds for the 108-pound hundredweight.

Originally wool was not weighed on the beam. The custom of wool was the king's prerogative and was weighed on the *king's tron* (balance) which from the description was a large steelyard suspended from a suitable support, later becoming a beam balance when the auncel was outlawed. These wool-weighing devices could handle loads of several hundred pounds. In 1298 the City of London was required to construct for Edward I a large steelyard for the weighing of wool in 'our town of Len', that is King's Lynn in Norfolk as we know it today. Besides the six woolmen 'Sworn to the Tron at Lynn' there was one other, Thomas Turgot, an 'ancermaker', who certified the scale as showing the true weight.[76,77] But as early as 1290, foreign merchants, notably Spanish wool-traders, petitioned the king[78] that wool should be weighed by balances, claiming that, in having their wool weighed at Southampton, they were losing half a pound in the clove. Presumably the weighers at Southampton were not using balances but were using an auncel calibrated on the 6¾-pound clove and not the 6¼-pound clove of the 100-pound hundredweight. One can readily imagine foreigners expecting a hundredweight to be 100 pounds and the clove therefore 6¼ pounds. With a loss of 28 pounds in a sack of 350 pounds it was little wonder the merchants complained, but no action was taken to regularize this matter for quite some time.

Even with beam balances there was a different technique employed as between weighing bullion (which was always weighed with the balance) and all other goods. For bullion the balance was always struck with the beam horizontal. For other goods it had long been customary to keep adding merchandise to its scale pan until it not only balanced the weights but tipped the scales towards the goods being weighed, 'quod statera debet trahere inclinando versus pecuniam . . .', as the Archives of London for 1256[79] tell us. This extra weight of goods given to the buyer was justified on the basis that there was likely to be wastage, spillage, or other loss in the transaction and this became a firmly established practice which the Mayor and Aldermen of London succeeded in maintaining. But, it being noted that the weigher can 'allow a greater [extra] weight to one than to another', it was decided that on the Saturday after the Feast of St. Nicholas (6 December) 1256, all goods weighed by the king's beam should be weighed as for bullion without the scales tipping and that, in lieu of this, the seller was to give the buyer an extra four pounds in every hundred. This was reiterated in 1260. The extra allowance was called *tret* or *cloffe*.[80] Its introduction did not pass without royal challenge but the mayor was able to maintain that if true weight was obtained with the beam horizontal then the buyer was entitled to the four pounds in lieu of having the scales tip. Everyone wanted the extra four pounds in the hundred, so the hundredweight of 100 pounds went to 104 pounds, that of 108 pounds to 112 pounds and we have two stones so generated, one of 13 pounds, the other of 14 pounds, and two new cloves, one of 6½ pounds, the other of 7 pounds.

In the 'Ordinacio facta de modo ponderandi per balanciam' of 1309[81] we are told of heavy goods such as wax, almonds, rice, copper, tin, 'and whatever has to be weighed by balance' being weighed by the hundredweight of 112 pounds. Light goods, ginger, saffron, sugar, mace, and other kinds of wares which are sold by the pound are to be weighed by the hundredweight of 104 pounds. Sometimes the hundredweight of 104 pounds is called the *spicers' hundredweight*. Pegolotti[82] writing a few years later confirms the sizes of the two hundredweights at 104 and 112 pounds and also reiterates the wares for which they may be used.

Goods for the lighter hundredweight would use the smaller beam; goods for the heavier would use the king's great beam. Figure 30 shows a fifteenth-century market scene centring on the great beam. To summarize, we have:

The Hundredweights

Original	*Later*	*Goods*
100 lb	104 lb	Spices, ginger, sugar, etc. Goods of high value.
108 lb	112 lb	General bulk goods, wax, copper, tin, alum, iron, rice, and nuts.

The Cloves

100 lb cwt — 6¼ lb	104 lb cwt — 6½ lb
108 lb cwt — 6¾ lb	112 lb cwt — 7 lb

The use of the clove of 6¼ pounds and therefore the 100-pound hundredweight at the time of Edward I is shown by his shield-shaped lead weight now in the Science Museum, London. It is shown in Figure 29. A recent weighing gave it as 6.337 lb Avoirdupois. It is a handsome article carrying on its surface 'Three lions passant'. The slot for the leather strap, a feature of many of the later woolweights, is clearly visible.

Thus we see that in the second half of the thirteenth century *cloffe* was being given for a large number of commodities and together with hundredweights of 100 and 108 pounds we find those of 104 and 112 pounds appearing. Dealers in wool were not to be left out for we find in the customs accounts for Winchelsea and Pevensey for 28 May — 14 September 1288[83] a list showing a total of 43 sacks and 14 cloves exported. The various entries include two involving pounds, one of three and the other of four, yet the total gives no pounds, only sacks and cloves. The inference is that here 7 pounds made the clove, 112 pounds the hundredweight, and 364 pounds the sack. *Cloffe* at 4 pounds the hundred was being applied to wool. By implication the balance was in use whereas at Southampton the steelyard was used as the Spanish traders found to their cost in 1290.

The custom of reckoning the sack of wool at 364 rather than at 350 pounds was something buyers would encourage but the practice does not seem to have been legalized until 1340 when, by the Statute 14 Edward III Stat 1 c 21, the sack was declared to be of 26 stones each of 14 pounds and this was repeated in Statute 2 of the same year. Further, balances were to be used. This did not prevent steelyards continuing in operation for we read of petitions the next year (1341)[84] and in 1343[85] asking that: 'wools be according to the weights ordained by Statute i.e. 14 lbs to the stone and 26 stones to the sack' and 'that the stone shall be 14 lbs and 26 stones to the sack according to the statute made at Westminster', while in 1351 a similar petition[86] reads:

> That while divers merchants are wont to buy and weigh wool and other merchandise by a weight called the auncel, to the great damage and deceit of the people, the Commons pray that this weight called the auncel be ousted and each one sell and buy by balances.

To this end the king assented, reaffirming the sack to be 26 stones each of 14 pounds. This petition passed into statute the same year almost verbatim, the Act (25 Edward III Stat 5 c 9, 1351/2) continuing:

> and that every person do sell and buy by the Balance, so that the Balance be even . . . so that the sack of wool weighed no more than 26 stones and every stone to weigh 14 pounds and that the Beam of the Balance do not bow more to the one part than to the other and that the weight be according to the Standard of the Exchequer. . . .

There was no longer any doubt. The 350-pound sack of the Tractatus had acquired its *cloffe* legally and was augmented by 3.5 × 4, that is 14 pounds, to become 364 pounds. The *clove* or half a stone was now 7 pounds Avoirdupois and being $\frac{1}{16}$ of a hundredweight there were now 112 pounds to the hundredweight. Moreover the sack was now 500 *libbrae* of the City of Florence, near enough.

We can sum up this development chronologically as follows:

*c*1250 The Tractatus (usually dated 1301/2) declares the sack of wool to be 350 pounds.

1256 London: Introduction of cloffe to compensate for a horizontal beam. The beginnings of the 104 and 112-pound hundredweights which include the 4 pound cloffe.

1288 The local use at Winchelsea and Pevensey of a 7-pound clove, a 112-pound hundredweight, and a 364-pound sack, that is cloffe introduced for wool trading.

1290 Spanish traders expect a 6¼ pound clove (100-pound hundredweight) rather than the 6¾ pound (108-pound hundredweight) in use at Southampton.

*c*1290 Fleta declares the sack to be 350 lb.

1298 The steelyard at King's Lynn calibrated on a 364 lb sack.

*c*1300 The 6¼ lb clove of Edward I. The 100-pound hundredweight still in use (it continued certainly into the fifteenth century).

1309 The 'Ordinacio' declares the 112-pound hundredweight to be used for heavy goods and the 104-pound hundredweight for spices. Both hundredweights include the 4-pound cloffe.

*c*1330 Pegolotti confirms the hundredweights as 104 and 112 pounds respectively.

1340 Statute 14 Edward III Stat 1 c 21 declares the sack to be 364 pounds and balances are to be used.

1352 The sack declared again to be 364 pounds.

The Abolition of the Auncel

The appeals of the traders of 1290 and subsequent years were answered in 1352 by the Act 25 Edward III Stat 5 c 9[87] in which the auncel was condemned.

> Item, Whereas great Damage and Deceit is done to the People, for that divers Merchants used to buy and weigh Wools and other Merchandises, by a Weight which is called Auncel; It is accorded and established that this Weight called Auncel betwixt Buyers and Sellers, shall be wholly put out; and that every Person do sell and buy by the Balance, so that the Balance be even, and the Wools and other Merchandises evenly weighed by rightweight so that the Sack of Wool weighed no more but XXVI stones and every stone to weigh XIV pounds and that the Beam of the Balance do not bow more to one part than to the other and that the Weight be according to the Standards of the Exchequer; and if any Buyer do contrary he shall be grievously punished as well at the suit of the party as at the suit of our Lord the King.

This Act condemned the auncel but the record shows it continued in surprisingly good health, notwithstanding a further Act the following year[88] proclaiming that wool was to be weighed in a balance and to show good faith the sum of £40 was expended the next year in providing twelve new trons (balances) for twelve ports and the old trons (auncels) were removed. The abolition of the auncel was confirmed in 1353[88] and again in 1360[89] and this called for additional legislation making provision for the weighing of other commodities which had also been the subject of this kind of weighing. Early in 1430 a petition[90] stated that in ancient times cheese was sold by the *wey*, then it came to be weighed by the auncel, but that these by the Act of Parliament had been destroyed and other '*poises cochanty*' (weights *couchants*, that is, lying in a scale pan) used but it was not known how many pounds by these weights the *wey* ought to be. It was therefore asked that it be 32 cloves each of 7 lb. This was agreed to in the same year. The wey of cheese was to be 32 cloves each of 7 lb 'according to the Standard'.

It will not go unnoticed that the wey of cheese of 224 lb is just one tenth of the Avoirdupois ton of 20 cwt, that is 2 cwt.

Matters continued much as they had been for a long time until positive action from an unexpected quarter finally started the rooting out of the auncel. The Archbishop of Canterbury, Henry Chicheley, in 1428[91] pronounced the penalty of the Greater Excommunication on all 'who use or cause to be used in any manner any weight notably disagreeing with the King's standard and especially the aforesaid weight, vulgarly called le Auncel, Scheft or Pounder or any of them . . .'. The terror of hell had greater effect than all the statutes put together, but just to make sure a Statute referring to previous enactments in this regard was issued the following year (8 Henry VI c 5):[92]

> Whereas . . . by a Statute made the 25th year of King Edward III it was ordained . . . that the weight called Auncel . . . shall be utterly left and set apart and the wools and all other manner of Merchandise and all other things weighable bought and sold shall be weighed by the Balance so that the tongue of the Balance do not incline more to the one party than to the other . . . Our Lord the King . . . hath ordained . . . that the Statutes shall be firmly kept. . . .

From 1428 the use of the auncel gradually faded away with the balance rapidly ascending in popularity and every city borough and town was ordered to have a common balance (11 Henry VI c 8) 1433.[93]

Although all this vastly reduced the number of auncels in use it was not a total extinction, for a petition of 1439[94] complains that 'the Schafte, otherwise called a Pondre, otherwise called a Hauncer . . . beth nowe used, the which hurteth the pouere clothe makers . . .'. Kingdon records the cases of two women reported in 1458 to the Dean and Chapter of St. Paul's Cathedral for having auncels,[95] while in Robert Thompson's *Catalogue of Abuses of False Weights and Measures* of about 1634[96] we read: '3. Item, hee hath seized many falce Beames and Scales . . . And likewise also the Ancell Beame which being altogether prohibited yet are used by many'. Indeed, as has already been mentioned, the bismar and the steelyard continued into the twentieth century in many parts of the world.

With the establishment of the wool sack at 364 pounds with its subdivisions, the *tod* $\frac{1}{13}$ (28 lb), the stone $\frac{1}{26}$ (14 lb), the clove $\frac{1}{52}$ (7 lb) and the pound ($\frac{1}{364}$), we find Pegolotti,[97] writing of the early decades of the fourteenth century: 'Wool is sold in London and through all the island of England by a sack of 52 cloves and each clove weighs 7 English pounds'. But the imagination was stirred. In a fifteenth-century document we read (in modern English):[98]

> For a sack of wool contains for everyday of the year (except 1 day and 6 hours) one pound of wool, for every week a nail (clove) of wool, for every fortnight a stone of wool, and for every month (lunar) a tod of wool and so the sack will contain 364 pounds.

We must not think however that the 100-pound hundredweight became 104 and the 108 became 112 within a few days or even a few years. Mixtures of all of them appear in the records from time to time. A manuscript, Ms Cotton Vespasion E IX ff 86–110, of the fifteenth century,[99] tells of the 112-pound hundredweight for heavy goods (called 'Grete Warys') but the spices, the 'Sotyll Warys' (fine goods), are to be reckoned by the hundredweight of 100 pounds. The manuscript further states that the pound contains 16 ounces, the clove 7 pounds, the stone 14 pounds, the tod 28 pounds, and the sack 364 pounds. It is written that the sack of wool is: 'IIIC 1 qrs and this C that pertaineth to wool is the C of Haburdy peyse and it amounteth to 112 lbs'. The symbols 'IIIC 1 qrs' means 3 cwt and ¼ cwt, that is, $3 \times 112 + 28$ or 364 lb just as before. All is measured on the Avoirdupois scale. The information in this document is reiterated in a manuscript of the first or second decade of the seventeenth century attributed to J. Lloyd, a Denbighshire landowner,[100] and addressed to King James I. The 112-pound hundredweight was destined to outlast all the others and to continue to the present. It had the advantage of division by a large number of factors, 2, 4, 7, 8, 14, 16, 28, and 56 which was always a convenience in earlier days when division into numerous factors was more important than ease of calculation. Nor should it be thought that once the cloffe of 4 pounds had been added to the hundredweight that from here on all weighings would in fact be by the balance with the level beam. As late as 1789 the Act 29 George III c 68 entitled 'An Act for repealing Duties on Tobacco and for Granting New Duties in lieu thereof' states in Section 54:

> Excise Officer . . . in weighing and taking account of all tobacco and snuff respectively . . . and by this Act directed to be weighed and taken an account of therein [shall] give the Turn of the Scale in favour of the Crown and in lieu thereof shall allow the Importers, Proprietors, and Consigners thereof 2 pounds weight

Avoirdupois upon every Hogshead, Cask, Chest or Case of such tobacco. . . . (Section 9 provided that no hogshead, cask, etc. would contain less than 450 pounds so the allowance is becoming miserly).

Similarly in 1795 (35 George III c 118, s 15), referring to the weighing of coffee and cocoa nuts with a view to levying duty on the same, we are told by statute that: 'Customs and Excise shall . . . give turn of the Scale in favour of the Crown and in lieu thereof shall allow Importers etc. 1 lb Avoirdupois on every 100 lbs weight of coffee and 2 lbs on every 100 lbs weight of cocoa nuts'. Similar provisions were enacted for the Duty on Glass in 1809:[101]

> the turn of the Scale shall be in favour of the Crown and in lieu, there shall be allowed to such maker or makers of such Spread Window Glass or Crown Glass respectively 1 lb weight upon each and every 100 lbs of such Spread Window Glass or Crown Glass respectively so weighed.

The allowances represent a sort of inverse of the cloffe of the Middle Ages and are smaller than those given in earlier centuries. The weighings were made in controlled warehouses. To consider tobacco first, it can be assumed that, if the turn of the scale was to be in the Crown's favour, on entering the warehouse more tobacco would be added to the pan of the balance than would bring it horizontal. It will 'incline towards the goods' and surely this extra would be more than the two pounds credited to the importer. On removal from the warehouse, duty was paid and one can be sure that the balance would then be level. The king not only obtained a duty on each pound of tobacco but he also accumulated tobacco to his credit in the warehouse, to be sold later for the support of the Privy Purse. It would be the same with coffee, cocoa, and glass, and indeed with other commodities, especially paper for which the allowance was 2 lb per cwt in 1795; glass bottles, 1 lb per cwt in 1798; salt, $\frac{1}{2}$ lb per cwt in 1803. Although the turn of the scale in favour of the Crown was abolished in 1809 it was reinstated for paper and glass in 1811.[102]

The *Oxford English Dictionary* of the 1930s defines cloffe as 'an allowance (now of 2 lb in 3 cwt or $\frac{1}{168}$) given with certain commodities in order that the weight may hold good when they are sold by retail'. The implication is that the practice of allowing cloffe continued into the present century.

Wool Weights

So important was wool that special weights were produced for weighing it. Many of these are objects of considerable beauty and it is fortunate that so many are extant. No shield-shaped bronze wool weights are known earlier than the reign of Henry VII. For the most part these weights are seven pounds Avoirdupois and were intended to be used in pairs, for they are usually slotted so that a leather strap may be passed through each and the combination placed across the back of a horse, one weight on each side, as the weigher moved from one location to another. Occasionally a wool weight is found with a ring for handling rather than a slot for a leather strap and one in ten of the surviving weights are of 14 rather than 7 lb.[103] The use of wool weights was discontinued in the reign of George III. An impression of these delightful objects, which are now collectors' items, may be gathered from the photographs in Figures 31–33.

A wool-weight of a very different type is the 91 pound Avoirdupois weight of Edward III now preserved in the Winchester City Museum and which has been described earlier in this chapter. This weight is the quarter sack. When weighed in 1927 it was 90 lb 14 oz and 6 drams Avoirdupois (see p.129). This is only one-eighth of one per cent short of 91 pounds Avoirdupois of 7000 grains of the present day. The weight further demonstrates the use of the Avoirdupois scale in the fourteenth century. There is no other known example of the quarter-sack weight. The production of woollen cloths rapidly increased, with exports rising markedly during the first half of the sixteenth century. The domination of the market by English cloth put several European cloth centres out of business. But there can be too much of a good thing. The Antwerp cloth market collapsed in mid-century[104] and this with other factors led to a severe reduction in the cloth business which continued far into the seventeenth century. Many measures were enacted to protect the domestic industry, mostly in the form of prohibitions of one sort or another, but they were unsuccessful in restoring trade to its former position. Soon desperate measures were adopted to provide support to the industry. One of the more remarkable was the Act 18 and 19 Charles II c 4 of 1666 entitled 'An Act for Burying in Woollen onely'[105] which read:

> For the encouragement of the Woollen manufacturers of this kingdom and prevention of the exportation of the moneys there of for the buying and importing of linnen, be it enacted . . . that from and after the 5 and 20th day of March in the year of our Lord 1667 no person or persons whatever shall be buryed in any shirt, shift, or sheete made of or mingled with flax, hemp, or silk . . . or other than what shall be made of Wooll onely . . . upon pain of the forfeiture of the sum of £5, to be employed to the use of the Poore of the Parish.

There was to be no penalty if the deceased died of the plague. The Act was not strictly followed, for in 1678 by the Act 30 Charles II c 3[106] it was decreed that the previous Act was not of 'sufficient remedy' and that on and after 1 August 1678 no burial was to take place except in wool. An affidavit was to be brought to the parson. If no such certificate was forthcoming the goods and chattels of the deceased were to be seized to the value of £5 and if no such distress goods were available the persons acting contrary to the Act were to be liable. A register of burials and affidavits was to be kept.

This Act was effective and was in force until repealed by 54 George III c 108 of 1814. The church registers of the Norfolk village of Worstead (which gave its name to worsted cloth) tell a part of the overall story:[107]

> 1678 In primis Robert Webster, labourer, was buried in woollen August 19 and a certificate brought August 26th'
> 1728 Sarah, the daughter of Charles Themylthorpe and Sarah his wife was buryed November 25th and in linen and the penalty paid according to Act of Parliament.

We are further told[108] that in 1752 Samuel Johnson's beloved wife 'Tetty' died and was buried in wool.

NOTES AND REFERENCES: CHAPTER VIII

1. HULTSCH, Friedrich Otto. *Metrologicorum Scriptorum Reliquiae* Vol.II, p.61 et seq. Stuttgart: Tuebner 1864-66, reprinted 1971. See also p.17 et seq.
2. Lists of these (and many other) weights and their subdivisions will be found in KISCH, Bruno. *Scales & Weights* Appendix 2, p.224 et seq. New Haven & London: Yale University Press 1965.
3. HOOVER, Herbert Clark and HOOVER, Lou Henry (eds. & trs.) AGRICOLA, Georgius, *De Re Metallica* p.253 et seq. New York: Dover Publications Inc. 1950.
4. This opinion I find to have been held by Philip Grierson from his Presidential Address, Weight and Coinage *Numismatic Chronicle* (1964) Vol.IV, p.ix and from his comments respecting a lead weight, *Libra* (1963) Vol.II, p.16, Eastbourne.
5. (a) URDANG, George (ed.) *Pharmacopoeia Londinensis of 1618* Madison, Wisconsin: State Historical Society of Wisconsin 1944. Facsimile edition. (see Introduction.)
   (b) WELBORN, Mary Catherine. The De Ponderibus et Mensuris of Dino di Garbo *Isis* (1935) Vol.XXIV, pp.15-36. Brussels.
6. HULTSCH, Friedrich Otto. Loc. cit. Vol. II pp.128-9 and 130-1.
7. HALL, Hubert and NICHOLAS, Frieda J. *Selected Tracts and Table Books Relating to English Weights and Measures*. Camden Miscellany Vol.XV, p.47. London: Camden Society 1929.
8. CHANEY, H. J. *Our Weights and Measures* p.19. London: Eyre & Spottiswoode 1897.
9. Report of the Lords' Select Committee Session 1823 (brought from the Lords) on petition from Glasgow, relating to the Weighs and Measures Bill. *Parl. Papers,* 1824 (HC94) vii, p.431. See p.435.
10. SKINNER, F. G. *Weights and Measures* p.86. London: Science Museum HMSO 1967.

11. ROBERTSON, E. William. *Historical Essays* p.20. Edinburgh: Edmonston & Douglas 1872.
12. Third Report of Commissioners appointed to inquire into the Condition of the Exchequer (now Board of Trade) Standards — On the Abolition of Troy Weight, *Parl. Papers* 1870 (C30) xxvii, p.81.
13. ZUPKO, Ronald Edward. *British Weights and Measures* p.29. Madison, Wisconsin: University of Wisconsin Press 1977.
14. RIDGEWAY, W. *The Origin of Metallic Currency and Weight Standards* p.384 et seq. Cambridge: C.U.P. 1892.
15. I am indebted to Ms Susan Wheeler for an extract from E. C. Streeter's inventory relating to this weight and for providing the photograph reproduced in the text.
16. SMITH, Reginald A. Early Anglo-Saxon Weights. *Antiquaries Journal* (1923) Vol.III, pp.122-9. Oxford.
17. I am indebted to Miss Susan Read, Archaeologist, Reading Museum, for this information and for reweighing all ten weights of the Silchester collection.
18. WHITELOCK, Dorothy (ed.) *English Historical Documents c 500-1042* Vol.I, No.130, p.548-50. London: Eyre & Spottiswoode 1955. This has been the subject of comment by Stewart Lyon, Historical Problems of Anglo-Saxon Coinage (3) Denominations & Weights *British Numismatic Journal* (1969) Vol.XXXVI, pp.204-222. See p.209.
19. ATTENBOROUGH, F.L. *The Laws of the Earliest English Kings.* Cambridge: C.U.P. 1922. See p.161 Aethelstan IV 6 2,
 'An ox shall be valued at a mancus', and p.165 Aethelstan IV 8 5 '. . . shall forfeit 30 pence or an ox'.
20. This question has been discussed by HARVEY, S., Royal Revenue and Domesday Terminology, *Economic History Review* (1967) Second Series Vol.XX, No.2, pp.221-8. London.
21. ROBERTSON, A. J. *The Laws of the Kings of England from Edmund to Henry I* p.79. Cambridge: C.U.P. 1925.
 IV Aethelred 9 2 '. . . the pound contains 15 ores'.
22. CHADWICK, H. Munro. *Studies in Anglo-Saxon Institutions* p.25. Cambridge: C.U.P. 1905.
23. LENNARD, Reginald Vivian. *Rural England 1086-1135* p.373. Oxford: Clarendon Press 1959.
24. See HARVEY, S. Loc. cit., for a good account of this problem.
25. FORBES, J. S. and DALLADAY, D. B. Composition of English Silver Coins (870-1300) *British Numismatic Journal* (1960) Vol.XXXI, pp.82-7. London.
26. CARSON, R. A. G. (ed.) *Mints, Dies and Currency, Essays dedicated to the memory of Albert Baldwin;* No. X LYON, C. S. S. *Variations in Currency in Late Anglo Saxon England* London: Methuen 1971.
27. GRIERSON, Philip. Presidential Address. Coin Wear and the Frequency Table *Numismatic Chronicle* (1963) Vol.III, pp.i-xvi.
28. HARVEY, S. Loc. cit. p.221.
29. Ibid p.224.
30. MADOX, Thomas. *The History and Antiquities of the Exchequer* Ch.IX, p.188. London: John Matthews 1711.

31. LARKING, L. B. (ed. & tr.) *The Domesday Book of Kent*, p.93 (line 14) and p.8. London: J. Toovey 1869.
32. JOHNSON, Charles (ed. & tr.); Richard (FITZNEALE), *Dialogus de Scaccario* p.70. London & New York: T. Nelson 1950.
33. PEGOLOTTI, Francesco Balducci. *La Practica della Mercatura* EVANS, Allan (ed.) p.255. Cambridge, Mass.: Mediaeval Academy of America Pub. No.24 1936.
34. Assisa Panis et Cervisie; *Statutes* Vol.I, pp.199-200.
35. RUFFHEAD, Owen (ed.) *Statutes at Large* 1786.
36. Tractatus de Ponderibus et Mensuris; *Statutes* Vol.I, pp.204-5. See also: HALL, Hubert and NICHOLAS, Frieda J. Loc. cit. p.9.
37. RICHARDSON, H. G. and SAYLES, G. O. (eds. & trs.) *Fleta* Vol.II, Bk.II, Ch.12, p.119. Selden Society Publication Vol.72 of 1953. London: The Selden Society 1955.
38. HALL, Hubert and NICHOLAS, Frieda J. Loc. cit. pp.11-12.
39. ELLIS, Sir Henry (ed.) *Registrum Vulgariter Nuncupatum: The Record of Caernarvon* p.242. London: The Record Commission 1838.
40. DOLLEY, R. H. M. (ed.) *Anglo Saxon Coins* Section XV. GRIERSON, Philip *Sterling*. London: Methuen 1961.
41. See for example:
    BERRIMAN, A. E. *Historical Metrology* p.22. London: J. M. Dent & Sons Ltd. 1953.
    ZUPKO, Ronald Edward. *British Weights and Measures* p.11. Madison, Wisconsin: University of Wisconsin Press 1977.
42. Assisa (Tractatus) de Ponderibus et Mensuris; *Statutes* Vol.I, p.204.
43. See for example:
    (a) MILLER, W. H. On the Construction of the New Imperial Standard Pound and its Copies of Platinum . . . *Phil. Trans. Roy. Soc.* (1856) Vol.CXLVI, p.755. London.
    (b) CHISHOLM, H. W. *Seventh Annual Report of the Warden of the Standards* p.17. London: HMSO 1873.
    (c) ZUPKO, Ronald Edward. Loc. cit. p.20.
44. See for example:
    (a) Report from the Committee appointed to inquire into the Original Standards of Weights and Measures (The Carysfort Committee) 26 May 1758 *Reports from Committees of the House of Commons* Vol.II (1737-65) p.420.
    (b) BERRIMAN, A.E. *Historical Metrology* p.22. London: J. M. Dent & Sons Ltd. 1953.
    (c) GRIERSON, Philip. Presidential Address. Weight and Coinage *Numismatic Chronicle* (1964) Vol.IV, p.iv, Note 1.
45. HALL, Hubert and NICHOLAS, Frieda J. Loc. cit. p.11.
46. RICHARDSON, H. G. and SAYLES, G. O. Loc. cit. p.119.
47. HALL, Hubert and NICHOLAS, Frieda J. Loc. cit. p.9 (thirteenth century) Ms Reg. 9A II f 170^{b}.
48. Ibid p.8 (fourteenth century) Ms Add. 24059 f 15 14C.
49. Ibid p.13 (fifteenth century) Ms Cotton Vesp. E IX.

50. CHISHOLM, H. W. Loc. cit. p.27; Harl Mss. Vol.698 f 64-5 (AD 1496).
51. HARRIS, Mary Donner (ed.) *The Coventry Leet Book* p.396. London: E.E.T.S. 1907-13.
52. See Reference 44(a).
53. Statute of Stamford 3 Edward III 1309. p.270; 8/9 Edward III Stat 1. c 1 1335. p.337; 27 Edward III c 10 1353; *Statutes* Vol.I, p.154.
54. LETTER BOOK, F. City of London Records f 106.
    KINGDON, J. A. *The Strife of the Scales* p.9. London: Rixon & Arnold 1905.
55. For example GRIERSON, Philip Ref. 44(c) p.iv.
    The idea of a 16-ounce pound goes at least as far back as the Tractatus in which we read at its beginning:
    'A sack of wool ought to weigh twenty-eight stone, that is three hundred and fifty pounds and in some parts thirty stone, that is three hundred and seventy-five pounds and they are the same according to the greater or lesser pound . . .'.
    The ratio 350/375 is almost exactly 15/16. The sack is therefore 350 heavier pounds or, what is the same thing, 375 lighter pounds. The thinking is clearly of 350 lb each of 16 oz and 375 lb each of 15 oz, for the two are the same number of ounces to within half of 1 per cent. These latter pounds are *librae mercatoriae*. We find the same concept of a pound of 16 Troy ounces occurring in the Norwich Assembly Roll for 1422 (see HUDSON, Rev. William and TINGEY, John Cottingham, *The Records of the City of Norwich* Vol. 2, Norwich & London: Jarrold & Son 1906, which gives us: 'It is ordained that the chandlers being sellers of tallow candles for the future shall sell a pound of candles by the old weight, viz by 4 marks of Troy weight and in no other way under penalty of losing those candles which are not of that weight and if anyone shall do contrary of this ordinance he shall lose his freedom'.)
    Four marks are equal to 32 Troy ounces, that is two 16-ounce pounds.
    Other fifteenth-century references have been given above (p.127) but the error was not continued in the next century; rather a different error emerges which is found in the declaration of Elizabeth's jury of 1574 where we are told (CHISHOLM, H. W. Loc. cit. p.19): 'and the lb weight of avoir de poiz weight dothe consiste of fiftene ounc troie . . .'. This lies nearer the truth but is still incorrect. The Avoirdupois pound of 7000 grains is in fact only 14.583 Troy ounces. Fifteen Troy ounces is the *libra mercatoria*. See this chapter p.125 et seq.
56. CHISHOLM, H. W. Loc. cit. p.19.
57. 24 Henry VIII c 3 1532; *Statutes* Vol.III, p.420.
58. WILDE, Edith E. Weights and Measures of the City of Winchester. *Papers and Proceedings of the Hampshire Field Club* (1931) Vol.X, Pt.3, pp.237-248. Winchester.
59. WILDE, Edith E. Loc. cit. p.239.
60. 31 Edward III Stat 1, s 2, 1357; *Statutes* Vol.I, p.350.
61. HALL, Hubert and NICHOLAS, Frieda J. Loc. cit. p.8.
62. BATESON, Mary (ed.) *Records of the Borough of Leicester* p.191. London & Cambridge: C. J. Clay & Sons 1899.
63. See for example:

(a) O'KEEFE, John Alfred *The Laws of Weights and Measures* 2nd ed. Sec. 5, App.1, pp.3-4. London: Butterworth 1978.
(b) SKINNER, F. G. The English Yard and Pound Weight *Bull. Brit. Soc. for the History of Science* (1952) Vol.I, p.184. London.
(c) SKINNER, F. G. *Weights and Measures* p.96. London: HMSO 1967.
(d) BERRIMAN, A. E. Loc. cit. p.6.
(e) MACHABEY, Armand. *La Mètrologie dans les Musées de Province* Paris: Revue de Mètrologie Pratique et Legale, and the Centre Nationale de la Recherche Scientifique (1962), pp.375, 384, 388, and 389.
(f) ZUPKO, Ronald Edward. Loc. cit. pp.25-27.

64. SKINNER, F. G. Ref. 63(b) pp.184-5.
65. DENT, Major Herbert C. *Old English Bronze Wool Weights* p.10. Norwich: H. W. Hunt 1927.
66. GREAVES, J. *A Discourse on the Romane Foot and Denarius* p.121. London: William Lee 1647.
67. KISCH, Bruno. *Scales and Weights* p.228. New Haven & London: Yale University Press 1965.
68. MACHABEY, Armand. Loc. cit.
69. BERRIMAN, A. E. Loc. cit. p.6.
ZUPKO, Ronald Edward Loc. cit. pp.25 and 27.
70. ROBERTSON, A. J. Loc. cit. p.29.
71. DUEMMLER, Ernest (ed.) *Monumenta Germaniae Historica* Epistolarum, Tomus IV Karolini Aevi II Berlin: Weidmannschen Verlagsbuchhandlung 1895. Alcuini Epistola 100 (pp.144-6).
72. 3 Edward IV cap 1, 1463; *Statutes of the Realm* II p.392.
73. KINGDON, J. A. Loc. cit. p.53.
74. COWELL, J. *The Interpreter* Cambridge: John Legate 1607.
75. See:
ZUPKO, Ronald Edward. *A Dictionary of English Weights and Measures* Madison, Wisconsin 1968, p.6, s.v. 'auncel'.
For woodcuts see:
SALZMAN, L. F. *English Trade in the Middle Ages* Oxford: Clarendon Press 1931.
76. For this paragraph see:
KINGDON, J. A. Loc. cit. pp.26-8.
77. Ibid p.75. This steelyard appears to have been constructed on the scale 364 lb = 1 sack.
78. STRACHEY, J. (ed.) *Rotuli Parliamentorum* Vol.I, p.47 (AD 1290) London: HMSO 1767-77.
79. STAPLETON, T. (ed.) *De Antiquis Legibus Liber* p.25. London: Camden Society 1846.
80. For this paragraph, see:
KINGDON, J. A. Loc. cit. pp.29 et seq.
81. HALL, Hubert and NICHOLAS, Frieda J. Loc. cit. p.42.
82. PEGOLOTTI, Francesco Balducci Loc. cit. pp.254-5.
83. PELHAM, R. A. Exportation of Wool from Winchelsea and Pevensey in 1288-9. *Sussex Notes and Queries* (1935) Vol.V (August), No.7, pp.205-6, Lewes.
84. STRACHEY, J. (ed.) Loc. cit. Vol.II p.133 Item 57 (1341).

85. Ibid p.142, Item 53, 1343.
86. Ibid pp.239-240, 1351.
87. 25 Edward III Stat 5 c 9 1351/2; *Statutes* Vol.I, p.321.
88. 27 Edward III Stat 2 c 10 1353.
89. 34 Edward III c 5 1360; Ibid p.365.
90. STRACHEY, J. (ed.) Loc. cit. Vol.IV, p.381, 9 Henry VI 1430.
91. KINGDON, J. A. Loc. cit. p.50 et seq.
92. 8 Henry VI c 5 1429: *Statutes* Vol.II, p.241.
93. 11 Henry VI c 8 1433: Ibid p.282.
94. STRACHEY, J. (ed.) Loc. cit. Vol.V, p.30, 1439.
95. KINGDON, J. A. Loc. cit. p.59.
96. HALL, Hubert and NICHOLAS, Frieda J. Loc. cit. p.51.
97. PEGOLOTTI, Francesco Balducci. Loc. cit. pp.254-5.
98. HALL, Hubert and NICHOLAS, Frieda J. Loc. cit. p.16.
99. Ibid p.12.
100. Ibid p.23.
101. 49 George III c 63 s 13 1809; *Statutes at Large* Vol.XVII, p.733.
102. See Appendix to the First Report of Commissioners on Weights and Measures 7 July 1819. *Parl. Papers* (HC 565) XI p.307.
103. DENT, Major Herbert C. Bronze Weights of England. *Apollo* (1929) (July), pp.25-33. London.
HUGHES, Bernard. Old English Bronze Wool Weights of the Period Edward IV to George II *The Connoisseur* (1969) (March) pp.153-9. London.
104. *Cambridge Economic History of Europe* Vol.V, p.254. Cambridge: C.U.P. 1941.
105. 18 and 19 Charles II c 4 1666; *Statutes* Vol.V, p.598.
106. 30 Charles II c 3 1678; Ibid p.885.
107. Reference 65 pp.11-12.
108. CLIFFORD, James L. *Dictionary Johnson* p.99. London: Heinemann 1979.

CHAPTER IX

Measures of Capacity

The laws of the Saxon kings tell us that there were measures in use in pre-Conquest England but none has survived, nor is there any description or comparison which would enable us to obtain an unequivocal value for their contents in Saxon times. The measures have names however. There was the *amber* used for Welsh ale, honey, butter, and meal; there was the *mitta* which contained two *ambers;* and there were measures called *sesters* and *coombs*. The Laws of Ine who ruled Wessex 688-725 mention in 70.1 'twelve ambers of Welsh ale',[1] while the land grant of Oswulf dated 804-29[2] says in part '30 ambers of good Welsh ale which are equal to 15 mittas' and a charter of Edward the Elder of 900[3] speaks of 'church-mittas' being, no doubt, a customary measure set according to a local unit in the keeping of a priory, monastery, or cathedral for tithes or charitable gifts. Later on, in the thirteenth century, we are told[4] that an *amber* is four bushels, so a *mitta* would be a quarter, though this relationship cannot be regarded as either precise or founded on a substantial body of evidence.

In Saxon times it is unlikely that the *amber* could contain as much as four bushels, otherwise the donations and land rents in kind of ecclesiastical foundations for both meal and ale would have been enormous. We really have little to go on as to the size of these measures but it has been thought likely that the *amber* was close in volume to the Roman *amphora* of about six gallons.[5]

The *sester* carried forward from Anglo-Saxon times, for it is mentioned on several occasions in the Domesday Book. This unit of capacity varied with the commodity being measured, but for some goods it would appear to have been close to the Roman *sextarius* (about a pint), and for others, for example for honey, the *sextarius* is given as 30 ounces.[6] Again in the thirteenth century this unit is declared to be much larger, for Fleta states that for wine a *sester* is four gallons.[7]

Richard I's Assize of Measures of 1196 gives no guidance[8] (see Ch.VI). His uniform measure for 'grain, beans, and suchlike' is 'a good horseload', (una bona summa equi), sometimes called a *seam,* which later apparently equalled 64 gallons or 8 bushels, that is a quarter[9] (see p.170). A good horseload meant the load a pack horse might be expected to carry. It can hardly be said to be either precise or definitive. The *seam* is an Anglo-Saxon (not Norman) term and was used as a capacity measure and as a weight. The Assize declares that measures shall everywhere be measured *stricken:* 'et haec mensura sit rasa tam in civitatibus et burgis quam extra'. This means that measures are to be filled to the brim and a rod is to be passed over the edges of the measure to flatten and make level the grain contained therein. There is, however, no evidence that so large a measure as the *seam* was constructed in those days but gallons and bushels were made relating to the *seam*. This implies that notionally at least the *seam* was so many bushels, the actual number being given a full century later in the Year Book for the 20th year of Edward I (1292) where we read:[10] '. . . 3 seams of wheat and 1 seam of peas. (This measure is the same as a quarter)', and a quarter is eight bushels (2.909 hectolitres). The seam was still equal to eight bushels in 1820,[11] but in the general Act of 1824 on Weights and Measures (5 George IV c 74) only the quarter is mentioned. The word '*seam*' is not used.

Thus the *seam* of wheat would weigh approximately 400-450 pounds Avoirdupois. It was sometimes regarded as 500 pounds, depending on the commodity being measured.

Though there may not have been a seam measure, the bushels and gallons were quite definite. The standards were not only made but were distributed throughout the kingdom. The Pipe Roll for the 9th Year of Richard I has this entry under Michaelmas (29 September) 1197 for London and Middlesex:[12]

> 'and a purchase to make measures and gallons and iron yards and scales and weights to send to all counties of England, eleven pounds, sixteen shillings and sixpence' (Et in emptione ad faciendum mensuras et galunos et virgas ferreas et trosnos et pondera ad mittendum in singulis comitatibus Anglie XI li et XVI s et VI d),

while under Hampshire for the same date we read: 'and for measures made at Winchester, five shillings' (Et pro mensuris factis aput Winton, vs). While Richard calls for uniform measures for wine, ale, and other liquids, they are to be 'according to their nature'. There is no suggestion of a single volume measure for all liquids, far less for dry measure as well, nor for a single system of weights.

King John's Magna Carta of 15 June 1215 says: 'one measure of wine . . . and one measure of ale and one measure of corn, that is the London quarter . . .' (una mensura vini . . . et una mensura cervisie et una mensura bladi, scilicit (quarterium) London . . .). This is repeated verbatim in the Magna Carta of Henry III 1225 and nearly so in that of Edward I 1297. There is nothing here to suggest a single measure for wine, ale, and corn, yet 25 Edward III Stat 5 c 10. 1351-2 states:[13]

> Whereas it is contained in the Great Charter that one measure shall be throughout England which Charter hath not been well kept and holden in this point in times past . . . it is accorded and asserted that all the measures that is to say bushel, half-bushel, gallon, pottle [half-gallon] and quart throughout England within Franchises and without, shall be according to the Kings standard and the quarter shall contain 8 bushels by the Standard and no more and every measure of corn

shall be stricken without heap saving the Rents and Ferms of the Lords which shall be measured by such measures as they were wont in Times past.

Here the composer of the Statutes has had the expectation of a single unit and he records the fact that a single unit has not been employed in the past. Little was he to know that centuries were to pass before his ideal was achieved.

The names of all the measures mentioned above, from bushel to quart, are French in origin and in all probability are no older than the Conquest.

The Gallon, Bushel, and Quarter

Indeed, there was little information on measures until the mid-thirteenth century when both the Tractatus de Ponderibus et Mensuris,[14]* and the Assize of Bread and Ale,[15]† after stating that 32 grains of wheat make a *sterling* (penny) and 20 pence make an ounce and 12 ounces make a pound, go on: 'and eight pounds make a gallon of wine and eight gallons of wine make a bushel of London which is the eighth part of a quarter'. In this declaration there are several ambiguities in the interpretation. First, is it eight pounds of wheat or eight pounds of wine which generate the volume to be known as a gallon? Secondly, as we have seen in the last section, 32 wheat corns did not make a sterling, so we must ask is the pound to be the Tower pound or the Troy pound, or the *libra mercatoria?* The date does not permit it to be the Avoirdupois pound which would appear to have originated in about 1280–1300 but measures dating from later than this might well be on the Avoirdupois scale of weight. Thirdly, if generated from wheat, to identify the volume of the gallon we must ask what is the bulk density of wheat? Fourthly, of what measure is the 'quarter' a quarter? (This last question will be deferred until we have sorted out the gallon.)

As we proceed into our inquiry as to the origins and development of the English system of capacity measures throughout the centuries, it should be stated now that there exists no body of universally accepted opinion on the topic. In what follows a study will be make of the extant documents, analysing them for the information they contain, correcting gross errors where they appear, and comparing the results of the inquiry with the actual volumes of standard measures that have come down to us. The two, we shall find, mutually support one another even if formal proof is lacking.

On the first point, the various manuscripts of the Tractatus vary a little; for example that printed in 'Statutes of the Realm' is from *Liber Horn* and simply says: 'and 8 lb make a gallon of wine' (Et VIII libre faciunt galonem vini); that printed in 'Statutes at Large'[16] is from Cotton manuscript Claudius D II and it says: 'and 8 pounds of wheat make a gallon' (Et octo libre frumenti faciunt

*For full text see Appendix A (d).
†For full text see Appendix A (a).

galonem). The Record of Caernarvon[17] and Ms Reg. 9A II f 170^b (*c*1300)[18] say the same, adding it is the gallon of wine, 'Et octo libre frumenti faciunt galonem vini'. Further, of the extant gallons, all are too large to represent 8 pounds Tower or Troy *of wine* so there can be little doubt really but that the official gallon measure was intended to be the volume of 8 pounds of wheat. We cannot ignore, however, the possibility that the declaration of the Tractatus was accepted by some as meaning 8 pounds of wine. We shall later revert to this possibility, but meanwhile shall continue to explore on the basis that the gallon was to be generated, in theory at least, from 8 pounds of wheat. But which pound was intended? Bearing in mind what has been said about Tower weight being restricted to mint operations we would not expect to find any extant model of a unit of capacity based on it. However the Exchequer had very close ties with the Mint so before dismissing it out of hand, it might be well to try to find the volume of 8 Tower pounds of wheat and see if there be any vessel of this capacity in the Exchequer or other official repository. The bulk density of wheat is a problem, for the weight of a cubic foot of wheat is at best an imprecise quantity and one which is likely to vary from year to year and also from sample to sample of grain of a given year.

We must therefore resign ourselves to the possibility of finding, after about 1300, two families of gallons, one based on dry measure (8 pounds of wheat) and the other on liquid measure (8 pounds of wine), each with no less than four distinct pound weights possible; Troy, Tower, *libra mercatoria,* and Avoirdupois. The number of possible permutations is no fewer than eight and this number is increased if a dry measure is defined as eight times another, for then the additional implication of *stricken* or heaped measures comes into the picture, as we shall see (p.156).

Norris[19] in 1775 stated that a cubic foot (1728 cubic inches)* of wheat weighed $47\frac{1}{2}$ pounds Avoirdupois,†[20] that is 61.57 Tower pounds. Hence 8 Tower pounds of wheat would occupy a volume of 224.5 in^3. On 25 May 1688 an old standard wine gallon found at the Guildhall, London, was tested by the astronomers Halley (of comet fame) and Flamsteed and found to contain 224 in^3.[21,22] A remeasurement[22] in 1819 gave the volume as 224.4 in^3. More recently a volume of 220 in^3 was found for an old undated copper gallon thought to be of Tudor design. It is now somewhat dented, thus reducing the volume, and it is thought that this is none other than that which was measured in 1688 and 1819.[23] Thus there does seem to be some evidence for the existence of a gallon based on 8 Tower pounds of wheat, although the likelihood of Tower weight being used is slight. We may yet find another explanation of the origin of the 224 in^3 gallon which is more appealing (p.158-9). The excellent agreement of the volume as measured and that

*Hereinafter cubic inches will be denoted 'in^3'.
†This is 59 lb Avoirdupois to the bushel.

calculated from Norris's density must be regarded as fortuitous, for his figure of $47\frac{1}{2}$ pounds Avoirdupois cannot be other than an average value from which there is likely to be a 6 to 8 per cent spread on either side.

The doubt as to whether the Tower or Troy pound was intended was ended in 1496 by the Act 12 Henry VII c 5[24] where we are told explicitly

> That the measure of a bushel contains VIII gallons of wheat and that every gallon contain VIII pounds of wheat *of Troy weight* and every pound contain 12 ounces of Troy weight . . . according to the old laws of this land.

Although this is not the first mention of Troy weight in the Statute Book, it is the first mention of Troy as a standard weight. Here again wheat is used to determine the volume but the pound is the Troy pound, not the Tower pound. There is further fifteenth-century evidence that at that time the gallon, in theory at least, was to be raised from 8 pounds of wheat, Troy weight. One such document, after declaring as usual that the penny weighs 32 wheat corns from the midst of the ear, reads[25] '1 pound makes a pint of 7680 wheat grains, [5760 Troy grains] . . . 8 pounds make a gallon, 61440 wheat grains' (Una libra facet 1 pynt 7680 grana . . . 8 libre faciunt 1 galon, 61440 grana). The information given here is crystal clear as to which pound is to be used for generating the capacity units. This is further exemplified in 1474[26] and in 1497[27] in the description of bullion (Troy) weight being used for gold, silver, bread, *and measure*.

Now, accepting again Norris's density, 8 Troy pounds of wheat occupy a volume of 239.5 in^3. This is close, but not very close, to the official measure of 231 in^3 used in the seventeenth century and long before, for the gauging of wine casks with a view to imposing customs and excise duty. The Carysfort Committee* was told in 1758[21] that a gallon measure of that volume had resided 'time out of mind' in the Guildhall, London. This may indeed be the theoretical origin of this unit of volume, but as with the gallon of 224 in^3 we may later find an alternative explanation[28] (p.159 et seq.). It must be said that there is no evidence to show that gallons or any other measures were constructed actually using wheat, but standards were always made according to some older standard residing in the Exchequer. A fine bronze wine pint inscribed 'A Wine Pynte Tryed by John Renolds at the Tower 1641' was in the World of Beer Museum at Tower Bridge, London.[29] It was measured in August 1980 by Mr M. Stevenson and the writer. The mean diameter was 7.291 cm and the mean depth 11.674 cm giving a volume of 487.4 cm^3 or 29.74 in^3. The corresponding gallon would be 237.94 in^3, clearly intended to represent the 231 in^3 gallon.

*A Committee of the House of Commons, under the Chairmanship of Lord Carysfort, appointed to 'Inquire into the Original Standards of Weights and Measures in this Kingdom', which reported to Parliament in 1758 and again in 1759.

Three Guildhall wine gallons were examined in 1819:[22] (1) that of 224 in^3 measured in 1688 as already stated; (2) that of 1733 which was a correct copy of the Exchequer standard but a dent in the vessel had reduced its capacity to 230.0 in^3; and (3) that of 1798 containing 230.8 in^3. Nor should we expect great accuracy in the gauging of measures of capacity for we find the Exchequer standards themselves at variance by several cubic inches when tested.

But we have also the important information contained in a manuscript of 1497.[27] The following extracts are of significance and will be put into modern English and numbered serially for ease of reference:

(1) After reciting the usual 32 wheat grains weigh a *sterling* and so on, it states: 'And 12 ounces make a pound Troy weight for selling gold, bread, and measure with ½ ounce which weight makes a pint of wheat and 2 pints make a quart and 2 quarts make a pottle and 2 pottles make a gallon and 8 gallons make a bushel of wheat . . .'.
(2) 'The same time ordains that 16 oz Troy make the Avoirdupois pound for to buy and sell spice with . . .'
(3) 'Twelve ounces and a half contain a pint of measure for wheat and wine . . .'.
(4) 'Two pottles make a gallon for wine and ale which contains 100 oz Troy'.
(5) 'Two half bushels make a bushel which contains in weight 50 pounds which is in measure 64 pints . . . 800 oz to the half hundred which contains XVIC pounds in gold. And the quarter of wheat contains in weight 4C weight'.

Upon these extracts the following comments can be made:

(1) Here we have a statement that Troy weight is the basis of units of measure (capacity) but that the pint does not contain 1 Troy pound of wheat but 12½ oz of wheat. This is contrary to previous declarations.
(2) The Avoirdupois pound never did contain 16 oz Troy (see Ch.VII).
(3) This repeats the 12½ oz pint.
(4) A gallon of 8 pints of wheat would weigh 100 oz with the pint at 12½ oz, but the ounces are Avoirdupois; see (5) next.
(5) Here we are told the bushel weighs 50 lb which amount to 800 oz and then we are told this contains 'XVIC pounds in gold'. The interpretation of this is as follows: 'The bushel of wheat weighs 50 lb or 800 oz, both Avoirdupois (i.e. 16 oz pounds) which weigh X^{VI} lb in gold, i.e. 60 lb of gold.' (There has been a scribal error in giving the number of pounds of gold for in XVIC the C is superfluous and the figure should read X^{VI}, that is sixty.) Fifty pounds Avoirdupois are exactly 60.76 Troy pounds, that is bullion weight. A certain rounding off has occurred to the figure sixty but the discrepancy is only a little over 1 per cent — little enough when it comes to capacity measures.

The quarter, we are told, contains four hundredweights (cwt). At eight fifty-pound bushels to the quarter, we arrive at 400 pounds. This is an example of the original 100 pound cwt (see Ch.VIII), and confirms the interpretation that Avoirdupois weight is intended for this description of the weight of the bushel. Troy weight was never reckoned in hundredweights; that was reserved for bulky goods or 'haverdepois'. Until displaced by metric weight, when using Avoirdupois measure, we spoke of a pound, a half pound, a quarter pound, seldom if ever in ounces, whereas Troy weight is almost always given in ounces. The Exchequer Troy weights of Elizabeth I were marked in ounces not in pounds to avoid the possibility of confusion with the Avoirdupois standards. The pound of the market-place in the fifteenth century was the Avoirdupois pound but the only ounce known was the Troy ounce, hence the confusion.

There is of course some mix-up in the document between extracts (1), (3), and (4) on the one hand and (5) on the other. The document declares the bushel to be 800 Troy oz. and at the same time '60 lb of gold', that is 60 Troy lb or 720 Troy oz. The corresponding gallons would weigh in terms of wheat 100 and 90 Troy oz. respectively and their volumes would be 249.5 and 224.5 in^3 respectively, and both of these have appeared as measures. The latter is just the volume of the old wine gallon in the Guildhall, and there is evidence of the production of standards of 250 in^3, for in the Combination Room of the University of Cambridge, together with other weights and measures of the University, is a wine measure inscribed 'A Wine Pottle Tryed by John Renolds at the Tower. 1641'. Previously in the possession of the Museum of Archaeology, Cambridge, it was loaned in 1939 to the Science Museum, London. Its capacity was then determined to be 125.0 in^3 from its dimensions. The equivalent gallon would therefore be 250 in^3, only 0.5 in^3 in excess of that computed from the document of 1497 (p.154) and only 2.5 in^3 less than that of $1\frac{1}{8}$ times the ancient gallon of 224 in^3. (see p.152) The evidence therefore leads to the belief that there were gallons of 224, 231 (or 239.5), and 250 in^3 in use in the Middle Ages. The old Guildhall gallon of 224 in^3 has already been mentioned as an artifact. We now turn to the oldest Exchequer capacity measures that have come down to us, namely the bushel and gallon of Henry VII (dated one year after the initiating Act of 1496) which are now located in the Science Museum, London.

The Measures of Henry VII

Fully expecting to find the gallon to be one or other of the three mentioned above it is with surprise that we find Henry VII's gallon as measured in 1931-2 to be 268.43 in^3 and his bushel 2144.81 in^3. The bushel we see is close to eight times the gallon (that is 2146.72 in^3) (Fig.34).

Reverting to Norris[19] again, a cubic foot of wheat is 47.5 lb. Avoirdupois which is 57.73 lb Troy. Wheat to fill the bushel would therefore weigh 71.66 lb Troy. This was intended presumably to be 72 lb Troy, so at 8 lb Troy to the gallon, using the traditional measure of the official gallon, Henry's bushel contained not 8 but 9 gallons. Likewise for the gallon, our calculation from the Act of 1496 yielded a gallon of 239.5 in^3 (p.153). Increase this by one-eighth and we have 269.4 in^3, which is very close indeed to the measured volume of Henry's gallon, being out only 1 in^3. We have already (p.155) observed that the 250 in^3 gallon is similarly related to that of 224 in^3.

How could this come about? The clue is to be found in that part of the Statute 25 Edward III Stat. 5, c 10, of 1351-2[13] quoted at the beginning of this chapter, where it is required that 'every measure of corn shall be stricken, without heap, saving the Rents and Ferms of the Lords which shall be measured by such measures as they were wont in times past'. The easiest way of getting a little more was to heap the bushel of grain and not to *strike* it or level it off at the brim with a rounded wooden stick. A modest amount of heaping would enable eight heaped to measure nine stricken (legal) bushels and for this to be accepted as a quarter, or eight heaped gallons measuring nine stricken gallons for a bushel. The idea of heaping the measure had endured for centuries in spite of innumerable statutes prohibiting the practice. The statutes themselves did not help for, as we have seen, the rents and farms of the lords were to be as they had been in times past, and not according to the Statute. Sometimes it was the lords' privileges that were to be protected, sometimes the king's, sometimes those of the two universities (Oxford and Cambridge), sometimes those of the Duchies of Lancaster or Cornwall, sometimes those of the burgesses of a given city to whom a charter of right had been given, sometimes those of a livery company of London.[30] Little wonder the standardization of weights and measures took centuries to be accepted by all.

The Statute for Bakers and Brewers (Statutum de Pistoribus)[31], probably of the reign of Henry III or Edward I, said that no grain was to be sold by heap (or *cantle*) except oats, malt, and meal. These exceptions wrecked any hope of uniform practice in the market-place. The custom grew and developed of expecting (and receiving) nine stricken gallons for each bushel, and of these large bushels eight heaped or nine stricken for the quarter. The law was obeyed to the extent that the measure was stricken but nine measures were taken, and not eight as the law provided. Not only was this the practice in the market-place. It appeared in writing long before the days of Henry VII. An extract from the leet roll of the city of Norwich for 21 Edward I, 1293 reads:[32] 'Presentments of Wymer and Westwyk, of Robert de Wymundham, leyner, because he refused the Lord King's measure when straked [stricken] and will not take it unless heaped up' (no penalty is recorded). The Customal of Sandwich in Kent,[33] written in 1301, states: 'The measure is computed by the pound in weight; thus a gallon contains 8 lbs of wheat; 8 gallons and 8 lbs make a bushel and 8 bushels make a London quarter'.

Here eight gallons plus eight pounds, that is nine gallons, make the accepted bushel while retaining the primary definition of the gallon as the volume of 8 lb of wheat. The gallon of Henry VII contained 9 lb of wheat with the bushel eight of these gallons. Thus the bushel was to be either 8 × 9 or 9 × 8 lb Troy of wheat and 72 lb Troy are 59.25 lb Avoirdupois, which is only 3 lb over $1\frac{1}{8}$ times the 50 lb Avoirdupois of the document of 1497 (p.154).

Nearly a century after the Customal was written the practice was continuing, for we read in the Act 15 Richard II c 4 of 1391:[34]

> Whereas it is ordained by divers statutes that one measure of corn, wine and ale should be throughout the Realm and that 8 bushels striked [stricken] make the quarter of corn; nevertheless, because that no Pain [penalty] is thereupon ordained in the said statutes, divers People of divers Cities, Boroughs, Towns, and Markets will not take neither buy in the said cities . . . ne [nor] in any other place, but [at] 9 bushels for the quarter . . . It is ordained and assented that the said statutes shall be firmly kept and holden, as well in the City of London . . . and that none from henceforth do buy in the City of London, or any other place, any manner of corn or malt but after 8 bushels to the quarter. . . .

A further century later matters were still the same, for while Henry's Act of 1496 said one thing, Henry's bronze measures of 1497 said another. In making his measures, the custom and practice of the market-place was more important and relevant to commerce than what was contained in the statutes. The measures were made each one-eighth larger than those provided by law so that the merchants could receive eight stricken gallon measures for a bushel and at the same time receive the time-honoured measure of nine. But the merchants and the King's Purveyors had exceeded even this.

In the Act I Henry V c 10 of 1413[35] it is stated that in times past the measure of corn was to be eight bushels to the quarter, one bushel of which could be heaped. The King's Purveyors however had for some time been exacting nine bushels for a quarter, not stricken and often with unstamped, uncertified measures. Furthermore, London merchants took nine bushels of wheat to the quarter and called it a *faat,* and they also took, in addition to the faat, another bushel, making ten, and likewise for a quarter of oats they took ten bushels. The Act goes on to order that the old ordinances are to be kept by all, including the King's Purveyors, and that in future no one was to buy corn except by eight stricken bushels to the quarter on pain of a year's imprisonment, a fine of a hundred shillings to the king, and payment of a like sum to the aggrieved party.

From this we gather that to take one heaped bushel in the eight making a quarter was not then frowned upon. There would be spillage, wastage, and loss here and there so this would ensure the buyer got a fair quarter. The king's agents had gone far beyond that, taking nine heaped bushels of uncertain and therefore (very likely) overly large measure. Nine heaped bushels would approximate to ten stricken. In London, the nine bushel quarter of wheat was so common that it was given a name, the *faat* (spelled variously *fatt, fat,* etc.), and on top of the *faat* they exacted a further bushel, making ten in all. The bronze gallon of Henry VII may have reflected the practice of the market-place but it stopped short of ten measures to the unit in question. If we increase the gallon of the Act of 1496, that is 239.5 in³, by a quarter we get 299.4 in³. This would be the volume of wheat weighing 8 *librae mercatoria* to within 0.4 in³. There is, however, no such measure extant, but if we take 1.25 times the Guildhall gallon of 224 in³ we get 280 in³.

The ale gallon described in the 'Memorial to the Commissioners of the Treasury from the Commissioners of Excise and Hearth Money' dated 15 May 1688[21] had a capacity of 282 in³. The wine gallon was given as 231 in³, and these were the two standard volumes used by excise officers in determining the duties payable on ale and on wine.

Eight pounds Avoirdupois of wheat would occupy 291 in³, larger than any extant gallon measure, so we can set aside any suggestion that a gallon was based on Avoirdupois weight, stricken or heaped.

Thus far we have progressed on the gallon based on pounds of wheat. Do we get any help by interpreting the gallons that have come down to us on the basis of pounds of wine? This was an interpretation opened up by the imprecise wording of the Tractatus (p.151). The following table tells the story:

Pounds of Wine:	*8 lb Tower*	*8 lb Troy*	*8 lb Avoirdupois*	*8 lb librae mercat.*
Vol. of gallon	172.3 in³	183.8 in³	223.4 in³	229.8 in³

As said earlier (p.152) there are no extant gallons of 172 or 183 in³, although we shall have occasion to refer to the latter in Ch.X (h), but we do see that eight Avoirdupois pounds of wine generate a volume very nearly equal to that of the old Guildhall gallon (224 in³) while eight *librae mercatoria* of wine generate a volume only 1.2 in³ short of the wine gallon of 231 in³. It may not be without its significance that these two old wine measures could have been generated from a volume of wine. They were both non-official, as we shall see, the latter being represented in the Exchequer only in the early eighteenth century, the former never. The gallons of 250, 268, and 282 in³ were obtained by heaping, so were based on wheat, not liquid, measure.

To sum up thus far, we find we have evidence for no fewer than five different gallons according to the table which follows:

Volume (in³)	*Volume (litres)*	*Description*
224	3.67	Ancient gallon at Guildhall, London, perhaps the volume 8 Tower pounds of wheat but more likely from 8 Avoirdupois pounds of wine (223.4 in³). No Exchequer standard.

Volume (in^3)	*Volume (litres)*	*Description*
231	3.79	The wine gallon, perhaps derived from eight Troy pounds of wheat, volume 239.46 in^3, but more likely from eight *librae mercatoria* of wine (229.8 in^3). Later the wine gallon of Queen Anne of 1707.
250	4.10	A gallon probably derived from taking nine measures stricken for eight heaped i.e. $1\frac{1}{8} \times 224$ (252 in^3). No Exchequer standard, but standards made and sealed by John Reynolds in the seventeenth century. Alternatively and equivalently, though less likely, it could have been the measure of nine Avoirdupois pounds of wine.
268.43	4.40	The corn gallon of Henry VII; one-eighth of his bushel (2144.8 in^3) generated from nine stricken gallons of wheat each of which was defined as eight Troy pounds of wheat, i.e. $2144.81 \div 8 = 268.1$ in^3. Alternatively and equivalently, $\frac{9}{8}$ of the volume of eight Troy pounds of wheat as given above = 269.4 in^3 (i.e. $1\frac{1}{8} \times 239.46$ in^3). Later 272 or $272\frac{1}{4}$ in^3, called Winchester measure (see below)
282	4.62	The ale gallon, probably derived from taking ten measures for eight, i.e. 1.25×224 in^3 (280 in^3) or $1\frac{1}{8} \times 250 = 281\frac{1}{4}$ in^3. The Science Museum, London, has an ale quart of 1689 of volume 71.089 in^3, and one of 1709 of $70\frac{1}{2}$ in^3. The corresponding gallons would be 284.4 in^3 and 282 in^3 respectively. The volume of the Science Museum's Elizabethan (1601) ale quart is 70.22 in^3 giving a gallon of 280.88 in^3 while the Museum of London's ale gallon (1601) contains 288.75 in^3.

The Measures of Elizabeth I

Elizabeth generated her Exchequer standards of capacity from those of Henry VII. They were issued in 1601-2. (See Fig.35) For comparison the measures of Henry and of Elizabeth are set down in the accompanying table, giving the results of the remeasurement carried out 1931-2.

Exchequer Standards

	bushel	gallon	quart	pint*
Henry VII (1496)	2144.81 in³	268.43 in³	—	—
Elizabeth I (1601)	2148.28	270.59 268.97	70.22	34.46 (of 1602)

Two gallons of Elizabeth were found in the Exchequer. The larger is stamped 'ER' with a crown, the smaller 'ER ER' with two crowns. Both are of 1601. Eight times the capacity of the smaller gallon closely matches the bushel, being only 3.5 in^3 in excess, while the larger would generate a bushel some 16.5 in^3 in excess. Strangely enough, the Carysfort Committee of 1758 were presented by Mr Farley, one of the Deputy Chamberlains of the Exchequer, with the larger gallon as the primary standard, and not the smaller which fitted the bushel a little better. Doubtless the selection had been made many years before, at the very beginning of the seventeenth century, and the 270.59 in^3 gallon is believed to have been used as the standard for dry measure (corn). It is close to the capacity of the Winchester gallon 272-272.25 in^3 long used for grain (see p.164-5).

The two smaller gallons were described by the Carysfort Committee[36] as containing 'less by two spoonfuls' than the standard (of 270.59 in^3). The spoonful of water supposedly weighed 3 *drachms* (see Ch.X (g)) so two would weigh $\frac{3}{4}$ oz Troy and have a volume of $1\frac{1}{8}$ in^3. The two gallons differed from the standard of 270.59 in^3 by 2.16 and 1.62 in^3 respectively, the average being 1.89 in^3 which is reasonably near 1.125 in^3 having regard to the imprecision in the size of the spoon used.

Four of Elizabeth's quarts would give a gallon of 280.9 in^3, and eight of these gallons a bushel of 2247.2 in^3, considerably larger than any of the extant measures of her day in the Exchequer. The quart is somewhat anomalous, being considerably greater relatively than that of any other Elizabethan standard. Indeed this would seem to be the quart of the ale gallon and not intended for use as a dry measure. We are told that in Elizabeth's Exchequer there was also a gallon of 282 in^3, used for ale.[37] No such gallon appears to be extant at the present day but we certainly have Elizabeth's Exchequer (ale) quart of 70.22 in^3 (Science Museum, London) and the Museum of London has an ale gallon of 1601. Its capacity is 288.75 in^3. Winchester also has an Elizabethan ale quart and pint of capacities 70.98 and 35.315 in^3 respectively. Derby's 1601 pint and its leather case are shown in Figure 36.

The pint dated 1602 of 34.46 in^3 would generate a gallon of 275.68 in^3 and a bushel of 2205.44 in^3, again in excess but not to the same extent as with the

*The pint of 1601 in Figure 35 (now in the Jewel Tower, London) was not measured in 1931-2 and its present condition is unlikely to yield a reliable measurement of its cubic capacity.

quart. The corresponding quart would be 68.92 in^3, smaller than the standard quart vessel by more than a cubic inch. It is quite common to find that the liquid measure generated from a number of smaller measures is larger than the actual artifact representing the larger unit. For example, eight measures of one pint filled are usually greater than the contents of the gallon measure, the idea being that in filling a larger vessel from a smaller one some allowance must be made for spillage.[38] The vessels ranging in size from pint to gallon were primarily for the measurement of liquids. The measures of the gallon and bushel were dry measures for grains, seeds, etc. One does not encounter a bushel of any liquid, but seedsmen of the twentieth century still sell small amounts of beans, peas, etc. to gardeners in units of pints and quarts. There is no allowance for spillage between the size of the gallon and the bushel. Presumably it was considered to be only liquids that were spilled.

These divergences among the Elizabethan primary standards of the realm were of great concern to the Carysfort Commission of 1758. Although they were not to know it at the time, these measures were destined to remain the primary standards until 1824. They concluded[39]

> and thus bushels, sized at the Exchequer, will greatly differ, if they are sized by the Standard bushel or by eight times the Standard gallon, or thirty-two times the Standard quart; and yet, as the law now stands each of these different measures must be understood to contain the like quantities, are equally lawful and may be indiscriminately used.

If there was difficulty with the primary standards of the realm it can be imagined that there might be further difficulties with the copies sent to the principal cities. Some of these are listed in the table on this page and page 162. Unless otherwise stated they are of the issue of 1601-2.

Ale Measure[40]

Place	*Measure*	*Capacity (in^3)*	*Equivalent Gallon (in^3)*
City of London	gallon	288.75	288.75
(Guildhall)	quart	70.33	281.32
	pint	34.89	279.12
	half pint (late 17th Century)	17.55	280.8
Bristol	quart	70.76	283.04
Newcastle-upon-Tyne	half pint (1649)	17.77	284.32
	quart (1700)	70.01	280.04
Winchester	quart	70.98	283.92
	pint	35.315	282.52
Hastings	pint (late 17th Century)	35.11	280.88
Bridport	gallon	288.75	288.75

Wine Measures

Newcastle	gallon (1649)	228.88	228.88
	pottle	116.17	232.5
	quart	60.69	242.76
Hastings	quart	58.96	235.82
	pint	29.91	239.29
	half pint	15.28	244.52
Bristol	pottle (*c*1680)	116.42	232.84
	quart (*c*1680)	59.46	237.84
	pint (*c*1680)	30.09	240.73
	half pint (*c*1680)	15.17	242.78
	quarter pint (*c*1677)	8.20	262.4

From this table one can see that for ale measures the computed gallon is remarkably consistent, averaging 283.04 in^3, or, if the oversize Bridport and London measures are excluded, 281.77 in^3. The greatest deviation from the mean is 2 per cent (Bridport and London) which if excluded, falls to 0.9 per cent, and we see little variation in the computed gallon from the various lower denominations. For wine measure the average is 240.01 in^3, or, if the oversize Bristol quarter pint is excluded, 237.8 in^3. Here the greatest variation from the mean is 9.3 per cent (Bristol quarter pint). If this is excluded it is 3.8 per cent, very much greater than the spread in the ale measures for the same cities. It will be noticed that for the two cities having *pottle* measures, it is these for which the computed gallon is closest to the measure employed (231 in^3) by the Excise authorities[41] and which later became legalized by statute in 1707 (see next section). Further, for the wine measures of the three cities given there is a notable reduction in volume of the computed gallon as we proceed upwards from the smaller measures. This does not appear in the ale measures. It remains a mystery why there is only spillage requiring correction by generating a larger volume from the smaller measures of wine, and not for ale, although it must be admitted that the alcohol content is greater for wine and the hand probably less steady. Moreover beer was of less value than wine.

The Ale and Wine Gallon

The Carysfort Committee of 1758 endeavoured to find out why the ale gallon was 282 in^3 while the wine gallon was 231 in^3 and applied to the Commissioners of Excise for an answer. The Commissioners responded by sending copies of their records, specifically a memorial dated 15 May 1688 which said that all beer and ale had been gauged at 282 in^3 for a gallon and other excisable liquors at 231 in^3. The memorial further stated that the Commissioners had been advised that the true standard wine gallon contained only 224 in^3, whereupon a measurement of the Exchequer standards had been undertaken the same month (May 1688). Three standard gallons were found, each declared to be 272 in^3 in volume.[21] (These are doubtless the three gallons given in the table on p.160 ranging from 268.43 to 270.59 in^3 when measured using modern techniques.) No gallon of 231 in^3 was in evidence, so reference was made to the Guildhall where a gallon was discovered and measured, yielding 224 in^3. This led to an order being prepared to designate this as the wine gallon but it was not proceeded with for it was never signed, and further, in the opinion of counsel, only Exchequer standards were recognized in law. No opinion could be given in 1688 as to 'from whence the 231 in^3 wine gallon had come' but it was felt unwise to change a unit so long established by precedent. There was no standard at the Exchequer 'but what the King will be vastly a Looser by'.[21] This unsatisfactory situation was not likely to endure for long. The excise duty was too important to play with. A test case was heard in 1700,[42] the defendant being one Thomas Barker, a wine importer. The Crown's case was that Barker had imported more gallons (at 231 in^3) than duty had been paid. Barker's case was based on the size of the gallon at the Exchequer which was the only legal repository.

Evidence showed the ale, corn, and wine gallons employed by the officers of the Crown to be 282, 272, and 231 in^3 respectively. The Crown claimed there was a gallon of 231 in^3 at the Guildhall and Barker claimed there was a quart of 70.5 in^3 yielding a gallon of 282 in^3 at the Exchequer or Treasury and that this was the legal gallon. Neither of these gallon measures was mentioned as being in these locations in 1688, but there is little doubt that measures of these two gallons were available and in use at the time, and indeed there were then measures at the Exchequer, in particular a corn gallon of 272 in^3 which had been used as a wine measure in times past, and an ale quart on which a gallon of 282 in^3 could be based.

The trial of 1700 ended after five hours with the Crown abandoning the case, and with the matter being left to Parliament to resolve by statute, which duly appeared in the form of the Act 5 Anne c 27, s 17 of 1706[43] which declared:

> And to the end that the contents of the wine gallon, whereby the Duties hereby granted are to be levied, may be ascertained and known to all her Majesty's subjects and that all disputes and controversies touching the wine measures, according to which any customs subsidies or other duties are from and after the first Day of March 1707, to be paid or payable to her Majesty . . . that any round vessel (commonly called a cylinder) having an even bottom and being 7 inches in diameter throughout and 6 inches deep from the top of the inside to the bottom, or any vessel containing 231 in^3 and no more shall be deemed and taken to be a lawful wine gallon; and it is hereby declared that 252 gallons consisting each of 231 in^3 shall be deemed a Tun of wine and 126 such gallons shall be deemed Butt or Pipe of wine and 63 such gallons shall be deemed a Hogshead of wine.

This Statute is important on a number of points. It settled and gave legal status to the gallon used by Excise Officers in times past though no one could give the authority for its use. Certainly this gallon had not been defined previously by statute and it was to remain in force as the Exchequer standard for wine till 1824 when a new gallon (that of the twentieth century to the onset of metrication) usable for all liquids and dry goods was enacted, with all others being abolished. A further important consideration is that this wine gallon of Queen Anne was adopted as the standard for liquids in the USA and has so continued to this day.

The stated measurements of the vessel given in the Act of 1706 would yield a volume of 230.9 in^3. Subsequent determinations of the actual measure deposited in the Exchequer in 1707 are as follows:[44]

1758 — 231.2 in^3
1814 — 230.8 in^3
1819 — 230.9 in^3
1871 — 230.93 in^3,

from which it will be apparent that the measure was very carefully made. From 1707 there was no further argument as to the size of the wine measures. A photograph of the Exchequer standard wine gallon of 1707 is shown in Figure 37.

One is forced to ask why it was that there was no Exchequer wine gallon until that of Queen Anne? Considering the long history of the wine gallon already outlined and all of its vicissitudes, one would have expected to find such a measure of the time of Henry VII or at least of Elizabeth, but neither monarch had a national wine standard nor were wine measures issued until 1707. Could it be that the confusion found in 1688 and again in 1700 was well known even then and it was decided that, so long as there was no challenge in the courts to the use of the 231 in^3 gallon for wine duties, it was better to let sleeping dogs lie?

The Winchester Bushel

The bushel and gallon of Henry VII and the Elizabethan bushel of 2148.28 in^3 and gallon of 270.59 in^3 were used for dry goods, seeds, malt, and the like and the name 'Winchester' was associated with these measures (the gallon of 282 in^3, the quart of 70.22 in^3, and the pint of 34.46 in^3 were all ale measures). The first *mention* of 'the Winchester bushel' in the statutes is found in 1670 in 22 Charles II c. 8 sl.[45,46] The Winchester bushel was *defined* by 8 and 9 William III c. 22 s 9 and s 45 of 1696–7,[47] on the occasion of the placing of a duty of 6*d* per bushel on malt:

> And to the end all His Majesties subjects may know the content of the Winchester bushel whereunto this Act refers and that all disputes and differences about measure may be prevented in the future, it is hereby declared that every round

> bushel with a plain and even bottom being eighteen inches and a half wide throughout and eight inches deep shall be esteemed a legal Winchester bushel according to the Standard in His Majesty's Exchequer.

The definition was repeated in 1701.[48] As we shall see, the hope that disputes about measure would cease was forlorn but the intention was good. The City of Winchester standard quart and pint of William III, 1700, is shown in Figure 38.

This bushel so defined would have a capacity of 2150.42 in^3. It is therefore a little larger than that of Elizabeth, but only by 2.14 in^3. The corresponding gallon ought to be 268.8 in^3, but this did not prevent subsequent statutes from stating that the Winchester gallon was 272.25 in^3.[49] This latter figure is close to the gallons of Henry VII and Elizabeth but it is not one-eighth of the bushel of William III. The Act of Union of 1706[50] which joined the parliaments of Scotland and England also provided that the weights and measures of England would now be the weights and measures of Scotland too. There were no doubt adequate grounds for the complaint that this provision could not be implemented as the weights and measures of England were so imperfectly known, but the Commissioners of Weights and Measures in their First Report of 1819[51] demonstrated the widespread use of the Winchester bushel by providing a long list of towns, the vast majority of which were using this measure. As to the difference between corn (Winchester) measure and that for ale, the Commissioners commented that the standard gallon, quart, and pint of Queen Elizabeth had apparently been used indiscriminately for adjusting measures of both corn and ale, the difference between them having arisen by accident, so that now the Excise Department used an ale gallon some 4½ per cent more than the corn gallon of 272 in^3, 'although we do not find any particular Act of Parliament in which this excess is expressly recognized'.[52] This is not surprising if the suggestions made here are correct. As we have seen, the measures themselves did not always match the statutory description and sometimes there would appear to be no documentary authority for the measure at all. Lord Carysfort's 1758 Committee considered recommending the adoption of either 280 or 282 in^3 as the volume of the gallon to be used for all commodities.[53] His second committee set up on 1 December 1758 did recommend on 11 April 1759 that the gallon be 282 in^3 with multiples and subdivisions of the usual proportions[54] but this was not acted upon.

The year 1824 by the Act 5 George IV c 74[55] saw a complete reorganization of British metrology (see Chapter XII). In particular the standard of measure from 1 May 1825 for liquids, as for dry goods not measured by heaped measure (see Water Measure, Ch.X),

> shall be the gallon, containing ten pounds Avoirdupois weight of distilled water weighed in air at the temperature of sixty-two degrees of Fahrenheit's

thermometer the barometer being at thirty inches . . .

The pint was to be one eighth and the quart one quarter of this gallon, two of which made a peck and eight made a bushel and eight bushels made a quarter.

Section 14 of this Act further provided that the capacity of the gallon so defined was to be 277.274 in^3, and a brass measure was to be constructed of this capacity. This was duly accomplished. At the 1931-2 examination of the standards this Exchequer gallon was found to contain 277.421 in^3.

All other gallons and bushels were abolished by the Act of 1824, with this new gallon alone to be used for all purposes. The corresponding bushel of eight new gallons according to the Act would hold 2218.192 in^3. The gallon defined by this Act is, for the first time, called the *Imperial* gallon and has been so designated ever since. This measure is considered further in Chapter XII. The Winchester bushel, though by this Act abolished for purposes of trade, continued to be used for the establishing of corn rents. It is recorded that in 1885 in Lincolnshire the average price of a Winchester bushel of wheat for the previous 21 years was used to determine the corn rents for the following 21 years.[56] Clearly the late nineteenth century was a period of considerable stability.

In the twentieth century liquid chemicals were still being supplied in bottles called 'Winchester quarts'. To further confuse the issue, these cylindrical, narrow-necked bottles did *not* hold a Winchester quart but held *two Imperial* quarts or just over $2\frac{1}{4}$ litres. Since metrication they have been 'standardized' at $2\frac{1}{2}$ l. No one seems to know the origin of this anomalous measure.[57]

After many centuries of confusion over the gallon, bushel, quarter, and peck it was good to have them rigorously defined and universally applied. This happy state of affairs lasted some 150 years before the 1963 Weights and Measures Act discontinued the use of the quarter and decreed that on 31 January 1969 the bushel and peck would cease to be lawful measures for purposes of trade.

NOTES AND REFERENCES: CHAPTER IX

1 .ATTENBOROUGH, F. L. *The Laws of the Earliest English Kings* pp. 59, 192. Cambridge: C.U.P. 1922.
(See also WATSON, Sir Charles M., *British Weights and Measures* p. 12, London: John Murray 1910.
and ROBERTSON, E. William, *Historical Essays* p. 68. Edinburgh: Edmonston & Douglas 1872.)

2. (a) THORPE, Benjamin. *Diplomatarium Anglicum Aevi Saxonici,* p. 460.

London: Macmillan 1865.
(b) HARMER, Florence Elizabeth. *Select English Historical Documents* pp. 1, 39, 73-4. Cambridge: C.U.P. 1914 (Sawyer No. 1188).
3. THORPE, Benjamin. Loc. cit. p. 144. (Sawyer No. 359).
4. HARMER, Florence Elizabeth. Loc. cit. pp. 73-4.
5. Ibid. pp. 79-80.
6. ROBERTSON, E. William. Loc. cit. p. 2, footnote 1.
7. RICHARDSON, H. G. & SAYLES, G. O. (eds.) *Fleta* Ch. 12, p. 120. Selden Society Publication. Vol. 72 of 1953, London: Selden Society 1955.
8. STUBBS, William (ed.) *Chronica Magistri Rogeri de Houedene* Vol. IV, pp. 33-4. London: (Rolls Series) Longman et al 1871.
9. (a) HALL, Hubert & NICHOLAS, Frieda J. *Select Tracts and Table Books relating to English Weights and Measures (1100-1742)* Camden Miscellany Vol. XV, p. 21. London: Camden Society 1929.
(b) Second Report of Commissioners on Weights and Measures 13 July 1820, Appendix A p.31 *Parl. Papers* 1820 (HC 314) vii, p.473.
'Seam or seem, sometimes a quarter of corn or malt'.
10. *Year Books of the Reign of King Edward the First, Years XX & XXI,* HARDWOOD, Alfred J. (ed. & tr.) 20 Edward I 1292 Pt. 1 p. 16. London: HMSO and Longmans 1866.
11. Reference 9 (b).
12. *Pipe Roll, 9 Richard I,* STENTON Doris M. (ed.) Vol. XLVI N.S. VIII pp. 160 & 17. London: Pipe Roll Society 1931.
13. 25 Edward III Stat. 5 c 10 1351-2; *Statutes* Vol. I, p. 321.
14. Ibid. Vol. I, p. 204.
15. Ibid. Vol. I, pp. 199-200.
16. *Statutes at Large* London Edns of 1763, 1769 Vol. 1 pp.148-9.
17. ELLIS, Sir Henry (ed.) *Registrum Vulgariter Nuncupatum - The Record of Caernarvon* p. 242. London: Record Commission 1838.
18. HALL, Hubert and NICHOLAS, Frieda J. Loc. cit. p. 8.
19. NORRIS, H. Ancient Weights and Measures Prior to Henry VII *Phil. Trans. Roy. Soc.* (1775) Vol. LXV, p. 48. London.
20. John Powell in his *Assize of Bread,* London 1684, states that a bushel of wheat weighs 56 pounds Avoirdupois and 12 Troy ounces of wheat makes a pint. A bushel of wheat is given as a half cwt (56 lb) by the Act 14 George III c5 s6 of 1774, and estimated as 57 lb in the Act 29 George III c58 s24 of 1789 yielding 45.0 and 45.8 lb respectively for the cubic foot. With the change to Imperial measure we find the 1921 Corn Sales Act equating a bushel of wheat to 60 lb Avoirdupois. Norris's equivalent would have been 60.97 lb.
O'KEEFE, J. A., *Law of Weights and Measures* 1966 states that native English wheat is now 504 pounds Avoirdupois per quarter. This is 63 lb to the larger Imperial bushel and 49.1 lb to the cubic foot. More recently the supplement to Weekly Bulletin of the Home Grown Cereals Authority, Vol. 14 dated 17 December 1979 gives the specific weight for the average of 288 wheat samples from all parts of the kingdom for the 1979 harvest as 74.4 kg/hl, which is equivalent to 46.44 lb per cubic foot or 59.6 lb to the Imperial bushel. Much depends on the moisture content and on the size and weight of the individual grains. (See Ch. 1).

21. Report from the Committee appointed to inquire into the original Standards of Weights and Measures in this Kingdom (Carysfort's Committee) 26 May 1758. *Reports from Committees of the House of Commons* Vol. II (1737-65) p. 432.
22. First Report of Commissioners appointed to consider the Subject of Weights and Measures 7 July 1819. *Parl. Papers* (HC 565) xi p.307. See particularly Appendix A of this Report.
23. STEVENSON, Maurice. The size of Liquid Measures in the 17th and 18th Centuries. *Libra* (1964) Vol. III, No. 2, p.12 (wrongly attributed in the Index to *Libra* to R. F. Homer). Eastbourne.
24. 12 Henry VII c 5 1496; *Statutes* Vol. II, pp. 637-8.
25. HALL, Hubert & NICHOLAS, Frieda J. Loc. cit. p. 36.
26. HARRIS, Mary Dormer (ed.) *Coventry Leet Book* Vol. 1, p. 396, London & New York: E.E.T.S: K. Paul, Trench, Trubner & Co. Ltd. H. Milford, O.U.P.
27. CHISHOLM, H.W. *Seventh Annual Report of the Warden of the Standards* p. 27 et seq London: HMSO 1873. (Harleian Mss Vol. 698 f 64-5; printed in part). The full document is available in photographic reproduction from the Science Museum, London. Negative No. 283/50. The document is entitled 'The Standard of Weights and Measures in the Exchequer. Anno 12 Henrici Septimi'.
28. See MOODY, B. E. The Origin of the 'Reputed Quart' and Other Measures *Glass Technology* (April 1960) Vol. 1, No. 2, pp. 55-68. Sheffield.
29. Unfortunately, the Beer Museum ('The World of Brewing') was closed on 30 September 1980. The present location of the wine pint of 1641 is not known to this writer.
30. A few examples of Statutes containing these exceptions or exemptions may be given:
    Exemption the Borough of Dorchester - 9 Henry VI c 6 1430-1.
    Exemption to the Counties of Devon and Cornwall and to 'water measure' on board ship - 11 Henry VII c 4 1494.
    Exemption of the County of Lancaster - 13 Richard II c 9 1389.
    Exemption of the Lords of the Manors - 14 Edward III c 12 1340.
    Exclusion of measures sealed by the Fruiterers Co. of London - 1 Anne c 9 s2, 1701. (Given in *Statutes* as 1 Anne c 9, s2.)
    Exemption of the Colleges and Halls of the Universities of Oxford and Cambridge - 11 William III c 15 s1, 1698/9.
31. Statutum de Pistoribus; *Statutes* Vol. 1, pp. 202-4.
32. HUDSON, Revd William & TINGEY, J. C. *The Records of the City of Norwich*. p. 371. Norwich & London: Jarrold & Sons 1966.
33. BOYS, W. *Collections for a History of Sandwich in Kent,* Vol. II, The Customal of Sandwich, p. 542. Canterbury: 1892.
    I am indebted to Mr Maurice Stevenson for directing my attention to this work.
34. 15 Richard II c 4, 1391; *Statutes.* Vol. I, p. 70.
35. 1 Henry V c 10, 1413; Ibid. Vol. II, p. 174.
36. Ref. 21, p. 431.
37. Ref. 27, p. 24.
38. This suggestion originated with Mr Maurice Stevenson.
39. Ref. 21 p. 433.
40. This Table is based on the series of articles by STEVENSON, Maurice, The Size of Liquid Measures in the 17th and 18th Centuries *Libra* (1964) Vol. III, pp. 2, 9, 16.

41. Ibid. p. 17.
42. (a) Ref. 21, pp. 432-3.
    (b) Ref. 27, p. 35.
43. 5 Anne c 27, s17, 1706; *Statutes at Large* Vol. IV, pp. 244-5.
44. CHISHOLM, H. W. Loc. cit. pp. 36-7; see also Ref. 22 Appendix A.
45. *Statutes* Vol. V, p. 662.
46. Note that the years of the reign of Charles II attached to Acts of Parliament date from the execution of his father in 1649. There are, of course, no statutes of either Charles for the years of the Commonwealth. The enactments of Charles II begin with the Restoration of 1660 in the twelfth year of his supposed reign.
47. 8 & 9 William III c 22, s9 and s45 1696-7; *Statutes* Vol. VII, pp. 248 and 256.
48. Ibid. Vol. VII, p. 744.
49. This discrepancy was commented upon in the First Report of Commissioners . . . (Ref. 22).
    See also ZUPKO, Ronald Edward. *A Dictionary of English Weights and Measures*, pp. 70-1. Madison, Wisconsin: University of Wisconsin Press 1968 for further examples of the gallon of $272\frac{1}{4}$in^3.
50. 6 Anne c 11, 1706 Art. 17; *Statutes* Vol. VIII, p. 570.
51. Ref. 22 Appendix B.
52. Ref. 22 p. 311.
53. Ref. 21 p. 433.
54. Report from the Committee appointed upon the 1st Day of December 1758 to inquire into the original standards of Weights and Measures in this Kingdom, 11 April 1759 *Reports from Committees of the House of Commons* Vol. II, (1737-65) p. 456.
55. 5 George IV c 74, 1824; *Statutes at Large* Vol. XXIII, p. 759.
56. O'KEEFE, John Alfred. *The Law of Weights and Measures* p. 10. London: Butterworth 1966.
57. (a) WILDE, Edith E. Weights and Measures of the City of Winchester *Papers and Proceedings of the Hampshire Field Club* (1931) Vol. X, Pt. 3, pp. 237-248. Winchester.
    (b) MOLL, P.D. The Winchester Quart. *The Industrial Chemist* Sept. 1954. London.

CHAPTER X

Particular Measures of Capacity

(a) The Tun of Wine

After this lengthy discussion of the gallon and the bushel, we perhaps need reminding that one of our original questions remains unanswered, that is, of what measure is the 'quarter' a quarter?

In reality there are two answers, one for dry goods and one for liquids as we might expect, for the smaller measures of capacity up to and including the gallon were usually but not exclusively used for liquids while the bushel and quarter were almost invariably used for dry goods such as grain. There could therefore be a conceptual unit of 4 quarters, that is, 32 bushels for grain and another of 4 times 64 gallons for liquids. We will consider dry measure first.

The quarter or *seam* was described in the late twelfth century as 'a good horse-load'. Four quarters would be a sizeable amount of grain. At 56 lb Avoirdupois per bushel* this would amount to 1792 lb. The only large units of capacity we encounter are the *wey* (pp.130, 139, 320) and the *chaldron* (pp.180, 257). Looking at the word *wey* we find its Old Teutonic root to be *weg-* or *wag-*, meaning to move, journey, or carry. This root is common to 'wain', 'weigh', and 'wagon'. Thirty-two bushels would make a wagon-load in the Middle Ages so we can think of the quarter as a quarter of a *wey*. The *chaldron* of coal was also 32 bushels originally so the quarter could be a quarter of a *wey* or a quarter of a *chaldron*. Just as in North America, where the quarter of a dollar is called simply a quarter, so the measure of capacity dropped the explanatory words which were soon lost sight of, only the word 'quarter' remaining. The *wey* soon disappeared from common usage and seldom were grain shipments described other than in quarters and bushels.

For liquids, especially wine, large volumes continued to be used not only as units of account but as actual casks in the importation of this commodity. Throughout history we have been told that eight gallons make a bushel and eight bushels make a quarter, so four quarters would generate the unit sought. This is 256 gallons. The ale barrel, from the fifteenth century (and probably much earlier) until the late seventeenth century, was 32 gallons in capacity[1], that is, just one-eighth of this amount, and this 256 gallons was the original *tun* of wine, etc. At a very early date however four gallons were withdrawn. Thus we read in the Statute Book for 1423[2] (although the change had occurred much earlier) that the *tun*, or ton, of wine was to be 252 gallons, with the pipe

*See Ref. 20 p.167.

126, the tertian 84, and the hogshead 63 gallons respectively. The lost four gallons are referred to in a document of 1517[3] thus: 'And a ton, beyng caske, shulde holde VIII barelles of Ale lacking IIII galons of old tymes'. The Carysfort Committee declared in 1758[4] that the records of the Tower of London for 1347 contained a complaint that the

> Ton of wine did not bear its right measure. And the King is prayed that the Ton might be gauged by the Verge according to the Standard of England; but the King not consenting to this petition they [the Commission members] are unable to explain this ancient Standard of England farther than that it appears by a subsequent Statute of 2 Henry VI c 11* that the ancient standard ought to contain 252 gallons.

The Committee may be excused for not being able to sort out this reference to the fact that the 'ton' was not according to ancient custom. The explanation to be offered now is that the ancient standard was 256 gallons and that in some way, now forgotten, it was reduced to 252 gallons without adequate written record being left. Edward III in 1347 was not likely to wish to lose the duty on four gallons for each tun of wine imported, so we should not be too surprised that the king did not consent. Had he agreed, the petitioners would be the gainers; he would be the loser. As the reduction had already occurred there would be little incentive for him to effect a change. The capacities of the variously sized wine casks remained unchanged for centuries. They are variously referred to as:†

The tun (or ton)	252 (12 score + 12) gallons
The pipe (or butt)	126 (6 score + 6) gallons
The firkin (or puncheon or tertian)	84 (4 score + 4) gallons
The hogshead	63 (3 score + 3) gallons
The tierce	42 (2 score + 2) gallons ($\frac{1}{6}$ tun)[5]
The barrel	$31\frac{1}{2}$ gallons ($\frac{1}{8}$ tun)
The rundlet	18 or $18\frac{1}{2}$ gallons ($\frac{1}{14}$ tun)[6]

Casks of the same capacities were used for oil and honey according to the Act 18 Henry VI c 17, 1439, but what exactly were these capacities if the size of the gallon was problematical? To government officials there was no doubt that the wine gallon was the measure in current use, that is one of 231 in^3, and the casks were gauged accordingly.

Naturally the tun maintained its gallonage when the wine gallon was

*This is the Act of 1423 already referred to, now given as 2 Henry VI c 14 1423 in the Chronological Table of Statutes (C.T.S.).

†The old measures for wine were recited in 2 Henry VI c 11 (now given in C.T.S. as c 14) of 1423: they were the *tun, pipe, firkin,* and *hogshead.* These were confirmed in 1439 (18 Henry VI c 17) and again in 1483 (1 Richard III c 13) which included the *barrel* and *rundlet.* In 1536 (28 Henry VIII c 4) the *tierce* was added.

legalized in 1707 at 231 in^3 by Queen Anne. No change was needed, for this had been the measure used by the Excise for a long time previously. With the advent of the Imperial gallon in 1824, being almost exactly 1.20 times the wine gallon of Queen Anne, the various casks retained their capacities but were rated by Imperial measure, for example the previous hogshead of 63 gallons was used as before but described as 52.5 of the new (Imperial) gallons, and this has continued to the present.[7]

As for the possibility that the quarter defined a unit of weight, it is true that the word was so used, but to denote a quarter *of a hundredweight*, that is 28 lb. As for four quarters of 63 gallons each, it has been taken by some to represent a ton of 2000 lb Avoirdupois weight. Certainly the weight of wine in a cask of 252 gallons of 231 in^3 would be 2084.4 lb Avoirdupois, taking the specific gravity of wine as 0.99. This rough and ready equivalent does not appear to have had solid foundation or wide currency. (See this Chapter section (e).)

(b) The Barrel

Throughout the history of English metrology we find one unit of capacity, the *barrel,* being used for an entire spectrum of commodities ranging from ale and beer to soap and tar. To be packed like herring in a barrel is an every-day expression and we hear of barrels of eels, honey, beef, and oil as well, to mention but a few. The evolution of some of the more important casks is given below.

(i) Ale, Beer, and Wine

The most frequently encountered units of ale and beer in bulk are the *barrel,* the half barrel or *kilderkin,* and the quarter barrel or *firkin,* while those for wine are the *tun,* the *pipe* or *butt* ($\frac{1}{2}$ tun), the *firkin* ($\frac{1}{3}$ tun), the *hogshead* ($\frac{1}{4}$ tun), and the *barrel* ($\frac{1}{8}$ tun). It will be noticed that the two firkins are very different in capacity, the one for malt liquors being a quarter of a barrel while the wine firkin is one third of the very large measure, the tun.

In the fifteenth century beer and ale were provided in barrels of distinctive and different sizes, the beer barrel being 36 gallons, the ale barrel 32 gallons[8] (see Ch.XI). This continued until 1688 when it was enacted[9] that the barrel of beer or ale, strong or small, was to contain 34 gallons according to the Exchequer ale quart (that is 1 gal. = 282 in^3) and in 1699 it was enacted[10] that the barrel of vinegar should likewise be 34 gallons. This situation continued until 1803 when the barrel of ale and beer was raised to 36 gallons.[11] Following the reorganization of all the measures in 1824,[12] the barrel was redefined as 36 Imperial gallons, a reduction of some 1.6 per cent from the previous capacity, at which value it has continued to this day. The original barrel of 32 gallons was $\frac{1}{8}$ of the ancient tun (see this Chapter, section (a)) and for wine the barrel has continued as $\frac{1}{8}$ of the tun, albeit the tun reduced by 4 gallons to 252 gallons. The wine barrel was therefore $31\frac{1}{2}$ gallons. Note that the wine gallon contained 231 in^3 while the ale gallon contained 282 in^3 prior to 1824.

We are repeatedly advised from the fifteenth to the seventeenth century that wine measure was also used for oil and honey,[13] although in 1581 we are told the barrel of honey was to be 32 wine gallons,[14] rather than the $31\frac{1}{2}$ gallons of the wine barrel.

(ii) Herrings, the Barrel, and the Cran

We read in the thirteenth century Tractatus[15] that a *last* of herrings contains ten thousand fish while the Statute Book for 1357 tells us:[16] 'and the Hundred of Herrings shall be accounted by six score and the Last by ten thousand'. In a document of the fifteenth century[17] we are told specifically that the hundred for herring is 5×20, that ten thousand make a last, and, as one cannot pack a thousand herrings into a barrel, there shall be twelve barrels to the last, yet in 1741[18] we read that 'herrings go by 120 to the c [hundred] and 12 hundred to the thousand which makes a barrel and 12 barrels make a last'.

Here we have 1440 herrings to the barrel, whereas our earlier source declared that 1000 would not go into the barrel. This is not totally absurd, for in the three centuries that elapsed between the two writings the size of the fish may well have diminished, so that in later years more fish would be needed to fill the barrel. This has a parallel in the marked reduction in the size of cod caught and offered for sale in this century. As a boy this writer remembers cod several feet in length; now they are of more modest size, no doubt as a result of more intensive fishing as the century progressed.

In the year 1423[19] we are told that the barrel of herrings (and of eels) is to be 30 gallons fully packed,★ while the *butt* of salmon is to be 84 gallons (hence the salmon barrel contained 42 gallons), yet in 1482–3[20] the barrel for herrings is to be 32 gallons while those for salmon and eels are to be unchanged at 42 gallons. But which gallon is intended?

The Statute Book contains a curious enactment of 1571[21] which reads more like a petition than an enactment. It states that for time out of mind, herring fishers had packed in barrels of about 32 *wine* gallons 'according to such usual brasse measure as is out of your honourable Court of Exchequer delivered to your said honourable Citie of London'. Now complaints had arisen that this measure was not being adhered to, 'which they never did', and it was requested that the assize of herring barrels should be about 28 gallons of the Standard, 'well packed and conteyning in every barrel usually a thousand full

★Bishop Fleetwood, *Chronicon Preciosum*, 2nd edition London: T. Osborne 1745, confirms this volume for 1444 (p.86) and speaks of a *cade* of herrings (p.82 et seq). One *cade* (Lat *cadus* = cask) held 720 fish, that is 6 long hundred or a half-barrel of 1440 fish.

herring at least . . . any ancient or former law or statute to the contrary notwithstanding'.

This remarkable declaration was commented upon in the 1758 Report of the Carysfort Committee,[22] who pointed out that the Act of 1423, still in force, had declared for 30 gallons but the Committee made no mention of the enactment of 1482-3 which had declared for 32 and which presumably superseded that of 1423. Be that as it may, the barrel of 1571 was declared to be about 32 wine gallons by Guildhall measure which would then be the gallon of 231 in^3, for a total of 7392 in^3. For such a volume to be deemed 'about 28 standard gallons', the standard gallon would be about 264 in^3 near enough to 268 in^3, the standard corn gallon of Henry VII and not so far away from those of Elizabeth herself (namely 268.97 and 270.59 in^3) as to pass for the measurement of a commodity so heterogeneous in size as whole fish. It cannot be said that any of these Acts are crystal clear, and those which followed were no better, for in about 1600 we were returned to the unit of 1423 when it was declared that the barrel of herrings and of eels ought to be '30 gallons in content, fully packed'.[23] This was repeated in 1615 but the next year we are informed that the barrel of herrings, butter, and soap is the same as that for ale,[24] namely 32 ale gallons, not wine gallons. The 1616 statement was reasserted in 1682[25] but in 1718 by statute[26] we are told the herring barrel is to be 32 gallons English *wine* measure, so we are back on course again. This Statute (Sections 14 and 16) also confirmed the salmon barrel at its 1423 value of 42 gallons, English wine measure, that is a gallon of 231 in^3. In 1721 a further enactment[27] placed a duty of 20*d* on every thousand (that is a barrel) of red herrings, while a separate Act of the same year[28] declared the duty on a barrel of 32 gallons of white herrings to be $\frac{3}{4}$*d*. The Statute 43 George III c 69 of 1803[29] confirmed the herring barrel as 32 gallons for both red and white fish, while the barrel of beef or pork was to be of the same volume, and it also advises us that the barrel of salmon is still at its 1423 value of 42 gallons. The herring barrel was again confirmed in 1808[30] and when we enter the twentieth century we find in 1913[31] that 'no [herring] barrel shall be branded unless it have a capacity of $26\frac{2}{3}$ Imperial gallons'. Now $26\frac{2}{3}$ Imperial gallons are exactly 32 wine gallons of 231 in^3. Clearly from 1824, when the Imperial gallon of 277.274 in^3 displaced all other gallons, the herring barrel continued as before but in a new guise. Remarkably enough the herring barrel would appear to have remained at the standard of 32 wine gallons from 1482-3 to the present, suffering along the way some rather harsh buffeting.

No indication as to the weight of fish in a barrel was given until the Herring Fishery (Scotland) Act of 1815[32] in which it was stated that, if packed with small salt, the weight was to be 235 lb (Section 31); if with large salt, 212 lb.

By this last-mentioned Act another measure, the *cran*, was recognized, though doubtless in use as a customary measure long before. Section 13 empowered the Commissioners for Herring Fishery to determine its contents. This was done by the Commissioners in 1832,[33] the cran basket being defined as 45 gallons of English Wine Measure. It was redefined in Imperial measure in 1852 as $37\frac{1}{2}$ gallons. In fact 45 wine gallons are exactly 37.49 Imperial gallons, so once again the move to Imperial units yielded a reasonable equivalent.

The Commissioners proceeded to give a detailed description of this rather substantial basket built of oak slats three quarters of an inch in thickness and between two and four inches broad, 'bound with six good iron hoops'. There were also half and quarter crans but the half cran fell into disuse while the full cran, although the unit of measure, was too large for practical use. The quarter was the actual measure employed at the herring ports. Very detailed instructions were given in 1898[34] for the construction of the quarter cran in either basket or box form. As a basket it was now normally made of wicker with a convex bottom and a stick was used to level off fish at the brim. As these quarter crans aged the basket stretched and it was not uncommon to find one 15 per cent over measure.[35] Naturally every buyer wanted his fish measured in old baskets. Some idea of this unit can be obtained from the inside measurements of the box form, namely 31 inches long, 7 inches deep, and $14\frac{3}{4}$ inches broad.[34] Even allowing for the corner blocks 7 inches long and $2\frac{1}{2}$ inches square the volume of the quarter cran is substantially in excess of $\frac{45}{4}$ wine gallons, being 3025.75 in^3 instead of 2598 in^3, perhaps to allow somewhat for the space between the fish in the box, or do we have here another example of the volume of a heaped measure being taken to represent the measure itself?

These regulations were further elaborated in 1908 and in this year, by the Cran Measures Act 8 Edward VII c17,[36] the use of the cran and quarter cran was extended to include England and Wales and formally made legal. The half cran was never legalized.

Not being a legal unit at the time, the Weights and Measures Act of 1878[37] makes no mention of the cran (or of the barrel for that matter) but Section 10 of the corresponding Act of 1963[38] provides that nothing in that Act shall affect the legislation authorizing the use of cran measures. This 'saving' clause was necessary, for the schedule listing permissible units of capacity does not include the cran. In fact the cran under this Act is strictly not a lawful measure but by allowing its enabling legislation to stand the cran continued in use. As it did not appear in the Weights and Measures Act the Board of Trade had no standard of this measure.

With the onset of metrication and the entry of Great Britain into the European Economic Community, the industry agreed to adopt a new 'unit', one of 100 kg divided into 4 'boxes' of 25 kg each. It will be noted in conclusion that 100 kg is very close indeed to the weights given in 1815 for the herring barrels with fine or coarse salt, so perhaps once again we have gone full circle.

(iii) Honey

As we have seen, the barrel of honey was to be 31½ wine gallons from the fifteenth to the seventeenth century except for the perturbation of 1581 which raised it to 32 gallons. In all likelihood the 1.5 per cent difference between the 31½ and 32 gallon barrels would not be noticed in practice. This seems to have been one of the more stable barrel volumes for the value of 32 gallons is given in 1820,[39] although it is then given the alternative value of 42 gallons of 12 lb each. It would have been simpler had the barrel been given just as 504 lb of honey, without ever involving a capacity measure.

(iv) Soap and Butter

These commodities have barrels of unusual definition. The barrel of soap was defined in 1531[40] as 32 gallons or more, the empty barrel weighing 26 lb and no more, the half barrel to hold 16 gallons, the container to weigh 13 lb and no more, and the firkin 16 gallons, the container weighing 6½ lb 'on pain of forfeiture for every barrel etc. hereafter made or used contrary to the Statute, 3*s* 4*d*'. The size of the barrel was reduced in 1711[41] when the barrel was redefined as 256 lb of soap net weight and this definition continued into the nineteenth century.[39] The density of ordinary household soap is about 1.023 so 256 lb would occupy a volume of almost exactly 30 wine gallons (231 in^3), a reduction of two gallons from the declaration of 1531.

As for butter, the barrel was never given as so many gallons. Rather, in 1662 an Act[42] stated that from time immemorial the *kilderkin* (half barrel) of butter should weigh 132 lb gross at least, consisting of 112 lb of butter in a barrel weighing 20 lb, the firkin at least 64 lb gross with 56 lb of butter and the barrel weighing 8 lb, and the pot of 20 lb consisting of 14 lb of butter and the pot itself 6 lb.

The barrel of butter, then, consists of 224 lb (2 cwt) of that commodity. With the density of butter being 0.9 we find the volume of the butter to be almost exactly 30 wine gallons, hence the change in the soap barrel in 1711 brought both barrels to the same cubic content, that is 30 wine gallons.

From all this information an anomaly will be seen in the containers themselves for both soap and butter. The containers are not very substantial and the Act of 1531, if followed exactly, would require the smaller casks to be made of wood of decreasing thicknesses, for the weight allowed decreases in exact proportion to its volume.

Imagine a barrel of soap cut evenly in two; both halves contain 16 gallons and the wood weighs 13 lb in each piece, but there is no cover for the sections of the cut. To create a half-barrel fully enclosed needs additional wood and, if the weight is to be fixed in strict proportion, the wood throughout must be thinner. For the quarter-barrel it must be thinner still.

The situation for butter is even worse. The *kilderkin* is allowed a cask of 20 lb but its half, the *firkin*, has to make do with an 8 lb cask. The 20 lb pot does rather better, being allowed a container of 6 lb, almost as much as the 56 lb firkin, but from its name we might conclude that the pot was not a wooden container but a dish of some sort, hence its relatively substantial weight.

(v) Oil

We have seen that from the fifteenth to the seventeenth century the barrel of oil had the same capacity as that for wine ($31\frac{1}{2}$ gallons). This value continued into the nineteenth century. In 1824, with the advent of one gallon for all commodities, it was specifically stated that 'all tuns, pipes, tertians, hogsheads, or other vessels of wine, oil, honey, and other gaugeable liquors . . . shall be gauged according to the standard of capacity for liquids directed by this Act', that is by the gallon of 277.274 in^3.

The barrel of oil of which we hear so much today is in reality an American measure consisting of 42 US gallons.[43] Remembering that the US gallon is none other than the 1707 wine gallon of Queen Anne of 231 in^3 we are back at a barrel of 42 wine gallons, the shade of the salmon barrel. This is taken as 35 Imperial gallons and indeed is very close, for the exact equivalent is 34.99 Imperial gallons.

(c) The Wirksworth Dish[44]

Lead has been mined in Derbyshire, particularly in the High and Low Peak districts, since Roman times and this activity continued through mediaeval times to the present century. In 1288 it was stated[45] that miners should have weights for their lead and measures for their ore, the suggestion being that only smelted and refined metal was weighed.

The measures used by miners up to the present century were oaken troughs, being copies of the bronze vessel known as the Wirksworth Dish, originally kept in the Moot Hall at that place. The dish dates from 1513 as the inscription, which covers both sides, proclaims:

> This dish was made the IIII day of October, the IIII year of Henry VIII before George Earle of Shrewsbury, steward of the Kyng's most Honourable Household and also Steward of all the Honour of Tutbery, by the assent and consent as well of all the Mynours as of all the Brenners within and adjoining the Lordship of Wyrkysworth percell of the said Honour. This dish to remayne in the Moote Hall at Wyrkysworth hanging by a cheyne so as the merchants or mynours may have resort to the same at all Tymes to make the trew measure after the same.

We have, however, two bronze dishes extant, both dating from the sixteenth century. One is in the Derby Industrial Museum, the other is in the

Science Museum, London. Chaney[46] says that the dish formerly in the Moot Hall, Wirksworth, is that (in 1897) in the Museum of Practical Geology, London. Berriman[47] speaks of a *copy* of the dish being in the Science Museum, formerly having been deposited in the Museum of Practical Geology. So it would seem that both bronze dishes have supporters for their claim to be 'the original'. An analysis carried out by P. T. Craddock of the British Museum Research Laboratory in September 1981 showed the Derby vessel (which weighs 79½ lb) to contain 0.45 per cent lead whereas that in the Science Museum, weighing 76 lb, contains 5.3 per cent lead. The metal used for the two dishes is therefore quite different.[48]

The measured volume of the Derby Dish is 13.39 Imperial pints or 464.40 in^3 while the one in the Science Museum contains 14.047 Imperial pints or 487.2 in^3. The document of 1288[49] says 14 pints and these would be Winchester pints of about 33.75 in^3 each for a total of 472.5 in^3 which nicely fits between the actual capacities of the two dishes. The rough internal dimensions, $4 \times 5 \times 21.5$ inches, given by Berriman,[47] yield 430 in^3.

The 1820 Report,[50] which gives 'Dish — Derbyshire, of lead ore, 14 pints of 48 cubic inches making 672, in the Low Peak hundred, weighing 58 lbs; but in the High Peak 16 pints' is very far from the mark. No pint was ever 48 in^3 in volume.

The 1824 Weights and Measures Act did not disallow this measure. It continued to be used for local trading. The High Peak Mining Customs and Mineral Courts Act (14 and 15 Victoria c 94) of 1851 declared the dish to contain fifteen pints in the High Peak district, while a similar Act of the next year (15 and 16 Victoria c. 163) states that if ever lost the Wapentake or local council of Wirksworth is to change to the fifteen pint measure of the High Peak. Fifteen Imperial pints would amount to 520.24 in^3, a volume considerably above either dish.

The wooden replicas used by the miners contained about 65 lb of ore. In early times every thirteenth dish went to the king, every tenth to the Church, and every ninth paid a fee for measuring and for the right to sell the ore outside the extent of the manor.[51]

The measure continued in use into the twentieth century. The Science Museum dish is seen in Figure 39.

(d) Water Measure and Heaped Measure

In the fifteenth century and probably long before that there was a customary measure recognized in law called *water measure,* applicable for the sale of dry goods, grain, fruit, etc. on board ship or in ports and maritime towns. Water measure is specifically excluded from the provisions of the Act 11 Henry VII c 4 of 1495[52] by Section IV and we are told that 'Water measure within the ship board shall onely contain V pecks after the said standard rasen and stricken'.

As 4 pecks equal 1 bushel, water measure amounted to $1\frac{1}{4}$ bushels.

The use of this measure continued, being confirmed in 1640 by statute,[53] but thirty years later this statute was modified by 22 Charles II c 8 s 1 in which we are told that with effect from 29 September 1670 that part of the 1640 Act dealing with water measure is repealed with respect 'to the measuring, selling or buying of corn or grain, ground or unground or salt'. For these only the Winchester bushel was to be used, stricken.

Water measure was redefined later by Queen Anne (1 Anne Stat. 1 c 15 1701 Sect. 1 & 2*) in a statute entitled 'An Act to ascertain the water measure of fruit' when it was stated that water measure for fruit (apples and pears) should be round, 18.5 inches in diameter, and 8 inches deep (in fact the bushel of William III of 1696) and 'shall be heaped as usually'. The penalty for not using it when appropriate was to be ten shillings. The Act went on to say that this enactment was not to extend to the measures sealed and allowed by the Fruiterers' Company of the City of London which were to be used in the City and for three miles round about. This exemption guaranteed that there would be at least two measures for fruit. So much for the age-old call for one weight and one measure throughout the land.

Thus the water measure of Henry VII was of five pecks, that is $1\frac{1}{4}$ bushels stricken (see Ch.IX), while that of Queen Anne was a bushel as defined in 1696 by her predecessor, William III, in the words quoted above, but heaped up. No doubt Anne's heaped bushel was intended to equal the bushel and a quarter of Henry, and heaped measure became accepted as $1\frac{1}{4}$ measures, rather than the original $1\frac{1}{8}$ (see Ch.IX). Anne's definition of the water measure continued into the nineteenth century, being given in 1819 as a heaped bushel,[54] and in 1820 as five pecks.[55] Water measure was abolished by the Weights and Measures Act of 1824,[56] which, though it completely reorganized the system of metrology in this country and introduced a totally new gallon and bushel, nevertheless retained the concept of heaping, going so far as to specify the shape and size of the heap on top of the new bushel. Sections 7 and 8 say:

> . . . coals, culm [coal-dust], lime, fish, potatotes or fruit and all other things commonly sold by heaped measure . . . shall be duly heaped up in such a bushel in the form of a cone, such cone to be of the height of six inches and the outside of the bushel to be the extremity of the base of such a cone. . . .

Section 7 of the Act obligingly gave the *outside* diameter of the bushel measure as 19.5 inches. (The inside diameter did not have to be specified, for the volume was to be defined not by mensuration but by the weight of water it contained, 80 Avoirdupois pounds). A little calculation shows that the

*Now given in C.T.S. as 1 Anne c 9 1702.

volume of the conical heap so defined would be a quarter bushel.

Further definition of the heaping of the measures was provided the next year when it was enacted[57] that all such measures should be cylindrical as from 1 January 1826 with the diameter at least double the depth and the height of the cone, three-quarters of the depth of the vessel, the edge of which was to be the base of the cone. Here again an elementary calculation shows that for a bushel measure of diameter equal to twice the depth, the conical heap specified would amount to a quarter bushel.

These definitions of heaping, and indeed the concept of water measure itself, is a throw-back to the earlier heaping arrangement whereby ten stricken measures were taken instead of eight or alternatively eight such heaped measures would have the volume of ten stricken. The Act of 1824 ensured the buyer that for heaped measure he would get one quarter more of the goods being sold.

Heaped measure was abolished by the Weights and Measures Acts of 1834 and 1835 (4 and 5 William IV c 49 s 6 1834;[58] 5 and 6 William IV c 63 s 7 1835).[59] Surprisingly, heaped measure was again mentioned, somewhat needlessly, in Section 16 of Victoria's Weights and Measures Act of 1878,[60] though not permitted for use:

> 16. A bushel for sale of any of the following articles, viz. lime, fish, potatotes, fruit or other things which before 9th September 1835 [the effective date of the 1835 Act] were commonly sold by heaped measure shall be a hollow cylinder having a plane base, the internal diameter double the internal depth and other measures are to be made the same shape.

This unfortunate section was repealed by the Weights and Measures Act of 1889 (52 and 53 Victoria c 21, 1889) and thereafter we hear no more of heaped measures in the Statute Book.

(e) Coal Measures

The Chaldron

The earliest enactment known to the writer respecting coal measures is that of 1421 when a duty of two pence a *chaldron* was imposed on sea-coal. This merchandise was to be carried in ships called 'keels' which ought to carry twenty chaldrons, as stated in the Act 9 Henry V Stat. 1 c 10 of that year.[61] The chaldron was not then defined but we are told it was originally 32 bushels.[62] This is quite likely, for in 1543[63] we are told that every quarter of coal shall contain eight bushels at least and the chaldron of 32 bushels would just be four quarters, each of eight bushels.* As might be expected the eight bushels

*Being four quarters, the chaldron would represent the *wey* or *tun* (ton). See this chapter (a).

became eight heaped bushels, then called nine bushels, stricken, to allow for modest heaping. Four nine-bushel quarters would yield a chaldron of 36 bushels stricken and of course these bushels would in time become heaped too. Sure enough we read in the Act 16 and 17 Charles II c 2 (1664–5) that

> All sorts of coale commonly called sea-coales brought into the River Thames and sold, shall be sold by the chaldron containing thirty-six bushels heaped up according to the bushel sealed for that purpose in the Guildhall. And that all other sorts of coales comeing from Scotland and other places commonly sold by weight and not by measure shall be sold by weight after the proportion of 112 lbs. to the hundred of Avoirdupois weight.

We note in passing the most unusual reference to the bushel at the Guildhall rather than to a bushel in the Exchequer.

In 1676, and again in 1694, in placing a duty of five shillings a chaldron on imported coal, this measure was defined as 36 Winchester bushels for all coal sold by measure with no mention of heaping and a similar tax on coal sold by the ton of 20 cwt.[64] This may be no more than identifying the chaldron with the ton, but if this is to be taken literally as meaning that 36 stricken Winchester bushels weigh roughly 20 cwt and therefore carry the same duty, with the density of coal being about 1.5 we see that only a little over half the volume is solid coal, the rest being air in the interstices. However the very next year, by 6 & 7 William III c 10 of 1695, we learn that the chaldron of coal is to be 53 cwt, with the wagon load $17\frac{1}{2}$ cwt and the cart load $8\frac{3}{4}$, the intent being that 1 chaldron = 3 wagon loads = 6 cart loads. Presumably $52\frac{1}{2}$ cwt was near enough 53 to pass muster, but this chaldron is anomalously large, for a block of solid coal of the above density having a volume of 36 *heaped* Winchester bushels weighs barely 47 cwt.

The explanation of all this would appear to be as follows. The London chaldron was usually defined as four *vatts* 'ringed and heaped'. The metropolis therefore imported coal and valued it by measure. Points north such as Scotland and Tyneside exported coal by weight. Whether traded by weight or volume, each chaldron was taxed so there was every incentive to increase gradually the size of the chaldron. The record shows that the Newcastle chaldron which in 1421–8 was 2000 lb (about 18 cwt) rose steadily through the years until it was 53 cwt in 1689, at which figure it plateaued. In the eighteenth century this 53 cwt chaldron was equal to *two London chaldrons.* (See J. U. Nef, *The Rise of the British Coal Industry,* London 1932 Vol.II, Appendix C p.367.) This makes the London chaldron equivalent to a weight of $26\frac{1}{2}$ cwt or thereabouts. For this to be 36 heaped bushels means that the volume of actual coal should be $\frac{26.5}{47}$ or roughly $56\frac{1}{2}$ per cent of the measure. But the chaldron measure continued through the nineteenth century, when it was again defined as 36 bushels, into the twentieth before becoming officially extinct. (See next section.)

The Coal Bushel

In the early years of the eighteenth century, Queen Anne introduced a new measure specifically for coal. This was the *coal bushel*, defined in 1713[65] as '19½ inches from outside to outside and to hold a Winchester bushel and one quart of water according to the standard of the Winchester bushel of 13 William III c 5'. The Act went on to provide that these oversized bushels were themselves to be heaped and the 1664–5 definition of the chaldron was repeated but the 36 bushels were now to be coal bushels.

Instructions were given for a standard coal bushel to be constructed according to this enactment but it was apparently delayed, only appearing in 1730*. This bushel should have contained 2217.62 in^3 if the definition is taken as meaning $\frac{33}{32}$ of a Winchester bushel, or 2218.48 in^3 if taken as a bushel of 2150.42 in^3 plus one quarter of a gallon of 272.25 in^3. A measurement made on 6 February 1872[66] gave its actual volume as 2222 in^3. This bushel of 1730 (George II) is shown in Figure 40.

In 1819[54] the coal bushel was described as a Winchester bushel and a quart of water with the coals heaped to form a cone. Like water measure, the coal bushel was abolished in 1824. By the Weights and Measures Act 5 and 6 William IV c 63 of 1835,[59] the Winchester bushel was abolished by name (Section VI) together with all local and customary measures including heaped measure, with the gallon and bushel to be as defined in 1824. It further decreed (Section IX) that all 'coal, slack, culm, and cannel' was to be sold by weight and not by measure with effect from 1 January 1836. (Here *slack* is dross, *culm* is coal dust or slack of anthracite, and *cannel* is bituminous coal.) This was re-enacted in the Weights and Measures Act of 1878.[60] The 1878 Act, while prescribing weight as the measure for coal, nonetheless defined the chaldron again as 36 bushels. The chaldron was finally abolished by the 1963 Weights and Measures Act,[67] which also reiterated that coal was to be sold by weight.

The Sack

But there had been another measure for coal — the *sack*. In 1552[68] we are told that every sack is to contain 'foure bushelles of good clean coles' and this is repeated in 1634.[69] With the increase already mentioned in the size of the coal bushel in 1713, we find a corresponding reduction in the number of bushels to the sack, for in 1730[70]† by Act of Parliament the sack was to contain three bushels and not four as previously. This Act also gave the dimensions of the sacks themselves as 50 inches long and 26 broad 'at least'.[71] With the English

*Reign of George II.
†Given as 1729 in C.T.S.

climate, the sacks were found to shrink in the rain so in 1758 the sack was lengthened by two inches, the breadths remaining the same.

If the new coal bushel was to be one quart greater than the Winchester bushel, that is 33 quarts, and then the measure was to be heaped, the volume of coal would be one quarter greater, that is $41\frac{1}{4}$ quarts. Three such heaped measures would be $123\frac{3}{4}$ quarts. Previously the sack had contained four Winchester bushels, that is 32 quarts. Four of them would yield 128 quarts. Presumably $123\frac{3}{4}$ was near enough 128 for the change to be acceptable.

The Act of 1824 (s. 7) defining the Imperial gallon and bushel also defined the sack of coal as three heaped bushels. In law, the Imperial bushel was 2218.192 in^3 so three heaped (that is $3\frac{3}{4}$) would have a volume of 8318.22 in^3. The quart of the previous paragraph was one-fourth of the gallon of William III, that is either 67.2 or 68.06 in^3 depending on the value taken for the gallon, either 268.8 or 272.25 in^3. Thus three heaped Imperial bushels would hold either 123.8 or 122.2 of these quarts, nearly enough to the previous $123\frac{3}{4}$ to be acceptable. One can infer that efforts were being made to give the customer the same volume of coal in the sack as traditionally, notwithstanding changes in the units themselves.

But the sale of coal by volume was decreasing, and following the Act of 1835,[59] it became mandatory to sell coal by weight. Those who can remember the days before smokeless zones, central heating, and oil-fired furnaces will also remember the familiar sight of the early decades of this century, of coal merchant's lorries with their sign declaring that each bag contained 112 lb net weight.

(f) The Imputed Bushel

The Statute Book of the late seventeenth, eighteenth, and early nineteenth centuries gives a few examples of a commodity being sold nominally by the bushel but actually by weight with the weight of a so-called bushel being given by Act of Parliament in Avoirdupois pounds.

By the Act 22 Charles II c 8 of 1670 certain sections of an Act of the previous reign (16 Charles I c 19 1640) were repealed in that water measure, as we have seen, was no longer to be employed for measuring corn or salt but all were to be measured by the Winchester bushel stricken. This was modified in 1694[72] to allow Chester salt to be sold by a weight of 56 pounds which was to be considered as a Winchester bushel while rock salt was to be sold at 120 ('six score') pounds to the bushel. These 'equivalents' were repeated in statutes of the years immediately following.

This arose because to fill a bushel often meant pulverizing the goods, to the loss of the seller, and reflected the difficulty of obtaining or making a true bushel measure. Thus we read in 7 and 8 William III c 31, s 43 of 1695/6[73] that

> whereas the measures for making the Winchester bushel at 8 gallons to the bushel . . . are various and unequal and have proved inconvenient . . . all salt at all salt works, except rock salt [shall be sold at the] rate of 56 lbs. weight to the bushel.

Rock salt was attended to shortly thereafter by the Act 9 William III c 44 s 30 1698 where we read:[74]

> And whereas salt rock or rock salt taken out of pits in such great lumps that cannot be measured without breaking the same to powder would be great loss to the Proprietors thereof, be it enacted that . . . 120 lbs. weight . . . shall be deemed . . . a Winchester bushel of 8 gallons Winchester and rated and taxed accordingly.

This bountiful rate of 120 lb of rock salt to the bushel did not last long. Within a year we read in 10 William III c 11 s1 1698 as follows:[75]

> . . . and being sensible the rock salt which in and by the said Acts . . . is chargeable . . . after the rate of 120 lb weight to the bushel, may be and is used in kind for curing fish, flesh and other purposes without being refined into white salt and all white salt made from brine or otherwise being chargeable with the said duties after the rate of 56 lb weight to the bushel and that 120 lb weight of rock salt used in kind will serve the use of almost 2 bushels of white salt reckoned at 56 lbs weight . . . that from and after 15 May 1699, every 75 lbs weight of rock salt . . . shall be deemed and taken to be a Winchester bushel . . .'

In 1697 a specific statute[76] was devoted to requiring retailers to sell salt by weight only.

In the first year of the reign of Queen Anne,[77] in the Salt Duties Act 1702, it was decreed that salt could only be bought and sold by weight and a penalty of ten shillings was to be imposed on those trading by measure for each bushel bought. Here we find that 84 lb weight of foreign salt is to be reckoned as a bushel while the weight to represent the bushel of rock salt was reduced from the above 75 lb to 65 lb. Section XI sums it up as follows:

> 56 lb weight make a bushel for English white salt
> 65 lb weight make a bushel for English rock salt
> 84 lb weight make a bushel for foreign salt.

This is repeated in 1718 (5 George I c 18 s 4).

A century later George III imposed additional duties on salt in Great Britain, Section IV of this Act[78] stating that 65 lb Avoirdupois weight of rock salt made a bushel but of all other kinds only 56 lb were needed. Salt for curing or preserving fish was not to be subject to duty.

It is strange, it having been decided that salt should be sold by weight, that the fiction should have continued that it was really being measured by volume. Presumably there was some correlation as the equivalence lasted so long but it cannot have been close.* The idea that salt was sold by measure continued into the twentieth century, especially in the countryside. The writer can recall his grandmother ordering 'half a peck of salt', while he as a boy went to the greengrocers for a 'forpit' (fourth of a peck) of potatoes. Sure enough, she was supplied with 7 lb in accordance with the rate 56 lb to the bushel; he received $3\frac{1}{2}$lb.

Section 12 of the Fifth Report of the Commissioners appointed to Enquire into the Condition of the Exchequer Standards, 1871,[79] tells of applications being made to have weights of $3\frac{1}{2}$ lb and $1\frac{3}{4}$ lb legalized, but they were refused as being too similar in appearance to 4 and 2 lb weights. Three and a half pounds of flour made a quartern (4 lb) loaf, $1\frac{3}{4}$ lb made a half-quartern loaf (2 lb), and the septenary scale of Avoirdupois weight above 1 lb fitted well the long established baking practice with 7 lb of flour making the gallon (8 lb) loaf. (See Ch.XI.)

(g) Apothecaries' Measure

For a long time it was customary to sell liquid drugs and pharmaceutical concoctions by weight. Liquid measures made of horn preceded the introduction in the late eighteenth century of glass vessels but the *London Pharmacopoeia* of May and December 1618 gives liquid measures converted into weight in the following magnitudes; the tablespoon, the wineglassful, the *cotlya,* the pound, the *sextarius,* and the gallon.

A tablespoon of distilled water weighed (supposedly) 3 drachms while the wineglassful was $1\frac{1}{2}$ oz, the *cotlya* 9 oz, and the pound, of course, 12 oz. The *sextarius* was 18 oz and the gallon, equal to six *sextaria,* was therefore 108 oz. This is clearly nine 12-oz (Troy) pounds. So once again we have an instance of nine measures being taken instead of the basic eight. Interestingly enough, attention has been drawn to the manuscript annotation in the British Library copy of this *Pharmacopoeia* by which we are informed that the *congius* is 'A measure of liquids containing sixe sextarios which is of our measure a gallon one pint; it may be taken for O^r^ [our] gallon'.[80] Here we have a complete description, recognizing the gallon or *congius* of six '*sextarios*' as nine pounds, that is a gallon and a pint, but saying also that the extra did not matter.

If we again adopt Norris's density for wheat, 47.5 lb Avoirdupois to the cubic foot, we find by a little calculation that the gallon given above (108 Troy ounces) would have a cubic capacity of 269.4 in^3 which is remarkably close to the Exchequer gallon of Henry VII (268.43 in^3) and to both gallons of Elizabeth (268.97 and 270.59 in^3). There can be little doubt that in the early

*Were a bushel of 2150 in^3 to consist of solid crystalline salt of specific gravity 2.163 (KAYE, G. W. C. and LABY, T. H. *Tables of Physical and Chemical Constants.* 12th ed. p.117 London: Longmans, Green & Co. 1959) it would weigh 168 lb. The conceptual bushel here described would therefore be $\frac{2}{3}$ air and $\frac{1}{3}$ granular salt.

seventeenth century the gallon of the druggist was the current Exchequer standard, with lesser measures in proportion. The annotation further indicates the correctness of the view that the measures of Henry VII, continued by Elizabeth in her standards, were based on nine measures and not eight. By the mid-eighteenth century there was a noticeable shift from Apothecaries' weight to Avoirdupois with Reynardson[81] commenting that apothecaries seldom use Troy weights above two drams (drachms). For all above that Avoirdupois weight was used. Chamberlayne[82] in 1741 stated that 'although Apothecaries make up their medicines by Troy weight they sell their drugs by Avoirdupois weight'.

In 1851 the *Pharmacopoeia Collegii Regalis Medicorum Londinensis* authorized measures as follows:

60 minima (m)	=	1 fluid drachma
8 fluid drachmas (ʒ)	=	1 fluid ounce
20 fluid ounces (℥)	=	1 octarius
8 octarios (O)	=	1 congius (C)

The *octarius* was the pint and the *congius* the gallon in this post-1824 declaration, as we shall see from Schedule 2 of the Weights and Measures Act of 1878.

Formally, Avoirdupois weight replaced Apothecaries' weight by the Medical Act of 1858, and in 1864 the General Medical Council, in the first edition of the *British Pharmacopoeia*, used the Avoirdupois pound instead of the Apothecaries' (Troy) pound, leading Squires in his *Companion to the British Pharmacopoeia* of the same year to treat of weights and measures extensively so as to reflect the changed weight.[83]

By the Weights and Measures Act 1878 (41 & 42 Victoria c.48)[60] only the Avoirdupois pound was retained, with the Apothecaries' ounce of 480 grains being kept for drugs on a permissive basis, but increasingly retail drugs were being dispensed on the Avoirdupois scale. This is reflected in the table of pharmaceutical measures given in this Act (Schedule 2):

20 minims*	=	1 fluid scruple
3 fluid scruples	=	1 fluid drachm
8 fluid drachms	=	1 fluid ounce
160 fluid ounces	=	1 Imperial (1824) gallon,

for the Imperial gallon was defined as the volume of ten Avoirdupois pounds of water at 62°F.

The 1963 Weights and Measures Act,[67] while retaining the fluid ounce, gave notice of the abolition of the fluid drachm and minim as legal units, 'on

*A *minim* was often equated with a *drop*, but as drops varied in size, it is clear this is only an approximation.

such a date not earlier than 31 January 1969 as the Board (of Trade) shall appoint by order'. This date turned out to be 1 February 1971 in accordance with The Weights and Measures Act (Amendment of Schedules 1 and 3, Order) 1970. (S.I.* 1970/1709).

(h) The Reputed Quart[84]

The standard wine or spirit bottle encountered in the United Kingdom was almost always one nominally of $26\frac{2}{3}$ fluid ounces (46.24 in^3). In Scotland the whisky bottle was affectionately known as the 'whisky quart'; and in legislation and government reports all bottles of this capacity are referred to as 'reputed quarts', notwithstanding the fact that the bottle is only two-thirds of an Imperial quart.

A study of the history of glass-making in England shows bottlemaking to have commenced in the seventeenth century, prior to which nearly all wine bottles in England were imports from the Rhineland and were made of earthenware.[85] At least as early as 1663 'quart glass bottles' were being sold at sixpence each. Being hand-made, these bottles would not be expected to show highly accurate and consistent contents but they are grouped around three standard values, namely the ale quart of 70.5 in^3 $(\frac{282}{4})$, the wine quart later legalized by Queen Anne of 57.75 in^3 $(\frac{231}{4})$, and the 'reputed quart' of 46.24 in^3. From 1800 onwards the great majority of wine bottles had a volume of between 46 and 47 in^3, and they would seem to have originated in the middle of the seventeenth century.

Why would such a unit of capacity come into use and from whence might it have come? It has been suggested,[84] and very likely correctly, that it arose as a true quart (quarter) of the wine gallon generated from eight Troy pounds of wine. In Ch.IX p.158 we showed this to be 183.8 in^3. The quart would therefore be almost exactly 46 in^3. The Statute 43 George III c 68 of 1803 speaks of wine bottles making about five to the gallon. The gallon in question would be that of Queen Anne, that is 231 in^3, one fifth of which is 46.2 in^3. (This being the gallon of the USA, the standard bottle would be one-fifth of their gallon and today liquor is sold in the States in 'fifths' or what is the same thing 'four-fifths of a quart', US measure.) With the arrival of Imperial measures, the reputed quart did not change except in definition for it then became one sixth of the gallon[86] of 277.274 in^3 (that is 46.21 in^3), near enough the old value for all practical purposes.

The term 'reputed quart' appears in the Act, 5 George IV c 54 of 1824 and in Section 17 of the Fifth Report of the Commissioners appointed to Enquire into the Conditions of Exchequer Standards 1871[79] as one sixth of an Imperial gallon. Its abolition was recommended in the Hodgson Report of 1951[87] but

*Statutory Instrument.

the Customs and Excise Act of 1952 (15 and 16 George VI and 1st Elizabeth II)[88] uses the term to define a wholesale sale as distinct from a retail sale, thus: 'Wholesale means a sale of not less than two gallons or one dozen reputed quart bottles of spirits; for beer $4\frac{1}{2}$ gallons or 2 dozen reputed quart bottles'. Retail means a sale involving not more than this definition. Section 59 of the Weights and Measures Act of 1963 modified the 1952 Customs and Excise Act by removing the words 'one dozen reputed quart bottles' and substituted 'one case', the case being defined as one dozen units each containing not less than 23 or more than 28 fluid ounces.

Now that metrication is upon us we see wine bottles of 75 centilitres in great profusion. This is close to but a little less than the old reputed quart, being 45.77 in^3. As the difference is less than half a cubic inch few will notice the reduction, but we do observe wines in 70 centilitre bottles in the shops now. This is only 42.7 in^3 and the bottles are observably smaller.

(i) Capacity Measures of the United States of America

As there was frequently in the past confusion between Imperial measures of capacity and those of the USA it might be well to set down the origins of the latter and show the relation between the two systems.

The Americans have only one bushel and this is for dry measure. Its volume is 2150.42 in^3 and is based on the gallon and bushel of William III of 1696–7. It is divided into 64 dry pints each of 33.6003 in^3. The only gallon is that for liquids which has a volume of 231 in^3 and is based on Queen Anne's wine gallon. It is divided into four liquid quarts and eight liquid pints. The liquid pint is therefore 28.875 in^3 in volume and is not the same as the dry pint of the USA. The liquid pint is further divided into sixteen fluid ounces. One US fluid ounce has a volume of 1.80 in^3 and therefore weighs 1.04 oz Avoirdupois.

	Bushel	*Gallon*	*Quart*	*Pint*
Imperial (1824)	2218.19 in^3	277.274 in^3	69.32 in^3	34.66 in^3
USA (dry)	2150.42	–	67.20	33.60
(liquid)	–	231	57.75	28.875

In Britain and those countries which used Imperial measure the gallon since 1824 was the volume of ten pounds Avoirdupois distilled water, that is it was 277.274 in^3 as defined in the Act of 1824. The British quart contained 2.5 lb of water and the pint 1.25 lb, their volumes being respectively 69.32 in^3 and 34.66 in^3. The British gallon weighed 160 ounces Avoirdupois and under the Imperial system a fluid ounce was the volume of one Avoirdupois ounce of distilled water at 16.7°C (62°F). The gallon contained therefore 160 fluid ounces, the quart 40, and the pint 20. The British fluid ounce has a volume of

1.733 in^3 and by definition one fluid ounce of water weighs one ounce Avoirdupois.

Twenty-five Imperial fluid ounces equal twenty-four US fluid ounces to better than one part in one thousand. The US gallon is five sixths of an Imperial gallon to better than one part in four thousand.

NOTES AND REFERENCES: CHAPTER X

1. See the Assize of Bread and Ale, Ch. XI.
2. 2 Henry VI c 14 (formerly c 11) 1423 *Statutes*, Vol. II, p. 222.
3. HALL, Hubert and NICHOLAS, Frieda J. *Select Tracts and Table Books relating to English Weights and Measures (1100-1742)*. Camden Miscellany XV, p. 49. London: Camden Society 1929.
4. Report from the Committee appointed to inquire into the Original Standards of Weights and Measures in this Kingdom (Carysfort Committee) 26 May 1758. *Reports from Committees of the House of Commons* Vol. II, (1737-65) p. 415.
5. In 28 Henry VIII c. XIV s 5 of 1536 the *tierce* is given erroneously as 41 gallons.
6. The *rundlet* is given as 18½ gallons by Statute, 1 Richard III c 13 1483-4 and again about 1600 in the *Common Place Book* of J. Lloyd. (See Ref. 3 p. 21) and also in S. Jeake's *Arithmetic* of 1696. By 1700 it had dropped to 18 gallons, being $\frac{1}{14}$ tun. (See: CHAMBERLAYNE, John *Magnae Britanniae Notitia,* London: Goodwin Timothy 1708, also the edition of 1741 at p. 157). In 1820, (Second Report of Commissioners appointed to consider the subject of Weights and Measures, 13 July 1920 Appendix A p. 31) it is given as 18 gallons. After the introduction of the Imperial gallon in 1824 this cask was described as 'about 15 gallons'. In fact 18 gallons of 231 cubic inches are just 14.996 Imperial gallons.
7. O'KEEFE, John Alfred. *The Law of Weights and Measures.* London: Butterworth 1966, and 2nd ed. 1978.
8. HALL, Hubert and NICHOLAS, Frieda J. Loc. cit. p. 50.
9. 1 William & Mary c 24, s 4 1688 *Statutes at Large* Vol. III, p. 406.
10. 10 & 11 William III c 21 s 15 1699 Ibid. Vol. IV, p. 16.
11. 43 George III c 69, s 12 1803 Ibid. Vol. XV, p. 848.
12. 5 George IV c 74 1824 Ibid. Vol. XXIII, p. 759.
13. HALL, Hubert and NICHOLAS, Frieda J. Loc. cit. pp. 15, 21, 29. See also *Statutes*: 18 Henry VI c 17 1439 and 1 Richard III c 13 1483-4.
14. 23 Elizabeth I c 8, s 54 1581 *Statutes*. Vol. IV, Pt. 1, p. 670.
15. Tractatus de Ponderibus et Mensuris 1266 Ibid. Vol. 1, p. 204-5.
16. Ordinacio facta de allece vendend';
    31 Edward III Stat. 2 1357 Ibid. Vol. 1, pp. 353-4.
17. HALL, Hubert and NICHOLAS, Frieda J. Loc. cit. pp. 16-17.
18. CHAMBERLAYNE, John. *Magnae Britanniae Notitia* 34th ed. London: D. Midwinter et al. 1741.

19. 2 Henry VI c 11* 1423 *Statutes* Vol. II, p. 222.
20. 22 Edward IV c 2 1482-3 Ibid. Vol. II, pp. 470-1.
21. 13 Elizabeth I c 11 s 4 1571 Ibid. Vol. IV, Pt. 1 p. 546.
22. Report from the Committee appointed to Enquire into the Original Standards of Weights and Measures in this Kingdom. (The Carysfort Committee) 26 May 1758, pp. 417 and 421. *Reports from Committees of the House of Commons* Vol. II, (1737-65) pp. 407-451.
23. HALL, Hubert and NICHOLAS, Frieda J. Loc. cit. p. 23.
24. ZUPKO, Ronald Edward. *A Dictionary of English Weights and Measures* p.16 Madison, Wisconsin: University of Wisconsin Press 1968.
25. HALL, Hubert and NICHOLAS, Frieda J. Loc. cit. p. 30.
26. 5 George I c 18, ss 6, 12, 13, 15. 1718 *Statutes at Large* Vol. V, p. 136.
27. 8 George I c 4 s 2 1721 Ibid. Vol. V, p. 252.
28. 8 George I c 16 s 2 1721 Ibid. Vol. V, p. 266.
29. 43 George III c 69; Schedule (c) p. 862, 1803 Ibid. Vol. XV, p. 846.
30. 48 George III c 110, ss 3 and 36 1808 Ibid. Vol. XVII, p. 498.
31. 3 & 4 George V c 9 s 1(3) 1913 *The Statutes Revised* Vol. XV, p. 71.
32. 55 George III c 94 1815 *Statutes at Large* Vol. XX, p. 260.
33. O'KEEFE, John Alfred. Loc. cit. p. 629.
34. Ibid. p. 631 et seq.
35. MACKENZIE, N. P. B. The Sale of Herring at Mallaig Fishing Port *The Monthly Review* (Dec. 1979) Vol. LXXXVII, No. 12, p.214.
36. 8 Edward VII c 17 1908*Public General Statutes* Vol. XLVI, p. 30.
37. 41 & 42 Victoria c 49 1878 Ibid. Vol. XIII, p. 308.
38. Elizabeth II c 31, 1963. London: HMSO *Weights and Measures Act 1963* p. 500.
39. Second Report of the Commissioners on Weights and Measures, 13 July 1820. *Parl. Papers* 1820 (HC314) vii, p.473. Appendix A p. 7.
40. 23 Henry VIII c 4 s 4 1531-2 *Statutes* Vol. III, p. 367.
41. 10 Anne c 18 s IX 1711 Ibid. Vol. IX, p. 596.
42. 13 & 14 Charles II c 26 1662 *Statutes at Large* Vol. III, p. 240.
43. WILLIAMS, Howard R. & MEYERS, Charles J. *Manual of Oil and Gas Terms* 5 ed. p. 52. New York & San Francisco: Matthew Bender 1981.
44. For much of what follows, see:
    TUDOR, T. L. The Lead Miners' Standard Dish or Measure, *Derbyshire Archaeological & Natural History Society Journal* (1937) Part I pp. 95-106, (1938) Part II pp. 101-116.
45. Ibid. Pt. 1 p. 99.
46. CHANEY, H. J. *Our Weights and Measures* p. 136. London: Eyre & Spottiswoode 1897.
47. BERRIMAN, A. E. *Historical Metrology* p. 156. London: J.M. Dent & Sons Ltd. 1953.
48. Private Communication, Dr D. Vaughan, Science Museum, London. June 1982.
49. TUDOR, T. L. Loc. cit. Pt. I p. 103.
50. Second Report of the Commissioners on Weights and Measures, 13 July 1820, *Parl. Papers* 1820 (HC314) vii, p473. Appendix A, p. 15.

*In some printings this is given as c 14.

51. TUDOR, T. L. Loc. cit. Pt. I p. 101.
52. 11 Henry VII 1495 *Statutes* Vol. II, p. 570-1.
53. 16 Charles I c 19, s 7, 1640 Ibid. Vol. V, p. 130.
54. First Report of Commissioners appointed to consider the Subject of Weights and Measures 7 July 1819. *Parl. Papers* (HC565) xi, p. 316.
55. Reference 50, p. 36.
56. 5 George IV c 74 1824 *Statutes at Large* Vol. XXIII, 1824, p. 759.
57. 6 George IV c 12 1825 Ibid. Vol. XXIV, 1826, pp. 20-21.
58. 4 & 5 William IV c 49 s 6 1834 Ibid. Vol. XXVII, 1834, p. 630.
59. 5 & 6 William IV c 63 s 7 1835 Ibid. Vol. XXVII, 1835, pp. 977-986.
60. 41 & 42 Victoria c 49 s 16 1878 *Public General Statutes* Vol. XIII, pp. 308-341.
61. 9 Henry VI Stat. 1 c 10 1421 *Statutes* Vol. II, p. 208.
62. ZUPKO, Ronald Edward. Loc. cit. p. 35.
63. 34 & 35 Henry VIII c 3 1543 *Statutes* Vol. II, p. 208.
64. 6 & 7 William & Mary c 18 1694 Ibid. Vol. VI, p. 603.
65. 12 Anne Stat. 2, c XVII 1713 *Statutes at Large* Vol. IV, p. 624.
66. CHISHOLM, H. W. *Seventh Annual Report of the Warden of Standards* p. 38. London: HMSO 1873.
67. Elizabeth II c 31 1963. London: HMSO. *Weights and Measures Act 1963* London 1964.
68. 7 Edward 7 c 7 1552-3 *Statutes*. Vol. IV, Pt. 1 p. 171.
69. Reference 3, p. 52.
70. 3 George II c 26 s 13 1730 (now given as 1729 in C.T.S.); *Statutes at Large* Vol. V. p.551.
71. Ibid. s 11 1730.
72. 5 William & Mary c 7 1694 *Statutes*. Vol. VI, p. 603.
73. 7 & 8 William III c 31 s 43 1695-6 Ibid. Vol. VII, p. 137.
74. 9 William III c 44 s 30 1698; ibid p.433.
75. 10 William III c 11 s 1 1698; Ibid. p. 508.
76. 9 & 10 William III c 6 1697 *Statutes at Large* Vol. III, p. 665-6.
77. 1 Anne c 15 s 6 & 9, 1702 *Statutes*. Vol. VIII, p. 61-2.
78. 45 George III c XIV 1805 *Statutes at Large* Vol. XVI, p. 270.
79. Fifth Report of Commissioners to Enquire into the Conditions of the Exchequer Standards; *Parl. Papers* 1871 (c 257) xxiv p.647.
80. MATTHEWS, L. G. *History of Pharmacy in Britain* p. 360-1, Note 3. Edinburgh & London: E & S Livingstone 1962.
81. REYNARDSON, Samuel. A State of the English Weights and Measures of Capacity . . ., *Phil. Trans. Roy. Soc.* (1749-50) Vol. XLVI, pp. 54-71. London.
82. CHAMBERLAYNE, John. Loc. cit. p. 154.
83. MATTHEWS, L. G. Loc. cit. p. 100.
84. A good discussion is given by MOODY, B. E. The Origin of the 'Reputed Quart' and Other Measures. *Glass Technology* (April 1960) Vol. 1, No. 2, pp. 55-68. Sheffield.
85. MOODY, B. E. Ibid. p. 60.
86. Report of Commissioners appointed to consider the Steps to be taken for the Restoration of the Standards of Weights and Measures, *Parl. Papers* 1842 (HC356) xxv, p.281. Recommendation 92. (Report dated 21 Dec. 1841).

'That the wine bottle be recognized as a measure containing one-sixth part of a gallon'.

87. Report of the Committee on Weights and Measures Legislation (The Hodgson Report) Cmnd 8219 London: HMSO May 1951. *Parl. Papers* 1950–51 (Cmd 8219) xx. p.913.
88. George VI and 1 Elizabeth II (1952) c.44; *Customs and Excise Act 1952*, c44 p.953.

CHAPTER XI

The Assize of Bread and Ale[1]

Bread and ale or beer have been with mankind since the earliest times of which we have record. Loaves or cakes have been found among the remains of the Swiss lake-dwellings of the Stone Age. Beer was known in Babylonia in the sixth millennium BC and in Egypt in the third. Both bread and beer are nutritional but bread holds pride of place in this regard. We are told by Jonathan Swift and Matthew Henry* that it is the staff of life and everyone seems to know intuitively that bread is good and wholesome. The high nutritional value of bread, white or brown, has been clearly demonstrated by tests on German orphans at the end of World War II when it was shown to be a complete food enabling growing children to thrive even in the absence of milk and vegetables.[2] While neither ale nor beer is mentioned in Holy Writ, we are told that Noah planted a vineyard after the flood and drank the wine thereof somewhat immoderately. In northern climes where the vine did not flourish the drink of the common people of historical times was ale, and bread with the occasional dish of fish, meat, or cheese was their staple diet.

We should not be surprised that the basic drink of the people was ale. In the England of the Middle Ages the water supply, whether from well or stream, was, as often or not, polluted. There was no sewage treatment. Privies emptied into the nearest waterway which served all other purposes. The alcohol of the ale served as an antiseptic and was of proper strength if the carbohydrates had for the most part been so converted in the brewing, hence the need for a simple test for sugar (see p.222).

Recognizing that bread and ale was the food and drink of the people, regulations were made centuries ago to ensure that wholesome bread and good ale was freely available at controlled prices which reflected the current price of grain and the costs of baking and brewing. These regulations were in the form of *Assizes* which equated so much bread (or ale) with so much money and thus enabled the local civil power to regulate the trade in these commodities. In addition, provision was made to check that the regulations were being carried out, that the people obtained a loaf of the weight prescribed according to local costs of baking, that the quality was up to standard, and that no unjust profits were made at the expense of the people. The same went for ale.

The Assize of Bread took a different form from the Assize of Ale so they will be treated separately.

*J. Swift, *A Tale of a Tub,* 1704; M. Henry, *Commentaries*

The Assize of Bread

The Assize of Henry II, 1202, 1266, and 1290

It would appear that no Saxon assize has survived to the present day in written form. One of the earliest to survive from continental Europe is that contained in Charlemagne's Frankfort Capitulary of 794, which reads in part:[3]

> When the measure of oats is sold for 1*d*, that of barley for 2*d*, that of coarse wheat 3*d* and that of wheat 4*d*, yielding 12 wheaten loaves of 2 lb each, the loaves shall be sold for 1*d*; if of coarse wheat, 15 loaves of the same weight for 1*d*; of barley 20; and of oats 25 of similar weight.

We see that Frankish bread was not baked exclusively from wheat flour and, whereas the size of the measure is not given specifically in the Capitulary, it was in all probability the new *modius* (roughly a peck), established by Charlemagne prior to 794 and referred to in the *Consuetudines Corbeienses* of 822[4] as follows:

> It is our wish that every year there should arrive 750 corbi of well winnowed and husked spelt, each corbus having twelve modii, well compared and standardized to the new modius which the Lord Emperor has set.

The earliest English assize known to the writer is one of the time of Henry II [5], who reigned 1154-89, but there may well have been earlier assizes. Matthew Paris[6] in his *Chronica Majora* records an assize of the year 1202. Further there is an assize given in the Annals of the Monastery of Burton of 1256, but this differs only in minor detail from the much better known Assisa Panis et Cervisie,[7] given in the Statute Book as of 'uncertain date' but presumed in earlier printings to date from 51 Henry III (1266). It is certainly of the thirteenth century. For present purposes we will accept the 1266 date and will so refer to the Assize. Because of its similarity to that of 1266, the Burton assize will not be considered further.

Fleta (1290)[8] in Chapter 9 has a section entitled 'De Assisa Panis et Cervisie' and it may be of interest to compare these four assizes, extending over a period in excess of a century, noting their differences and their similarities. Of early assizes we therefore have that of Henry II, and those of 1202, 1266, and 1290.

The two earlier assizes give the weights of farthing loaves, that is loaves costing $\frac{1}{4}$*d*, for white bread *(wastel)* and wholemeal *(panis de toto blado)* for various prices of the quarter of wheat ranging in steps of 6 pence (6*d*) from 6 shillings (6/-) down to one shilling and 6 pence (1/6)*. The two later assizes

*In what follows in terms of money we shall use the abbreviations, common in the earlier part of the twentieth century, of the *solidus* to denote shillings, for example 6/3 means 6 shillings and 3 pence, and '*d*' (for *denarius*) to denote pence, that is 6*d* means 6 pence. The pound as a weight was 20/-, that is 20 shillings. The shilling weight *(s)* is that of 12 pennyweights (dwt).

give the weights of loaves of wastel bread only, according to a much more extensive table which runs, again in steps of *6d*, according to the price of grain, from 1/- to 20/- in the assize of 1266 and from 1/- to 12/- in Fleta. It will be noted that the price order is reversed from that found in the two earlier documents.

The 1266 assize begins:*

> When a quarter of Wheat is sold for XII*d*, then Wastel bread of a farthing shall weigh VI pounds and XVI *s* [6 pounds and 16 shillings]. But Bread cocket of a Farthing of the same Corn and Bultel [Sieve] shall weigh more than a wastel by II *s*. And Cocket Bread made of Corn of lower price shall weigh more than Wastel by V *s*. Bread made into a Simnel shall weigh II *s* less than Wastel. Bread made of whole Wheat shall weigh a Cocket and a half, so that a Cocket shall weigh more than a Wastel by V *s*. Bread of Treet shall weigh II Wastels and bread of common Wheat shall weigh Two great Cockets†.

Wastel is best white bread. Cocket is second-quality bread, but appeared in two grades according to the quality of the grain used. Simnel bread was more expensive than wastel for it was twice baked or boiled then baked; it was used at the principal church festivals such as Christmas, Easter, etc. Wholemeal or wheaten loaves were made from whole wheat from which the bran had been removed. Treet is bread made from coarse flour. An identical set of interrelations is given in *Liber Albus*[9] as of the time of Edward I (1272-1307) and by Fleta.[8].

The assize continues, that

> When a quarter of wheat is sold for 18 pence then Wastel bread of a Farthing, white and well baked shall weigh 4 pounds 10 shillings and 8 pence. When for 2 shillings; 3 pounds and 8 shillings . . . when for 2 shillings and 6 pence; 54 shillings $4\frac{3}{4}$ pence . . . when for 3 shillings; 48 shillings weight . . .

and so on, the weight of the loaf being adjusted for every rise or fall of sixpence in the price of a quarter of wheat. From the four assizes, a table can be extracted showing the weight of the loaf as the price of wheat varies.

It will be observed from the table on page 196 that all four give the same weight of bread when wheat is moderately priced (4/- to 3/- and almost so at 2/6) but when corn was dear or cheap the later assizes give the purchaser a relatively heavier loaf.

We also see that the weight of the farthing loaf changed in rough inverse proportion to the price of grain (see also p.204). The farthing was the smallest unit of money exchangeable:[10] finer divisions of the penny were not

*For full text see Appendix A (a)

†This great variety of bread would not be generally available. Villages and small towns might provide two or three. Our later information indicates three principal kinds; white, wheaten, and household. The nomenclature also might change from place to place with such variants as 'tourte' and 'blanc peyne' appearing.

Wheat Price per Quarter	*Weight of Farthing Wastel Loaf*			
	Henry II	*1202*	*1266*	*1290*
6/-	16 shillings	16 shillings	22*s* 8*d*	21*s* $8\frac{1}{4}$*d*
5/6	20 shillings	20 shillings	24*s* 8*d*	24*s* $8\frac{1}{4}$*d*
5/-	24 shillings	24 shillings	27*s* $2\frac{1}{2}$*d*	27*s* $2\frac{1}{2}$*d*
4/6	30 shillings	32 shillings	30*s*	30*s*
4/-	36 shillings	36 shillings	36*s*	36*s*
3/6	42 shillings	42 shillings	42*s*	42*s*
3/-	48 shillings	48 shillings	48*s*★	48*s*
2/6	54 shillings	54 shillings	54*s* $4\frac{3}{4}$*d*	54*s* $4\frac{3}{4}$*d*
2/-	60 shillings	60 shillings	£3 8*s* (68*s*)	68*s*
1/6	66 shillings	77 shillings	£4 10*s* 8*d* (90*s* 8*d*)	90*s* 8*d*
1/-	–	–	£6 16*s* (136*s*)	136*s*

practicable, hence the variation in weight rather than in price.

To obtain a scale of relative weights of loaves of different kinds of bread we note that when grain was 12*d* a quarter, the 1266 and 1290 Assizes require the farthing wastel to weigh 6 pounds 16*s*, that is 136*s*. Based on this the following table can be made from the description (on page 195):

Type of Bread	*Weight of Farthing Loaf (wheat at 12d a quarter)*
Simnel	134*s*
Wastel	136*s*
Cocket	138*s* or if of inferior corn 141*s*
Treet	272*s*
Whole Wheat	$211\frac{1}{2}$*s*
Common Wheat (mixed grain)	282*s*

We assume here that the 'great cocket' is the one weighing 5*s* ($\frac{1}{4}$ lb) more than wastel, and this establishes the weights of treet, whole, and common wheat loaves.

The implied exclusive use of wheat is surely unlikely especially as the document *Judicium Pillorie,*[11]† also of uncertain date but again said to be of 1266, states that when the price of bread is being settled, account should be taken of the price of wheat *and oats*. The *Statutum de Pistoribus,*[12]‡ again stated to be of 1266, gives the punishment for the adulteration of oatmeal, so presumably oats were also incorporated into the loaf of the thirteenth century. The *Statutum de Pistoribus* reads, in part,

★Given correctly in another version as 45*s* 4*d*.
†For full text see Appendix A (b)
‡For full text see Appendix A (c)

And if any presume to sell the Meal of Oats adulterated or in any other deceitful manner, for the first offoence he shall be greviously punished; for the second he shall lose all his Meal; for the third he shall undergo the judgement of the Pillory; and for the fourth he shall abjure the Town.

Troy Weight for Bread

It will be observed that the weights used to define the farthing loaf are given in pounds, shillings, and pence, that is in currency units, but the pound is that of the Troy scale. Being so prime a necessity it is not unreasonable to weigh it by Troy weight rather than by the 15 oz *libra mercatoria* or the Avoirdupois pound. Indeed we have suggested earlier that Avoirdupois weight did not enter England's commerce until the late thirteenth or early fourteenth centuries, so this could not be the pound intended in any case. The Coventry Leet Book for 1474[13] says that:

> . . . XXXII graynes of wheat take out of the mydens of the Ere makith a sterling otherwise called a penny and XX sterlings maketh an ounce and XII ounce maketh a Pounde for sylver, golde, *bred* and measure. . . .

From quotations such as the one just given, and from the fact that bread weight was given in pounds, shillings, and pence, it might be thought, as some have done,[14] that bread was weighed by Tower weight, as was the coinage; but this cannot have been the case, for it would be most unlikely to find Tower weight in use, far less in common or everyday use, outside the premises of the moneyers. By 1500 this question is completely resolved, with the weights of loaves being given in ounces and pennyweights, that is in Troy weight (see *Arnold's Chronicle*,[15] below), nor is there any indication of a change from another weight to that of the Troy system at this time. That bread is to be weighed by Troy weight is stated more than once in a document of 1496 (Harl Mss Vol.698 f 64–5),[16] for example 'And XII unces make a li of troy weight for selling golde *breade* and measure' and again 'Twenty pence the unce of troy is for sylver and golde and *brede* . . .'.

The use of Troy weight for bread is again made quite plain when we read in the 'Declaration of the Several Waightes and Measures according to the Standard' (1588):[17] 'Troy weight in the Queen's Exchequer. Troy weight is most used by the ounce for gold and silver, *bread*, electuaries which is muske and civett and such like things.' Throughout, bullion weight is in use for bread, a situation to be changed only at the beginning of the eighteenth century, as we shall see.

Payments and Penalties

All four assizes give the payments and allowances due to the baker for his out-of-pocket expenses and his profit. For ease of reference they are shown in the table on this page, from which we see considerable consistency over a period exceeding a century. Only the Assize of 1266 shows much fluctuation, the baker receiving 4*d* for every quarter baked instead of 3*d*, but the allowance for firewood is 2*d*, not 3*d*, and only in the Assize of the time of Henry II is the payment for servants less than ½*d* each.

PAYMENTS AND ALLOWANCES
Per quarter baked

	Henry II	*1202*	*1266*	*1290*
To the Baker	*3d + bran + 2 loaves*	*3d + bran + 2 loaves*	*4d + bran + 2 loaves*	*3d + bran + 2 loaves*
Servants	For 4 – 1½*d*	For 4 – 2 *d*	For 3 – 1½*d*	For 3 – 1½*d*
Boys	For 2 – ¼*d*	For 2 – ¼*d*	For 2 – ½*d*	For 2 – ½*d*
Salt	½*d*	½*d*	½*d*	½*d*
Yeast	½*d*	½*d*	½*d*	½*d*
Wood	3*d*	3*d*	2*d*	3*d*
Candles	¼*d*	¼*d*	¼*d*	¼*d*
Sieve (sifting)	½*d*	½*d*	½*d*	½*d*

The concluding paragraphs of the 1266 Assize specify that a baker who offends by selling light

> in the weight of a farthing loaf of bread not above 2*s* weight that then he be amerced [fined] . . .; but if he exceed 2*s* then he is to be set upon the pillory without any redemption of money.
> In like manner shall it be done if he offend often and will not amend, then he shall suffer the judgement of the body, that is to say, the pillory if he offend in weight of a farthing loaf under two shillings weight as is aforesaid.

Cunningham[18] gives the price of wheat in 1289 as 6 shillings the quarter*, hence in that year the weight of the farthing wastel loaf would be (from the table in the Assize) 22 shillings and 8 pence. The baker would thus escape the pillory if no more than 9 per cent underweight.

The Baker's Dozen

To meet such narrow tolerances in the thirteenth and fourteenth centuries would be no mean feat. It would be difficult today without modern methods and technology. Each baker was required to place his mark on the loaves so that they might be identified. It is little wonder that the baker's dozen (13) came into being under these circumstances. In *Liber Albus*[19] compiled in 1419,

*It would appear that a figure of about 6 shillings the quarter was the average price of wheat from 1261 to 1400.

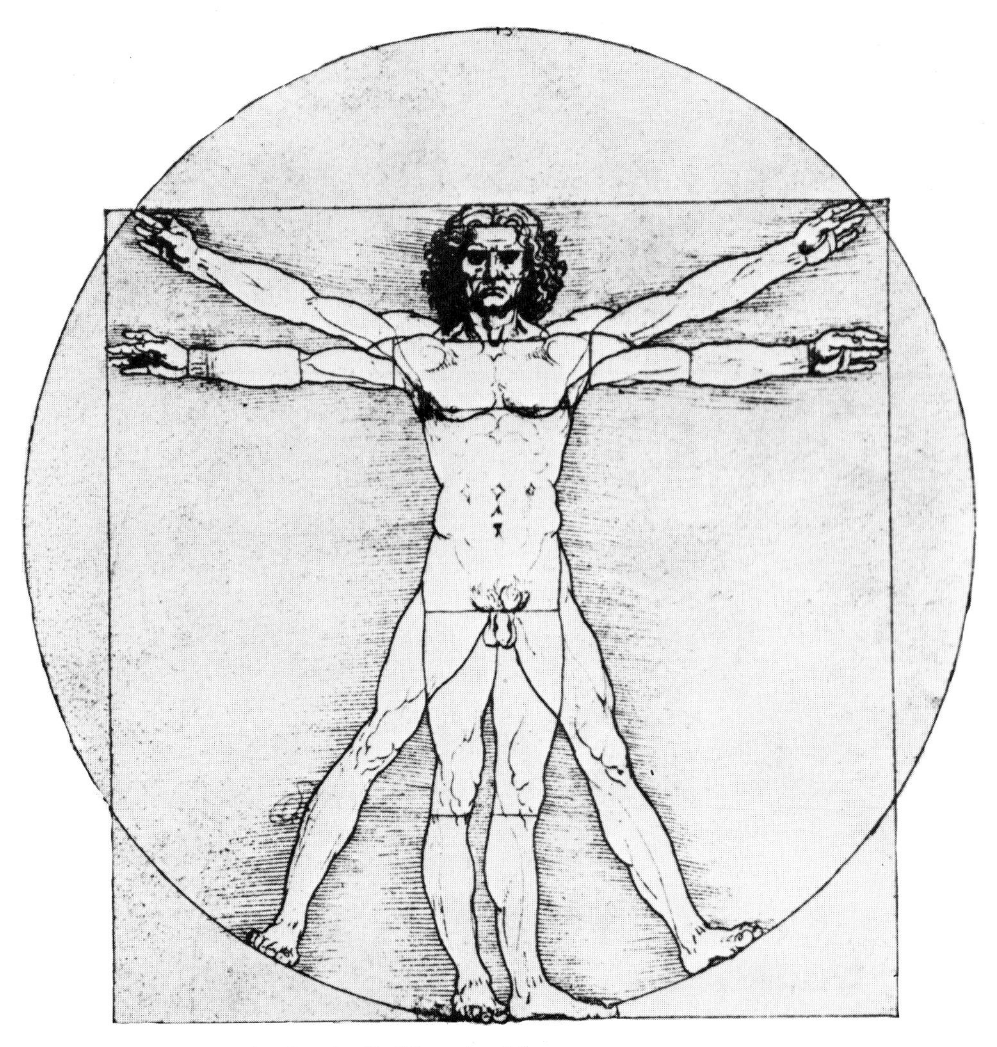

Fig 1 Leonardo da Vinci's Vitruvian Man.

Fig 2 Egyptian workmen measuring land with a rope knotted at intervals. From a wall painting *c.*1400 BC.

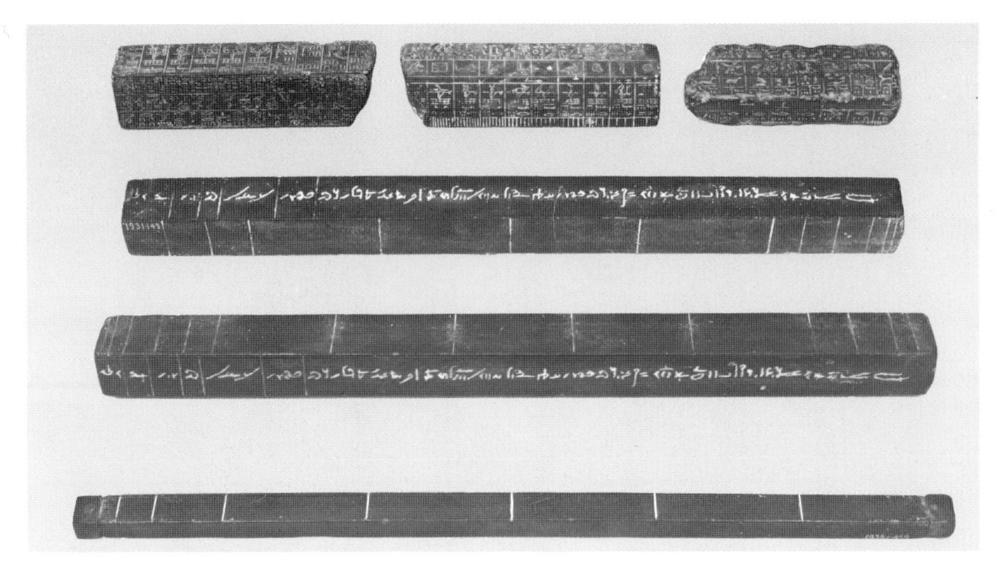

Fig 3 Replicas of Egyptian royal cubits.

Top: Basalt fragments *c.*2500 BC.
Centre: Two views of a complete basalt cubit *c.*250 BC.
Bottom: Wooden cubit with bronze-capped ends. First century AD (Roman Occupation).

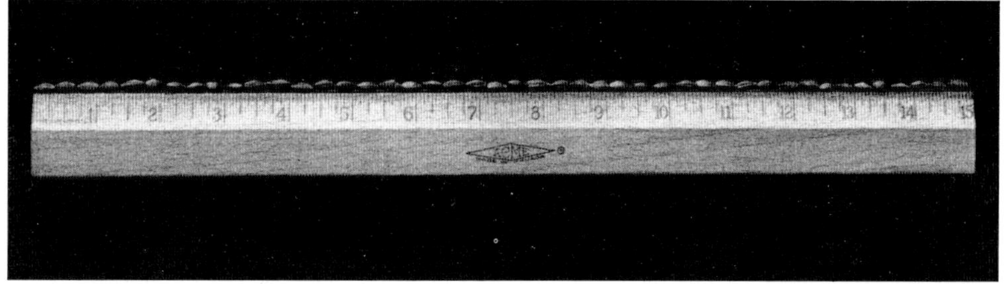

Fig 4 A 15-inch ruler along which are laid 45 barley corns from the 1936 crop showing their agreement in length to within ½ per cent.

Fig 5 A collection of Egyptian stone weights of the peyem, qedet, and bequa systems *c*.2900 – 2000 BC.

Fig 6 Bronze 'ox-hide' ingot from Eukomi, Cyprus *c*.1200 BC. Weight 37.02 kg.

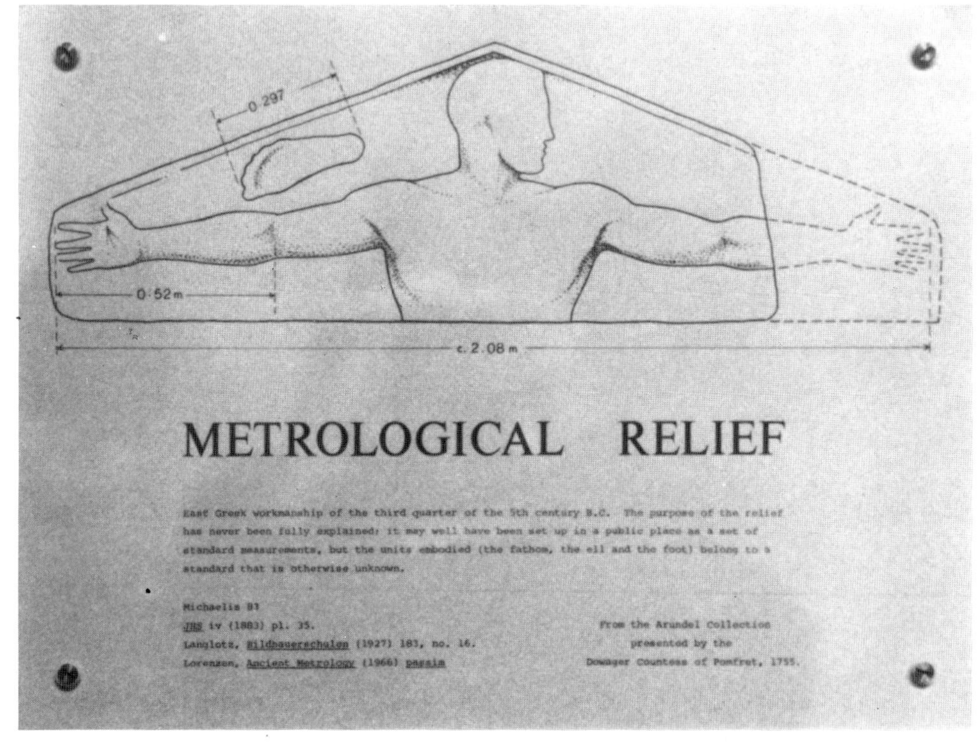

Fig 7 The Arundel relief in the Ashmolean Museum, Oxford, *c*.450 BC.

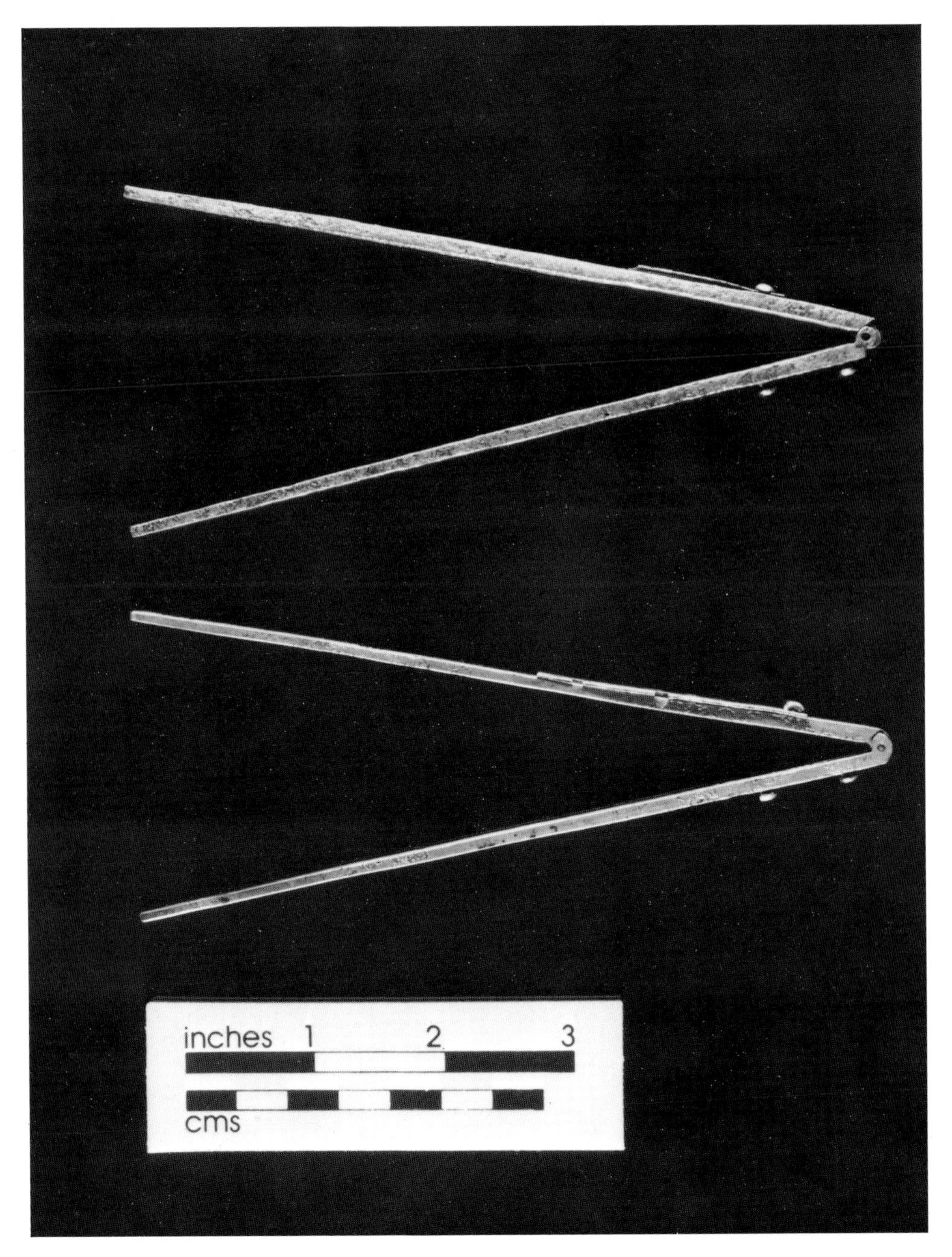

Fig 8 Roman footrules of the second century AD.

Fig 9 The Roman *denarius* (a) showing the symbol X to the left of the bust indicating its value as 10 asses; (b) showing the symbols XVI indicating its value as 16 asses; and (c) showing the symbol ★ to the lower right of the bust.

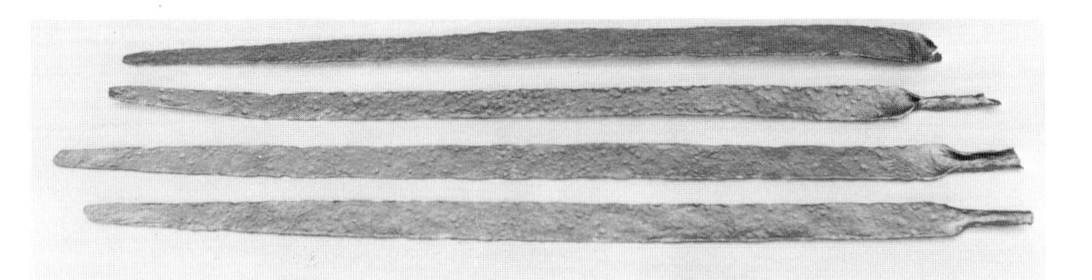

Fig 10 Four currency bars from the Winchester area.

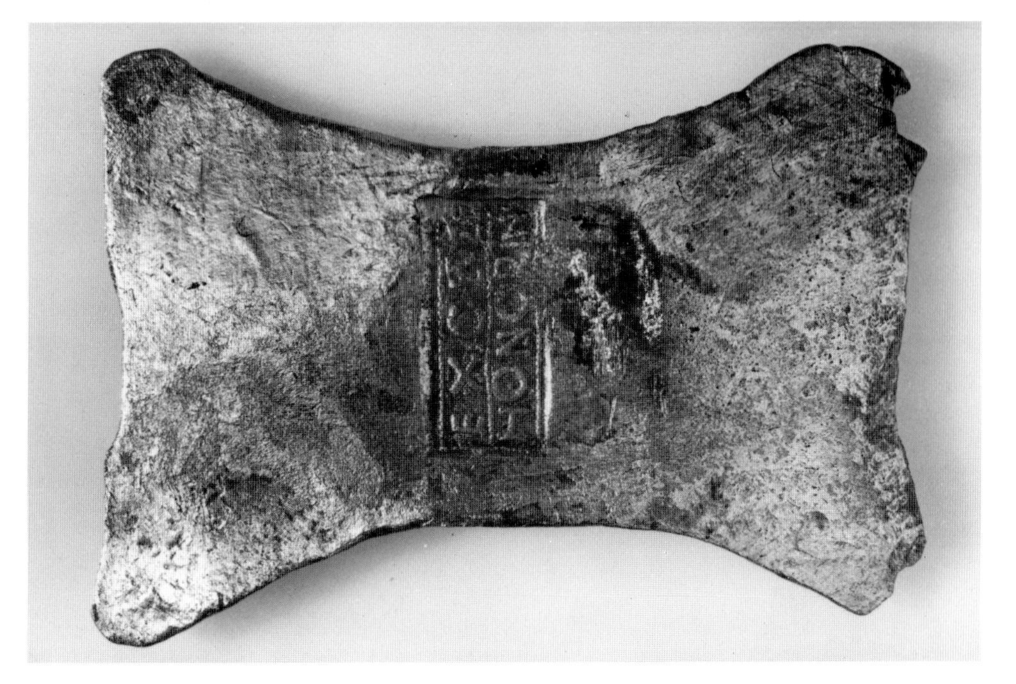

Fig 11 Silver 'axe-head' ingot of the time of Emperor Honorius (AD 395–423). Its weight was only 4 g short of the Roman pound of 327.5 g.

Fig 12 Roman scales and bronze weights from Pompeii AD 79.

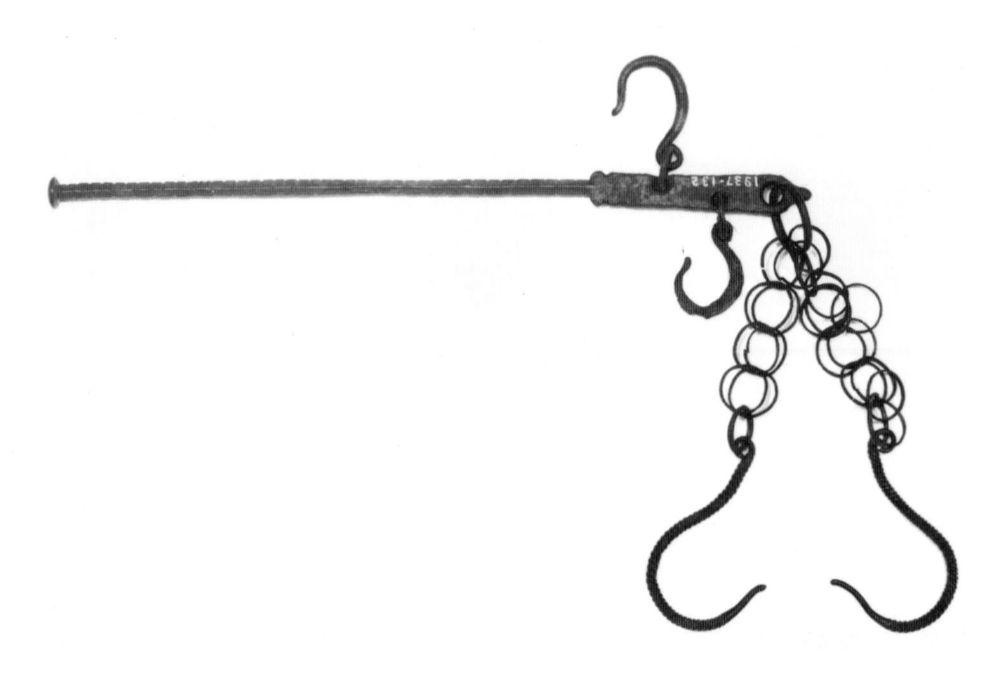

Fig 13 Roman-Etruscan steelyard.

Fig 14 (a) A sixteenth-century Swedish woodcut of a scene depicting a bismar in use (front, left).

Fig 14 (b) A Norwegian bismar similar to that in *Fig 14* (a) and of the same century.

Fig 15 A Malaysian bismar (steelyard) of the twentieth century.

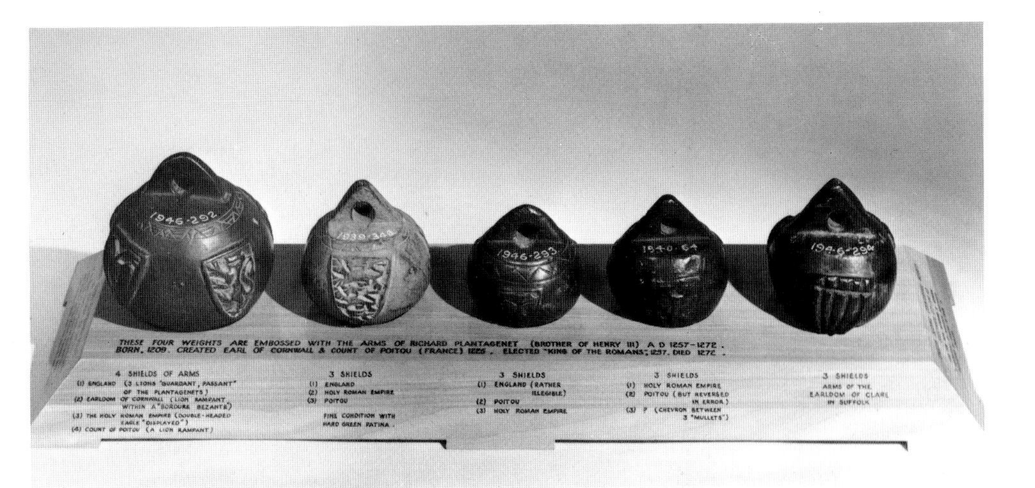

Fig 16 A group of five ornamental counterpoise weights for steelyards.

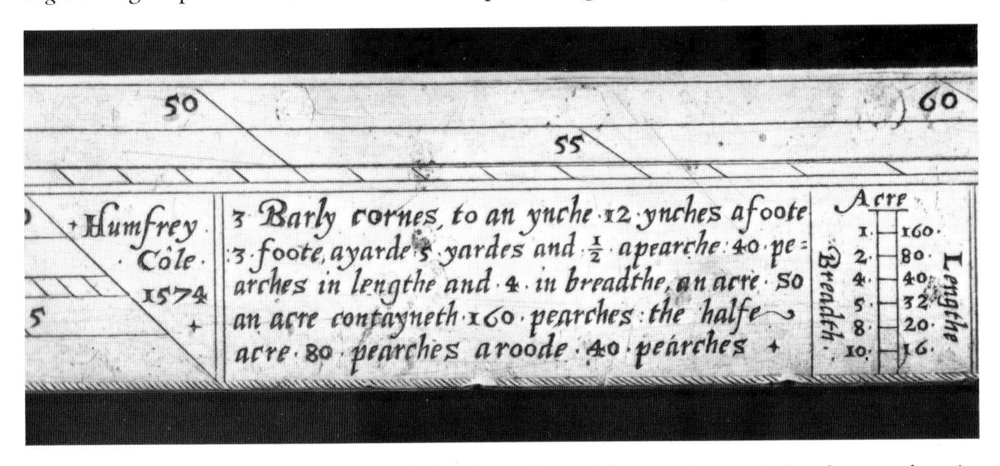

Fig 17 Inscription on a surveyor's rule by Humfrey Cole (1530?–1591) dated 1574, showing '3 barley corns = 1 inch' (detail).

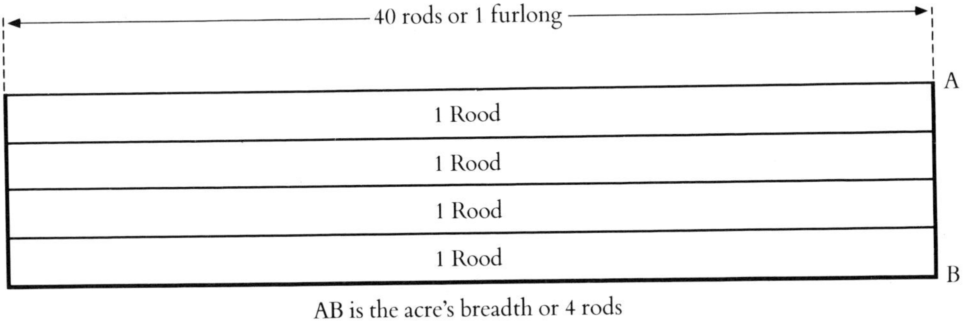

40 rods or 1 furlong

A

1 Rood

1 Rood

1 Rood

1 Rood

B

AB is the acre's breadth or 4 rods

Fig 18 Diagram of the acre and rood showing the acre 40 rods (1 furlong) by 4 rods with the rood 40 rods long and 1 rod broad.

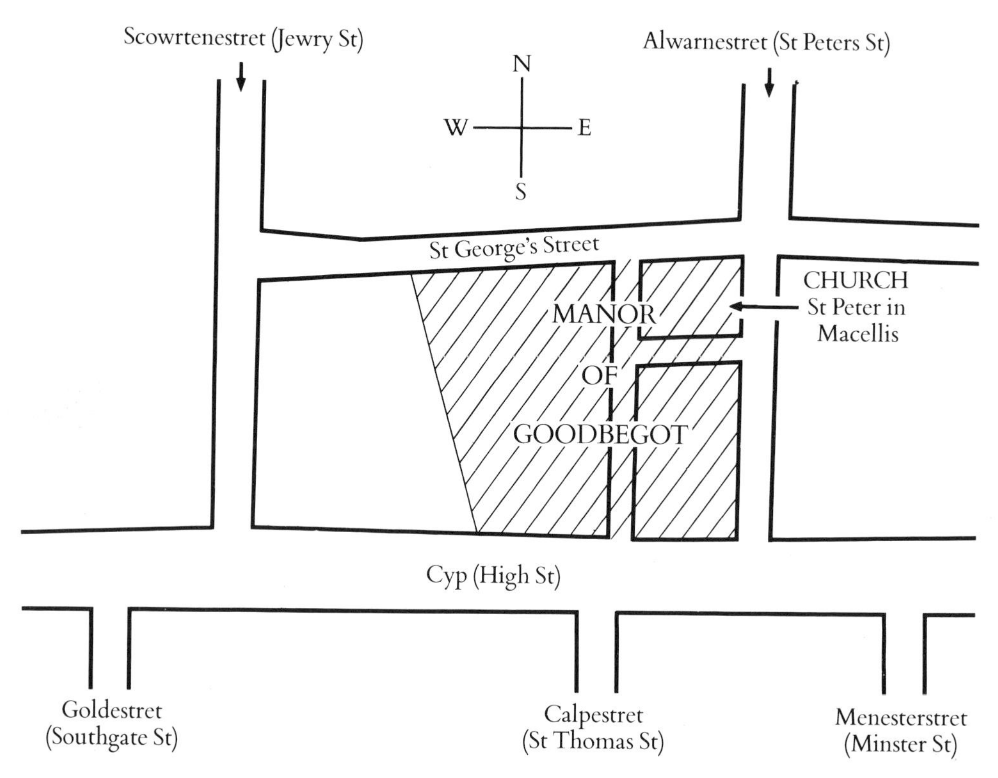

Fig 19 Tarrage map of the manor of Goodbegot (Godbegot), Winchester, 1416, (redrawn). Modern street names are in parentheses.

Fig 20 The Manor of Godbegot in 1980.

meyn Meßrůte / darmit mann das Feldt meſſen ſoll/ Vnnd ge-
ſchicht in geſtalt wie in nachfolgender Figur angezeygt wirdt.

So nun wie oben gelehrt vnnd angezeygt/ die ſechtzehen Per-
ſonen nach einander / ieder einen fůß fůrgeſetzt hat/ vnnd die Růt

Fig 21 A woodcut from Jakob Kobel's *Geometrie* of 1531, showing how to determine the land rod by lining up 16 men, heel to toe, in soft leather shoes, as they emerged from church.

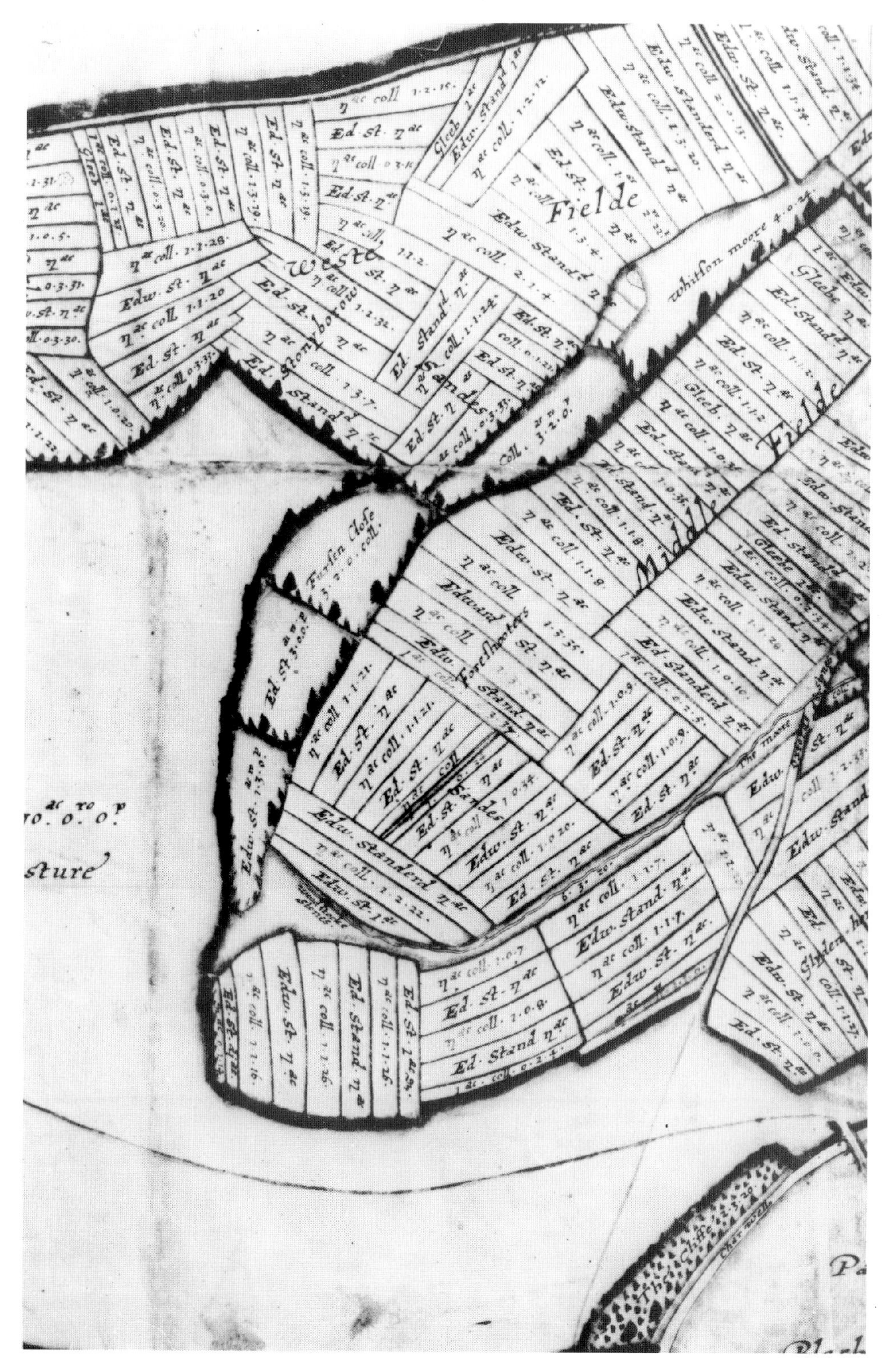

Fig 22 Map of the village of Whitehill in Oxfordshire 1605, showing Mr Standerd's nominal two-acre strips and the college's surveyed strips (detail).

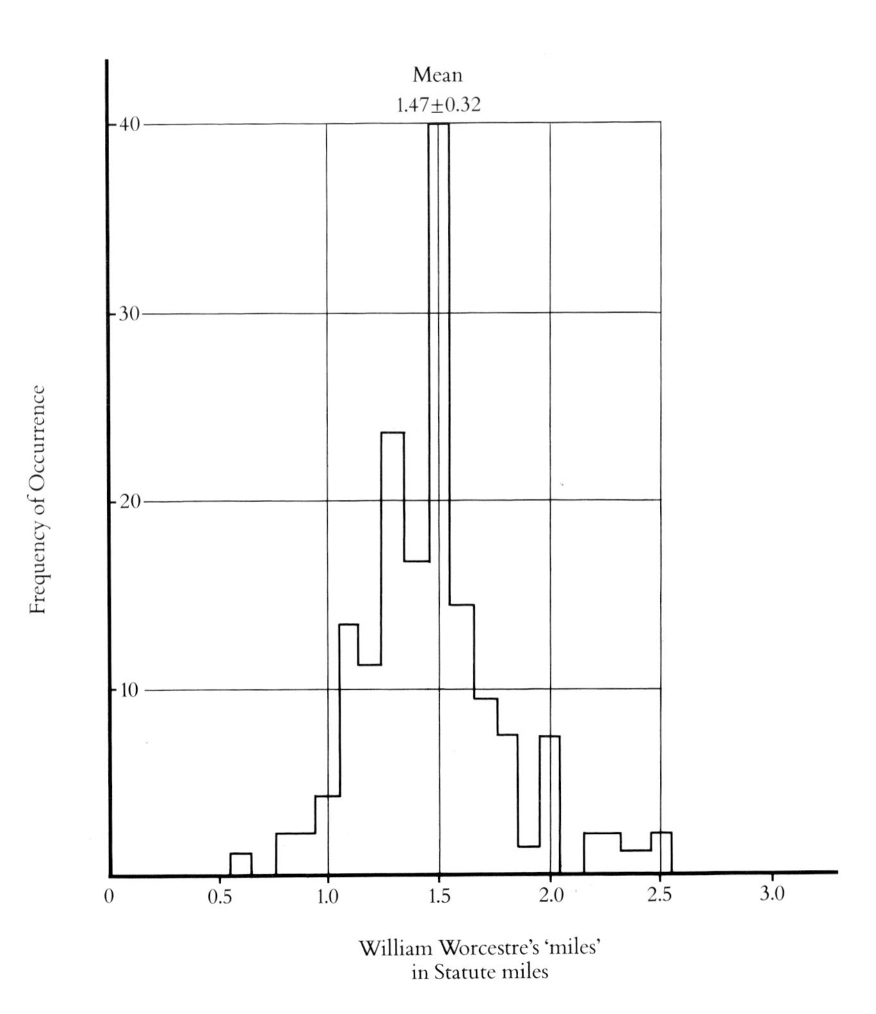

Fig 23 A histogram of the length of William Worcestre's mile in terms of the statute mile, for 161 of his distances, showing a mean value 1.47 ± 0.32 statute miles.

Fig 24 Bronze plaque of linear measures for public display and use.

Fig 25 A *sceatta* and a penny.

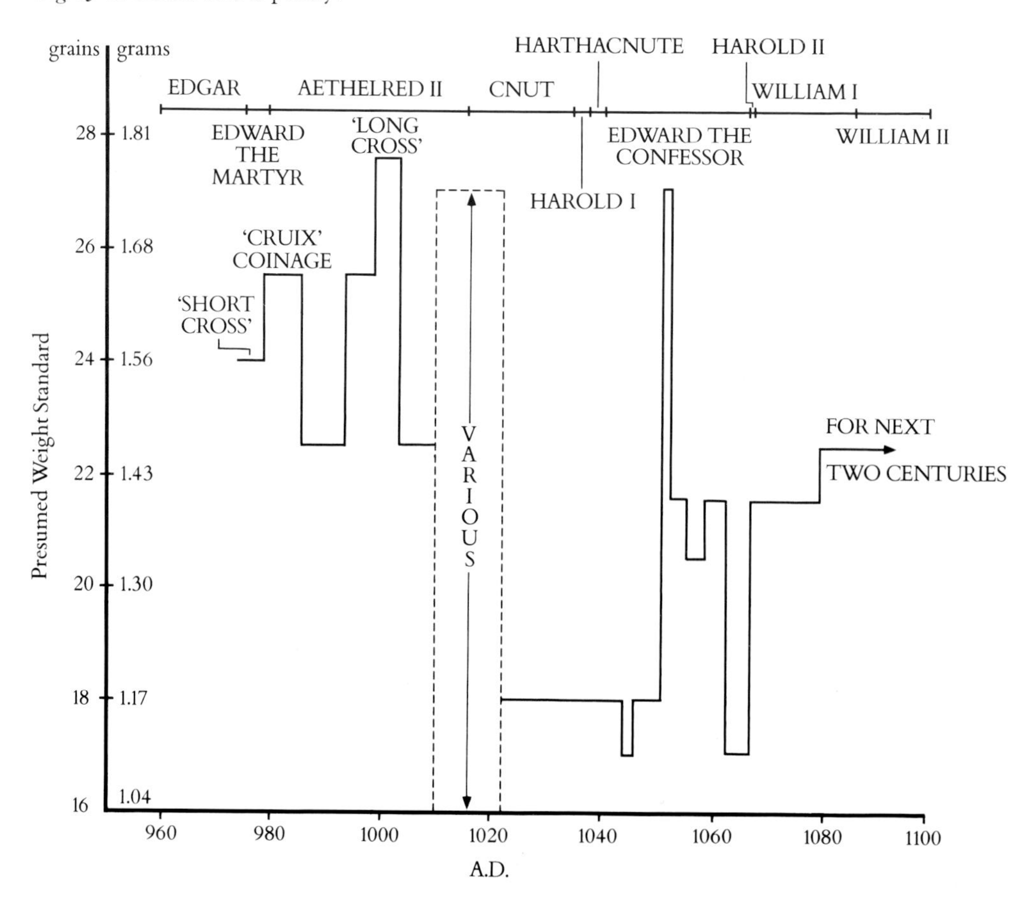

Fig 26 Diagram showing the variations in the average weight of the silver penny from *c.*AD 960 to *c.*1090, i.e. from Edgar to William II.

Fig 27 Lead weight of 8 Troy ounces, surmounted to a bronze gilt disk engraved in a basket-weave pattern said to be eighth- or ninth-century (Streeter Collection of Weights and Measures, Medical Historical Library, Yale University).

Fig 28 (a) The City of Winchester's bronze Avoirdupois weights of Edward III, *c.*1357. The weights are of 91, 56, 28, 14, 7, and 7 lb respectively; (b) Detail of 56-lb Edward III weight.

Fig 29 A 6¼-lb Avoirdupois clove of Edward I.

Fig 30 A window in Tournai Cathedral, Belgium, showing a fifteenth-century market scene with the great beam in operation.

Fig 31 Seven-pound wool weight of Henry VII.

Fig 32 Twenty-eight pound wool *tod* of James I.

Fig 33 Fourteen-pound wool weights of George I and III.

Fig 34 Exchequer bushel and gallon of Henry VII, 1497.

Fig 35 Exchequer bushel, two gallons, quart, and two pints, Elizabeth I, 1601-2.

Fig 36 Derby City's pint of 1601 and its leather case.

Fig 37 Exchequer wine gallon of Queen Anne, 1707.

Fig 38 Standard quart and pint of William III, City of Winchester, 1700.

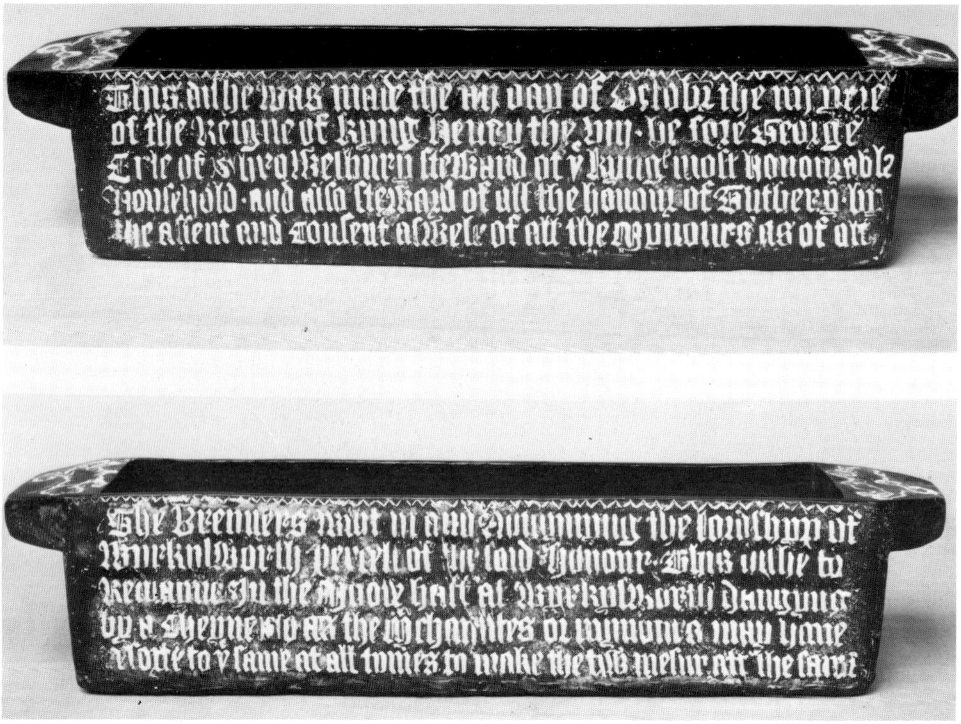

Fig 39 The Wirksworth Dish.

Fig 40 Exchequer standard coal bushels of George II, 1730.

Fig 41 Drawing from *Liber Albus* showing: (a) a baker at work; (b) a baker being drawn on a hurdle through the streets with the faulty loaf about his neck; (c) drawings of the pillory.

Fig 42 A weight for the quartern loaf (4 lb 5 oz 8 dr), 1804.

Fig 43 The churchyard wall at Great Wishford, Wiltshire, showing the price of the gallon loaf from 1800.

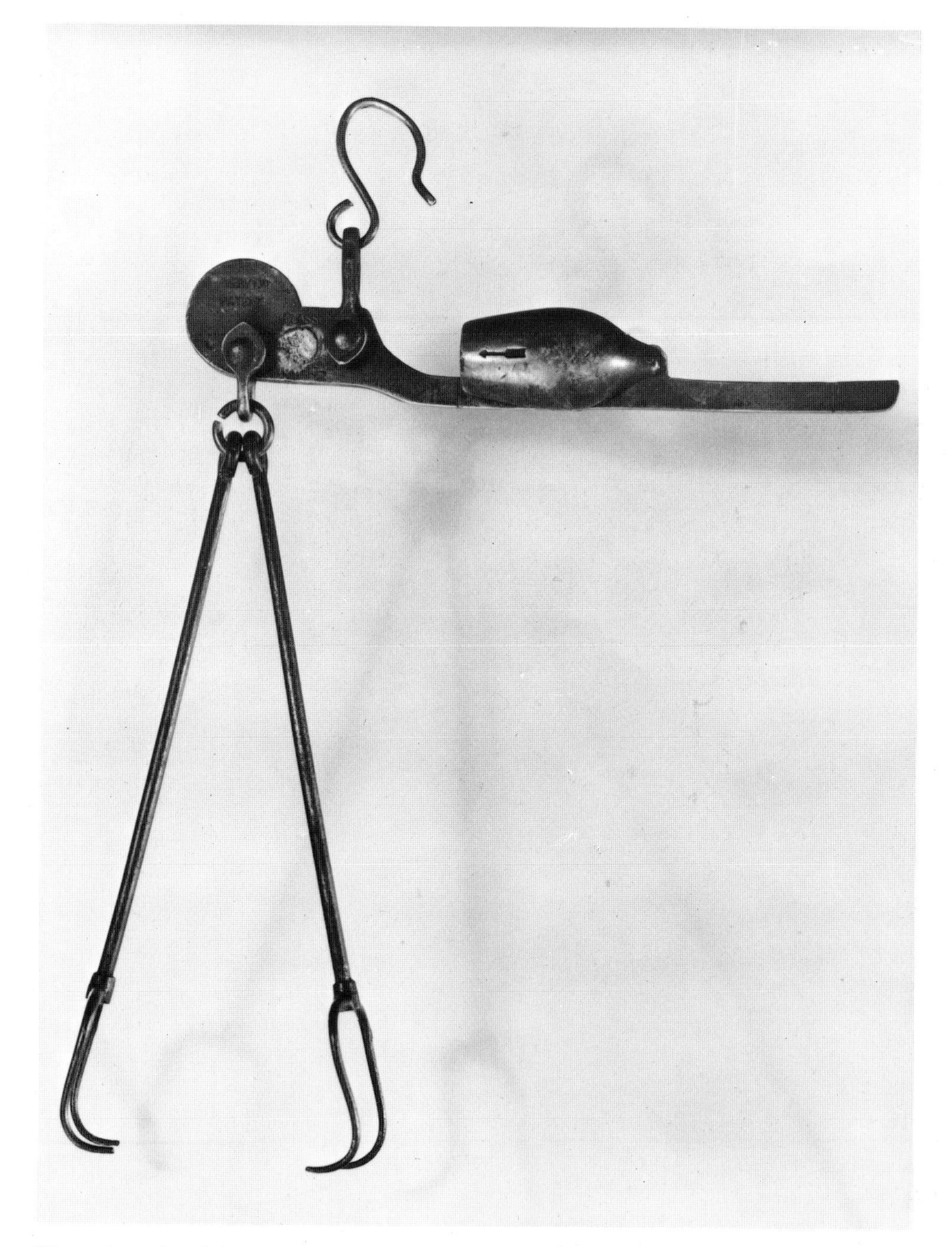

Fig 44 A steelyard for weighing 4-lb or 2-lb loaves.

Fig 45 The silver yard of the Merchant Taylors' Company being carried by the beadle Mr L. C. Oakman before the master Mr A. T. Langdon-Down as they proceeded to 'make search' at the Cloth Fair, Olympia, London, in 1979.

Fig 46 (a) Winchester's bronze yard of Henry VII, restandardized in the reign of Elizabeth I; (b) detail of the 'h'-end; (c) detail of the 'E'-end.

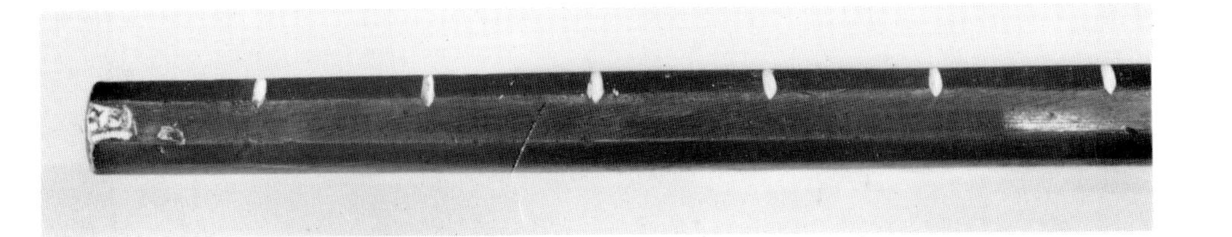

Fig 47 (a) Exchequer standard yard of Henry VII, showing the inches; (b) King Henry VII presiding over a trial of weights and measures. Unjust measures are being burned in the right foreground (From Harleian manuscript, British Museum, MS 698 f 64–5) 1496.

Fig 48 The Avoirdupois weights of Henry VII from 112 lb to 2 lb, 1497.

Fig 49 Exchequer standard Avoirdupois weights of Elizabeth I. First Series, 1558.

Fig 50 Standard Avoidupois weights of Elizabeth I, 1588. Third Series.

Bell-shaped. 56 lb – 1 lb
Disk-shaped. 8 lb – 2 drams

Fig 51 Standard Avoirdupois (disk) and Troy (cup) weights of Elizabeth I, 1588. Third Series.

Disk-shaped 8 lb – 2 drams (Avoirdupois)
Cup-shaped 256 oz – 1 oz and 10, 5, 2, and 1 dwt Troy.

Fig 52 Conical standard measures, George IV, 1824. The quart, gallon, and pint.

Fig 53 Imperial standard bushel, 1824.

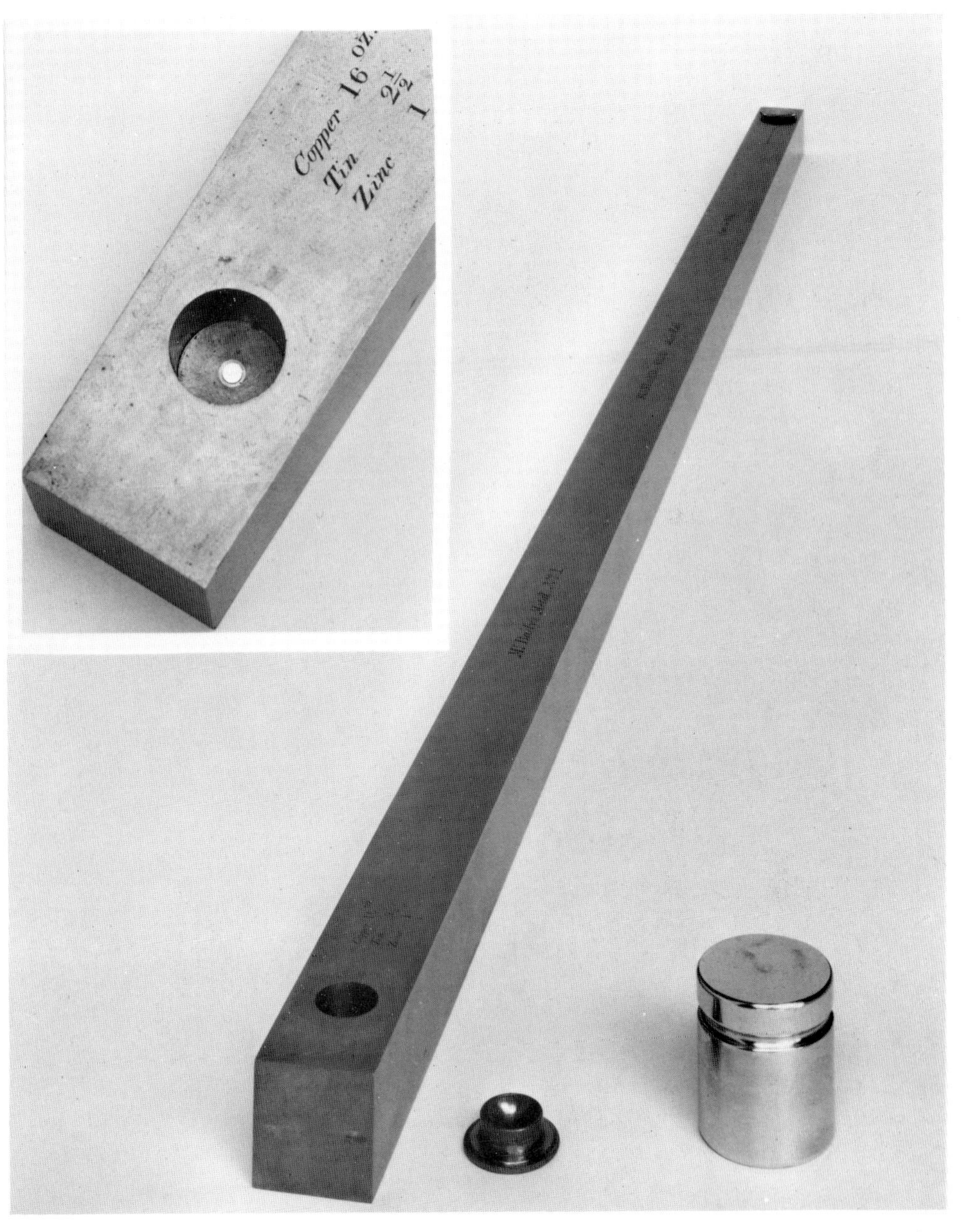

Fig 54 Imperial standard yard (1845) and pound (1844) in bronze and platinum respectively – Board of Trade, London, now at the National Weights and Measures Laboratory, London.

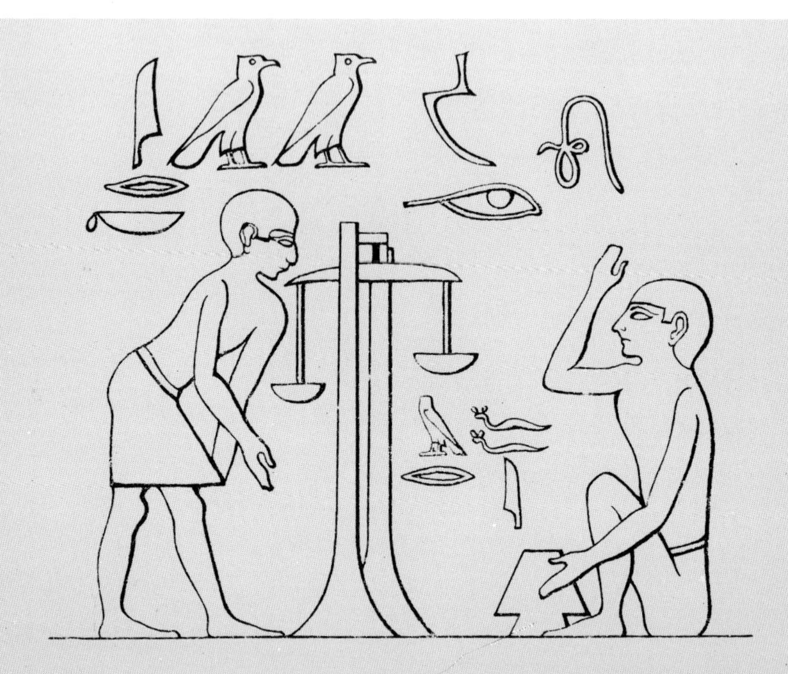

Fig 55 (a) Butter scales (nineteenth century); (b) A drawing showing Egyptian scales of *c.*2500 BC which are not very different in design from the butter scales.

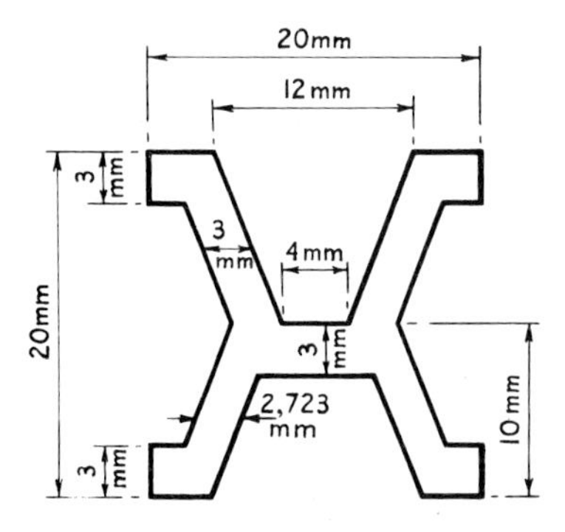

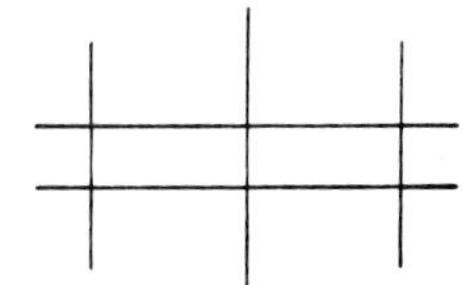

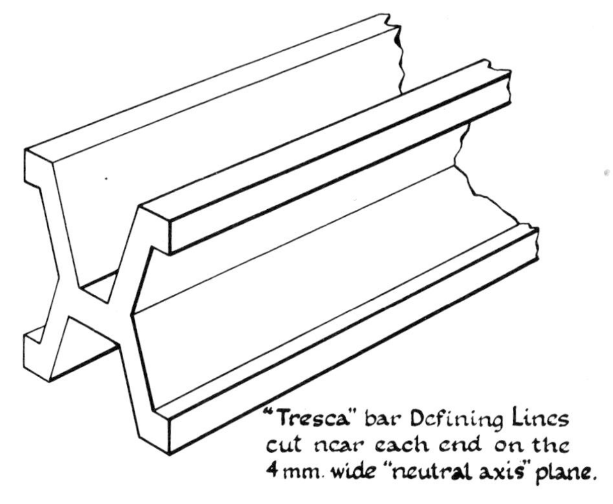

Fig 56 Sectional diagram with dimensions of Tresca's winged section for the metre.

Fig 57 Standard kilogram British copy No. 18.

Fig 58 French *cadil* (litre), made in 'An2' i.e. 1793–4.

covering events in London from roughly 1250 to 1380 with additional documents to the mid-fifteenth century, we read under 'Regulations for Bakers' as follows: 'Item, that they [bakers] shall sell unto the Hucksters only 13 loaves as 12 without gift or other courtesy'. This is repeated in 1522 for the City of Coventry:[20] 'Ordained that the bakers shall bake no horse-breds* but of good stuffe and that they sell no more for a penny but III lovis and XIII to the dezen . . .'. A similar provision for London is made in 1684 in the 'Orders and Articles made by the Privy Council on information from John Powell, Clerk of the Market of his Highness' most Honourable Household',[21] namely, that 13 penny's worth of horse-bread shall be given for 12 pence.

We would note that to give an extra one in twelve (8.33 per cent) would compensate nearly enough for being light 2 shillings in $22\frac{2}{3}$ shillings (8.29 per cent), but in so doing the profitability was substantially impaired and there is little doubt that short weight was given on many an occasion by accident or design. Part of the baking operation became clandestine, with some unmarked loaves appearing and being bought each day.

The 'Statute concerning Bakers and Brewers' *(Statutum de Pistoribus)*[12] already mentioned, attributed variously to 51 Henry III (1266) or to 13 Edward I, (actually of uncertain date but most probably belonging to the thirteenth century) begins by declaring that the Assize of Bread is to be kept and that a baker who is short in weight in the farthing loaf by not more than 2*s* 6*d* shall be fined but if over in deficiency he shall suffer punishment of the pillory, 'which shall not be remitted to the offender either for gold or silver'. Here the deficiency warranting the pillory is slightly greater than that noted previously, but in 1474, two centuries later, we are told[22] that:

> . . . if he lack 1 ounce in the weight of the farthing loaf he is to be merced [fined] at 20*d*, if he lack $1\frac{1}{2}$ oz. he is to be merced at 2*s* in all manner of bread so lacking and if he bake over that assise he is to be judged unto the pillory.

Previously he was put in the pillory if he were 2*s* ($\frac{1}{10}$ lb) or 2*s* 6*d* ($\frac{1}{8}$ lb) light. Now he goes to the pillory if $1\frac{1}{2}$ oz ($\frac{1}{8}$ lb) light, which is the same as 2*s* 6*d* and his fines are written down beginning at one ounce short weight. At first sight this seems very little but with wheat then close to 6/6 a quarter[18] the farthing (wastel) loaf would weigh only 8 oz or so (see pp.196, 203), which means that his short weight is one part in eight.

Enforcing the Assize

The document *Judicium Pillorie*[11] mentioned earlier provided that:

> Six lawful men shall be sworn truly to gather all measures of the town . . . [and] one

*Horsebread: a coarse bread made from peas and beans intended as food for horses but no doubt eaten by the poor. By the Statute, 22 Edward IV 1482, it could be bought at 9 or 12 lb for 1*d* or about half the price of wheaten bread.

loaf of every sort of bread . . . and also upon every loaf the name of the owner distinctly written. . . . And the bailiffs shall be commanded to bring in all the bakers and brewers with their measures . . .'.

Fleta in 1290[23] says the same thing:

> And the bailiffs are immediately to send before him [the coroner] at least six of the more law-worthy men of the whole township who, when they have been sworn are instructed by the coroner to make an honest collection of all the measures and all the weights in that town. . . . Moreover they shall cause all and sundry who make and offer bread for sale in that bailiwick to send at least one average loaf of every kind of bread. And these measures, weights and loaves are to bear the names of their owners. . . . And in the meantime the local bailiff is to send before him all the bakers, ale-wives, and innkeepers and all the owners of the aforesaid weights and measures.

The Customal of Sandwich in Kent,[24] written in 1301 (though extended later to the commencement of the reign of Edward IV), sets out the Assize of Bread, the manner in which the Assize is to be conducted, and the penalties for short weight. Not surprisingly, the weights given for the farthing wastel loaf are very similar to those of Fleta of 1290, the only variants being as follows:

Price of Wheat (qr)	*Weight of Loaf* (1301)	*Weight of Loaf* (1290)
18*d*	4lb 10*s* 6*d*	4lb 10*s* 8*d*
6/-	22*s* 8*d*	21*s* 8¼*d*

The Customal adds that for wheat at 6/6 a quarter, the loaf should weigh 20*s* 11*d*. The baker's profit is to be 4*d* plus the bran and two loaves for each quarter baked. He also gets cash allowances as follows: 1½*d* for 3 servants, ¼*d* for 2 boys, ½*d* for salt, ½*d* for yeast, ½*d* for candles, 3*d* for wood, and ½*d* for sifting the flour. These are a little different from Fleta, the overall difference being that here the baker receives 1*d* more. Relative to the Assize of 1266 his cash gain is again 1*d*, though made up differently.

Bakers were not held in high esteem in Sandwich at this time. When an assize was to be made (and it was made three times a year), the mayor summoned the jurors 'secretly on account of the fraudulent disposition of the bakers'. Some went to one end of the town, some to the other, and 'in the most cautious manner' took samples of all the kinds of bread offered for sale. It is entirely consistent with what has been said earlier with respect to the weight being Troy weight that we read in the Customal that the 'person that weighs the bread is commonly some goldsmith'.

As for penalties, for the first two offences the baker was fined 21*d* and the whole baking forfeited. For a third offence in addition to these amercements the oven was to be pulled down and the baker had to leave the trade for a year and a day. 'This is the custom in our liberty because none here ever suffer the pillory or tumbrel or other such like punishments as are inflicted in the county'. But elsewhere the pillory held sway.

Fourteenth-century London

In the fourteenth century the penalties imposed on a baker in London whose loaves were defective were even more stringent than those of the preceding century;[25]

> And that two loaves shall be made for 1*d* and four loaves for 1*d* [i.e. halfpenny and farthing loaves] and that no loaf shall be coated with bran. . . . And that each baker shall have his own seal as well for brown bread as for white; so that it may be the better known whose bread it is. . . .
> And if any default shall be found in the bread of a baker of the City, the first time, let him be drawn on a hurdle from the Guildhall to his own house, through the great streets where there may be most people assembled and through the midst of the great streets that are most dirty, with the faulty loaf hanging from his neck. If a second time he shall be found in the same transgression let him be drawn from the Guildhall through the great street of Chepe, in form aforesaid, to the pillory and let him be put upon the pillory and remain there for at least an hour in the day. And the third time that such default shall be found he shall be drawn and his oven shall be pulled down and the baker made to forswear the trade within the City forever.

To make things even clearer the manuscript carries a drawing (Fig.41) of an unfortunate baker, loaf about his neck, being drawn by a horse through the streets on a hurdle.

In Edward II's reign, 1307-27, we read of two of the assistants of the bread assayers stealing two penny loaves. They were sent to the pillory with the loaves hanging about their necks.[26] And again a baker was sentenced to the hurdle for selling $\frac{1}{2}$*d* horse-bread loaves weighing 5*s* 11*d* and 7*s* 2*d* instead of the regulation 8*d* $7\frac{1}{2}$*d*.[27] Yet another offended with horse-bread and on 26 January 1379 was drawn through Chepe to Temple Bar.[27]

No doubt the later Statute of 13 Richard II Stat 1 c 8[28] of 1389-90 was an endeavour to stop this giving of short weight:

> . . . and that no hosteller make horse bread in his hostry nor without, but bakers shall make it; and the Assize thereof shall be kept that the weight be reasonable after the price of corn in the market . . .

Greater definition is given in 1482:[29]

> . . . as long as the price of beyns beyn at iiii *s* or above that every baxter of this city shall sell thre hors loffys for 1*d* and that every loffe shall weye thre pound; and if the price of beyns be undyr iii s that then every baxter of thys cite shall sell thre hors loffys for 1*d* and that every lofe shall weye four pound weght. . . .

Probably the last sentence to the hurdle occurred on 10 October 1438, when a baker was convicted of having a 1*d* loaf 10 ounces light in weight and, furthermore, made of unclean wheat and 'unseasonable dough'.* Thereafter the punishment was simply the pillory. The difficulties and penalties confronting the baker were considerable. Little wonder that even in the seventeenth century the term of apprenticeship was seven years, in common with that of other trades.[21]

These rules, regulations, penalties, and rewards should not be thought of as being of national application. The price of wheat varied from place to place and it was up to the local authority to set up its own weights for the farthing loaf. (By the end of the thirteenth century there was a $\frac{1}{2}d$ loaf too and later still a loaf costing 1*d*). The London Assize was the master plan which many cities and towns adopted or adapted to their own needs, usually with only minor modification.

Setting the Assize

The weights of the various loaves were determined in London by direct experiment. Wheat was bought, milled, and baked and the number of loaves derived from a quarter was ascertained. After assessing the various payments and allowances, the weight of the loaf to be given for a farthing was calculated and promulgated by the magistrates. Trials of the bread in 1311 and in 1517 are recorded.[21] The Customal of Norwich dating from the early fourteenth century states:[30]

> Also for keeping the Assize of Bread in the city let four honest and lawful men be chosen yearly and let them be sworn in the presence of the community, to wit, two from the office of baker and two others from the lawful men of the city. These four to buy corn, cause it to be ground and bolted and baked and the bread sold and the baker has for his trouble the ancient common assize used in the city. The Bailiffs are to supply the cash to buy the corn until the bread is sold. The Assay to be held twice a year, at the feast of St. Michael and after Easter.

The Assize of c1500

For the end of the fifteenth century an assize in some detail was published in *Arnold's Chronicle* (1st Edition 1502).[15] This gives the weight of the various loaves according to the price of wheat in the range 3/- to 20/- a quarter but omits any note of the baker's allowances. These can however be found elsewhere as we shall see. In Arnold's table the weights are given in standard Troy units, that is, ounces and pennyweights (there being 20 dwt to the

*The statement in the introduction to Vol.I of *Liber Albus* (p *ci*) that the hurdle was abolished in the reign of Edward II is inconsistent with the statements quoted in Vol.III of that work and given above.

ounce), and not money units as heretofore. Three entries will serve to indicate the scope of the table given by Arnold:

AD 1500			
Price of wheat per quarter	*3/-*	*6/6*	*19/6*
Wt of farthing symmell	$15\frac{3}{4}$ oz	$8\frac{3}{4}$ oz + $\frac{1}{2}$ dwt	2 oz + 2 dwt
Wt of $\frac{1}{4}d$ white coket	$17\frac{3}{4}$ oz	$10\frac{3}{4}$ oz + 1 dwt	5 oz
Wt of $\frac{1}{2}d$ white	25 oz + 1 dwt	$21\frac{1}{2}$ oz + 2 dwt	10 oz
Wt of $\frac{1}{2}d$ wheaten	$52\frac{1}{2}$ oz + $1\frac{1}{2}$ dwt	$32\frac{1}{8}$ oz + $\frac{1}{2}$ dwt	15 oz
Wt of 1*d* wheaten	$105\frac{3}{4}$ oz + $\frac{1}{2}$ dwt	$64\frac{3}{4}$ oz + 1 dwt	30 oz
Wt of $\frac{1}{2}d$ wheat 'of all greynes'	70 oz + 2 dwt	$43\frac{1}{8}$ oz + $1\frac{1}{2}$ dwt	20 oz

By comparison with 1266 we find a considerable price rise. With wheat at 3/- a quarter taken in both tables, instead of 576 dwt of white bread for a farthing we now get 501 dwt for *two* farthings ($\frac{1}{2}d$), an increase in price of a factor of 2.3.

There is an entry in another place of the allowances for the baker per quarter baked in 1495, nearly enough contemporaneously with Arnold's figures.[21]

Furnace and wood	*6d*	
Miller for grinding	*4d*	
2 men + 2 boys	*5d*	
Salt, yeast, candle & sackbands	*2d*	
The baker, his house, wife, dog + cat	*7d*	Total 2/-

(The part to be played by the dog and cat is not given.)

Whereas 2 men and 2 boys used to cost $1\frac{1}{2}d$ in baking a quarter this now amounts to 5*d*. The baker's wage goes up from 4*d* to 7*d*. Taken together, the wage bill for these people has risen by a factor of 2.2, almost exactly in line with the increase in the cost of bread. This increase in price thus appears to have been due to increased labour costs for the *Chronicle of the Grey Friars of London*[31] states that wheat was sold for 4/- a quarter ('VI d a bushelle') in the eighth year of Henry VII (1493), cheaper than it had been two centuries earlier.

The Flaw in the Assizes

It has been mentioned (p.195) that the weights of the farthing loaf in the thirteenth century changed in rough inverse proportion to the price of grain and this was true also in the fourteenth century. The pivotal entry in the Assize table is found by processing grain into bread and from the resulting weight and the known costs and allowances, determining how much can be given for a farthing. Thereafter the table was set up with each entry representing a change in the weight of the loaf for each rise or fall of 6*d* in the price of grain.

This is only valid, however, if grain is the only item of cost. We see it is not, for the baker has to have his cash allowance and the two loaves as well as his expenses for assistants, materials, fuel, etc.

Let us examine the 1266 Assize in detail, working the calculation for wheaten bread. With grain at 6/- a quarter (p.198) from 1261 to 1400, the farthing *wastel* loaf is to weigh $22\frac{2}{3}$ shillings (p.196). The weight of the *wheaten* loaf is, from the table on p.196, $\frac{22\frac{2}{3} \times 211.5}{136}$ shillings, that is 1.76 lb Troy. To pay for the grain alone 288 loaves must be sold. To meet the cash requirement ($9\frac{3}{4}d$), a further 39 loaves must be sold. Add in the two loaves for the baker and we require a total of 329 loaves each of 1.76 lb Troy (a total of 579 lb) from a quarter of wheat.

In 1800 we are told that a quarter of wheat weighing only 55 lb Avoirdupois per bushel, instead of the usual 59, was baked into 433 lb Avoirdupois of wheaten bread and 25 lb Avoirdupois of household bread.[32] These weights are 526.2 and 30.4 lb Troy respectively. In 1266 the bread would be coarser than in 1800, and further the poor quality of the grain used in the test of 1800 would yield fewer loaves than normally expected. Nonetheless we see a reasonably close fit between the 579 lb Troy required by the assize and the 526.2 + 30.4 lb Troy generated by the actual test.

However, if we repeat the calculation for the cheaper grain we find for grain at 3/6 the quarter some 682.6 lb Troy of bread is needed from the quarter to meet the cost of grain and the baker's allowances etc.* For grain at 1/- a quarter the impossibly large figure of 941 lb is demanded.†

Equally for grain at the high price of 12/- a quarter, the 1266 Assize gives the weight of the farthing wastel loaf as 11 shillings and 4 pence. The corresponding weight of the farthing wheaten would be $\frac{11\frac{1}{3} \times 211.5}{13.6}$ shillings or 0.88 lb Troy. To pay for the grain needs the sale of 576 loaves, the $9\frac{3}{4}d$ needs a further 39, and there are the baker's two loaves for a total of 617 loaves weighing in all 543 lb Troy. Such a weight of bread can be made with ease from a quarter of wheat and there would in fact be an excess.

Clearly high prices after the setting of the assize would favour the baker. If prices fell rapidly thereafter the baker simply could not meet the assize. Fortunately for the baker, prices tended to rise.

This defect in the assize was noticed prior to 1500 for in the assize given in the previous section as AD *c*1500 we no longer find the rough inverse proportionality, but for cheaper grain the weight of the loaf is much less. In

*The weight of the farthing wheaten loaf is now 42*s*.
†The weight of the farthing wheaten loaf is now 136*s*.

fact, the pendulum has swung the other way now and is equally faulty, for using the values for the 1*d* wheaten loaf in the table for 1500 we find for grain at 19/6 a quarter, 645 lb of bread is needed to pay for the grain and the baking costs (now 2/-), while 550.4 lb are needed with grain at 6/6 and 528.8 lb with grain at 3/- a quarter. It is now only when grain is dearest that the baker has difficulty in creating the necessary weight of bread to meet all of his outgoings and his own salary.

Of course, between assizes the price of grain did not fluctuate anything like as widely as the extremities of the table. The implication is that the assize was taken whenever price fluctuations warranted it, and not just at the normal times of autumn and spring, and the weight of the current loaf was determined by direct experiment with a new table being set thereafter. Modest fluctuations in the price of grain could be handled by the baker in meeting the requirements of the table and before any major shift in grain prices had occurred a new experiment would have determined the pivotal weight of the loaf about which a new table would be constructed.

The Assizes of 1600 and 1684

An important work referred to earlier, which first appeared in 1600 but with several successive editions (including one of 1684 from which the following information has been gathered) extending into the early years of the eighteenth century, is the book entitled *Assize of Bread* written by John Powell, Clerk of the Market of the Royal Household* in the reigns of Elizabeth I and James I.[21] The book is for the guidance of magistrates and other local officers as to the rules and procedures for setting the assize of bread. It is brief, condensed, and unpaginated. In earlier editions of this book no allowance was given in the tables for the baker,

> either in respect of their baking or yet for any needful charges belonging unto the same mystery, but the same was referred unto the discretion of all such magistrates and officers as should have authority to deal therein . . . and divers officers by reason of their unskillfulness and want of knowledge do not afford the baker such sufficient allowances therein as are answerable unto all the charges of baking at this day. (Edn. of 1684)

After repeating the provisions of the Assize of 1266, he refers to the allowance for bakers of 2*s* mentioned above, of date 1495, that is 'when wheat is 12/- the quarter, the baker should bake 14/- the quarter', but now (1684) things are much more expensive.

*For a record of the fines imposed by Powell in Middlesex in 1581 as Deputy Clerk, see the Appendix (pp.85-89) of N. Williams, Sessions of the Clerk of the Market of the Household in Middlesex; *Transactions of the London and Middlesex Archaeological Society* (1957) Vol.XIX, Pt.2, pp.76-89. London.

> I do therefore think it very good and necessary that there be set down in the Book of the Assizes of Bread, that bakers of cities, boroughs and corporate towns where the three sorts of bread viz. white, wheaten and household are usually baked and sold, shall have 6/- in allowance for the baking of every quarter of wheat, over and above the second price of wheat in the market [this is the 'corn of lower price' mentioned in the Assize of 1266], because of their great and daily charges and in bearing Scot and Lot [municipal taxes] within the said cities . . . which 6/- is to be allowed according to the same former allowance [2/-] in the said 12th year of Henry 7th (1495) aforesaid.

When second wheat is 12/- per quarter the baker is to bake at 18/-, that is his return from which his expenses and living is to be met is 6/-.

Powell further states the ½*d* and 1*d* cocket loaf is made of second priced wheat and the wheaten and household loaves take their assize from the farthing coarse cocket given in the 1266 Assize as weighing £7 1*s*, that is the weight when wheat is 12*d* a quarter with cocket bread weighing 5*s* more than wastel. From this we gather that the finer breads — simnel, wastel, and white — are made from first-quality wheat as might be expected. Household bread was brown bread. His assize table runs from wheat at 12*d* per quarter to £3 0*s* 6*d*, a table surely wide enough to cover every eventuality with a price range exceeding a factor of 60! Had the assize not been set frequently, the problems outlined in the previous section would have been insurmountable.

Looking into the table for wheat at 6/6 a quarter as before we find:

Kind of Bread	*Weight of Loaf*
½*d* white of fine cocket	25¼ oz + 4½ dwt
1*d* white of fine cocket	50¾ oz + 4 dwt
½*d* wheaten	39 oz
1*d* wheaten	78 oz
1*d* household	104 oz + 2 dwt

Here more bread is being given in 1684 for 1*d* than in 1500 and with grain at 6/6 a quarter and the baker's allowance of 6/-, the impossibly large figure of 975 lb of wheaten bread is required from a quarter. Clearly the price of grain was more nearly the figure in his example, namely 12/- (p.205), a figure which makes the assize workable.

The penalty for breaking the assize in 1684 is written down: a fine for the first two offences; a heavier fine and a warning for the third; and for the fourth, the pillory without redemption,

> But if the Baker do exceed in breaking the Assize of Bread the full weight of 2/6 which is 1½ oz in his farthing white loaf [not given in the tables] then he shall suffer the judgement of the pillory without any fine or admonition given unto him.

This 1½ oz is just over 11.5 per cent in the weight of the assize. This gives a little more leeway than the 9 per cent of the 1266 assize.

Again the price of bread was to change with each *6d* increase in the quarter of wheat and having said Troy weight was to be used which '. . . serveth only to weigh Bread, Gold, Silver, precious jewels and electuaries of which weight there is but 12 oz 1 lb,' Avoirdupois weight was to be used for 'butter, cheese, flesh, tallow wax, and every other thing which beareth the name Garbol and whereof issueth a refuse or waste.' White, wheaten, and household [brown] breads are mentioned

> yet nevertheless horse-bread hath had his continuance for so long time and the use thereof so profitable and necessary for the commonwealth that it standeth rightly with the Law to have his continuance and the Assize hath been and must continue: That three horse-loaves be sold from the baker for 1*d*, 13*d* [worth] for 12*d*, and that every one loaf shall weigh the full weight of the 1*d* white loaf according to the price of wheat.

Only bakers are to bake bread for sale and only those breads allowed by statute, such as simnel, wastel, white, wheaten household, and horse-bread, and there must be ¼*d*, ½*d* and 1*d* white loaves; ½*d* and 1*d* wheaten and 1*d* and 2*d* household loaves. No baker is to sell

> any spice-cakes, buns, bisket or other spice-bread (being bread out of use and not by Law allowed) except it be at Burials★ or upon the Friday before Easter or at Christmas upon pain of forfeiture of all such spice-breads to the poor.

So much for Powell and the seventeenth century.

The Assize of 1710

At the beginning of the eighteenth century large-scale changes were underway which were reflected in the 1710 Act 8 Anne c 19.[33]†

First, the 1266 *Assisa Panis et Cervisie* of 51 Henry III is to be repealed insofar as bread is concerned. Mayors or magistrates are now to set the assize and weight of bread according to the price of grain, meal, or flour, although in the chart accompanying the statute only the cost of the bushel of wheat (and bakery costs) are shown. We can assume that if flour is bought, the cost of milling will be known and hence the price of the original grain, so that the table need only show this latter. Only white, wheaten, and household bread is to be offered for sale and such others as are licensed to be made. Prices have risen drastically. Wheat prices are given per bushel now and a table is given ranging from 2 to 15 shillings a bushel including the costs of baking a bushel of wheat. Previously the tables showed only the price of wheat and the corresponding loaf weights in Troy measure; now we find Avoirdupois

★In North Staffordshire, the custom of eating a specially baked spiced bread at funerals continued into the twentieth century.

†Given as c 18 of 1709 in the 'Common Printed Edition'.

weight in use. The worked example of the Statute showed how the mayor and magistrates were to arrive at the price of bread. If wheat is 5/- a bushel and the magistrates allow 1/6 to the baker for his trouble, look into the table given in the Statute for 6/6 and we find that white loaves are to weigh half and wheaten three-quarters that of household loaves.

The Act states that bread is to be weighed by Avoirdupois weight and the table states 'Note: That 16 drams make one ounce and 16 ounces one pound'.

	White	*Wheaten*	*Household*
1*d* loaf	7 oz 2 dr	10 oz 11 dr	14 oz 4 dr
2*d* loaf	14 oz 4 dr	1 lb 5 oz 6 dr	1 lb 12 oz 8 dr
6*d* loaf	–	4 lb+ 3 dr	5 lb 5 oz 9 dr
12*d* loaf	–	8 lb+ 5 dr	10 lb 11 oz 2 dr
18*d* loaf	–	12 lb+ 8 dr	16 lb+ 11 dr

The use of Troy weight for bread has been discontinued and the much more commonly used Avoirdupois is to be employed for bread, a change that remained to the last quarter of the twentieth century.

But like most changes the shift to Avoirdupois weight for bread was neither instantaneous nor universal. In 1741 we read in John Chamberlayne's *Magnae Britanniae Notitia:*[34]

> Bakers who live in corporation towns make their bread by troy weight but those who live not in corporations are to make it avoirdepois weight for freemen are allowed 3*d* in the bushel more profit than those who are not free. For instance when the market price of middling wheat is 5/- a bushel a freeman baker must make a penny wheaten loaf to weigh 11 oz troy weight . . . but they that are not free men must make it as heavy as when the market price is but 4/9 a bushel . . .

Now 11 oz Troy is 12 oz 1 dr Avoirdupois. Extracting from the tables of 1710 and 1757 which bracket the date of this edition of Chamberlayne's book we find the weights given in the table below.

Weight of penny loaf 1710			*Weight of penny loaf 1757*		
Price of bushel and baking	*Wheaten*	*Household*	*Price of bushel and baking*	*Wheaten*	*Household*
4/9	14 oz 10 dr	19 oz 8 dr	4/9	12 oz 12 dr	17 oz 1 dr
5/-	13 oz 14 dr	18 oz 9 dr	5/-	12 oz 1 dr	16 oz 6 dr
5/3	13 oz 4 dr	17 oz 10 dr	5/3	11 oz 9 dr	15 oz 7 dr
5/6	12 oz 10 dr	16 oz 14 dr	5/6	11 oz 2 dr	14 oz 10 dr
5/9	12 oz 1 dr	16 oz 2 dr	5/9	10 oz 8 dr	13 oz 9 dr
6/-	11 oz 9 dr	15 oz 7 dr	6/-	10 oz 2 dr	13 oz 9 dr
6/3	11 oz 2 dr	14 oz 13 dr	6/3	9 oz 11 dr	13 oz 1 dr
6/6	10 oz 11 dr	14 oz 4 dr	6/6	9 oz 4 dr	12 oz 10 dr
6/9	10 oz 5 dr	13 oz 12 dr	6/9	9 oz 0 dr	12 oz 1 dr

From these data it would appear that when Chamberlayne says 'the market price of middling wheat is 5 shillings a bushel' he must mean 'the market price of middling wheat, including baking costs, is 5 shillings a bushel', for it is under that value that we find his weight for a wheaten loaf of 12 oz 1 dr Avoirdupois. If we assume the Assize in 1741 was the same as in 1757 we see that between 1710 and 1741, some 31 years, the weight of the penny loaf has diminished some 12.5 per cent. In other words the price of bread has risen this much in 31 years.

He tells a strange story of two classes of bakers, one working on Troy the other on Avoirdupois weight (and according to statute), with the latter bound to give a heavier loaf than the former by one unit in the assize table. If at all widespread the situation he describes was one destined to produce dissension where the two systems touched.

The Assize of 1757

The Act of 8 Anne of 1710 was intended to be in force for three years but was in fact continued annually till 1757 when the Act 31 George II c 29 of 1757[35] repealed it, together with the 1266 Assize, with effect from 1758 and promulgated a completely new set of bread regulations.

From 1758 bread could be made from rye, barley, oats, beans, or peas. The weight is to be on the Avoirdupois scale and new types of loaves are to be baked, namely those derived from baking a peck, a half peck (gallon), or quarter peck of flour, to be called the peck, half-peck, and quartern loaf. Wheaten and household bread are to be weighed and sold in proportion to the price of a peck loaf. Here the *price* of the loaf is to vary with the price of grain, meal, or flour and not the *weight* of a loaf costing a certain sum as heretofore but bread of 1, 2, 6, 12, and 18 pence per loaf is still to be baked. These loaves costing a fixed sum but of variable weight are the *loaves of assize*. Peck, half-peck, and quartern loaves are called in the Statute '*prized bread*', and are to be of a standard weight which is priced according to costs. Prized bread may not be sold where bread of the assize is offered for sale. Prized bread is an innovation, not found in the 1710 Act of Anne. Here again, though the price of bread is to vary with the cost of grain, meal, or flour, the accompanying tables use only the price of the bushel of wheat. No doubt, as before, if this was known, the prices of meal and flour of wheat and the other grains mentioned were also known. If bread is offered for sale defective in weight in relation to the assize the penalty shall not be more than 5/- or less than 1/- for each ounce so deficient. For less than one ounce underweight the penalty shall not exceed 2/6 or be less than *6d*. No person is to expose or offer for sale any except wheaten and household (brown) bread or such others as shall be authorized by the magistrates. The price of bread is to change only when there is a change of *3d* in the price of the bushel of wheat.

Section XXI is of interest:

> No alum . . . or any other mixture or ingredient whatever (except only the genuine meal or flour which ought to be put therein, common salt, pure water, eggs, milk, yeast, and barm* and such leaven as shall at any time be allowed to be put therein

In the table accompanying the statute the weight of the assize bread and the price of prized bread is given for wheat of 2/9 to 14/6 the bushel, including baking costs. In the worked example given in the table, if wheat is 5/- a bushel and the magistrates allow 1/6 to the baker, look up 6/6 in the table and that gives a pennyloaf wheaten to weigh 9 oz 4 dr, and if of household bread, 12 oz 10 dr. The peck loaf is to cost 2/6 wheaten and 1/10 if of household quality and is to weigh 17 lb 6 oz Avoirdupois. The sack of flour is to weigh 2 cwt 2 qrs nett (that is $2\frac{1}{2}$ cwt) and from every sack on average there should be a yield of twenty such peck loaves.

A second table is provided for breads from grains other than wheat. The weight of the 1, 2, 6, and 12*d* loaf and the price of the peck loaf is given for rye, barley, oats, beans, or mastin ($\frac{2}{3}$ wheat, $\frac{1}{3}$ rye) for a total cost of grain plus baking ranging per bushel from 1/- to 7/-.

Wheaten loaves are to be three quarters the weight of household and white three quarters the weight of wheaten. White bread is not shown in the table. Its weight is given only through this comparison and presumably would be baked only when authorized by the local authority. The mid-eighteenth century was a time of poor harvests and great scarcity. It would be important to extract as much flour from the bushel of grain as possible, giving a darker loaf. Permission to bake with white flour would be given only on rare occasions, so it is anomalous to see white bread shown as nine sixteenths the weight of household instead of one half as given in 1710, especially when white bread is under a virtual interdict.

The Baker's Allowance, Eighteenth Century

Neither the Act of 1710 nor that of 1757 mentions the 'two loaves' that traditionally were the perquisite of the baker having baked a quarter of wheat, probably because the assize is worked now in bushels rather than quarters. Instead, the magistrates are to allow the baker a fixed sum for his trouble over and above the cost of the bushel of wheat. But there is a perquisite built into the tables in that the baker can generate a sufficient number of loaves to meet his costs of materials and his allowance and still have some loaves left over which he could either sell or divert to his own use. We can check the calculation using the information of the 1757 Act which states that a sack of 280 lb Avoirdupois of flour will make twenty peck loaves[36] (wheaten) each weighing 17 lb 6 oz and each to be sold at 2/6. For a loaf of this price the tables appended to the Act give the cost of a bushel baked as 6/6. Now a bushel bakes

*Old word for yeast or the froth of fermentation.

into four peck loaves by definition, if not in practice, and these four loaves are together worth 10/-. Thus the baker meets his grain costs (5/-), has his allowance (1/6), and over and above has 3/6 worth of bread as a perquisite. His total advantage per bushel is his allowance of 1/6 and his perquisite of 3/6 worth of bread, of total value 5/-, that is the worth of two peck loaves. This is equivalent to 34 lb 12 oz of bread. If the magistrates allow more, his gain is further increased.

Evidence placed before the Select Committee on Laws relating to the Manufacture, Sale, and Assize of Bread (Report dated 6 June 1815)[32] indicated that more than twenty peck loaves could be obtained from a sack of flour and that the figure is more like 20½ or 21 (82–84 quartern loaves), hence the gain is the greater.

It should be remembered that each district had its own price for wheat and the magistrates did not act in concert throughout the land in deciding what were reasonable allowances; consequently the price of the 17 lb 6 oz peck loaf varied from place to place. Sometimes, the returns to the baker were hardly sufficient to keep him in business notwithstanding his 'free' bread.

The 1757 Act was explained further by 3 George III c 11 of 1762.[37] Here it was stressed that assize bread must not be offered for sale in the same place as prized bread. Local justices were to decide which was to be offered. The weights of the loaves are as given in 1757 and mention is made of peck, half-peck, quarter-peck, and half-quarter-peck loaves, with the peck loaf again weighing 17 lb 6 oz Avoirdupois. Loaves were to be stamped 'W' for wheaten and 'H' for household and if made from grain other than wheat that was to be signified too.

Events at the Turn of the Century

High prices and general scarcity forced down the weight of the assize loaf. Whereas the average price of a quarter of wheat in the decade 1750–60 was £1 19*s* 3¼*d* it rose in the next decade to £2 11*s* 3¾*d*. The average for the twenty years 1773–93 was a little lower, at £2 6*s* 3*d*, but the poor harvests of 1795 and 1797 forced the price up. In 1801 the price averaged £5 19*s* 6*d*, in 1803 £2 18*s* 10*d*, in 1805 £4 9*s* 9*d* and in 1813 £6 8*s* 8*d*. Such a considerable price rise could only be reflected in the price of bread, with draconian measures being taken in an attempt to ensure the provision of a subsistence level for the lower classes. The penny wheaten was down to 6 oz 9 dr in 1767 and in 1796 the 2*d* wheaten was only 8 oz 13 dr. The poor could hardly survive and other materials besides wheat had to be used in breadmaking. Recourse was made to legislation.

The Statute 36 George III c 22 1795[38] begins:

> Whereas it is expedient in order to diminish the consumption of wheat that bakers should be permitted to make and sell in all places various kinds of mixed bread and such kinds of wheaten bread as they cannot now sell in places where an assise is set;

> and whereas it is not expedient to apply to such sorts of bread the restrictions contained in the tables of the Assise and Price of Bread . . . it is now lawful to sell . . . peck loaves, half peck loaves, quartern loaves, and half quartern loaves made of the whole produce of wheat deducting only 5 lbs weight of bran per bushel or made of any sort of wheaten flour mixed with meal or flour of barley, oats, buckwheat, indian corn, peas, beans, rice, or any other kind of grain whatever or with potatoes in such proportions and at such prices . . . as the maker and seller thereof shall deem proper and reasonable, any law, custom, or usage to the contrary notwithstanding. . . .

The proportions of the ingredients were to be specified and displayed in the shops, loaves were to be marked with an 'M' where mixed grains were used (Section 2), and the weights of the loaves were to be in accordance with the tables of the assize. The bread was to contain the mixture of materials declared and not some other mixture but no alum was to be incorporated (Section 3). The situation was becoming desperate. The Act 37 George III c 98 of 1797[39] set up for London a detailed accounting system of prices and sales of grain, with weekly returns being made on Mondays to the Cocket Office in the Mansion House of the City of London and an assize to be held every Tuesday. A substantial table of prices was appended giving the price of prized bread and also of bread of the assize for wheaten and household loaves according to the price of wheat running from 27/- to 104/4 per quarter or correspondingly the price of flour running from 25/- to 98/4 a sack with an allowance of 14/- per quarter of wheat for milling, baking, etc. or 11/8 for baking a sack of flour. The price of bread was to change with every advance of $2\frac{1}{2}$*d* on the bushel of wheat or 1/8 on the sack of flour. Here again the tables enable us to calculate the baker's advantage. When the peck loaf (wheaten) costs 2/6 the cost of a sack of flour (280 lb) and its baking was 50/-, that is a bushel (56 lb) baked is 10/-. We are told the sack of flour costs 38/4 so the bushel costs 7/8 and the allowance for baking is 2/4 now (11/8 ÷ 5).* But the bushel produces only four peck loaves, sold for a total of 10/-, so the baker's advantage has fallen drastically since 1757 to 2/4 or 16.2 lb of bread, that is his cash allowance only, with no 'free' bread, and we are back to Powell's concern for the welfare of the baker.

Five years later (41 George III c 16 1800)[40] the same assize is being adhered to. This Act further decreed that the manufacture of any fine flour from wheat or any grain whatsoever was to be prohibited and the making of bread solely from fine wheat flour was forbidden. It went on to specify that from and after 17 January 1801 in the City of London or within forty miles thereof, and from and after 21 January 1801 in all other parts of Great Britain no meal dressed or bolted finer than with a sieve with thirteen wires on each side of a square inch was to be produced. From 31 January and 7 February respectively there was to

*There are 5 bushels of flour, each 56 lb in weight, in the sack of 280 lb.

be no baking for private use or for sale with any flour finer than this. The penalties for disobedience were increased and of the five pages of the Statute more than half relate to penalties and proceedings to be taken if violations occur. By 39 and 40 George III c 74 1800[41] it was pointed out that the table in 31 George II c 29 of 1757[35] stopped at a price of 14/6 a bushel and a new table of weights and prices was to be produced in proportion to that of 1757 as the price rose above 14/6 per bushel. Again an assize was to be held every Tuesday with the magistrates checking bread by weighing within 48 hours of its having been baked.

By the Act 41 George III c 17 1800[42] the sale of new bread was prohibited. Only bread at least 24 hours old could be sold and it was not to be kept warm nor was its moisture to be conserved. This was found either unacceptable or unworkable (probably both) for it was repealed the next year by 42 George III c 4 of 1801.

In 1801 the extraction rate which had up till then been over 90 per cent was raised to 100 per cent. The Act 41 George III (U.K.) c 12 1801[43] permitted flour of the whole produce of wheat to be made (without the exclusion of the 5 lb of bran per bushel) or indeed flour of bran only or 'of bran and pollards' (bran flour or bran mixed with flour) 'at any price at which any person may be willing to purchase the same but the price is to be less than the assise for wheaten bread . . . notwithstanding any Act to the contrary'.

Section 2 removed the requirement to display the mixture of materials used in the bread being sold (that is it repealed 36 George III c 22 section 2), but Section 3 provided that wheaten loaves of inferior quality be marked H and a loaf of mixed grains marked X.

This flurry of legislation around the year 1800 gives a good picture of the great dearth prevailing in the land. The Act 53 George III 1813[44] has a very extensive table listing the prices of wheat and of flour, the prized bread and the weights of the bread of assize, but whereas the tables of 1797 begin with the price of a bushel of wheat as 3/4½*d*, running to 13/0½, the table of 1813 begins at 4/11½ running to 22/6¾ per bushel. The allowance for grinding and baking rose to 15/10 per quarter, or for baking a sack of flour, to 13/4, but the price of the peck loaf for a given total cost of baking a sack of flour did not change.

The tables list three different kinds of bread, 'wheaten, standard wheaten, and household'. The tables further enact that the number of pounds of bread of these categories to be made from a quarter of wheat are respectively 413 lb, 434 lb, and 468 lb or per sack 347½ lb for all three kinds of bread. This rate per sack is precisely the same as that given in 1757 where the sack was to generate 20 peck loaves each of 17 lb 6 oz, a total of 347½ lb.

We note that a quarter of wheat is to make 413 or 434 lb of wheaten or standard wheaten bread, figures very close to those given in the test of 1800 mentioned earlier (p.204).

Standard weights of 17 lb 6 oz, 8 lb 11 oz, and 4 lb $5\frac{1}{2}$ oz were made for the convenient weighing of these peck, half peck, and quartern loaves. Some are to be seen in the Museum of London and elsewhere dating from the first quarter or so of the nineteenth century. (See Fig.42). Although mentioned in statutes they had no legal sanction.

At Great Wishford (Wishford Magna) in Wiltshire stone plaques were set in the boundary wall of the church of St. Giles recording the price of the gallon loaf (half peck) of 8 lb 11 oz which in 1757 had cost 1/3. The first tablet reads: '1800 Bread 3/4 per gallon'. Thereafter we have 1801 — 3/10; 1904 — 10*d;* 1920 — 2/8; 1946-8 2/1; 1965 — 5/4 and 1971 — 8/- or in decimal currency 40*p*. The photograph Figure 43 shows the plaques as they were in May 1980.[45] Looking into the table for 1800 we find the price of the gallon (half peck) loaf to be beyond its scope, for it stops at the price of a sack of flour at 105/- with the half peck (gallon) loaf costing 2/11. The table provides that for higher grain prices the cost of prized bread is to be pro-rated. Doing so leads to a remarkable price in 1800 for a sack of flour, namely 120/-.

We can trace some of the troubles of the late eighteenth and early nineteenth centuries to the corn laws created in the reign of William and Mary in the late seventeenth century with a subsidy for wheat exports. This ensured that there would be every effort to export grain rather than keep it for home consumption. When this was coupled with heavy duties on imported grain, as in 1670, the problem became doubly serious. By the mid-eighteenth century the duty on imported grain was much reduced but the subsidy on exports continued, with as much as £324,000 being paid out by the Exchequer in 1749. By the last decade of the eighteenth century the crisis was such that bounties were being paid on grain imports in an endeavour to offset several years of poor harvests, but this did not prevent the great rise in the price of grain in the early years of the nineteenth century.

The Nineteenth Century

The assize continued in many places into the nineteenth century though many towns and cities had abandoned it, finding that where this was done bread prices tended to fall. London kept the assize and bread cost more there than in many other locations. The explanation is that if prices are fixed according to the price of flour or grain there is no incentive for bakers to seek the best, that is the lowest, price. This appears to have been the cause of London's problems.

Standard-sized loaves of variable price were popular but did not become so overnight. The people were accustomed to the penny and two-penny loaves and we find weights for loaves of a given price being set until the second decade of the nineteenth century. The fixing of the price of bread was abolished in 1815. Indeed, the Act 55 George III c 99 1815[46] repealed a long list of bread Acts relating to London and ten miles round about, declaring 'there shall be no longer any assize of bread within the same city . . .', but the regulations governing the weights of the various loaves were retained. Section 2 provided for bread to be made from flour or meal of a large variety of grains and from potato flour together with salt, pure water, eggs, yeast, milk, barm, leaven, and potato yeast 'mixed in such proportions as the makers and sellers shall think fit, any law custom or usage to the contrary notwithstanding'. Section 9 listed the weights of the loaves as we have had them before (17 lb 6 oz, etc) but added the pound loaf of 16 ounces. Section 10 provided for scales and weights to be kept in bakers' shops. Section 11 provided penalties for light weight; for each ounce light a fine not exceeding 10/- and in proportion for a deficiency less than an ounce, the weighing to take place within 24 hours of baking. The final comment in the First Report of Commissioners 'appointed to consider the Subject of Weights and Measures' of 1819,[47] is that it is doubtful if the public gained or lost by the provisions of this Act of 1815.

Finally, in 1822, by Act 3 George IV c cvi (Local Act),[48] the assize was totally abolished in London and for a radius of ten miles round. This was extended in 1836 by 6 and 7 William IV c 37[49] to all other parts of the country except Ireland. All Acts relating to the regulation of the price of bread were repealed.

Of the latter Statute, which is very similar to that of 1822, Section II provides that bread may be made from

> Flour or meal of Wheat, Barley, Rye, Oats, Buckwheat, Indian Corn, Peas, Beans, Rice, or Potatoes or any of them with common salt, pure water, eggs, milk, barm, leaven, potatoe, or other yeast and with no other ingredients whatsoever.

Section III makes it lawful for bread to be made of such size and weight as bakers think fit. Section IV provides that bread shall be sold by weight. Section V states the weight is to be Avoirdupois of 16 ounces to the pound. Section VI requires bakers to have scales and weights in their shops whereby bread may be weighed, and Section VII makes similar provision that scales and weights shall be carried on any cart or carriage conveying bread. The Act added a proviso that a baker could sell French or fancy bread or rolls without previously weighing them.

In an appeal in 1869 against a conviction for selling bread unweighed which in 1836 was regarded as fancy bread but which in 1867 was 'a common article of consumption', the seller's appeal was dismissed, it being held that the Act defined conditions at the time of selling, not at the time of the Act coming into force, Justices Lush and Hayes concurring, Mr Justice Hannen dissenting. Mr Justice Lush in his judgement gave the intent of the Acts thus:[50]

> The object of the Legislature in passing the Act was to liberate the trade from all restrictions of the Assize Act and leave the baker at liberty to make bread of any size

> and shape he pleased and to charge his own price for it, but in order to protect the customer from imposition, it required the baker to sell by weight. He is no longer at liberty to sell at so much per loaf, he must sell at so much per pound, and the customer is to be supplied with so many pounds of bread, unless he chooses to have an article of exceptional quality, something that is not ordinary bread and if he buys that, the baker is to be at liberty to sell it without reference to its weight. But unless it is of an exceptional character, if it is the common article of consumption, the baker must sell it as such.

In 1822 the quartern loaf shed its odd $5\frac{1}{2}$ ounces and it became exactly 4 lb in weight.* It then became customary to make bread in loaves of 1, 2, and 4 pounds, a practice which continued into the twentieth century, as we shall see presently.

The Twentieth Century

The Bread Act of 1836 (6 and 7 William IV c 37)[49] was altered by the Bread Order of 1917[51] which in Article 7 required that 'No loaf of bread shall be sold or offered for sale unless its weight be one pound or an even number of pounds', and was further amended by the important Sale of Food (Weights and Measures) Act of 1926.[52] Section 6 of this latter Act (1) forbade the sale of bread other than by weight, (2) stipulated that a loaf had to be one pound net weight or an integral multiple of pounds (this repealed Article 7 of the Bread Order 1917), and (3) excluded fancy bread or loaves not exceeding twelve ounces in weight from the foregoing provisions. These provisions were to apply to pan loaves, French loaves, etc. save that 'in Scotland any such loaf of the net weight of $1\frac{3}{4}$ pounds may be sold and delivered for sale if the weight is clearly marked by an impression made in the baking or on any printed band or wrapper surrounding or containing the loaf'.

These formal provisions take the writer back to the days of his childhood in Edinburgh when he remembers buying 'high pan' loaves with the legend '1 pound 12 oz' standing in relief on the crust, having been formed by this inscription on the pan containing the dough. Another familiar loaf was 'plain bread' which weighed 2 lb Avoirdupois baked in a batch with the bottom crust of the loaf flat and the top crust rounded. One asked for 'a half loaf' at the baker's shop. Perhaps few of the shopkeepers and even fewer of the customers realised that the request was for a half-quartern loaf.

*The first edition of the *Sunday Times*, 20 October 1822, tells of the 4 lb loaf being sold for $7\frac{1}{2}$*d*, with 86 such loaves being required from a sack of flour. The baker's advantage is calculated for an 8*d* loaf, indicating this to be the more usual price, with the sack of flour costing 40/-. The profit is 17/4. A footnote states; '86 loaves at 4 lbs each is 334 lbs (*recte* 344), three lbs and a half under the weight of 80 quartern loaves ($347\frac{1}{2}$ lb as before), in favour of the baker per sack'. The indication is that the quartern loaf is still being reckoned as 4 lb $5\frac{1}{2}$ oz for profit calculations while the actual loaf weighs 4 lb.

'Fancy bread' was exempt from this Act of 1926. Being undefined by statute no satisfactory description ever emerged. There were court cases in abundance yet the question was never formally settled.

Since 1836 a legal requirement for the carrying of scales and weights or a weighing instrument suitable for weighing bread on any vehicle carrying bread for sale or delivery had been in force. Various appliances had been introduced to comply with this requirement, including beam-scales, spring balances, and a special kind of steelyard designed to weigh loaves of 1 or 2 lb, or 2 and 4 lb. An example of this steelyard is shown in Figure 44, where the position of the pivoted metal poise determined whether a 2 or 4 lb loaf was of the correct weight. These were deemed to be 'not of a suitable pattern' under Regulations introduced in 1907. The degree of lightness especially had to be known exactly so we find steelyards of this model ceasing to be manufactured about 1912.

Section 15(3) of the Sale of Food (Weights and Measures) Act, 1926, repealed the part of Section 7 of the 1836 Act which required the carrying of scales on vehicles selling or delivering bread, but under Section 6(4) of the 1926 Act the requirement for the maintenance of scales in shops selling bread was continued as follows:

> Every person selling bread or having in his possession any bread for the purpose of sale, shall provide in some conspicuous part of his shop or premises a. correct weighing instrument of a pattern suitable for weighing bread. . . .

This section of the 1926 Act continued in force until repealed by Section 63(1) (b) of the 1963 Act with effect from 31 July 1965.

The advent of World War II led to a number of changes in the weights and sizes of loaves and in the composition of the flour, for it was obviously central to the aims of the day to make available as much food as possible from a given quantity of grain and at the same time to have individuals eat a little less. By the Bread (Restriction of Sales) Order 1940,[53] the Ministry of Food restricted the loaf to a weight of 1 lb or an even number of pounds (save the Scottish loaf of 1 lb 12 oz); rolls were not to exceed 2 oz nor the Vienna loaf 8 oz.

The Bread (Control and Maximum Prices) No. 2 Order of 1943 (S.R. & O.* 1943, No. 1653) provided that batch bread in Scotland should be 75 per cent National flour† and other breads should be 87½ per cent. Later, by S.R. & O. 1945, No. 1149, this latter figure was reduced to 75 per cent. This did not last, for in 1947 the problems of feeding Britain and a devastated Europe were enormous. S.R. & O. of 1947 No. 1867 returned bread to 87½ per cent National flour.

*'Statutory Rules and Orders'
†That is of high extraction

Part II Section 3 of the 1943 Order restricted the oil and fat content of the dough to not more than 2 lb per 280 lb of flour (this is precisely the weight of the sack of flour given in the 1757 Act mentioned earlier), and 'improvers of the nature of Yeast food' were not to be used beyond that normally employed by the trade in producing such bread.

Section 4 specified all 1 lb loaves were to be of the same shape.

Section 5 of this 1943 Order specified all breads except a Vienna loaf to be 1 lb or an integral number of pounds. The Vienna loaf was to weigh 8 oz and no roll was to exceed 2 oz. An exception was made again for the $1\frac{3}{4}$ lb loaves of Scotland and Northern Ireland, provided they were so marked. The First Schedule to this Order lists the materials which may be included in the baking of National bread besides National flour, that is white flour, oils and fats, water, salt, yeast, 'improvers of the nature of yeast food', any acid or acidic substances suitable for regulating the acidity of the dough, potato or potato flour, and barm. The Second Schedule reproduced below gave the price of loaves of various weights, an interesting return to an assize involving a modification of 'prized bread':

Maximum Price

	4 lb loaf	*2 lb loaf*	*1 lb 12 oz*	*1 lb loaf*
1. *Northern Ireland*	$9\frac{1}{2}d$	$4\frac{3}{4}d$	$4\frac{3}{4}d$	$2\frac{3}{4}d$
2. *Scotland* (a) *north & west of the Caledonian Canal*	10	5	5	$2\frac{3}{4}$
(b) *the Hebrides, Orkney, & Zetland*	11	$5\frac{1}{2}$	$5\frac{1}{2}$	$2\frac{3}{4}$
(c) *the rest of Scotland*	9	$4\frac{1}{2}$	$4\frac{1}{2}$	$2\frac{3}{4}$
3. *Depopulated areas (specified separately)*	$9\frac{1}{2}$	$4\frac{3}{4}$	–	$2\frac{3}{4}$

Elsewhere a 1 lb loaf shall cost $2\frac{3}{4}d$, any other loaf at the rate of $2\frac{1}{4}d$ per pound.

By 1946 the war was over but scarcity remained. In an endeavour to save flour by having people eat a little less, the weight of the loaf was reduced from 1 lb (16 oz) to 14 oz by S. R. & O. 1946, No. 626, and all loaves were to be of the same shape. No one was to offer a loaf for sale unless it weighed 14 oz or a multiple thereof, except for the Vienna loaf of 8 oz and a roll not exceeding 2 oz in weight. It will be noted, though, that a 28 oz loaf is just the old loaf of $1\frac{3}{4}$ lb of Scotland and Northern Ireland.

Order No. 626 of 1946 is important also in that the Schedule gave the maximum prices for bread, not very differently from the 1943 Order. The prices varied according to locality and therefore reflected availability, transport costs, and the like just as before. The $1\frac{3}{4}$ lb loaf could cost then 5*d*, $5\frac{1}{2}$, or $4\frac{1}{2}d$ in Scotland depending on whether the loaf was sold (a) north and west of the Caledonian Canal, (b) in the Inner and Outer Hebrides and the counties of Orkney and Zetland, or (c) in the rest of Scotland including the islands of the Firth of Clyde, respectively.

Apart from Northern Ireland, the Holy Isle, and sparsely populated parts of the kingdom, all of which had individual prices listed, the rest of the land was to pay $2\frac{3}{4}d$ per 14 oz loaf and for loaves of other sizes at the rate of $2\frac{1}{4}d$ per pound, whereas the 1943 Order had these prices for the 16 oz loaf. A further Order (S. R. & O. 1946 No. 1136) confirmed the above but reduced the weight of the Vienna loaf to 7 from 8 oz.

Bread rationing in Britain was introduced by S. R. & O. 1946 Nos. 1100 and 1385, and bread was returned as mentioned earlier to $87\frac{1}{2}$ per cent National flour by S. R. & O. 1947 No. 1867 which also reiterated the loaf's weight to be 14 oz or multiples thereof except Vienna bread of 7 oz and rolls not exceeding 2 oz. The components that could be added to national flour to make bread remained the same as those listed in the 1943 Order. Prices were to be as given earlier in S. R. & O. 1946 No. 626 but restrictions on fats and improvers were removed.

Bread rationing ended on 25 July 1948 and the trade slowly returned to normal.

The Weights and Measures Act 1963 c 31 specified (Schedule IV pt. IV)[54] that whole loaves exceeding 10 oz made for sale shall be of 14 oz or multiples of 14 oz while Statutory Instrument (the new name for Statutory Rules and Orders) 1976 No. 1297 specified that a whole loaf of bread exceeding 300 grams shall be made for sale only if it is of a net weight of

(1) 14 oz or a multiple of 14 oz
(2) 400 grams or a multiple of 400 grams.

Here 300 grams replaces the previous 10 oz. (10 oz. = 283.5 g) and 400 g = 14.1 oz, so we see after nearly a millennium of traditional weights for bread the presence of an approximate equivalent in metric units creeping into the regulations.

Many Statutes and Orders have in past times laid down what may go into bread. Some of these have already been quoted. This section concludes with an item from the *Daily Telegraph* of 9 June 1980.

> In France, law limits bakers to soft wheat flour, water, yeast, salt, and one preservative, ascorbic acid, better known as vitamin C. Britain's regulations permit about 26 chemical substances, most of them banned in other EEC countries. Analysts say that 15 or 16 additives go into a sliced white loaf. Maturing and bleaching agents artificially age and whiten the flour; persulphates increase water content and improve the elasticity of dough.
> Mr O'Neill [President of the Bakery Workers Union] said: 'Twenty years ago, I could have taken you round a bakery and anyone could recognise all the ingredients — lard, butter, yeast, flour, etc. Today you would have to be a chemist to appreciate what goes into dough.

But this is not restricted to the United Kingdom. A recent letter to the Editor of *Agenda,* the quarterly bulletin of the Science Council of Canada (DUFOUR, F. *Agenda* Autumn 1982, Vol.V, No.2, p.2), quoted a bread label thus: 'INGREDIENTS: made with enriched white flour (thiamine mononitrate, riboflavine, niacine), reduced iron, water, sugar (and/or dextrose, and/or glucose), whey solids, salt, shortening (modified and/or hydrogenated, vegetable and/or palm oil), yeast, yeast nutrients (calcium sulfate, ammonium chloride, starch, potassium bromate), protease, emulsifier (mono and diglycides), dough conditioner (calcium sulfate, dicalcium phosphate, diammonium phosphate, calcium dioxide, calcium propionate (a preservative).) *Contains* 0.24 mg thiamine (B1), riboflavine 0.17 mg (B2) 2.2 mg niacine, 1.8 mg iron per 100 g'.

We have thus surveyed the metrology of the making of bread from the earliest times of which we have extant records to the present day. In so doing we have been able to identify the times of plenty and the times of dearth, the tranquil periods and those of conflict and turmoil, the times of inflation and the times of stable economy. We have seen the trials of the baker and the quandaries of governments, but throughout it all with a few exceptions the people, by the sweat of their brow, have been able to eat bread as was foretold to Adam at the expulsion from the Garden. (Gen. 3.19)

The Assize of Ale

Besides meaning a malt liquor, the word 'ale' also signified a festival at which ales were provided. There were church-ales and parish-ales, audit-ales (which were dispensed when students at some of the colleges of Oxford and Cambridge settled their accounts), and lamb- and midsummer-ales which were held at the appropriate seasons. There were bid-ales (benefit festivals) and the word 'bridal' comes from bride-ale or wedding feast. Monastery records show brewing to have been one of the more active occupations of the brethren during the twelfth and thirteenth centuries.[55] From their custom of marking the barrels with a cross came the practice of brewers marking their barrels X to denote weak ale and XX to denote strong.[56] Commercial brewing had been a thriving industry much earlier and continued so to be. The control of this enterprise is presented in the Assize of Bread and Ale,[7] whose date we accepted as 1266, wherein it is stated:

> When a quarter of wheat is sold for 3 shillings or 3 shillings and 4 pence, and a quarter of barley for 20 pence or 2 shillings and a quarter of oats for 16 pence, then brewers in cities ought and may well afford to sell 2 gallons of beer or ale for a penny and out of cities to sell 3 or 4 gallons for a penny. And when in a town 3 gallons are sold for a penny, out of a town they ought and may sell four and this Assize ought to be holden throughout all England.

Here we see the quantity of the ale obtainable for a penny changing with location for a given price of grain. When wheat is in the range 30 to 40 pence a quarter and barley similarly in the range 20 to 24 pence, brewers in cities should give 2 gallons, in towns 3 gallons, and outside towns and cities — that is in rural England — 4 gallons for 1*d*. The grain was available in rural areas. Transportation costs were high and the cheapening effect of having a centre of population had not yet been experienced, hence country ale was cheaper than town or city ale. In early times ale was distinguished from beer in that the latter was made with hops, a distinction which has been abandoned today.

The document entitled *Judicium Pillorie*[11] of uncertain date but attributed to 1266 shows that the price of ale was to change only with a step change of six pence per quarter in the price of grain. The document mistakenly speaks of quarts instead of gallons throughout. In the following excerpt the error has been corrected:

> When a quarter of barley is sold for 2 shillings then 4 gallons of ale shall be sold for a penny; when for 2 shillings and 6 pence then 7 gallons of ale shall be sold for 2 pence; when for 3 shillings then 3 gallons for a penny; when for 3 shillings and 6 pence then 5 gallons for 2 pence; when it is sold for 4 shillings then 2 gallons at one penny. And so from henceforth the prices shall increase after the rate of 6 pence.

Sometimes this portion of *Judicium Pillorie* is printed separately under the title 'Assisa Cervisie'. It is given correctly and verbatim as above by Fleta of 1290.[57] The Customal of Sandwich[24] written in 1301 reads very much as the 1266 Assize. It begins:

> When a quarter of wheat is sold for 3 shillings or 40 pence and barley for 22 pence or 2 shillings and oats for 16 pence then brewers in cities ought . . . to sell 2 gallons of beer for one penny . . .'.

It concludes: 'And this Assize is holden throughout the kingdom by ordinance of King Henry III'.

In the *Statutum de Pistoribus*[12] (Statute concerning Bakers) ostensibly of 1266 also but actually of uncertain date like the others, we are told that the brewer may not raise his price except there be a sixpenny rise in the price of malt and if he break the assize he shall be fined for the first three offences but for the fourth 'he shall suffer judgement of the pillory without redemption'. The document states that the measures are to be sealed with the iron seal of our Lord the King and kept safe under pain of £100. No measure was to be used unless it was agreeable to the king's standard measure.

Innkeepers at the time of Edward I could resell ale which they had bought at a penny or penny-halfpenny a gallon at the rate of two pence a gallon measured by a sealed gallon, *pottle* (half gallon), or quart.[58] No pints or measures smaller than a quart are mentioned.

Some of the pressure to have nine bushels to the quarter instead of eight (see

Ch.IX) came from the brewers, who complained in 1390 that after cleaning nine bushels of malt there only remained eight for brewing, but this led to the order that malt should be cleaned before selling,[59] with the quarter continuing as eight bushels. To take nine bushels of malt to the quarter had been commonplace throughout the fourteenth century.

The Ale-Conners

Ale inspectors, the *'gustatores cervisiae'*, were appointed on oath:[60]

> You shall swear that you shall know of no brewer or brewster, cook, or pie-maker in your ward who sells the gallon of best ale for more than 1½ pence or a gallon of the second for more than a penny, or otherwise than by measure sealed and full of clear ale,

an oath which originated in the days of Henry V.[61] They were known variously as ale-inspectors, ale-tasters, ale-founders, or ale-conners, and it has been claimed that the office goes back as far as the days of Alfred.[62] Shakespeare's father held such an office in 1557 in Stratford-upon-Avon. Their duties were to call for a measure of ale, pour it on a bench, and sit in the puddle in their leather breeches for a while. If on attempting to rise the breeches had stuck to the bench the ale was demonstrably sugared and therefore unwholesome. Ale brewed from pure malt was not supposed on evaporation to be adhesive or sticky. The conner could also check the brew for clarity and flavour, and frequently did.

We can imagine that the office was not without its compensations. It would be a churlish innkeeper who would not provide adequate refreshment for the conner beyond his taste-sample as he sat at his allotted task. But it also had its disadvantages. Prolonged or frequent sampling was often called for, as the brewer had to notify the conner when each brew was ready for testing. With some conners having over a hundred establishments in their district to visit, it is not surprising that the ale-tester was on occasion presented at court charged with being drunk and disorderly.

Further, he could not shield a brewer who defaulted, for to do so in London left him liable to imprisonment for eight days and to be fined at the discretion of the Lord Mayor.

But during the sixteenth and seventeenth centuries a loosening was experienced in the regulatory function respecting ale, with some reluctance to press prosecutions for assize violations, and this progressed further in the eighteenth, so that the role of the ale-conner virtually disappeared, although officers continued to be appointed. The business of brewing and retailing ale and beer was becoming self-regulating, at least in part, and was much more responsible and ethical, although at the end of the seventeenth century innkeepers were still required by law to sell ale and beer by the Exchequer ale quart as a standard measure. Mayors were to compare all ale quarts and pints with the standard in their custody, but this was not to extend to the colleges and halls of the Universities of Oxford and Cambridge. By the second half of the eighteenth century it was deemed to be no longer necessary to have official sanction to change the price. The City of London appointed ale-conners annually each midsummer's-day into the present century but, of course, the modern office was entirely ceremonial.

The Price of Ale

In 1474 for Coventry it is recorded[63] that a brewer can make 48 gallons of ale from a quarter of malt* and, further, when a quarter of malt costs two shillings he shall sell a gallon of his best ale for a halfpenny; when it costs three shillings the gallon is to be sold for $\frac{3}{4}d$ and so on. No ale was to be sold till tested by the ale-conners.

This price was for 'off-sale', for the record goes on to say that a quart of best ale 'with-Inn hyms upon his tabull' shall cost $\frac{1}{2}d$, that is for consumption on the premises a gallon would cost *2d,* as opposed to the $\frac{1}{2}d$ or $\frac{3}{4}d$ off-licence price. Then as now it was more costly to drink in establishments than at home. But this record of Coventry shows that the brewer cannot hope to survive if he follows the 'off-sale' prices. From a quarter of malt costing him 2 shillings, that is 24 pence, he recoups only 48 half pence, that is 24 pence, if he must sell at $\frac{1}{2}d$ a gallon. There is nothing for the other ingredients or his labour or his expenses. We can assume that the assize was frequently broken, for the brewer could not always rely on his sales for consumption on the premises to generate the necessary overhead costs and still yield a profit. Yet 'if he sell not after the price of malt' he is to be fined 12 pence the first time, 20 pence the second, and for a third time he is 'to be judged to the Cukkyngstole' (that is, placed in a tumbrel and ducked in the pond or river) and afterwards put in the pillory.[63] Prices varied with the cost of grain which itself varied locally. The ale vendor could always be found to be breaking the assize. Fines became a recognized business expense, with some brewers making a yearly payment of $2\frac{1}{2}$ marks (33*s* 4*d*) to avoid endless court appearances.[65] These prices are recorded at Coventry in 1474 for brewers. Innkeepers of that time did rather better. They are to sell a pot of three pints of the best ale 'within hym' for one penny but if an innkeeper was also a brewer he was to sell as other brewers.[66] The pint was now a recognized measure for malt liquor.

The brewers of ale and beer were members of separate crafts with no great love the one for the other. Beer was ale flavoured with hops. The first mention

*This must have produced a very substantial ale for we are told[64] that in 1512 a quarter of malt produced 83 gallons and prior to World War II 76 gallons of strong beer and 150 gallons of mild ale were produced from a quarter.

of hopping occurs in 1391 in the Letter Books of the City of London.[67] In 1464 the beer brewers petitioned for recognition and the granting of rules for their trade, while in 1483 the ale brewers counter-petitioned that it be forbidden to put into ale any hops, herbs, or anything besides liquor, malt, and yeast.

By the beginning of the seventeenth century prices had risen but not markedly. We learn that in 1603[68] not less than a full ale quart of the best beer or ale or two quarts of small ale was to be given for a penny. Complaint was made about 1634[69] that

> many victualers and others that lett out their beere to tapsters at 14*s*, 15*s* and 16*s* the barrell whereby those tapsters by their juggs, blackpotts, canns, and bottles and other deceiptful measures doe much deceive and defraud the subject, they having but a bare wine pint of beer for a penny.

This was fraud indeed, for the ale quart was almost two and a half times as large as the wine pint. The Act 11 and 12 William III c 15 of 1700[70]* endeavoured to stop these practices. Liquors were to be sold

> by the full ale quart or pint or in proportion thereto in a vessel made of wood, earth, glass, horn, leather, pewter, or of some other good and wholesome metal, made, sized, and equalled unto the Exchequer standard and signed, stamped, or marked to be of the content of the said ale quart or ale pint.

The penalty for failure was not to be less than 10 nor more than 40 shillings but nothing in the Act was to extend to any of the colleges or halls of the Universities of Oxford and Cambridge.

The Thurdendel

During the seventeenth century steps were being taken to ensure that the customer received his due. John Powell's book *The Assize of Bread* in the editions of 1671 and 1684[21] contains the following:

> *The Assize and Order of Innholders and Victuallers*
> And for that ale and beer are not in themselves perfect liquors, but being filled into a small measure, the Yeast and Froth thereof will ascend by working very speedily, requiring a time in settling thereof again, there is also to be used and to be allowed within this Realm, sundry measures of lesser content for Brewers, Inholders, and Victuallers, selling their ale and beer by retail unto the Subjects: the which are named and called hooped quart and pint measures, thurdendels and half-thurdendels, being a small quantity somewhat bigger than the aforesaid standard, in respect of the working and ascending of the Yeast and Froth: by which quarts and pints the Inholders shall retail their Ale and Beer being after the rate of

*Given as 11 William III c 15 1698 in C.T.S.

> fourpence the gallon and by the same thurdendel and half-thurdendel the Victuallers shall retail their drink after the rate of three pence the gallon.

The price difference (4*d* and 3*d*) a gallon between the 'Inholders' and the 'Victuallers' reflects the fact that the former offered more services rather akin to those of an hotel, whereas the latter represented the pub or snack-bar operation. The use of the *thurdendel* enabled the customer to get a full measure of a quart of liquid with the froth floating on top. This is exactly analagous to the use of glass line measures today[71] where the true fluid measure is marked by a horizontal line on the vessel about 1 cm below the lip of the vessel. This extra room is to accommodate the froth as before, although the modern measures are of the pint and half pint.

Powell goes on to forbid the use of false or unsealed measures of lesser capacity than the Exchequer standards, with Justices of the Peace to set the price of ale and beer within each shire. The penalty for overcharging was to be 6/- on each barrel, 3/4 on each half-barrel or *kilderkin,* 2/- on a *firkin,* and for every vessel containing a greater number of gallons than the barrel (32 gallons for ale, 36 gallons for beer) the sum of 10/- but for quantities less than the firkin the sum of 12 pence, with one half of the fine going to the king and the other to the complainant. No musty malt or malt infested with weevils was to be used nor was any ale or beer to be put into 'musty or corrupt vessels'.

The Size of the Barrels

Not only were the two crafts separate; there were barrels of different sizes for the two malt liquors. (See Ch.X (b) (i).) The more ancient brew was ale but the size of the barrel is not given in any of the assizes of the thirteenth century. Beer as so defined would appear from the above to be more a late fourteenth- or early fifteenth-century development. Certainly in 1454 the barrel of ale is given as 32 gallons, that of beer 36 gallons.[72] In the Act of 1531 (23 Henry VIII c 4)[73] this was restated with the half-barrels (kilderkins) and the quarters (firkins) being defined in proportion with all measurement by the king's standard gallon. In 1660 the same sizes of barrels were being maintained, as we shall see.

In 1688, doubtless in an endeavour to bring some semblance of uniformity to the capacity of the barrel of the malt liquor trade, it was enacted that the barrel of beer or ale, strong or small, was to contain 34 gallons according to the Exchequer ale quart.[74] This striking of the happy mean lasted till 1803 when, by 43 George III c 69, s 12,[75] the capacity of the barrel of both liquors was increased to 36 gallons. This value for the barrel was restated in 1820.[76] In 1824[77] all measures were to conform to the new Imperial gallon of 277.274 in^3: The old ale or beer gallon was 282 in^3 and therefore the two were less than 2 per cent different. In practice today, therefore, we find the barrel of ale or beer continuing as 36 Imperial gallons.

The Levying of Duty on Ale and Beer

Whereas wine had been taxed for centuries, a duty was placed on beer for the first time in 1643.[78] Indeed there were two ordinances in that year affecting beer or ale, one of 22 July, the other of 8 September affecting only the six associated counties of Norfolk, Suffolk, etc. The substance of these edicts was that the former required 2/- to be paid per barrel of eight shilling ale or upwards by the first retailer, with an additional 1/- to be paid by the first buyer, while the brewer was to pay 6*d*. Home brewing was taxed at the rate of 1/- the barrel. The latter ordinance required the brewer to pay the 2/-, the householder paying 1/- for the home brew, but for weaker beer selling above 4/- but below 6/- a barrel, the brewer was to pay 6*d*.

In 1660, as a *quid pro quo* for the king relinquishing tenure by knight's service (Ch.VI) and all charges incident thereto, together with the Court of Wards and Liveries and Tenures which brought him a considerable sum each year, Parliament granted Charles II a duty on a whole variety of drinks with beer or ale leading the list (12 Charles II c24 s14 1660):[79]

s.14-25	For every barrel of beer or ale (sold above 6 shillings the barrel . . .	1/3*d*
	. . . at 6/- or under . . .	3*d*
	A hogshead of cider or perry . . .	1/3*d*
	A gallon of metheglin or mead . . .	1/2*d*
	A barrel of vinegar-beer . . .	6*d*
	A gallon of strong water or aqua-vite . . .	1*d*
	A barrel of beer imported from overseas . . .	3/-
	A gallon of coffee made and sold . . .	4*d*
	A gallon of chocolate, sherbet, or tea* made and sold. . .	8*d*

Section 21 provided, as already mentioned, that 36 gallons of beer and 32 gallons of ale made the barrel (as in 1454 and 1531), but all other liquors were to be according to the wine gallon. The duty on beer was removed in 1830, but replaced again in 1880 as a substitute for a duty on malt. Beer duty was normally imposed on the barrel and often was quite small, as when in 1894 the budget required an extra sixpence on a barrel. At the consumer level it was difficult for the producer to pass on these small increases, this last one amounting to no more than 0.02 of a penny per pint. The multiplicity of changes in the overall taxation of beer is a study in itself and goes beyond the scope of the present work, so will not be pursued further.

The End of the Assize

From the above the Assize of Ale will be seen to have been seldom a

*To twentieth-century eyes, a duty on tea eight times that on a similar volume of brandy is surprising. It reflects the comparative rarity of these non-alcoholic drinks relative to the stronger more traditional ones, and the high costs of freightage from India.

parliamentary enactment but rather a local town or city policy, laid down for the information of traders and citizens alike. There was no formal abolition of the assize. It just faded away into the mainstream of living under market influences and changes in trading practices. For the twentieth century, the cost of a pint of ale was to be controlled by taxation on the one hand and the price of grain and other market or production costs on the other. Charitably, they could be considered as having an equal influence on the price.

NOTES AND REFERENCES: CHAPTER XI

1. The only extensive treatment covering part of what follows is that by:
   STEVENSON, Maurice. The Assize of Bread. *Libra* (1966) Vol.V, pp.3, 10, 18, 26 Eastbourne: Weights and Measures Office.
   Ibid. (1967) Vol.VI, pp.2, 10, 17.
   Ibid. (1968) Vol. VII, p.13.
   for the Assize of Bread.
   For the Assize of Ale see:
   ROBINSON, G. J. A. The Duties of the Ale Taster (1) *Libra* (1968) Vol.VII, p.10. Eastbourne.
   Ibid. Some Malpractices of Inn Keepers & Brewers (2) *Libra* (1968) Vol.VII, p.21. Eastbourne.
   STEVENSON, Maurice. The Assize of Ale (3). *Libra* (1969) Vol.VIII, p.20. Eastbourne.
   Ibid (4). *Libra* (1970) Vol.IX, p.54. Eastbourne.
   Brief treatments are to be found in CHISHOLM, H. W. *Seventh Annual Report of the Warden of the Standards* pp. 51-2 London: HMSO 1873 and, ZUPKO, Ronald Edward. *British Weights and Measures* p.20, Madison, Wisconsin: University of Wisconsin Press 1977.
2. MacCANCE, R. A. and WIDDOWSON, E. M. *Breads White and Brown* Philadelphia: J. B. Lippincott 1956.
3. PERTZ, G. H. (ed.) *Monumenta Germaniae Historica* Leges Vol.I, cap.4. Capitulare Francofurtense, Stuttgart: Kraus Reprint 1965. The capitulary is also to be found in translation in CUNNINGHAM, W. *Growth of English Industry and Commerce — Early and Middle Ages* Appendix p.501, Cambridge: CUP 1890 (Also 5th ed. 1910, p.507).
4. HORN, W. and BORN, E. *The Plan of St. Gall* Vol.III, p.106, Berkeley: University of California Press 1979 and the notes and references given therein.
5. British Museum M.S. (Add. 14252 f85b)
   See CUNNINGHAM, W. (Ref. 3) for a printed version, 1890 ed. p.502, (1910 ed. p.568).
   An assize of the time of Richard I (1189-99) has been recorded giving the weights of the farthing wastel and simnel loaves. See ROBERTSON, Revd W. A. Scott,

The Assize of Bread *Archaeologica Cantiana* (1878) Vol.XII, p.321. Maidstone, Kent, but the weights in general do not agree with those of the Assize of Henry II nor with that of 1202. Either the Assize is anomalous or it reflects a period of hardship, for the loaves here are much lighter yet the range given for the price of grain is quite modest.

6. PARIS, Matthew. *Chronica Majora,* Luard, H. R. (ed.) Vol.II, pp.480-1. London: Rolls Series, Longman, Green, Longman & Roberts 1874.
7. *Assisa Panis et Cervisie. Statutes* Vol.I, pp.199-200.
8. RICHARDSON, H. G. and SAYLES, G. O. (eds. & trs.) *Fleta* Vol.II, Book II, Ch.9, p.117. Selden Society Publication Vol.72 of 1953, London: Selden Society 1955.
9. RILEY, Henry Thomas (ed.) *Munimenta Gildhallae Londoniensis: Liber Albus,* Vol.III, p.411. London: Rolls Series, Longman et al. 1859-62.
10. Prior to the first minting of silver farthings ($\frac{1}{4}d$) by Edward I in 1279, a penny was literally cut into quarters, each one of which would purchase the farthing loaf of the assize. Copper farthings were first minted in 1672 and ceased to be legal tender in 1961.
11. *Judicium Pillorie Statutes* Vol.I, pp.201-2.
12. *Statutum de Pistoribus Ibid. pp.202-4.*
13. HARRIS, Mary Donner. *The Coventry Leet Book* Vol.I, p.396. London & New York: E.E.T.S.; K. Paul, Trench, Trubner & Co. Ltd., H. Milford O.U.P. 1907.
14. STEVENSON, Maurice. The Assize of Bread, *Libra* (1966) Vol.V, p.4. Eastbourne.
ZUPKO, Ronald Edward. *British Weights and Measures* p.20. Madison, Wisconsin: University of Wisconsin Press 1977.
CHISHOLM, H. W. *Seventh Annual Report of the Warden of the Standards* p.51. London: HMSO 1873.
15. ARNOLD, Richard. *The Customs of London, commonly called Arnold's Chronicle* p.49 et seq. London: F. C. and J. Rivington 1811 (Reprint of 1st ed. of *c*1502).
16. CHISHOLM, H. W. Loc. cit. pp.27-9 for part of the document which includes the two quotations.
17. HALL, Hubert and NICHOLAS, Frieda J., *Select Tracts and Table Books relating to English Weights and Measures.* Camden Miscellany Vol.XV, p.45. London: Camden Society 1929.
18. CUNNINGHAM, W. *Growth of English Industry and Commerce* Appendix IV, 2 c, p.458. Cambridge: CUP 1882.
19. RILEY, Henry Thomas (ed.) Loc. cit. Vol.III, p.267; see also Vol.I p.705.
20. HARRIS, Mary Donner (ed.) Loc. cit. p.682 et seq.
21. POWELL, J. (Clerk of the Market in the time of Queen Elizabeth and James I) *Assize of Bread* 1684 (unpaginated).
22. HARRIS, Mary Donner (ed.) Loc. cit. p.397.
23. RICHARDSON, H. G. and SAYLES, G. O. (eds.) Loc. cit. p.120-1.
24. BOYES, W. *Collections for an History of Sandwich in Kent: The Customal of Sandwich* Vol.II, p.541 et seq. Canterbury: Printed for the Author 1892 (recte 1792).
25. RILEY, Henry Thomas (ed.) Loc. cit. Vol.III, p.82-3. See also Vol.I, pp.264-5.
26. Ibid. Vol.III, p.420.
27. Ibid. Vol.III, p.424.

28. 13 Richard II Stat 1 c 8 1389-90; *Statutes* Vol.II, p.63.
29. SELLERS, Maud (ed.). *York Memorandum Book* Pt. 1 (1376-1419), Pt.2 (1388-1493) p.170. Durham: Surtees Society Vols. 120 & 125 1912.
30. HUDSON, Revd. William and TINGLEY, J. C. *The Records of the City of Norwich* p.174. Norwich & London: Jarrold & Sons 1906.
31. NICHOLS: John Gough (ed.) *Chronicle of the Grey Friars of London* London: Camden Society Publication No.53 1852.
32. Report of Select Committee on the Laws Relating to the Manufacture, Sale and Assize of Bread: 6 June 1815. *Parl. Papers* (HC 186) V. p.1341-1490, 1814-15.
33. 8 Anne c 19 1710 *Statutes* Vol.IX, p.248.
34. CHAMBERLAYNE, John. *Magnae Britanniae Notitia,* 34th ed. London: D. Midwinter et al 1741.
35. 31 George II c 29 1757 *Statutes at Large* Vol.VII, p.217.
36. This is correct. The sack of 280 lb of flour contains 20 pecks each of 14 lb. The imputed bushel weighs 56 lb.
37. 3 George III c 11 1762 *Statutes at Large* Vol.VII, p.426.
38. 36 George III c 22 1795 Ibid. Vol.XIII, p.231.
39. 37 George III c 98 1797 Ibid. Vol.XIII, p.591.
40. 41 George III c 16 1800 Ibid. Vol.XIV, p.568.
41. 39 and 40 George III c 74 1800 Ibid. Vol.XIV, p.390.
42. 41 George III c 17 1800 Ibid. Vol.XIV, p.575.
43. 41 George III (U.K.) c 12 1801 Ibid. Vol.XV, p.39.
44. 53 George III c 116 1813 Ibid. Vol.XIX, p.237.
45. I am indebted to Mr M. Stevenson for directing my travels to this Wiltshire Village. See also his letter to the Editor, *Daily Telegraph,* London 4 December 1965 p.13 in reply to that of Miss M. Cardwell of 27 November 1965.
46. 55 George III c lxcix (Local Act) 1815 *Statutes at Large* Vol.XXII, p.677.
47. First Report of Commissioners appointed to consider the Subject of Weights and Measures 1819, *Parl. Papers* 1819 (HC565) xi, p.323 (printed page 17).
48. 3 George IV c cvi (Local Act) 1822 *Statutes at Large* Vol.XXII, p.1036.
49. 6 and 7 William IV c 37 1836 Ibid. Vol.XXVIII, p.108.
50. *The Law Reports. Court of Queen's Bench 1868-9* BULWER, James Redfoord (ed.) The Queen v. William Wood Vol.IV p.559 et seq 1869.
51. OWEN, George Alfred (Revised by POOLE, A. W.) *The Law Relating to Weights and Measures* 2nd ed. p.268. London: C. Griffin 1947.
52. 16 and 17 George V c 63 Sale of Food (Weights and Measures) Act 1926 *The Statutes Revised* (3rd edn.) Vol.XIX, p.217.
53. *Statutory Rules and Orders* No. 1457 of 1940.
54. *Weights and Measures Act 1963,* c31.
55. E.g. Baking and brewing are reported throughout the text of HALE, W. H. (ed.) *Domesday of St. Paul's 1222* London: Camden Society 1858.
56. WATNEY, J. *Beer is Best — A History of Beer,* p.132 et seq. London: Peter Owen 1974.
57. RICHARDSON, H. G. and SAYLES, G. L. Loc. cit. Ch.11, p.118.
58. RILEY, Henry Thomas (ed.) Loc. cit. p.lvii.
59. *Rotuli Parliamentorum* Vol.III, pp.281, 323, AD 1390. London: HMSO.
60. RILEY, Henry Thomas (ed.) Loc. cit. Vol.III, p.128; see also Vol.I, p.316.

61. KING, Frank Alfred. *Beer has a history* p.21. London & New York: Hutchinson's Scientific and Technical Publications 1947.
62. ROBINSON, G. J. A. The Duties of the Ale-Taster *Libra* (1968) Vol.VII, p.11. Eastbourne.
63. HARRIS, Mary Donner (ed.) Loc. cit. p.398.
64. KING, F. A. Loc. cit. p.58.
65. KING, F. A. Loc. cit. p.24.
66. HARRIS, Mary Donner (ed.) Loc. cit. p.399.
67. KING, F. A. Loc. cit. p.42.
68. 1 James I c 9 1603/4 *Statutes at Large* Vol.III, p.7.
69. HALL, Hubert and NICHOLAS, Frieda J. Loc. cit. pp.52–3.
70. 11 and 12 William III c15 1700 (now given as 11 William III c15 1698 in C.T.S.); *Statutes at Large* Vol.IV, p.48.
71. STEVENSON, Maurice. Editorial *Libra* (1969) Vol.VIII, p.26. Mr M. Stevenson comments further:

    'The use of line measures, although increasing, is not compulsory and the majority of public houses continue to use brim measures. In recent years several appeal cases have been heard in the High Courts of England and Scotland against convictions for serving beer and stout with such a head of froth as to cause a deficiency in the liquid content. In all of these, beginning with the case of Marshall v Searles (*Criminal Law Reports* 667, 1964) followed by Dean v Scottish and Newcastle Brewery Ltd and others (*Scots Law Times, Notes of Recent Decisions* 1978 p.24) and Bennet v Markham and King (*All England Law Reports Annotated,* Vol.III, 1982 p.641) it was upheld that the 'head' was part of the measure. The reasoning behind these decisions was, apparently, that as long as it was legal to serve beer and stout in brim measures it was impossible to give the full measure of liquid unless the beer or stout was served 'flat'. These decisions have caused considerable concern amongst Inspectors, members of CAMRA*, and beer drinkers generally'.

    (Private Communication May 1983).

    We note however a provision of the 1979 Weights and Measures Act (Statutes in Force 1979 c 45) where in s 19 it is stated: 'In ascertaining the quantity of any beer or cider for any of the purposes of Part VI of Schedule 4 to the Principal Act (i.e. the Weights and Measures Act 1963) . . . the gas comprised in any foam on beer or cider shall be disregarded'.

    This section has not been implemented. Section 24 (3) (b) states: '. . . s 19 . . . shall come into force on such a date as the Secretary of State may appoint by order made by Statutory Instrument . . .'. If it were ever implemented this Section would outlaw brim measures and line measures would take over completely. Under these circumstances beer and cider drinkers would receive a full measure of liquid plus the froth on top and we would be back to the days of the thurdendel.

    This situation has not improved with time. In 1983 one third of the beer sales tested in the metropolitan area were between one and two fluid ounces short of the pint; in 1984 this had risen to one half, with barely one in twenty giving full liquid measure. This led to a call to implement Section 19 of the 1979 Act already quoted. By early 1986 this had not occurred. Indeed the Eden Committee

*Campaign for Real Ale

Report on the Metrological Control of Equipment for use for Trade (Cmnd 9546, June 1985) recommended the implementation (pp 43-44). This was not accepted by the Government in its published response to this Report (Cmnd 9850, July 1986), it being stated that: 'the Government remains strongly in favour of a self-regulatory solution to the issue of serving a fair measure of beer. Such a solution is being explored at present and the implementation of this measure (now s.43 of the 1985 Act) will not be considered while that remains a good possibility'."

72. HALL, Hubert and NICHOLAS, Frieda J. Loc. cit. p.50.
73. 23 Henry VIII c 4 1531 *Statutes* Vol.III, p.366.
74. 1 William & Mary c 24 s 5 1688 *Statutes at Large* Vol.III, p.406.
75. 43 George III c 69 s 12 1803 Ibid. Vol.XV, p.848.
76. Second Report of Commissioners on Weights and Measures, 13 July 1820, *Parl. Papers* 1820 (HC314) vii, p.473, Appendix A.
77. 5 George IV c 74 1824 *Statutes at Large.* Vol.XXIII, p.759.
78. *Acts and Ordinances of the Interregnum 1642-1660* Vol.I, March 1642-January 1649 pp.208-9. London: HMSO 1911. (Excise Ordinance 22 July 1643.)
    See also p.273, An ordinance for levying monies by a weekly taxe upon the Sixe associated Counties of Norfolk, Suffolk . . . 8 Sept. 1643.
79. 12 Charles II c 24 s 14 1660 *Statutes* Vol.V, p.261.

CHAPTER XII

The Physical Standards

Thus far in our inquiry we have been relying mainly on the written record as it has come down to us from the many and varied sources to which reference has already been made. But the written record must be supported and illuminated by the artifacts which have been preserved and are now located in museums throughout the country. Here the artifact and the record go hand in hand, and although we shall concentrate on the Exchequer standards as the primary objects of reference we shall frequently illustrate the development of weights and measures by reference to collections in the care of cities or in private hands.

It is convenient to break up the discussion into various time groupings, as follows:

(a) To the accession of Henry VII
(b) The Issues of Henry VII
(c) The Issues of Elizabeth I
(d) From 1588 to 1760
(e) Events 1760 to 1834
(f) The Standards Restored

In this way we shall be able to review in this chapter the construction and development of the standards up to the onset of metrication.

(a) The Establishment of English Standards to Henry VII

Measures

Although it is known from documents that standard measures were made in Saxon and Norman times, it would appear that none has survived. This is not too surprising because the measures of capacity were in all probability made of wood and decay is likely.

We learn from *Liber Custumarium*[1] of an issue of measures by the Treasurer in 1321 and a comparison thereof with existing measures. The bushel of the City of London matched the new bushel, but the 'king's bushel' was in excess 'per cyathos circa decem' (by about ten *'cyathos'*). This word has several interpretations including 'a ladleful' or a 'cupful'.[2] This 'king's bushel' may have been the ancestor of an Exchequer bushel of Henry VII, for in 1758 the Carysfort Committee measured two bushels of Henry VII then residing in the Exchequer. One was found to hold 'near three pints of water'[3] more than the other, the standard corn bushel. 'Ten cupfuls' might well be 'near three pints'. Henry's standard bushel measured 2144.81 in^3 (Ch.IX p.160). Three pints more would give a total volume of 2245.35 in^3. Eight ale gallons, each of 282 in^3, have a volume of 2256 in^3, near enough that of the bushel to indicate the basis of these thirteenth-, fourteenth-, and fifteenth-century bushels, but as the bushel was not used as a liquid measure it is likely they were constructed in error, using the ale instead of the corn gallon. We have to wait till the reign of Henry VII before we have an extant measure, fabricated in bronze. We do rather better for weights, however.

Weights

Few weights have come down to us which can be dated earlier than the reign of Edward I, and of that reign there would appear to be only one extant item of lead made in the form of a shield. It weighs just over 6.3 lb Avoirdupois, presumably intended to be 6¼ lb, this being the *clove* or one sixteenth of the hundredweight of 100 lb (see Chapter VII). A photograph of this interesting weight is shown in Figure 29. It is to be seen at the Science Museum, London.

Winchester is fortunate in having in the City Museum (previously in the Westgate Museum) a unique set of the weights[4] of Edward III (see Fig.28). These are the 91 lb quarter sack mentioned in Chapter VIII together with 56, 28, 14, and two 7 lb weights and are all Avoirdupois as we have seen. The use of Avoirdupois weights was common in the fourteenth and later centuries. The 91 lb weight carries six embossed shields: the 56 and 28 lb weights each display four shields, and the remaining three weights carry three each. Three shields carry the quarterings of Old France on the Royal Arms and this quartering was only used in the period 1340–1405. It would be of interest to try and narrow down the dates of their probable manufacture.

The Act 14 Edward III Stat. 1 c 12 of 1340[5] describes an issue by the Treasurer '. . . of certain standards of bushels, galons, *de poys dairesme,* and send the same into every county where such weights be not sent before this time'. The Statute Book translates the italicized words as weights of auncel and a footnote indicates these to be weights of brass *(poids d'airain)*. Both are correct, though actually of bronze. An example of such a weight is that of Sandwich[6] which weighs 12 lb 4 oz and is a typical steelyard weight, but carries four shields, two carrying the arms of Old France and two those of the Cinque Ports, of which Sandwich was one. Winchester's weights are not of this type, being intended for balance weighings, so we must look beyond the issue of 1340 for their origin. Further evidence that Winchester's weights are not of 1340 can be inferred from the statute itself when it describes this issue going to 'every county where such weights be not sent before this time'. The county of Hampshire, with the important city of Winchester, would surely have been among the first to receive standards. Winchester could never have been relegated in anyone's mind to the status of a town lacking standards prior to this time. This implies, of course, the presence of standards at Winchester prior to 1340, of which there are now no trace.

An Act of 1357 (31 Edward III c 2) orders that: 'Certain Balances and Weights of Sack, half sack, and quarter, pound, half pound, and quarter according to the standard of the Exchequer be sent to all Sheriffs of England'. It is not to be supposed that literally only six weights were issued to each sheriff, for the gap between the quarter sack (91 lb) and the 1 lb weight is too great. Others would be needed to fill it. Moreover the difficulty of moving weights of 364 and 182 lb would be considerable. In all probability weights of these magnitudes were never produced. None is extant. Thus, together with the evidence of the quartering, it is believed that Winchester's set dates from this issue of 1357, for there are no other issues known between 1340 and 1405.

Although the set may be incomplete, missing the postulated intermediate weights and the lighter ones mentioned in the Act, we are fortunate to have what we do, especially when it is remembered that with each issue went out a call to destroy the old standards. For once, disobedience is seen to be a virtue.

It may be surprising to find two issues of weights so close together, in 1340 and 1357. The explanation may be that the former was exclusively of auncel weights for specific steelyards like that at Sandwich and the latter weights for the pans of balances. It may be recalled (Ch.VIII p.138 et seq) that the auncel was condemned by statute in 1352, 1353, and 1360, admittedly without effecting the desired end, but by 1360 more pressure was being applied. The issue of balance weights in 1357 allowed three years for their assimilation before the 1360 prohibition of auncel weights. We know that, subsequent to its acquisition of the steelyard, Sandwich appointed an official *tronager* (operator of the tron or public weighing machine) which means that a balance was then in use.[6]

Yards

Records exist from the century following the Assize of Measures of Richard I (Ch.VI) showing that the custody and use of yard measure for the purposes of trade was jealously guarded by the cloth guilds. In the Records of the Borough of Leicester[7] there is an entry for 5 February 1265 reading:

> On the day of St. Agatha, virgin, in the aforesaid year [1265] Roger Siveker of Knighton, Walter and Peter of Beeby, charged that against the liberties of the Gild they brought a yard [measure] into Leicester and cut linen-cloth and sold it by the yard and half [yard] and they abjured this kind of died [sic, deed?] in future.

Merchant Taylors' Yard

As for yard measures themselves, it is thought that the silver yard in the possession of the Merchant Taylors' Company, one of London's liveried companies, dates from 1445.[8,9] If this date is correct, then it is the oldest English yard measure, being of the reign of Henry VI.

The Master and Wardens of the Company were given, in their fourth royal charter dated 1439, the right of 'search throughout the whole city of London and also at St. Bartholomew's Fair', an annual cloth fair instituted in 1133 by Henry II which opened at Smithfield on St. Bartholomew's Eve (23 August) and which continued to be held until abolished in 1854. The Fair initially lasted for two weeks but was reduced to three days in 1691. It was a meeting place of suppliers and dealers in cloth from all over the kingdom.

The record of 1439 reads in part;[10]

> . . . when the said Maister shall think beneficial to the said Mysterie, make search throughout the whole cittie and the suburb thereof and also at St. Bartholomew's Faier during the tyme of that Faier amongst all the occupiers and freemen of their owne mysterie, as well for waights, measures, yardes, and ells . . .

The silver yard is a handsome plain cylinder encasing an iron bar of hexagonal cross-section. The silver has engraved upon it at each end the coat of arms of the Company and a crowned lombardic 'h' denoting Henry but, as we shall see, Henry VII, not Henry VI. Mr H. J. Chaney, Superintendent of the Standards Office, examined the yard in April 1892 and again in 1895. On the latter occasion its length was determined as 36.001 inches.

The yard is marked out by double lines about 1 mm apart into four 9 inch sections. One of the end sections is subdivided into one part of 4½ inches and two 'nails' of 2¼ inches each, divisions clearly indicative of its connections with the cloth trade. The bar terminates in two small caps of silver soldered on to the main part of the yard.

The company had sealed (that is verified) standard yards of iron at an early date as the account books show:[11]

1419/20	Item paie pur 1 standard messore seled, XIIII d
1455/56	Et in denariis solutis pro tynnyng of the standard metyerd, IIII d
1462/64	Et in denariis solutis pro stannacione virge ferri stannati . . . VI d

It is of interest to note that a certified yard measure could be obtained for 14 pence, while tinning the measure cost 4 pence, a figure destined to rise to 6 pence only six years later.

It was Mr Chaney's opinion that the silver yard was adjusted in the Exchequer sometime after its construction by comparison with the standard yard of Henry VII, when it was then stamped or engraved with the lombardic crowned 'h'.[11] This 'h' has the same form as that on the bushel and gallon of Henry VII. If the yard truly dates from 1445, the coat of arms of the company must also have been a later addition, for the arms were not granted until 1480.

Certainly the silver yard is mentioned in the 1512 inventory of the property of the Merchant Taylors; 'Item, a yerd of sylver and another of iron'. The inventory of 1609 states: 'One silver yard with an iron bar in the middle of it having the Company's armes engraven upon it and weighing 36 oz'. The inventories of 1709 and 1803 both mention the silver yard. More recently[11] the yard has been described as of length 36 inches, diameter five eighths of an inch and of gross weight 35 oz 12 dwt. It carries no hall-mark or maker's mark.

The dating of this yard cannot be said to have been established with certainty, but it is not unlikely that the date given is correct, in which case it is the oldest extant standard of the kingdom. That the yard still performs its original function today, albeit ceremonially, is shown in Figure 45. Here we see the Master, Mr A. T. Langdon-Down, in his robes, preceded by the Beadle, Mr L. C. Oakman, carrying the silver yard at the Cloth Fair at Olympia in 1979, when they appeared at the Fair to 'make search', in order to ensure the correctness of the measures in use.

The Taylors had begun their 'searching' at least as early as 1428 and their right of search was established, as we have seen, in their fourth charter of 1439 which was confirmed in 1465 by Edward IV and again by Henry VII in 1502. The latter confirmation states in part: 'None to search any liege members or their goods or wares, woolen cloths, ells or measures, save only the Master and Wardens . . .'.[12]

The Mercers' and Drapers' Company

The Merchant Taylors were not the only livery company of the City of London to claim the right of inspection of the measures used in the cloth trade. In the fifteenth and sixteenth centuries the alnage of certain cloths was the province of the Mercers' Company of London. Not only did the Mercers measure silk, which was their principal commodity, but also canvas and linen. The measurements were made by means of cords or ropes of approved lengths and care was taken to ensure that the measures were true. It is recorded[13] that in 1502 two Mercers, William Wydowson and Richard Janys, were appointed to hold the office or occupation of mesurage. On 22 June 1503, just nine months after William Wydowson's appointment, his rope was found longer by half of a quarter of a yard, 'the which was cut off by Master Warden . . .'. A little later a 'Common Meter' was appointed as we learn from the Acts of the Court of the Mercers' Company for 20 September 1520;[13]

> The officer called the Comen Meter of Sylkes should mete as well all sylkes as all maner of canvas lynyn cloth both white and brown . . .
> . . . no maner of clothe of gold, sylver nor sylkes, lynyn clothe and canvas, lokeram, dowlas [both are types of coarse linen] nor any maner of lynyn cloth whyte or brown made in Bretayn, France, Flaunders, Holand, Seland, Braband, Henygo . . . be brought into this city of London by the said [Merchant] Strangers or forreiyns

to be sold except the same be mette by the Comen Meter . . .

and not only the Mercers' but even more importantly the Drapers' Company claimed the right of search. The Taylors and the Drapers agreed to share this right, largely ignoring the Mercers, and each searched their own members for correct measure, but towards the middle of the fifteenth century the Taylors extended their search to others beyond their own ranks. This was strongly opposed by the Drapers, who in 1441 complained to the Lord Mayor of London who decided in favour of the Drapers. The King overruled the Lord Mayor in favour of the Taylors but in 1447 gave the Drapers the right of search also.[10]

Although relations between the Taylors and the Drapers were often strained they greatly improved by the end of the fifteenth century. This was destined not to last long. The Taylors' charter of 1502, their sixth, confirmed their right to buy and sell goods of every kind, and their right of search was again proclaimed. The Drapers were highly incensed and for the next fifty years the two companies were great rivals. But by the mid-sixteenth century the arm of civil authority was intruded into the affairs of the companies with magistrates appointing overseers to supervise the search. From then on, although both companies claimed the right of search, this right now had to be shared with others. With the passage of time and the provision of readily accessible measures, the need for routine searches passed and the searches themselves became more acts of ceremony than anything else.

(b) The Issues of Henry VII

In the Statute Book we find two Acts of Henry VII of considerable importance. The first, 11 Henry VII c 4 1495,[15] followed the petition of 7 Henry VII c 3 1491[16] which stated that divers Acts proclaim one measure and weight, meaning that at the Exchequer, '. . . which statutes have not yet been put into execution, some [standards] are more, some are less, because the said standard is not well known'. The king is asked '. . . at his cost and charge to make weights and measures of brass★ and to deliver them to the citizens . . .'. The Act of 1495 which followed stated that previously enacted statutes establishing a single weight and measure by which all men should buy and sell are not being observed, so the king has caused to be made standards of brass,★ and the knights and citizens of every shire attending Parliament shall convey or cause to be conveyed into these boroughs and towns a set of these, to remain in the keeping of the mayor or 'bailie', and the inhabitants are to make standards from those so sent. The copies are to be verified and sealed by the mayor or bailie and all defective measures are to be broken and burnt, the

★Actually bronze.

penalty for having a defective measure to be: first offence, a fine of 6 shillings and 8 pence; second offence, 13 shillings and 4 pence; third offence, one pound and the pillory.

All this was done. In accordance with the provisions of the Act, standards were issued to 42 named towns. Alas, the king's standards were themselves defective and were so found to be shortly after their receipt by these towns. Within a year a new Act, the second of those alluded to above as being of importance, was proclaimed (12 Henry VII c 5 1496)[17] which recited the former Act and continues '. . . which weights and measures upon more diligent examination had since the making of the said statute been proved defective and not made according to the old laws . . .'.

The standards which apparently were the most defective were those of capacity, as these are specifically mentioned. The recently issued bushels and gallons were not be used but were to be returned at the cost of the cities and towns upon pain of £10 to the king, to be broken, and with the metal* other new bushels and gallons were to be made and sized at the cost of the cities and towns. Meanwhile, the standards used prior to 1495 were to be used until the new and corrected measures are issued. This distribution took place in 1497, and the magnificent bronze yard now in the Winchester City Museum dates from this issue.†

The Winchester Yard

The Winchester yard is a well made bronze bar of hexagonal cross-section, five eighths of an inch thick over the hexagonal edges at the mid division. It is like no other standard measure in that it has at each end a cap; on one, made of bronze, there is a Lombardic 'h', denoting Henry VII, and on the other, made of iron, there is a Roman 'E', denoting Elizabeth I. The yard had therefore been standardized in the days of Henry and restandardized in the days of Elizabeth. The yard is shown in Figure 46 with details of the end-pieces which are soldered to the main bar. The rod is divided in binary fashion from one end at the 18 inch point, at 9, $4\frac{1}{2}$, $2\frac{1}{4}$, and $1\frac{1}{8}$ inches, again indicating its intended use by the cloth industry. This yard is not divided into feet or inches. Its overall length was determined on 24 May 1927 as 0.004 inches short of the Imperial standard. Measuring from the centre (18 inch) mark, the actual length to the end of the 'h' cap is 0.047 inches short of 18 inches, while to the 'E' end it is 0.043 inches too long. The adjustment of the soldered caps to such fine tolerances could not have been easy when we consider the amount of metal which had to be heated to make the solder run and the problem of adjusting the length while still hot.

*All the weights and measures were of bronze, not brass.

†In June 1953, Mr F. G. Skinner of the Science Museum gave the date as '*c*1487', but there would seem to be no basis for such a dating.

Henry VII's Exchequer Yard

The oldest *Exchequer* yard is that of Henry VII of 1497. It is to be seen in the Science Museum, London (Fig.47a). It has been described[18] as

> a roughly made octagon of brass, the yard being an end measure. Close to each end it is stamped with an 'h'. A very coarse line in the middle marks the half yard. From the end of the bar on the left hand side there are lines denoting $\frac{1}{16}$, $\frac{1}{8}$, $\frac{1}{4}$ and $\frac{1}{3}$ yard and from the end at the right hand side a foot is marked divided into 12 inches.

It measures 0.037 inches less than the present standard. Remarkably enough the Exchequer yard does not figure in the text of the Act of 1496, and the yard itself does not compare in quality or condition with the bronze yard of Winchester of the same date. (Indeed, why the Winchester yard was made so vastly superior to others, including the king's Exchequer yard, is not known.) The divisions on the Exchequer bar are so broad that they appear to have been made by an ordinary triangular file or similar instrument and the bar itself is thinner than the Winchester yard, making it less rigid and so more liable to bending.

Measures of Henry VII

The corn bushel and gallon of 1497 are also in the Science Museum. They are shown in Figure 34 and as their capacities have been discussed in some detail in Chapter IX, repetition here is not necessary. The measures are in good condition and show little sign of deterioration as they approach their fifth centenary. The inscriptions on the bushel and gallon are respectively: 'Henricus Septimus Dei Gratia Anglie et Francie' interspersed with Tudor roses, a portcullis, and a leashed greyhound and 'Henricus [Tudor rose] Septimus [Portcullis]'. Figure 47(b) shows Henry at an Assize of Measures.

The Weights of Henry VII

There is little documentation of these weights apart from a manuscript of 1496[19] entitled 'Standard Weights and Measures of Henry VII' and the Elizabethan report of 1574 which is discussed in the next section. The manuscript gives no details of the weights at all but gives the misinformation that 16 ounces Troy make an Avoirdupois pound, and Troy and Tower weight are declared to be the same, which they are not. However, in 1574 it appears that even then there were no Exchequer Troy weights, for recourse was made to those at the Goldsmiths' Company[20] as standard*, but the Avoirdupois weights of the cities of London, Exeter, Worcester, and Norwich, and of the Clerk of the Market, all of Henry VII's date, were compared, together with Winchester's set of Edward III weights. The 56 lb

*This Troy pile of the Goldsmiths' Company disappeared in the seventeenth century, probably lost in the Great Fire of London, 1666.[21]

weights* of London, Exeter, and Worcester agreed to within one grain of each other, that of Norwich being 3 oz heavier while that of Winchester was 6 oz lighter,[21] but the London, Exeter, and Worcester weights were all 3 oz heavier than the Exchequer standard 56 lb weight. This means the corresponding Winchester weight was 3 oz lighter than that in the Exchequer in 1574.

The 14 lb weights of Exeter, Winchester, Norwich, and the Clerk of the Market were 0.75 oz heavier and 2.0, 0.75 and 0.45 oz lighter respectively than that of the City of London.[20] The 14 lb Exeter weight was examined in 1873.[22] Although by then it had lost its iron ring and most of its staple, the weight could be approximated by making a new ring and staple of iron and weighing the whole together when it appeared to be 1.1 oz heavier than 14 Imperial Standard Avoirdupois pounds.

The weighing in 1574 of the other city weights revealed discrepancies of a greater or lesser magnitude so that the report declares 'they could not vouche any sort of the same for the just standard of England'.[23] A number of examples of the weights of Henry VII are to be seen in the Science Museum (Fig.48). Further, Professor Miller in 1856 mentioned a weight of 7 lb less 650 grains, probably of the time of Henry VII, then owned by a Fellow of St. John's College, Cambridge and one of 2 and another of 4 lb of the same monarch in the library of that University, but no details as to their actual weights were given.

(c) The Issues of Elizabeth I

The Exchequer linear standards of Elizabeth I (1588) are also in the Science Museum. They form a composite measure, consisting of a brass bed some 49 inches long into which the yard and the ell rods fit. The yard is half an inch square in section and bears a Roman 'E', crowned. Like that of Henry VII, it is an end measure. As with previous yard standards it is divided in binary fashion, ending in two $2\frac{1}{4}$ inch 'nails'. When examined in 1742 it was observed that the end faces were 'neither flat nor parallel'.[25]

Remarkably enough, this standard yard had been broken sometime between 1760 and 1819 and crudely mended with a dovetail joint.[26] This uncertainty of almost sixty years is an indication of the frequency of inspection then of the primary standards of the realm. In 1836 it was described as little better than a kitchen poker, but this condemnation is altogether too severe. True, it is of crude workmanship. Its divisions could hardly be described as finely engraved but, despite its trials and tribulations, it is now only 0.01 inches less than the standard yard of the twentieth century.

As stated earlier, the Elizabethan ell is the oldest Exchequer ell which is

*These weights were all of bronze with an iron ring and staple.

extant and in all probability was the first Exchequer ell ever to be constructed. It is a brass rod $\frac{6}{10}$ inch square in section and $1\frac{1}{4}$ yards long (5 quarters). It, like the yard, carries the crowned 'E'. It is divided along its length into sixteenths and actually measures 45.04 inches.[26] The bed, interestingly enough, not only carries the yard and ell in recesses but is itself marked off by coarse lines into three feet, the first foot being divided again into twelve inches and the first two inches subdivided into halves.

These measures of Elizabeth formed the primary standards of length of England from 1588, when they were deposited in the Exchequer, until 1824, when new standard measures were accepted and the ell abolished altogether as a standard of length.

Weights

If Henry had his problems with the capacity measures, they were nothing to those encountered by Elizabeth in producing her standards of weight. The first issue was that of 1558, the first year of her reign. These weights of Troy and Avoirdupois measure were found to be too heavy and not agreeable to the earlier standards. In all probability the Avoirdupois weights were based on the erroneous description of the pound as being 16 Troy ounces (but also see below). Seven of the weights of the first issue, two of 112 lb and one each of 28, 14, 7, 4, and 2 lb, are now in the Science Museum, London. (See Fig.49.) In 1574, complaint having been lodged that the weights throughout the realm were uncertain and that no true and just standards were then extant for the sizing of the weights, a jury of nine merchants and twelve goldsmiths of London was convened to re-establish both Troy and Avoirdupois weights. The result was equally unfortunate. This jury of 1574 made no mention of finding any Troy standards in the Exchequer. Their report shows that they regarded the Troy weights of the Goldsmiths' Company as the standard:[20] 'The goldsmithes great pile to us delivered is theonlie [sic] patern extant as far as we can find for the sizing of all other of the sorte . . .'. Happily the set of standard Troy weights at the Mint agreed very well with the Goldsmiths' pile when compared in that year. Those of the Clerk of the Market and the weights actually in daily use at the Mint in the Tower of London also agreed well but to wider tolerances with those of the Goldsmiths.

As for Avoirdupois standards of Henry VII, none is mentioned in the Report as being in the Exchequer. Rather the jury of 1574 took the standards of the City of London as the true measure of Avoirdupois weight in the realm.

Three of the seven weights now in the Science Museum were weighed by Elizabeth's jury against the London weights and all were weighed relative to the Imperial standard in 1873, with the results shown on the following table:

Elizabeth I; First Series[27]

	Weight in 1574 relative to London weights	*Weight in 1873 relative to the Imperial standard*
112 (No. 1)	–	112.659
112 (No. 2)	–	112.278
28	28.805	28.893
14	14.367	14.389
7	7.195	7.134
4		4.063
2	–	2.009

Of these seven weights, the 1574 comment was 'It semeth to the said jurors y'all the said weightes . . . were never sized', a comment which was repeated in 1873.[21]

With all this confusion it is little wonder that the jury decided to start afresh and produce new standards, but where to seek the official weights of the country? It will therefore be no surprise that the new weights so produced by them and issued in 1574 (Elizabeth's Second Series) were condemned eight years later as not being of the old standard. The jury had confused the weight of a penny with the Troy pennyweight and the Avoirdupois pound of 16 ounces totalling 7000 grains with the *libra mercatoria* of 15 Troy ounces totalling 7200 grains, declaring (erroneously) at the beginning of their report that the Avoirdupois pound was 15 Troy ounces.[28] A number of examples of the Second Series are in the University Museum of Archaeology and Ethnology, Cambridge.

A new jury of eighteen merchants and eleven goldsmiths of London was empanelled in 1582 for the construction of what came to be Elizabeth's Third Series of 1588.

The standards of 1574 were for the most part destroyed in 1583. A new Troy set of Exchequer weights based on the Goldsmiths' weights was prepared in 1582, together with two sets of Exchequer Avoirdupois weights based on the ancient 56 pound weight of Edward III found in the Exchequer, as already mentioned. The Troy weights are nested cup weights running from 256 to $\frac{1}{8}$ oz and the Avoirdupois weights were of two designs, flat discs and bell-shaped. The former ranged from 8 lb to 1 dram, the latter from 56 lb to 1 lb. They are shown in Figures 50–51. Copies were prepared and sent in 1588 to all the towns and cities which had received the Henry VII standards and to additional centres as well. A total of 58 sets, including the set for the Exchequer, was produced, marked '1588, 30 Eliz', at a cost of £547 17*s* 2*d*. There is no doubt that the Third Issue consisted of handsome, well-made weights, and they acted as the primary standards of the kingdom until 1824. These Exchequer standards are preserved in the Science Museum.

Measures of Capacity

The Elizabethan measures of capacity were based on those of Henry VII and agree very well (see Ch.IX p.160). They were issued in 1601-2 (Fig.35) and will not be discussed further except in one particular. The Exchequer had apparently no ale gallon of this issue. This is probably why statute after statute declares the ale and beer measures are to be gauged by the Exchequer ale quart rather than by a gallon measure. But the City of London has such a gallon dated 1601 on view at the Museum of London.[29] It is a handsome vessel, and is said to contain 288.75 in^3* 'very very nearly', and is therefore 2.3 per cent over the intended 282 in^3.

(d) *Progress from 1588 to 1760*

For all but the last few years of the quarter millennium following 1588, the primary standards of the realm were those of Elizabeth of 1588 and 1601-2, but a good deal of work was being done on the development of accurate standards by private and semi-private groups as well as by Government. Some of the more important will be considered in what follows.

Yards

The Science Museum, London, has a Tower of London end-standard brass yard of octagonal section inscribed 'The yarde is examined by me, John Reynols and sealed at the Tower of London, Aprill 30 Anno 1659'. It measured 35.97 inches in 1937.

Next was the construction of a yard based on the Exchequer standard for the Worshipful Company of Clockmakers of the City of London which was delivered to them from the Exchequer by indenture on the 4 September 1671.[30] This yard was an octagonal brass rod, about half an inch in thickness, sealed with the Exchequer seal and with 'C. R. crowned' at both ends. It was a bed measure, in that the distance of one yard was that between two upright posts, filed away to give the correct length. The present location of this yard is not known.

The next yard of importance was that made for the Board of Ordnance at the Tower of London, based on that of Elizabeth, by a Mr John Rowley, a noted instrument-maker, in 1720. This measure has been described as:[31]

> . . . a solid brass rod, about seven-tenths of an inch square and about 41 inches long; on one side of which is laid off the measure of a yard, divided into three feet, and each foot into 12 inches: The first foot has the inches divided into tenths, the second into twelfths and the third into eighths of an inch, and the first inch of all is divided into a hundred parts by diagonal lines.

*This is exactly the capacity of the Bridport 'Beare or Ale Gallon' sized by John Renolds at the Tower in 1659. (See Ch.IX p.161)

It was sealed with the Exchequer seal, twice stamped 'GR, crowned' and with a broad arrow.

In 1742 the Royal Society of London proposed to exchange copies of the English standards for those of the Paris standards. To this end Mr George Graham, a Fellow of the Royal Society, procured from Mr Jonathan Sissons, a London instrument-maker, 'two excellent brass scales of six inches each, on both of which one inch is curiously divided by diagonal lines and fine points into 500 equal parts', together with two brass rods, about 42 inches in length and $\frac{1}{2}$ inch broad with squared ends. On each of these rods Graham engraved three longitudinal lines. On one of the lines on each rod he laid off the length of the Tower yard of Rowley*, mentioned above. This was marked with the letter 'E'. The two brass rods and one of the six-inch scales were sent to the Royal Academy of Sciences in Paris together with one of two sets of weights made by Mr Samuel Read, a London scale- and weight-maker. These weights consisted of two single Troy pounds and two sets of Troy weights each running from 8 oz to $\frac{1}{4}$ oz, two sets of pennyweights from five to one-half pennyweight, and grain weights running from 6 to $\frac{1}{4}$ grain. In addition, two single Avoirdupois pound weights were included. The set of the weights and the six-inch scale transmitted were to be kept in Paris, and the Academy was to mark on each of the two 42 inch rods the length of the half-toise of Paris, the Academy to retain one and to return the other to the Royal Society together with a standard weight of 16 Paris ounces (2 marks), the Academy, having prepared two identical weights, retaining one for their own use.

All this was duly performed.[32] The measure of the half-toise was marked by the letter F on the brass bar. On the return of the bar now carrying the Tower yard and the half-toise, both marked on one of the three longitudinal lines, Graham instructed Sissons to subdivide the yard and the half-toise into 3 parts each in order to determine the relative lengths of the Paris and English foot.

Before depositing the Paris weight and the rod in the archives of the Society they were both examined carefully, whereupon it was found that the half-toise measured 38.355 English inches. The Paris foot was therefore 12.785 English inches. The Paris 16 ounce weight weighed 7560 Troy grains, giving a Paris ounce of 472.5 Troy grains. The ratio of the Paris ounce to the English Troy ounce was therefore exactly 63/64. Checking the Avoirdupois pound made by Read, it was found to weigh 7004 Troy grains.

In April 1743, at the request of the Royal Society, Mr Graham undertook a comparison of the Exchequer yard and ell of Elizabeth and the matrix into which they fitted, the Exchequer yard of Henry VII, the Clockmakers' yard, and the Tower yard, (E).[33] The chairman of the Supervisory Committee was

*Inquiries at the Tower of London, the Royal Society and the Science Museum, London, have failed to trace the present location of this important measure.

Lord Macclesfield. Although the ends of the Elizabethan yard rod were not parallel the length of the standard was taken to be the greatest length between the two ends.[25] The results were as follows:

1) The matrix of the yard exceeded the Elizabethan standard by 0.0102 inches.
2) The yard, (E), inscribed by Graham on the Royal Society's rod, exceeded the standard by 0.0075 inches.
3) The Henry VII yard was short of the standard by 0.0071 inches.
4) The Elizabethan ell exceeded 45 inches of the standard by 0.0494 inches.

The Guildhall measures were also compared but as they are not significant for our inquiry they shall not be discussed here.

At this point Graham laid off the length of the Elizabethan Exchequer standard yard on the central longitudinal line of the three which had been engraved on the Royal Society's bar. It was marked 'EXCH'. The third line carries three barely perceptible dots defining a yard and the half-toise. Which yard is depicted is not clear (Baily Ref.45). This bar became known as the Royal Society bar No.41 and is so inscribed today. The yard called the 'Royal Society's yard' is that marked E on this bar and was said to be equal to the Tower Yard of Rowley, but E is given as 0.0075 inches longer than the standard while Ref.21 gives Rowley's as 0.0111 inches in excess, a difference of 0.0036 inches.

The comparison of the Clockmakers' yard with that now laid down on the middle line (the standard) found the Clockmakers' yard to be short by 0.021 inches.

At this time the weights lodged in the Exchequer as the standards consisted of a set of nested Troy weights running from 256 oz to $\frac{1}{16}$ oz, but no pennyweights were to be found. As for Avoirdupois weights, there were: one 14 lb weight, 7, 4, 2, and 1 lb bell weights and a set of flat weights running from 14 lb to $\frac{1}{64}$ lb or $\frac{1}{4}$ oz. These weights together with those of the Royal Society, the Founders' Company, and the Mint, were compared one with another by Mr Read, who supplied pennyweights and grain weights as needed, the results being as follows:

The 14 lb and 1 lb Avoirdupois Exchequer bell weights together weighed 218 oz Troy 13 dwt $23\frac{1}{4}$ grains, which yielded an Avoirdupois lb of 6998.35 grains. Likewise the 7 lb and 1 lb Avoirdupois bell weights yielded a pound of 7000.7 and 7002 grains respectively. The bell-shaped 1 lb weight was found to be heavier than the flat 1 lb weight by $2\frac{1}{2}$ grains, hence the flat 1 lb weight weighed 6999.5 grains.

The Royal Society's Avoirdupois pound was lighter than the Exchequer 1 lb bell weight by 1 grain while their Troy pound was lighter than the Exchequer $8+4$ oz weight by 0.5 grain.

Further comparisons of the weights of the Founders' Company and those at the Mint need not occupy our attention.

These comparisons of the various yards and pounds have been given here in some detail because of the central importance they came to acquire in the light of subsequent events.

The Carysfort Committee

More than a decade after these events involving Mr Graham and the Royal Society the Government set up a committee 'to inquire into the Original Standards of Weights and Measures in this Kingdom and to consider the Laws relating thereto' under the chairmanship of Lord Carysfort, hence the name of the Committee. It had no fewer than 63 members. Their report is dated 26 May 1758,[34] from which it is clear that after reviewing the statutes from early times to date, commenting thereon, and interviewing a considerable number of people directly involved with the making or the use of measures, they proceeded to determine directly the cubic capacities of the various Exchequer standards* as follows:[35]

The Elizabethan bushel 1601	—	2124 in^3
The Standard Gallon (corn) 1601	—	271 in^3
The Standard Quart (ale) 1601	—	70 in^3
The Standard Pint 1602	—	34.8 in^3
The Standard Wine Gallon (Queen Anne 1707)	—	231.2 in^3

Those in charge of these measurements were Mr Bird, instrument-maker, and Mr Joseph Harris, HM Assaymaster at the Mint.

The Committee examined next the Exchequer standards of length and declared 'These are all very coarsely made, the Divisions that are upon them not exact, and the Rods appear to be bent, and therefore very bad standards'. Being advised that the Royal Society possessed the yard made by Graham in 1742 they secured it for examination. Mr Harris was of the opinion that the Exchequer standards of length and the yard of the Royal Society (Marked 'E' on rod No.41) were rather too short (even though 'E' was 0.0075 inches longer than the standard, see p.245) but that:[36] 'the Royal Society's Standard was made so accurately and by Persons so skilful and exact that he did not think it was easy to obtain a more exact one'.

Mr Harris stated that the yard should be a straight bar of brass some 38 or 39 inches long, about an inch broad and thick. Near each end a fine dot should be engraved, the distance between the dots being the measure of the yard. He suggested that two gold studs fixed in the brass rod should carry the engraved dots. He further asked the Committee whether such a rod should be made by comparison with the Royal Society's yard. This suggestion was approved by

*These are the standards which were remeasured in 1931/2 (see Chapter IX).

the Committee, and two rods were made by Mr Bird under the supervision of Mr Harris, one according to Mr Harris's specification as already described, the other a bar of rectangular cross-section with plane vertical end pieces exactly one yard apart. Into this bed measure a yardstick under test could be inserted if it were shorter than the standard and its degree of shortness determined. If the trial yard were precisely the standard length or in excess thereof it could not be inserted. The bed measure was convenient, therefore, for detecting short yardsticks.

After these rods and the Royal Society's yard[36,37] had lain together overnight to equalize the temperature, their respective lengths were compared by beam compasses and 'found to agree as near as it was possible'. The one, made to Mr Harris's specification, was marked 'Standard Yard 1758' with a crown over the symbols 'GII'. It was 39.06 inches long and 1.05 inches square in section with a fine point engraved on each of two gold studs or pins which were themselves about one tenth of an inch in diameter and were sunk into the upper surface of the bar. It was not subdivided at all. The bar was deposited with the Clerk to the House of Commons, with the suggestion that it might be considered as a primary standard in the future; the other being a bed measure was more suitable for quick comparisons at the Exchequer on a daily routine basis. To this end the bed measure was divided into three equal parts (feet), one of which was divided into twelve inches and one of these inches subdivided into tenths, thus facilitating comparisons.

The Committee then embarked on a very detailed comparison of the Troy and Avoirdupois weights in the Exchequer, noting discrepancies where these arose. Recourse was again made to Mr Harris as to how the true Troy and Avoirdupois pounds might be ascertained. As there could only be one standard, Mr Harris was of the opinion that it should be the Troy pound, as it had been longest in use, was the basis of the coinage, and was best known in the rest of the world. Avoirdupois weight, on the other hand,[38]

> is of doubtful authority, and though unfit to be made a Standard, yet the frequent use of it renders it necessary to ascertain and declare how many ounces, pennyweights and grains Troy the pound Avoirdupois ought to weigh'.

The Committee then directed that three Troy pounds should be made under the direction of Mr Harris, suitably marked 'lb T 1758' and with 'G2 crowned'. This was done and the weights were sized by Mr Harris.[39] These Troy pounds were of gunmetal,[40] one of which was designated as the standard.[41] There being insufficient time for the production of ounce weights and pennyweights, the desired relationship between Troy and Avoirdupois weight could not then be determined.

The Committee attempted to get the Goldsmiths' Troy pile for examination, having regard to the role it had played in the determination of

Elizabeth's Troy standards, only to find it had disappeared, probably destroyed in the Great Fire of London of 1666. The 56 lb Avoirdupois weight of Edward III upon which Elizabeth's Third Series had been based was also missing.[42] These two losses, though serious, were not fatal to the determination of the original standards.

The 1758 Report concludes with a number of resolutions or recommendations, the more important of which are:

(1) 'That the distance between the two points in the gold studs in the brass rod described in this report and delivered herewith ought to be the length called a yard'.

(2) 'That the gallon ought to contain 282 such inches (cubic inches)'. (Here the Committee is in favour of abolishing all gallons except the ale gallon).

(3) 'That the standard of weight ought to be the pound herewith delivered, described in this report', namely the Troy pound.

These were agreed to by the House of Commons on 2 June 1758.

The Second Report[43] from an enlarged Committee of 68 members dated 11 April 1759 records how Mr Harris was empowered to make the fractions of the Troy pound necessary for determining the true proportion of the Avoirdupois pound. This determination was undertaken. The Committee discussed the possibility of removing the Avoirdupois pound altogether or, where such weights existed, engraving them with their Troy equivalents so that they could continue in use. Thus a 1 lb Avoirdupois weight would be marked 1 lb 2 oz 12 dwt.[44] It was conceded that there would be difficulties were the Avoirdupois pound to be abolished but a detailed table giving Avoirdupois weight in terms of Troy weight was provided.

The 1759 Report (p.457) shows that Mr Harris succeeded in 1758 in procuring for the Committee two sets of brass weights, all multiples of the standard pound, that is 2, 4, 8, 16, and 32 lb Troy, which were adjusted and presented to the House of Commons. The 2 lb weights are specifically referred to in the Reports of the Commissioners of 1821 and of the Committee of the same year as we shall see. The recommendations at the end of the 1759 Report reiterated the Committee's views respecting the gallon, the pound, and the yard as given in the 1758 Report.

The Committee continued its work in 1760, having Mr Bird prepare a second yard bar identical, as far as could be ascertained, to that of 1758. This standard was marked 'Standard Yard 1760'.* A companion bed measure was also made as in 1758. The drafting of legislation commenced.

It is not known whether it was the novel nature of some of these recommendations which required further research on the part of parliamentary committees and which occasioned the long delay before any legislation

*It was a little longer overall than its 1758 counterpart being 39.73 inches long.

appeared, or whether it was fate, which, following the death of George II in 1760, caused Parliament to be prorogued just as legislation introduced by Lord Carysfort in the form of two Bills was in the committee stage. Certainly some of the proposals needed further thought even if Parliament had approved them.

(1) The yard gave no difficulty. It merely continued with better workmanship and precision that which had existed in the days of Henry VII and Elizabeth.
(2) The suggestion that all gallons should have the capacity of the ale standard was a difficulty. The wine gallon had only been *legally* established at 231 in^3 some fifty years earlier and the corn gallon was some ten cubic inches less than that proposed. Further the subdivisions of the gallon proposed (282 in^3) would lead to fractional parts of a cubic inch, the quart would be 70.5 in^3, the pint 35.25 in^3. To effect such a change was likely to bring confusion if not conflict.
(3) The suggested abolition of the Avoirdupois pound was even more serious for it was the unit employed almost without exception in commerce.
(4) The further recommendation making the Troy pound the primary standard of weight for all commodities, coupled with the suggestion that the Avoirdupois pound could be made but engraved in the equivalent Troy weight, that is 1 lb 2 oz 12 dwt, and used in that fashion, was not likely to command enthusiastic support.

In the final recommendations, the Committee had made no reference at all to the Avoirdupois pound but in their proceedings consideration was given to raising the Avoirdupois ounce from 437.5 to 438 grains, giving a pound of 7008 grains.

Nothing came of these two reports by way of legislation as we have just seen but, having said this, the impact of the work of the Committee was nevertheless considerable and led eventually to the establishment of the Troy pound as the primary weight standard, with the Bird measure of 1760, not the recommended yard of 1758, as the primary standard of length in 1824.

(e) *Events 1760 to 1834*[45]

General Roy's 42-inch Scale

The first observations for the survey of the kingdom that was later to be known as the Ordnance Survey were made in 1785, with the detailed survey beginning in 1791. The primary base-line was laid down on Hounslow Heath under the superintendence of General Roy (1726-90). The 42-inch brass measure, purchased and used by Roy and which came to be associated with his name, had been made by Sissons and divided by Bird in 1742 and had previously been in the possession of Graham. It was overall about 43 inches long and had been derived from the Royal Society yard. The length of the first 36 inches of Roy's scale was compared in 1785 with the Royal Society yard. They were found to agree exactly at 65°F. That which came to be called the 'Ordnance Yard' was in fact nothing more than these first 36 inches of Roy's scale.

A standard plank of wood, approximately 30 ft long, 9 or 10 inches broad, and 3 inches thick, was prepared on which six lengths of the first 40 inches of Roy's scale was laid off making a standard length of 240 inches or 20 feet. Initially the surveyors used wooden rods of deal 243 inches long carrying near their ends ivory pieces with engraved lines set 20 ft apart according to the standard scale on the plank. The survey was, unhappily, plagued with rain. The wooden rods appeared to vary in length according to the humidity and, though they stabilized, they were replaced with tubular glass rods, 240 inches long weighing 61 lb each. Using these, the base-line was successfully measured.

Because of the central role played by the first 40 inches of the scale, some later commentators have called the brass scale 'Roy's 40-inch scale' but there is no doubt that 42 inches are engraved upon it. Roy's article (Ref.45) so declares it to be and the actual scale can be seen in the Science Museum, London.

The Shuckburgh Scale

Another very important standard of length was constructed in 1796.[46] This was the 60-inch scale made by Troughton (a well known instrument-maker) for Sir George Shuckburgh after one made for himself in 1792 or 3 and subsequently referred to as 'Troughton's scale'.* The scale itself consisted of a brass bar 67.75 inches long, 1.4 inches wide, and 0.42 inches thick, divided into five feet each of twelve inches, and every inch divided into tenths. The inches were numbered, 0 to 60. It was mounted on a mahogany beam 75 inches long, 6 inches deep, and 5 inches wide and the scale was held in place on the beam by three finger screws. A second scale of 60 inches similarly divided called the 'beam' could be moved relative to the main scale by means of an accurately calibrated screw.

The Shuckburgh scale was the first to be constructed using micrometer microscopes instead of beam compasses. It was compared with a number of other measures, most notably the Royal Society yard (E), the Exchequer yard 'EXCH' laid down on the same bar, General Roy's scale, and both the 1758 and 1760 scales of Bird. It was found that the yard 'E' was 0.0130 inches longer than 36 of Shuckburgh's inches at 60.8°F while the 'EXCH' yard was 0.0067

*These scales can be confused since Shuckburgh referred to *his* scale as 'Mr Troughton's scale', after its maker. Troughton's bar was shorter, something over 62 inches, and carried 621 dots denoting $\frac{1}{10}$ inches but un-numbered. Though 62 inches long it is frequently referred to as 'Troughton's 60 inch scale'.

inches shorter at 60.6°F. Shuckburgh tells[45] of locating, in the custody of the Clerk of the Journals and Papers at the House of Commons, Bird's 1758 yard and in a box the 36-inch bed measure. Also located were the yard of 1760 and its companion bed measure. As these bars were too thick to be examined by his micrometer microscopes, their lengths were transferred to the Shuckburgh scale by beam compasses. The Bird yards appeared each to be of exactly the same length, 'at least [they] did not differ by more than 0.0002 inches', and the Bird yard measured 36.00023 inches at 64°F on the Shuckburgh scale, the error being so slight that the two could be taken as exactly the same. There is little doubt but that the Shuckburgh scale was regarded as the principal scientific yard measure of the time.*

Troughton also at this time made three sets of weights:

(1) binary grains ranging from 1 to 16,384 (that is 2^{14}) grains
(2) decimal grains ranging from 1 to 20,000 grains and
(3) Troy weights from 1 to 12 ounces.

Shuckburgh compared two of the three Troy pounds made by Harris in 1758 with those of Troughton with the following results:

> The Harris ('Standard') Troy pound was 5763.745 grains by Troughton's weights. The Harris duplicate pound was 5763.685 grains, giving a mean value of 5763.715 grains.

As the eighteenth century closed we see the accuracy of the various standards changing by several orders of magnitude. Precision could be and was demanded. A major change in direction could now be contemplated. The nineteenth century opened with a flurry of committees or commissions to consider one or more aspects of English metrology.

The Committee Report of 1814

The Report of the Select Committee 'to inquire into the Original Standards' was published on 1 July, 1814.[47] This Committee had the two Carysfort Committee reports referred to it and the recommendations contained therein were carefully considered.

Dr Wollaston, the Secretary of the Royal Society, and Professor Playfair of Edinburgh were examined and they advised the Committee of the recent determination of the density of pure water, namely that a cubic foot at 56.5°F weighs exactly 1000 ounces Avoirdupois and that 27.648 cubic inches was the volume of an Avoirdupois pound of pure water. This led the Committee to express the view that there should be but one gallon though not that

*Especially by Capt. Henry Kater (1777-1835), physicist, distinguished for his work in determining the length of the seconds pendulum and for his careful verification of standards. (See later, p.253 et seq.)

recommended by the Carysfort Committee (282 in^3). They gave it as their opinion that:

> the Standard gallon, from which all the other measures of capacity should be derived, should be made of such a size as to contain such a weight of pure water of the temperature of $56\frac{1}{2}$°F as should be expressed in a whole number of Avoirdupois pounds and such also as would admit of the quart and pint containing an integer number of ounces without any fractional parts. If the gallon is made to contain ten pounds of water, the quart will contain forty ounces and the pint twenty.

This recommended gallon would have a cubic capacity of 276.48 in^3. The corresponding bushel would be 2211.84 in^3 instead of the then current 2150.42 in^3,[48] and this new bushel would hold 80 Avoirdupois pounds of water. The quart would have a volume of 69.12 in^3 and weigh 40 oz, the pint would be 34.56 in^3 and would weigh 20 oz, and the half pint would be 17.28 in^3 or exactly one-hundredth of a cubic foot and would weigh 10 oz.

The Committee was led to this size of gallon because of the near-coincidence of this pint of 34.56 in^3 with Elizabeth's standard pint of 1602 of 34.8 in^3. Further, $\frac{51}{50}$ of the proposed gallon is very close (282.01 in^3) to the beer gallon of 282 in^3, and $\frac{10}{12}$ is very close (230.40 in^3) to the wine gallon of 231 in^3. This bushel of 2211.84 in^3 when multiplied by $\frac{35}{36}$ (2150.40 in^3) was very close to the Winchester bushel of 2150.42 in^3, and the pint multiplied by three (103.68 in^3) was very close to the Stirling jug of Scotland of 103.40 in^3. The Report stated:

> your committee are of the opinion that this departure from the corn measure which is employed in the collection of the Malt Tax and is supposed to be the most generous throughout the kingdom is justified by the advantages which they anticipate from the change.

The Committee went on to give its views as to the standard of weight. While agreeing that the Troy pound was the only weight established by due process of law, they admitted that the Avoirdupois pound was in use everywhere and for so many commodities that 'your committee cannot hesitate to recommend it in preference to the Troy lb'. The Troy pound, it was recommended, should remain as the bullion weight with Apothecaries' weight based on the Troy grain for drugs as heretofore. It was further stated that their proposed standard pound should be determined by weighing a solid cylinder of brass containing 27.648 in^3, first in air and again when completely immersed in water. The difference would be just the Avoirdupois pound,[49] for if 20 oz of water have a volume of 34.56 in^3 as we have seen, then 16 oz would have a volume of 27.648 in^3.

The Committee resolved that the Bird standard described in the 1758 Report and presently in the custody of the Clerk of the House of Commons ought to be the length of the yard. It further gave as its opinion, on the advice of its scientific consultants, that the length of a pendulum which vibrated 60 times a minute[50] in the latitude of London had been ascertained to be 39.13047 inches of which the standard yard contained 36.

Although the Committee did not make any recommendations with respect to the pendulum there was the clear hint that to use such a device would give a natural unit of length, for the Committee wrote:

> In this manner the standard of length is kept invariable by means of the pendulum, the standard of weight by that of the standard of length, and the standard of capacity by that of weight.

The actual recommendations of the Report included the Bird yard as standard, the gallon of 10 lb of pure water or 276.48 in^3, with the standard weight as the Avoirdupois pound being the weight of 27.648 in^3 of pure water at 56.5°F. The pound should be 16 ounces, each ounce of 16 drams; 'the third part of a dram is to be the scruple and the tenth part of the scruple to be one grain'.*

Hence we have some really new thoughts on the subject. There was no immediate government action, but a Royal Commission was now set up consisting of J. Banks. G. Clerk, D. Gilbert, W. H. Wollaston, T. Young, and H. Kater, 'to consider the Subject of Weights and Measures'. Their first Report was published in 1819[51] and again we find novel ideas.

The Commission Reports of 1819, 1820, and 1821

The Commission saw no reason to alter the standard of length but they recommended that for a legal definition of the yard, the scale employed by General Roy should be used,

> a duplicate of which will probably be laid down on a Standard scale by the Committee of the Royal Society appointed for the assistance of the Astronomer Royal in the determination of the length of the pendulum, the temperature being supposed to be 62°F when the scale is employed.

They proposed

> for purposes of identification or recovering the length of this standard should it ever be lost or impaired that the length of a pendulum vibrating seconds of mean solar time in London at the level of the sea and in vacuum is 39.1372 inches of this scale

be used. From this it will be noted that in the five years since the Committee

*This recommendation would have made the ounce contain 480 grains, but not Troy grains, for the Avoirdupois pound would contain 7680 of these grains.

report of 1814, there has been a noticeable shift in the alleged length of the pendulum when temperature and other effects are taken into account. Moreover the Commissioners felt that for capacity measures it was better to measure from a weight of water than from cubic content. They stated that 19 in^3 of distilled water at 50°F must weigh exactly 10 *Troy* ounces and they endorsed the 1814 suggestion of a gallon containing 10 *Avoirdupois* pounds of pure water, recommending that the gallon for ale and corn should contain exactly 10 Avoirdupois pounds of pure distilled water at 62°F. This would be 'nearly 277.2 in^3'* and would agree with the standard pint in the Exchequer, a consideration which had also influenced the 1814 Committee. But they were undecided as to whether the wine gallon should remain or whether it, too, should change to the 10-pounds-of-water gallon.

The Second Report of 1820[52] was quite brief. It stated that the Commissioners had discovered an error in the instruments used in measuring the base at Hounslow Heath and that in consequence they were reverting to the originals, preferring to the Roy scale the Bird standard of 1760, to which they had not previously had access. The length of the seconds pendulum was in consequence slightly modified to 39.13929 inches.

The Third Report, dated 31 March 1821,[53] begins by recommending the adoption of the Bird standard for length, identifying it by declaring that 39.1393 of its inches gave the length of the seconds pendulum. They go on to recommend the Troy pound in the form of the 2 lb weight (made by Harris in 1758 with other Troy weights, see p.248, not his 1 lb 'Standard') to remain unaltered as the weight standard and that the gallon generated by 10 lb (Avoirdupois) of water should be the '*Imperial*' gallon. There was no mention of a separate wine gallon. This was the first use of the word 'Imperial' to describe an English measure.

It will be observed that in 1820 and again in 1821 the Bird standard is now back in favour replacing that of General Roy. The story is that it was discovered that only the first 36 inches of Roy's scale had been verified by comparison with the Bird scale. The remaining 4 inches which had been used in setting off the base-line at Hounslow Heath had never been verified.[54]

The 1821 Report of the Select Committee

Two months after the Third Report from the Commissioners, the Select Committee on Weights and Measures reported.[55] Here the Committee concurred with the Commissioners as to the inexpediency of changing any standard. They recommended that the Bird standard yard of 1760 be accepted as the authentic legal standard of the British Empire and that the standard brass weight of 2 lb in the possession of the House of Commons (see pp.248 and

*A change from the 1814 figure of 276.48 in^3 owing to the temperature now being 62°F rather than $56\frac{1}{2}$°F.

256 footnote) and made by Harris in 1758 be considered authentic, one half of which to be the legal Troy pound of the British Empire, weighing 5760 grains with 7000 such grains being an Avoirdupois pound.

This Committee was of the opinion that a cubic inch of distilled water in a vacuum at 62°F weighed 252.72 grains so that the cubic foot would weigh 62.386 Avoirdupois pounds.

In their review of the various gallons in use throughout the kingdom's history they expressed their belief that the original gallon of England had been used for all commodities and that some variations arose by accident, others by fraud, until the present multiplicity was arrived at. They agreed that hereafter there should be only one gallon and they concurred with the pint as the volume of 20 oz Avoirdupois of distilled water, that is the gallon of 10 lb.

We must note in passing an important series of comparisons of various yard measures by Capt. Kater in 1821 because of the role they were destined to play in the future. The yards were those of the Royal Society ('E'), Roy, Shuckburgh, Bird's 1760 yard, a 40-inch scale known as the 'Ramsden scale',* and one made by a Mr Cary for Col. Lambton for the triangulation survey of India and known as 'Col. Lambton's scale' which was 61 inches long and carried Bird's name. Kater records the deviations of these measures from 36 inches of Lambton's as:

Shuckburgh	*1760*	*Roy*	*Royal Society*	*Ramsden*
+0.000642	+0.000659	+0.001537	+0.002007	+0.003147

Lambton's scale is therefore somewhat shorter than all the others. The Ramsden scale is described in Kater's paper of 1821[56] as 'no longer extant'. Its fate is unknown.

The Weights and Measures Act, 1824

A Bill was introduced into the Commons in 1822 by Sir George Clerk and again in 1823 with a few amendments. Finally the piece of legislation, which had been looked forward to as long ago as 1758, appeared in 1824 in the form of 'An Act for ascertaining and establishing Uniformity of Weights and Measures' 5 George IV c 74 of 1824. It was virtually identical to the 1823 version introduced by Clerk.

In framing the Act of 1824 the legislators had stepped carefully among the recommendations of the various committees and commissions, coming up with a composite statement which most nearly would achieve the desired result, namely the removal of the 'weights and measures, some larger, and some less, [which] are still in use in various places . . . and the true measure of the present standards is not verily known, which is the Cause of great Confusion and of manifest Frauds', as Section 1 of the Act has it.

*Jesse Ramsden, instrument-maker, 1732-1800.

From 1 May 1825 the Imperial standard yard was to be the distance between the points on the gold studs of Bird's brass yard, 'now in the custody of the Clerk of the House of Commons whereon the words and figures 'Standard Yard 1760' are engraved'. It was declared to be 'the original and genuine Standard of that Measure of Length or lineal Extension called a Yard' and it was to be the 'Unit or only Standard Measure of Extension wherefrom and whereby all other Measures of Extension whatsoever, whether the same as lineal, superficial, or solid shall be derived . . .'. If ever lost or destroyed the yard was to be recovered by reference to the length of the seconds pendulum in the latitude of London at sea level in a vacuum in the proportion of 36 inches to 39.1393 inches (Sec.III). This length of the pendulum is slightly greater (by 0.0021 inches) than that given in the Report of 1819 but almost identical to that of the 1820 Report.

From 1 May 1825 the Imperial Standard Troy pound was to be the one pound brass weight made in 1758 (by Mr Harris)* and now in the custody of the Clerk of the House of Commons. This standard was declared to be 'the original and genuine Standard Measure of Weight' and it was stated to be the 'Unit or only Standard Measure of Weight from which all other weights shall be derived . . .'. (Sec.IV). It was further declared to be 5760 Troy grains and the Avoirdupois pound was to be 7000 such grains.

The Avoirdupois pound was not abolished but it was no longer a standard weight, being defined only through this relationship of the grains. Provision was further made that if the Imperial standard pound was ever lost or destroyed it was to be regained by reference to the weight of a cubic inch of distilled water, which, at 62°F, with the barometer at 30 inches (of mercury) weighed 252.458 grains when balanced against brass weights in air. (Sec.V).

From 1 May 1825 the standard measure for capacity was to be, for liquids, as for dry goods, the Imperial standard gallon, which contained 10 lb Avoirdupois of distilled water weighed in air at 62°F the barometer being 30 inches (of mercury), and a measure of brass was to be made forthwith of this standard. Section XIV of the Act gave the gallon in cubic measure as 277.274 in^3, a more precise volume than that given in 1819.

For obvious reasons there was no claim as with the other two primary standards that this was the original and genuine standard, but this was to be the unit and only standard measure of capacity for 'wine, beer, ale, spirits, and all sorts of liquid as for dry goods not measured by heaped measure . . .' (Sec.VI).

Section VII enacted 'that the standard measure of capacity for coals, culm, lime, fish, potatoes, or fruit, and all other goods and things commonly sold by

*It is clear that in choosing the Harris 'Standard' pound Troy here, the Government had not accepted the 1821 proposals from the Commissioners and the Select Committee to use as the standard Harris's 2 lb weight but had gone back to the original and had adhered to Troy weights rather than Avoirdupois.

heaped measure shall be the aforesaid bushel containing 80 lbs Avoirdupois of water . . . being 19½ inches, outside diameter'.

Section VIII prescribed that for goods sold by heaped measure the bushel as described, heaped up, shall be used. The cone of the heap was to be of height six inches, the extremities of the bushel itself forming the base of the cone.* (See Ch.IX)

We are further told that three bushels make a *sack* and twelve sacks make a *chaldron*.

Section XI decreed that copies of the yard, pound, and gallon and 'the said standard for heaped measure i.e. the bushel, as judged expedient, shall, within three calendar months after the passage of the Act, be made and verified, and be deposited in the Exchequer, with the Lord Mayor of London and the Chief Magistrates of Edinburgh and Dublin and such other cities and places as shall be directed'.

Copies of the new Imperial standards were to be issued within six months to the counties, shires, and cities and purchased by HM Justices of Peace, and were to be verified against the Exchequer models. The expense was to be met out of the Rates (Property Tax). An entire page of the document is given up detailing how the money was to be raised.

This piece of legislation repealed all previous Acts from the thirteenth century onwards insofar as they related to establishing standards of weights and measures. It was a watershed in that measures after 1824 were 'Imperial' or 'post-Imperial'; measures defined previously were 'pre-Imperial'.

Wine casks, tuns, pipes, etc. were to be measured by the gallon of this Act. For example the hogshead, previously 63 gallons each of 231 in^3, became 52.5 Imperial gallons (strictly 52.486 Imperial gallons). The difference was insignificant.

Construction and Adjustment of the Principal Standards

In accordance with Section XI Capt. Kater undertook the supervision of the construction and adjustment of these principal standards.[57] Troughton was approached but because of age declined the task which then fell to one Dollond, for linear measure, and to one Bate, for weights and measures of capacity. Bate sought a new alloy to resist decomposition in the atmosphere of London, having the same or equal hardness and workability as hammered brass. His final choice was an alloy 576 parts copper, 59 parts tin, and 48 parts zinc†. The weights so produced were largely spherical but with a flattened bottom. Each had a small button on top which could be unscrewed, revealing

*It is remarkable that, after centuries of legislation prohibiting heaped measure and requiring the bushel to be stricken, we now find specific instructions in a Statute detailing how the heaping was to be carried out.

†Reference 57 says '48 parts of *brass*', clearly an error.

a cavity for holding small pieces of wire for final sizing of the weight. Five standard Troy pounds were made. The final errors in each when compared with the Imperial standard were +0.0005, −0.0015, +0.0021, +0.0022, and +0.0010 grains respectively, which attests to the skill and care lavished on their production. They were destined to go respectively to the Guildhall, Edinburgh, the Royal Mint, Dublin, and the Exchequer. The Mint as a repository for standards is not mentioned in the Act, but it is clear that it required the weight standard in view of its operations.

Metal containers of a triangular vertical cross-sectional pattern were made for gallon standards (see Fig.52). The bushels required a little more thought, for the weight of the vessel plus its water content would be about 250 lb. A special balance was made,* the beam of which was of mahogany 70 inches long, 22 inches deep in the middle, tapered towards both ends, and 2¼ inches thick. With 250 lb in each scale pan the beam would turn noticeably with one grain excess in one pan. The bushel under test was placed on one pan with weights amounting to 80 lb Avoirdupois inside and was counterbalanced with brass weights. The 80 lb were removed and water syphoned in to fill the bushel. A circular glass plate with a central hole of about ¼ inch diameter could be slipped over the top of the bushel and air bubbles could be removed through the hole. Equally, water could be added drop by drop through the hole as required.† Of the four bushels so prepared the errors were +6.55, +1.73, +6.47, and −1.07 grains of water respectively, an accuracy of better than one part in one hundred thousand. The standard bushel of 1824 is shown in Figure 53.

The yards first prepared by Dollond in 1824 were of brass, one inch square in section with one-inch high steel pillars at each end to form a bed measure, the yard being the distance between the opposing parallel faces of the two pillars. Also prepared were brass rods one yard long of fairly slight construction which could be lifted by two wooden handles. These yards were made to coincide with the first 36 inches of Shuckburgh's scale which Kater regarded as the same as the Bird standard, now the Imperial standard, and, being readily portable, were for routine trade comparisons and verifications at the Exchequer. Of greater significance is a set of four standard yards of brass, of the same approximate width and thickness as the Shuckburgh scale. The length of the yard was defined as the distance between two dots, one on each of two gold pins placed flush with the surface and mounted axially on the bar. One of the pins could be rotated and its dot was slightly off centre. By rotating this pin the two dots could be made precisely the requisite distance apart

*This balance, known as 'Kater's balance', is in the custody of the Museum of London.
†This method of testing local metal standard capacity measures of cylindrical form continued in use until 1963 when these measures were gradually replaced by the glass flask or burette type.

whereupon the pin would be permanently fixed in place. Again these were made according to Shuckburgh's scale.

Once the yards had been sized Kater proceeded to recheck them. His horror can only be imagined when he discovered that *all* were too short! Kater was too experienced a scientist to make any gross error so the cause was likely to be something quite subtle, as indeed it was. After some time had been spent examining the possibilities it was discovered that disagreements of about $\frac{1}{600}$ inch could be caused in the standards of length by 'almost undiscoverable inequalities in the surface of the table on which they were laid'.[57] In making the comparisons the various measures had been laid on a seemingly flat mahogany table but minor irregularities in the surface were transmitted through the measure to the inscribed surface causing it to flex. In the words of the Astronomer Royal[58]

> When the supporting surface is nearly plane a bar of any moderate thickness will accommodate its form to the small irregularities and the errors of divisions on the upper surface will be greater for a thick bar than for a thin one.

Kater's immediate solution was to cut down the ends of his yard measures to half thickness and to mount the gold studs at that level, that is in the neutral plane. Once this was done the problem disappeared[59] but of course the Shuckburgh scale could not be so treated as it was engraved throughout on the top side of the bar. The fact that this scale was liable to deformation led to its later rejection as a reliable standard. In the light of this experience, Kater recommended the use of very thin scales mounted on a substantial base.

The problem solved, the standards of length, weight, and capacity were issued, in accordance with the Statute, to the Exchequer and Guildhall in London, to Edinburgh, and to Dublin. These were to be the legal standards. The primary standards were ordered to be immured in the Palace of Westminster as we shall see and included not only Bird's 1760 yard but also his yard of 1758.

However, the provisions of this Act of 1824 did not come into force on 1 May 1825 as prescribed, for by further statute[60] the changeover was delayed until 1 January 1826.*

The Imperial standard Troy pound was further compared in 1829 by Captain Nehus with four others, two of brass and one of platinum in private hands,[61] and with the platinum Troy pound which had been made for the Royal Society by William Cary† in 1829.

Kater's next task was the creation in 1831 of a new standard of length for the

*The delay was probably due to the difficulty experienced by local authorities in obtaining their sets of the new standards, some having to wait till the 1830s.

†William Cary (1759-1825), a maker of scientific instruments and a dealer in platinum, a metal greatly favoured at the time in the construction of sulphuric acid stills.

Royal Society,[62] to become known later as 'Royal Society No.46'. The base on which the scale was laid was of brass, 40 inches long, 1.75 inches wide, and 0.6 inches thick. The scale itself was a brass strip 70 thousandths of an inch thick which was made to slide freely in a dove-tailed groove on the base. The strip could be locked in position by a screw. The divisions (inches) were in the form of dots on small gold disks let into the brass. One inch to the left of zero was subdivided into tenths. The scale was made by Dollond, and the mean error shown by 36 comparisons with the Imperial standard itself was −0.0006204 inches. The Royal Society No. 46 was therefore 35.99938 inches long.

From what has been said above it will be clear that there had been a wealth of inter-comparison data established prior to the standards having been legalized. This proved to be absolutely vital in the light of future events. A reference standard of the Avoirdupois pound was made and verified in 1824. When weighed in 1867 by the Warden of the Standards (H. W. Chisholm) it was only 0.0066 grain in excess of the current standard,[63] indicating the quality now being achieved in the production of these weights and measures. Copies of the standards of 1824 were issued to the cities, towns, and counties and many sets of yards, weights, and measures are to be seen in various town halls and civic museums throughout the land.

The 1824 Act, though of great moment, was not without its shortcomings. For instance, the measures in use up to that date were not made illegal. Indeed Section XVI provided that

> it shall and may be lawful for any person or persons to buy and sell Goods and Merchandise by any weights or measures established by local customs, or founded on special agreement; provided . . . the ratio of proportion which such customary measures or weights shall bear to the said weights and measures shall be painted or marked upon all such customary weights and measures respectively.

The inevitable happened. The old measures continued to be used, though on an ever diminishing scale,[64] for the same section went on to specify that weights and measures made after 1 May 1825 must be in conformity with the standards of the Act.

It was not until the Acts of 4 and 5 William IV c 49 of 1834[65] and 5 and 6 William IV c 63 of 1835[66] that all local and customary measures, including heaped measure, were abolished. Specifically, by name, the Winchester bushel and the Scottish ell were abolished and the penalty for failing to use Imperial measures was set not to exceed 40/- for each offence.

The primary standards of 1824 (the 1760 yard and the Troy pound of 1758) together with Bird's 1758 standard yard were ordered to be immured in the Palace of Westminster for safe keeping. Every ten years they were to be taken out for comparison with the copies and other standards with which they had been compared prior to being sealed up.

Alas, the frailty of human designs was to be high-lighted all too soon, for on the evening of 16 October 1834 a great fire destroyed both Houses of Parliament. The provisions as to what to do if the standards were ever lost or destroyed may have been more prophetic than had been imagined at the time.

(f) The Standards Restored

The fire, which apparently originated in the lobby of the House of Lords and demolished the Houses of Parliament, broke out shortly before 7.00 p.m. on the 16 October 1834.[67] The tide in the Thames was low and there was a lack of water. Only one or two fire engines were available and they were not particularly well placed to deal with so large a blaze. There was no general superintendent of the firemen so that concerted efforts were unlikely. Within a few minutes of the blaze being detected, the police were augmented by fifty members of the 1st Regiment of the Grenadier Guards who kept back onlookers. By 9.00 p.m. three regiments of Guards were in attendance and, for security, papers were being removed under guard in wagons, cabs, etc.

The fire was caused by overstoking the furnaces which fed the heating flues in the House of Lords.[68] Old papers and wooden receipt tallies from the Exchequer had been ordered to be burned. The woodwork surrounding the flues had ignited and smouldered for some time before bursting into flame. The destruction was quick, with fire-fighters concentrating on preventing a spread of the conflagration, and was so complete that the Imperial Troy pound was not even retrieved from the ashes.[69] Both of Bird's standards of length of 1758 and 1760 were recovered but in a ruinous state. The left hand gold plugs or studs had completely melted from both yards. The primary standard yard was 'somewhat bent and discoloured in every part like No. 1 which is marked 'Standard Yard 1758 G II''.[70]

The pound of 1758 and the yard of 1760 had had to wait 66 and 64 years respectively for legal recognition. It was a catastrophe, to put it mildly, that their official lives were so short — a mere decade — but like most disasters some good emerged from it, for a new look could now be taken at our system of metrology.

Steps to reconstruct standards were commenced in 1838 with the appointment of Commissioners on 11 May charged with that task. They were the Astronomer Royal (G. B. Airy), Mr F. Baily, Mr J. E. D. Bethune, Mr D. Gilbert, Sir J. F. W. Herschel, Mr J. S. Lefevre, Mr J. W. Lubbock, the Revd G. Peacock, and the Revd R. Sheepshanks, and it is clear from their report of 21 December 1841[70] that they had addressed the task in most serious fashion, bringing the latest scientific knowledge to bear on their deliberations.

The Report of 1841[70]

The Commissioners were far from being constrained by the 1824 Act which laid down how the yard was to be re-established (by reference to the seconds pendulum) and the Troy pound (by reference to the weight of a cubic inch of water). They pointed out that research subsequent to the passage of the 1824 Act had shown that, in determining the length of the pendulum reduced to the stated temperature in a vacuum and at sea-level at London, certain elements were doubtful, if not actually erroneous, and that to follow the procedure laid down would not necessarily recover the original length of the yard. The Committee in their 1841 report listed a number of papers, mainly in the Philosophical Transactions of the Royal Society, which showed where doubt and error lay with the pendulum, not least in the declared length of the pendulum itself, for the Shuckburgh scale had been used in its determination. Following Kater's work it could only be concluded that this scale was under a cloud and the length of the pendulum uncertain. On the other hand, there were several measures available which had been compared in detail with the primary standard. It was recommended therefore that the yard be regained by reference to these measures. This was accepted. No change was proposed in the unit of length itself.

Further, respecting the regaining of the pound, the weight of a cubic inch of water was not known in international circles to better than 1 part in 1200, for that was the spread in results obtained in the laboratories of several countries, whereas the actual weighing could be carried out with a precision of one part in a million. Here again a considerable number of extant standards had been compared with great precision with the Imperial standard pound, now destroyed. It was recommended and agreed that, in regaining the unit of weight, recourse be made to these. The multitude of inter-comparisons among the standards was now to be of great utility. But the Commission went further.

Evidence placed before them showed that Avoirdupois weight was much more commonly used than Troy. The ratio of sets of Avoirdupois weights to sets of Troy weights was variously stated to be anywhere from 2000/1 to 5000/1. They were told that 'the Troy pound is comparatively useless even in the few trades or professions in which Troy weight is commonly used and that to the great mass of the British population it is wholly unknown . . .'. Section III of the Report stated:

> The Avoirdupois pound on the other hand is universally known throughout this kingdom and moreover being now made equal to 7000 grains it is well adapted to subdivision by the decimal scale, an object which we think ought never to be placed out of view in considering the changes (in other respects producing no inconveniences) which may be made in the weights and measures of the country. We feel it our duty to therefore recommend that the Avoirdupois pound be

> adopted instead of the Troy pound as the primary standard of weight.

It was further recommended that the Troy pound be abolished and that the Troy ounce and pennyweight be retained but confined to the weighing of gold, silver, and precious stones. The Avoirdupois dram, they recommended, should be abolished, with the pound being defined as 7000 grains of which the Troy pound contained 5760. The word 'ounce' was now to mean one sixteenth of an Avoirdupois pound unless designated as a Troy ounce. Section III, Recommendation 30, asked for the retention of the gallon as defined in 5 George IV c 74 1824, so that no new standard of capacity need be constructed. It is clear from the record that the Commissioners spent considerable time considering the advantages and disadvantages of effecting a change in the fundamental units themselves but ultimately decided against any such change, recommending ('though not with absolute unanimity') to the Government that: 'No change be made in the value of the primary units of the weights and measures of this kingdom . . .', but it is clear that they did favour a decimalization of English metrology, although they regarded the chain of 22 yards (the acre's breadth) as 'the greatest anomaly in the whole English system of weights and measures', (notwithstanding the fact that it was introduced to decimalize the acre!). The mile, they contended, was given in inconvenient numbers and they advocated a measure of 1000 or 2000 yards. A thousand yards might be termed a *milyard*. In Section VIII, Recommendations 48, 49 and 50, it was proposed that weights of 14, 28, 56, and 112 lb or any of their multiples except those occurring on a decimal scale be not legal for any goods or merchandise whatever and that legal weights above 1 lb be in multiples of 1 lb not exceeding 10 lb and thereafter in multiples of 10 lb not exceeding 100 lb. To the unit of 100 lb the name *centner* or some other might be given. (The name later adopted for this unit was *cental*.)* Recommendation 17 asked for a committee of scientists to be set up to superintend the construction of new Parliamentary standards.

The introduction of a decimal coinage was strongly advocated, for in Section V we read:

> In our opinion no single change which is in the power of a government to effect in our monetary system would be felt by all classes as equally beneficial with this when the temporary inconveniences attending the change had passed away.

It was shown with what ease a new coin worth two shillings ('to be called by a distinctive name') could be introduced as one tenth of a pound; the farthing could be made £1/1000 instead of the current £1/960 and a coin worth £1/100 could be established, all of which would fit well into the present

*In use for grain from 1859, legalized by Order in Council 4 Feb. 1879, removed as an authorized unit by Statutory Instrument No. 484 of 1978.

scheme, including the shilling (£1/20) and the 6 penny piece (£1/40).[71] They

> felt it imperative on us to advert to it, because no circumstance whatever could contribute so much to the introduction of a decimal scale in weights and measures, in those respects in which it is really useful, as the establishment of a decimal coinage.

While some of these recommendations were not accepted most of the more important were. In particular, a new Commission was set up consisting of all the Commissioners of the 1841 Report save Mr Gilbert (now deceased) but including Professor W. H. Miller. The Earl of Rosse was subsequently added. Their assignment was to 'Superintend the construction of the new Parliamentary standards of length and weight', and they continued their work till 1854 when their report was submitted. It was felt by Government that those proposing the various steps to be taken would be the best group to ensure their being carried out.

The production of the length standard was entrusted to Baily and the weight standard to Miller, the recommendation to accept the Avoirdupois pound as standard having been approved. Some idea of the care that went into his work and the precision demanded at each stage can be gathered from the fact that over 200,000 micrometer readings were taken in producing the yards[72] and that Professor Miller's account[24] of the construction of the new standard pound which appeared in the Philosophical Transactions of the Royal Society for 1856 ran to 193 pages of print.

The Commission lost no time in getting down to business. Appointed by Treasury Minute dated 20 June 1843,[73] its first meeting was held on 11 July of the same year. Their Report, dated 28 March 1854,[74] formed the basis of the 'Act for legalizing and preserving the restored Standards of Weights and Measures, 1855', of which more later.

The Production of the Yard

The following is a summary of this work. Yard measures available to the Commission which had been compared with the lost standard consisted of two iron bars denoted A1 and A2, each three feet in length, that had been made for the Ordnance Survey*, the tubular 5 ft scale of the Royal Astronomical Society made in 1833† and compared with many others in 1834, Kater's Royal Society yard made in 1831 designated RS No. 46, and the Shuckburgh scale, of which we have heard so much. One further yard in

*These two bars carried defining points on the neutral plane and, to avoid the difficulties Kater had encountered, the measurements were performed microscopically with the bars supported, not on a table, but each on two rollers, placed one quarter of the length from each end. (Ref.54 p.627). The yard of the Royal Astronomical Society was similarly supported. This yard and A1 and A2 had been compared with the Imperial standard in 1834.

†This yard has been fully described by Baily (Ref.45).

private hands had been compared with the Imperial standard but to this yard the Committee was refused access. It was decided that only those measures which had been directly compared with the lost standard could be used in the production of the new. Thus the 62-inch scale of Troughton, though available to the Commission, could not be used as it had not been so compared.[74]

Considerable research went into the determination of the alloy to be used for the new yard measures. Bars of various composition, 40 inches long and 1 inch square in section, were tested for sag and breaking load when subjected to stress at the centre. The final choice was a bronze, known as 'Baily's metal No.4', 16 parts copper, $2\frac{1}{2}$ parts tin, and 1 part zinc, and the yard measures themselves were made 38 inches long and 1 inch square in section.

Mr Baily studied carefully the problems of flexure, confirming the observations Kater had made respecting the influence of unbelievably small irregularities of the supporting table. As the flexures in the Shuckburgh scale could not be overcome Baily recommended it be 'laid aside entirely', although this did not happen, as we shall see below. To eliminate these difficulties with the new yards Baily devised the method of boring a cylindrical hole in the bar to the mid level, that is to the neutral plane, placing the gold plugs which carried the defining marks flush with this level. To protect the gold plugs, loose push-fitting bronze caps were provided. Mr Baily thought little of dots as the defining marks on standards. When examining previously the now destroyed yards of 1758 and 1760 he had found the dots much enlarged, eroded, and blurred by beam compasses having been placed actually on the dots when a comparison was being carried out. Repeated measurements had led to considerable wear and the dots on the gold plugs were described by Baily as resembling 'the craters of miniature volcanos'. Now that microscope cross-wires were being used to define a point rather than the needle point of a compass, an engraved set of lines was deemed more appropriate. The final recommendation which was carried out in the actual production was to engrave the gold studs with

> three fine lines at intervals of about the one-hundredth part of an inch transverse to the axis of the bar and two lines at nearly the same interval parallel to the axis of the bar; the measure of the length is given by the interval between the middle transversal line at one end and the middle transversal line at the other end, the part of each line which is employed being the point mid way between the longitudinal lines; and the said points are herein referred to as the centres of the said gold plugs or pins.[75]

Unfortunately Mr Baily died on 30 August 1844, with by far the greater portion of the work of regaining the standard of length still to be done. The Revd Mr Sheepshanks undertook to complete the work, and embarked on a vast number of comparison readings among the several standards already mentioned (pp.250, 255) and a number of prototype yards which he had had made to facilitate the production of the final standards. His prototype bar called No.2 was finally chosen by him as a very near approximation to the lost standard. It compared as follows:[76,77]

Bar No.2 = 36.000084 inches when compared to Shuckburgh's scale 0–36 inches.
= 36.000280 inches when compared to Shuckburgh's scale 10–46 inches.
= 36.000229 when compared to R.S. 46.
= 36.000303 when compared to A1.
= 36.000275 when compared to A2.

The mean was 36.000234 which Mr Sheepshanks declared to be near enough 36.00025 inches as a true value.[78] The Commission accepted the scale on Bar No.2 as representing this length and production of the final standards went ahead.

With measurements being carried out to accuracies of better than one part in ten million it will be clear that the expansion of the standard yards was a matter of great importance, requiring temperatures to be known to $\frac{1}{100}$ or $\frac{2}{100}$ of a degree Fahrenheit. The final Report[74] of the Commissioners of 1854 (Section 17) refers to this matter, pointing out that no thermometer in England at the time could be read to such fine limits. Consequently Mr Sheepshanks had spent considerable time in making thermometers of the required accuracy.

Mr Sheepshanks prepared an *aide-memoire* for the Commission on 18 July 1848, a paraphrase of which follows: 'that all yard measures made in the last 100 years could be traced to that which Graham had laid down for the Royal Society in 1742 from the Tower standard. It was marked 'E' and was known as the 'Royal Society yard'. The yard 'E' was 0.0075 inches longer than the Elizabethan standard also laid down on the same bar and marked 'EXCH'. Bird's scales of 1758 and 1760 were 0.001 inch shorter than 'E'. When Troughton laid down his own (5 ft) scale it is certain that he used the Bird scale'.

> Hence all scales having Bird or Troughton for their author will be found to be about 0.001 inch shorter than Graham's 'E'. . . . The Shuckburgh scale made by Troughton is of the same family. . . . Another family of scales viz General Roy's and Ramsden's are said to have been carefully compared with Graham's Tower Yard 'E' and to have agreed exactly with it. . . . Ramsden's scale is lost, Roy's scale was made for Graham.

Forty line standards were prepared in 1845 and carefully compared with one another. One of these was selected as the new Imperial standard,

> . . . not only from its representing with the greatest precision, the assumed length of

the lost Standard but also from the clearness of the defining lines and from its general good workmanship.[79]

From the rest the four next best were selected as Parliamentary copies* and the temperature at which they were precisely one yard was noted. The primary standard was engraved: 'Copper 16 oz, Tin $2\frac{1}{2}$, Zinc 1. Mr Bailys Metal. No.1. Standard Yard at 62°.00 Fahrenheit. Cast in 1845 Troughton & Simms London'. The primary standard was deposited at the Exchequer.

The Parliamentary copies were engraved and located as follows:

'No.2 Standard Yard at 61°.94 Fah^{t}'	The Royal Mint
'No.3 Standard Yard at 62°.10 Fah^{t}'	The Royal Society
'No.4 Standard Yard at 61°.98 Fah^{t}'	Immured in the new Palace of Westminster
'No.5 Standard Yard at 62°.16 Fah^{t}'	The Royal Observatory, Greenwich

A sixth bar, marked No.6, was also selected. It was correct at 62°.00 F and, though not legally defined as a Parliamentary standard, it was retained for comparison purposes. The remaining 34 bars were distributed to London, Edinburgh, and Dublin and to the principal countries and cities throughout the world.

The primary standard and its Parliamentary copies were legalized in 1855, Section 2 of 18 and 19 Victoria c 72 1855,[75] reading:

> The straight line or distance between the centres of the two gold plugs or pins in the bronze bar deposited in the Office of the Exchequer as aforesaid shall be the genuine Standard of that measure of length called a Yard and the said straight line or distance between the centres of the said gold plugs or pins in the said bronze bar (the bronze being at the temperature of sixty-two degrees by Fahrenheit's thermometer) shall be and be deemed to be the Imperial Standard Yard.

Despite the problem which Kater stumbled upon (p.259) no guidance was given by the Act of 1855 as to how the yard measure was to be supported when measurements were being undertaken, but the question is alluded to in the Act of 1878 as we shall see. Mr Sheepshanks became ill while engaged in the comparisons but completed the work, dying the day before Royal Assent was given to the Act. It therefore fell to the Astronomer Royal to prepare for the Royal Society the paper on the re-establishment of the yard.[77]

The Production of the Pound

It will be recalled that Harris had made three standard Troy pounds in 1758, of

*A fifth Parliamentary copy designated 'Parliamentary copy No. 6' was cast in 1878 for the Standards Department, Board of Trade. It was correct at 62°F (See p.296 for Schedule 2 of the 1963 Act).

which one became the standard lost in the fire of 1834 (p.261). The two remaining pounds were located by the Commission and re-examined by Professor Miller. While they probably had been compared with the standard there was no evidence for this, so they were deemed unsuitable for regaining the lost weight. But Kater had compared the standards for the Exchequer, the Royal Mint, and the cities of London (Guildhall), Edinburgh, and Dublin with the lost standard, and Capt. Nehus had done likewise with two brass and one platinum pound, all the property of Professor Schumacher, and with the platinum Troy pound of the Royal Society. There were therefore standards to which reference could be made.

The following were among those gathered together for the purposes of regaining the lost standard:

(1) The brass Troy pound constructed by Kater in 1824 for the Exchequer (see p.260) and used thereafter as the legal standard, the primary weight being immured and therefore unavailable for normal comparison.
(2) The three brass pounds of London, Edinburgh, and Dublin.
(3) The three pounds already mentioned in the possession of Professor Schumacher.
(4) The Royal Society platinum pound of 1829 (see p.259).

When comparisons commenced it became clear that all the brass weights were displaying quite unacceptable discrepancies owing to oxidation. It was therefore resolved to use only the two platinum weights, that of Schumacher being denoted 'Sp', that of the Royal Society 'RS', and it was decided, in order to avoid this difficulty in future, to make the new primary standards in platinum.

But the problem was not to reproduce the Troy pound but to create a new Avoirdupois pound weighing 7000 grains *in vacuo*, not in air. This meant creating a weight 1240 grains in excess of a Troy pound and it was decided that there should be a primary standard Avoirdupois pound and four copies. To achieve this end a number of platinum weights were manufactured, from a total of 101 Troy ounces of the metal provided by Messrs Johnson and Cock, a company which later evolved into Johnson, Matthey & Co. of London.[80] The weights, made by a Mr Barrow who also provided the Commission with accurate scales, were as follows:

a prototype Troy pound designated 'T'
4 weights each of 1240 grains nominally
1 weight each of 800, 440, and 360 grains
4 weights of 80 grains
2 weights of 40 grains

As the weights 'Sp' and 'RS' were accurately known in terms of the lost standard[81] a comprehensive series of weighings of 'T' against 'Sp' and 'RS' yielded the results

> T = Standard − 0.52857 grains *in vacuo*
> T = Standard − 0.00745 grains *in air* at 65.66°F, the barometer being 29.753 inches of mercury. In a vacuum, then, T was equal to 5759.47143 grains.

The mean weight *in vacuo* of the four 1240-grain weights was found to be 1239.88601 grains, so this mean, added to 'T', yielded 6699.35744 grains *in vacuo*, only 0.64256 grains short of the desired 7000.[82]

Construction commenced on the primary standard itself. When the work was finished its weight was determined as 7000.000238* grains[83] of which the desired pound would contain 7000 and the lost standard 5760. The Commission agreed to accept this as the new Imperial standard and declared it to contain 7000.00000 grains and described it 'P.S.'. Four other platinum pounds were made as Parliamentary copies of the Imperial Standard. The standard and the copies had characteristics and were deposited according to the following table:[74]

	Density†	*Weight*‡	*Deposited at*§
PS	21.15702	7000.00000 grains	Exchequer
PC1	21.16634	PS + 0.00052 grains	Royal Mint
PC2	21.16334	PS − 0.00088 grains	Royal Society
PC3	21.16128	PS − 0.00178 grains	Royal Observatory, Greenwich
PC4	21.15549	PS − 0.00314 grains	Immured with Yard No.4 in the New Palace of Westminster[86]

The primary standard weight was deposited at the Exchequer on 17 September 1853 by Airy and Herschel, together with the primary standard of length.

The pound denoted 'RS' was returned to the Royal Society. Professor Schumacher's pound 'Sp', together with his two brass pounds, was returned to him at the observatory, Altona, and 'T' went to the Royal Observatory, Greenwich. In addition to the platinum pounds mentioned above 36 gilt bronze pounds were issued together with the standard yards to London, Edinburgh, and Dublin and to an array of countries and cities around the

*Thus in the official report[74] but W. H. Miller[84] gives 7000.00090 grains and also 7000.00093 grains while H. W. Chisholm[85] gives 7000.00093 grains.

†Miller gives the densities as follows PS − 21.1572, PC1 − 21.1671, PC2 − 21.1640, PC3 − 21.1615, PC4 − 21.1556,[84] and again PC4 − 21.1516[24] which are all sufficiently different from the values given in the official report as to warrant notice.

‡Miller gives the weight deviations from PS as follows: PC1 = PS + 0.00051, PC2 = PS − 0.00089, PC3 = PS − 0.00178, and PC4 = PS − 0.00316[84] and again PC4 = PS − 0.00314 grains.

§A fifth Parliamentary copy, 'PC5. 1879' was made in that year from a platinum-iridium alloy for the Standards Department, Board of Trade. (See p.296 for Schedule 2 of the 1963 Act.)

world.

The description of the Imperial standard weight is as follows:[75]

> And whereas the Standard Pound Avoirdupois so constructed as aforesaid and the Copies thereof are of Platinum the form being that of a cylinder nearly 1.35 inch in height and 1.15 inch in diameter with a groove or channel round it whose middle is about 0.34 inch below the top of the cylinder for insertion of points of the ivory fork by which it is to be lifted; the edges are carefully rounded off. . . .
> III. The said Weight of Platinum marked 'P.S. 1844, 1 lb' deposited in the Office of the Exchequer as aforesaid, shall be the legal and genuine Standard Measure of Weight and shall be and be denominated the Imperial Standard Pound Avoirdupois and shall be deemed to be the only Standard Measure of Weight from which all other Weights and other Measures having Reference to Weight shall be derived, computed, and ascertained and One Equal Seven Thousandth Part of such Pound Avoirdupois shall be a Grain and Five Thousand seven hundred and sixty such grains shall be and be deemed to be a Pound Troy.

The standard yard of 1845 and platinum pound of 1844 are shown in Figure 54. From this we see that the Commission's recommendation to abolish the Troy pound was not acceded to in this Act, nor were the suggestions to abolish the traditional weights of 14, 28, 56, and the cwt of 112 lb. No action was taken in respect of the chain and the mile, either. Indeed, no change at all occurred in the system save the recognition of the new yard and the adoption of the Avoirdupois pound as the standard with the Troy pound as 'a derivative Standard computed with reference to the Pound Avoirdupois' (Section IV). The last section of this Act provided that, if lost or destroyed, the Imperial standards may be restored by reference to or adoption of any of the Parliamentary copies.

One might have thought that things were now stabilized, with British metrology on a firm, universally accepted basis. Far from it. A Committee on Weights and Measures set up on 8 April 1862 reported on 15 July of the same year.[87] What they found was staggering. It was pointed out that the kingdom had no less than ten different systems of weights and measures, most of them established by Act of Parliament. As for weights, they listed ten systems, ranging from wool weight (using 2, 3, 7, and 13 as factors) and weights for hay and straw to Troy and Avoirdupois weight. There were three fathoms: that of a warship, 6 feet; that of a merchantship, 5½ feet; and that of a fishing-vessel, 5 feet. For capacity there were reported to them no fewer than twenty different bushels. When the President of the Board of Trade was giving evidence he was told that the Committee had evidence of fourteen different stone weights. He replied, 'I have not the least doubt of it. I should not be surprised if you had evidence that there was 60.' Obviously further clearing up was necessary, and this was done some sixteen years later by the Act of 1878 as we shall see.

The standards were now restored and legalized but their abode was changed in 1866 when, by Act of Parliament, 29 and 30 Victoria c 82, all the standards were transferred to the Board of Trade from the Exchequer[88] which from time immemorial had been the only legal repository.

The Weights and Measures Act, 1878

This Act, 41 and 42 Victoria c 49,[89] was the last major Act of its kind of the nineteenth century. Basically it reiterated the provisions of the 1855 Act, but went into great detail specifying the units of length, weight, and capacity which were to be legal for trade. All multiples and subdivisions were stated with their equivalents. These and only these could be used for trade. The tidying up had been accomplished.

One important change was the abolition of the Troy pound* to avoid confusion with its Avoirdupois counterpart, while the Troy ounce was elevated to primary status for the weighing of precious stones and metals, as recommended in 1841.

The Second Schedule provided that the brass gallon made in 1824 (see p.256 and below), then in the custody of the Warden of the Standards, should be deemed a Board of Trade standard for the gallon.

Section 10 of the Act declares the Imperial yard to be the measure between the engraved lines at 62°F when it is 'supported on bronze rollers placed under it in such a manner as best to avoid flexure of the bar and to facilitate its free expansion and contraction from variations of temperature . . .'.

A system of eight rollers had been prescribed[74] with the weight of the bar equally divided among them. However, it was found that the difference in length of the bar when supported by eight and when supported by two rollers, placed so as to make the length a maximum, is quite negligible, so in practice two rollers are used for routine comparisons.

The Report of 1854[74] Section 14 recommended that the yard measures be supported on eight rollers *when not in use* and stated that during the comparisons undertaken by the Commission the bars were actually floated in quicksilver (mercury). This would ensure no flexural distortion.

We note in passing that, by Order in Council dated 28 November 1889, the cubic inch of water was stated to weigh 252.286 grains and hence the capacity of the Imperial gallon is 277.463 in^3, a small change from the 277.274 in^3 of the 1824 Act. The new value so declared is remarkably close to the actual capacity of the new gallon constructed in 1824 and found in 1931-2 to hold 277.421 in^3.

The Board of Trade was directed by the Act to undertake a comparison of the Imperial standards every ten years excepting the immured standards but every twenty years they, too, were to be included in the check.

*Recommended by the Third Report of the Commissioners appointed to Enquire into the Condition of the Exchequer Standards, 1 Feb. 1870. c 30 *Parl. Papers* 1870 xxvii pp.81-217 (See Appendix B).

Heaped measure was made illegal by the Weights and Measures Acts of 1834[65] and 1835[66]. Section 16 of the corresponding Act of 1878 stated that, for goods sold by heaped measure prior to the passing of the 1835 Act (September, 1835), the shapes of the vessels used were to be cylinders whose diameters were double their depths. This requirement was repealed by Section 5 of the 1889 Act although the Model Regulations of 1890 which followed (see Appendix B, p.335) prescribed measures of this form or having diameters equal to the depth. These were repeated in the 1907 and 1963 Regulations and continued to the present.

Present Location of the Standards

By far the vast majority of the standards referred to in this chapter are on display in the Science Museum, London, derived either from their own collection or on loan from the Royal Society. The Exchequer Standards of Henry VII and Elizabeth are to be found there, as are also a number of linear measures of great interest, such as a yard sealed at the Tower by John Reynolds dated 30 April 1659, Royal Society No.41, General Roy's scale, Kater's Imperial yard (an end standard, made in 1824 after the Bird standard, brass 1 inch square with iron end cheeks stamped with a crown and the words 'Imperial Standard Yard 1825'), and the Shuckburgh scale, to mention but a few.

The collection is also rich in weights from early times to the present and in measures of capacity. It contains the 1824 Imperial standards of volume, gallon, quart, and pint marked 'Verified by Capt. Henry Kater FRS, Bate London', a bushel of 1824 verified by Kater but not so inscribed, Queen Anne's wine gallon of 1707, a coal bushel of 1730 (George II), wine quarts, Winchester bushels, etc., etc., in short a wealth of riches too numerous to describe adequately here.

Municipalities held custody of their own copies of the Imperial standards until very recently when, with the move to metrication and the abolition of ancient measures, some authorities decided to dispose of their now obsolete standards, especially bushels. Happily these have been purchased by museums and by private collectors and have not been lost or destroyed.

The stability of the standard yard and pound is discussed in Chapter XIV.

NOTES AND REFERENCES: CHAPTER XII

1. RILEY, Henry Thomas (ed.) *Munimenta Gildhallae Londoniensis: Liber Custumarium* Vol. II, Pt. 1, p. 382. London: Rolls Series, Longman et al. 1859-62.

2. SALZMAN, L. F. *English Trade in the Middle Ages*, p. 57. Oxford: Clarendon Press 1931.
   We note that a modern cup has a volume of about six fluid ounces so ten would be exactly three pints, but ten Roman *cyanthos* would only approximate to one pint.
3. Reports from the Committee appointed to inquire into the Original Standards of Weights and Measures in this Kingdom (Lord Carysfort's Committee) 26 May 1758. *Reports from Committees of the House of Commons* Vol. II (1737-65), p. 431. This bushel is said to be that on view in the Jewel Tower, London, situated opposite the west end of the Houses of Parliament.
4. WILDE, Edith E. Weights and Measures of the City of Winchester, *Proceedings of the Hampshire Field Club* (1931) Vol. X, Pt. 3, p. 237, Winchester, at p. 246 describes these weights as being of iron, while SKINNER, F.G. *Weights and Measures* p. 96 London: HMSO 1967 says they are of bronze. In fact the weights are of bronze, fitted with lifting rings and staples of iron, with lead on the bottom for sizing.
5. 14 Edward III Stat. 1 c 12 1340 *Statutes* Vol. I, p. 285.
6. STEVENSON, Maurice, A Note of the Sandwich Steelyard Weight. *Libra* (1963) Vol. II, pp. 15-16. Eastbourne.
7. BATESON, Mary (ed.) *Records of the Borough of Leicester* p. 106. London and Cambridge: C. J. Clay & Sons, 1899.
8. CLODE, C.M. *Early History of the Guild of Merchant Taylors of the Fraternity of St. John the Baptist,* London: Privately Printed 1888.
9. SKINNER, F. G. The English Yard and Pound Weight *Bulletin of the British Society for the History of Science* (1952) Vol. I, p. 179, on page 182 suggests a date of about 1497, i.e. the same as the Exchequer Yard of Henry VII (See this chapter, Section (b)).
10. CLODE, C. M. Loc. cit. p. 128.
11. FRY, F. M. and TEWSON, R. S. *Illustrated Catalogue of Silver Plate* - Merchant Taylors' Yard; London: Privately Printed 1929.
12. For much of the information given here concerning the Merchant Taylors' Yard I am indebted to references 8 and 11 and to discussion with officers of the Company who kindly allowed me to examine the silver yard.
13. LYELL, L. and WATNEY, D. *Acts of the Court of the Mercers' Company 1453-1527* Cambridge: C.U.P. 1936.
14. JOHNSON, Revd. A. H. *A History of the Worshipful Company of Drapers of London* Vol. I, p. 116 et seq. Oxford: Clarendon Press 1914-15.
15. 11 Henry VII c 4 1495 *Statutes* Vol. II, pp. 570-1.
16. 7 Henry VII c 3 1491 Ibid. Vol. II, p. 551.
17. 12 Henry VII c 5 1496 Ibid. Vol. II, p. 637-8.
18. CHISHOLM, H. W. *Seventh Annual Report of the Warden of the Standards* p. 34. London: HMSO 1873.
19. Harleian Mss. Vol. 698 f 64-5; printed in part by Chisholm (above) p. 27.
20. CHISHOLM, H. W. Loc. cit. p. 13.
21. Ibid. p. 19 & pp. 31-2.
22. Ibid. p. 32.
23. Ibid. p. 14.

24. MILLER, W. H. On the Construction of the New Imperial Standard Pound *Phil. Trans. Roy. Soc.* (1856) Vol. CXXXXVI, pp. 753-946. London.
25. Anon. An Account of a Comparison lately made by some Gentlemen of the Royal Society of the Standard of a Yard *Phil. Trans. Roy. Soc.* (1742-3) Vol. XXXXII, pp. 541-556. London. See p. 544.
26. CHISHOLM, H. W. Loc. cit. p. 25.
27. Ibid. p. xii and p. 19.
28. Ibid. p. 12.
29. I am obliged to Mrs Rosemary Weinstein of the Museum of London for showing me this gallon and providing information about it.
30. Reference 25 see p. 550.
31. See Reference 25, p. 549.
A brief outline for the life and work of John Rowley (? - 1728) is given in the pamphlet *At the Sign of the Orrery* by E. W. Taylor and J. S. Wilson, undated but published after 1950, being a history of the firm Cooke Troughton and Simms Ltd. In this booklet the 36 inch yard measure of Rowley is mentioned (p. 7) but it is not listed in the table of his extant instruments.
32. For much of this information and further details see Anon. An account of the Proportions of the English and French Measures and Weights, from the Standards of the Same, kept by the Royal Society. *Phil. Trans. Roy. Soc.* (1742-3) Vol. XXXXII, p. 185. London.
33. See Reference 25 for further details.
34. Report from the Committee appointed to Inquire into the Original Standards of Weights and Measures in this Kingdom ... (The Carysfort Committee) 26 May 1758. *Reports from Committees of the House of Commons* Vol. II, p. 420, 1737-65.
35. Ibid. p. 433.
36. Ibid. p. 434.
37. CHISHOLM, H. W. On the Science of Weighing and Measuring and the Standards of Weight and Measure *Nature* (1873) Vol. VIII, p. 367. London.
38. Reference 34, p. 436.
39. Reference 34, p. 437.
40. Gun metal - a bronze containing 5 - 11 lb of tin per 100 lb of copper, called 'brass' on occasion in official papers.
41. It will be recalled that at this time there was no standard Troy pound in the Exchequer, only a pile of ounce weights as described earlier.
42. Reference 34, p. 438.
43. Second Report from the Committee appointed to Inquire into the Original Standards of Weights and Measures in this Kingdom . . . (The Carysfort Committee) 11 April 1759 *Reports from Committees of the House of Commons* 1737-65, Vol. II, p. 456.
44. Ibid. p. 459.
45. The literature on what follows is rather scattered. The more important references are:
BAILY, F. Report on the New Standard Scale of this Society. *Memoirs of the Royal Astronomical Society* (1836) Vol. IX, pp. 35-184. London.
ROY, Major-General William. An Account of the Measurement of a Base of Hounslow Heath, *Phil. Trans. Roy. Soc.* (1785) Vol. LXXV, pp. 385-480. London.

SHUCKBURGH, Sir George. An Account of some Endeavours to ascertain a Standard of Weights and Measures *Phil. Trans. Roy. Soc.* (1798) Vol. LXXXVIII, pp. 133-182. London.
KATER, Capt. Henry. An Account of the Comparison of Various British Standards of Linear Measure. *Phil. Trans. Roy. Soc.* (1821) Vol. CXI, Pt. I, p. 75. London.
KATER, Capt. Henry. An Account of the Construction and Adjustment of the new Standards of Weights and Measures of the United Kingdom and Ireland, *Phil. Trans. Roy. Soc.* (1826) Vol. CXVI, Pt. II, p. 1. London.
KATER, Capt. Henry. On the Error in Standards of Linear Measure arising from the Thickness of the Bar on which they are traced. *Phil. Trans. Roy. Soc.* (1830) Vol. CXX, p. 359. London.
KATER, Capt. Henry. An Account of the Construction and Verification of a Copy of the Imperial Standard Yard made for the Royal Society *Phil. Trans. Roy. Soc.* (1831) Vol. CXXI, p. 345. London.
CHISHOLM, H. W. On the Science of Weighing and Measuring and the Standards of Weight and Measure. *Nature* (1873) Vol. VIII, (Pt. III) p. 327, (Pt. IV) p. 367. London.
GLAZEBROOK, Sir Richard. Standards of Measurement: Their History and Development. *Supplement to Nature* (1931) Vol. CXXVII, July 4th, p. 17. London.
ANON: Weights and Measures in History, . . ., *Chemist and Druggist* (1969) Vol. CX, p. 816 et seq.

46. The date is given as 1796 by Chisholm and as 1792 by Glazebrook (above). The actual scale is engraved 'Troughton London 1796'.
47. Report of the Select Committee to Inquire into the Original Standards 1 July 1814. *Parl. Papers* 1813-14 (HC290) iii, p. 131.
48. The Winchester bushel, defined by 8 and 9 William III c 23, 39 and 45, 1696.
49. The Principle of Archimedes states that when a body is immersed in a fluid it experiences an upthrust (loss of weight) equal to the weight of the fluid displaced by the body.
50. This pendulum is known to physics students as the 'seconds pendulum' in that it beats seconds, that is, the time from one extremity of the swing to the other is one second. The period (T), which is the time for a complete oscillation, there and back, for a mass point at the end of a weightless string of length l in a vacuum is given by
$T=2\pi\sqrt{\frac{l}{g}}$ where 'g' is the acceleration of gravity. Here T = 2 seconds.
51. First Report of the Commissioners appointed to consider the Subject of Weights and Measures, 1819 (7 July). *Parl. Papers* 1819 (HC565) xi, p. 307 et seq.
52. Second Report of the Commissioners appointed to consider the Subject of Weights and Measures, 13 July 1820. *Parl. Papers* 1820, (HC314) vii, p.433. This report is valuable for its Appendix A which in 32 pages lists the various measures, customary and statute, then in use throughout the realm.
53. Third Report of the Commissioners appointed to consider the Subject of Weights and Measures, 31 March 1821. *Parl. Papers* 1821 (HC383) iv, p.297 et seq.
54. AIRY, G. B. An Account of the Construction of the New National Standard of

Length and of its Principal Copies. *Phil. Trans. Roy. Soc.* (1857) Vol. CXXXXVII, p. 621 et seq.

55. Report from the Select Committee on Weights and Measures 28 May 1821. *Parl. Papers* 1821 (HC571) iv, p. 289 et seq.
56. KATER, Capt. Henry. An Account of the Comparison of Various British Standards of Linear Measure, *Phil. Trans. Roy. Soc.* (1821) Vol. CXI. Pt. 1, p. 75. London.
57. KATER, Capt. Henry. An Account of the Construction and Adjustment of the New Standards of Weights and Measures of the United Kingdom and Ireland. *Phil. Trans. Roy. Soc.* (1826) Vol. CXVI, Pt. II, p. 1. London.
58. AIRY, G. B. Loc. cit. p. 626.
59. It is well known that the neutral plane, that is the plane in a solid body which neither increases nor decreases in length when the body bends, is that located in the mid plane.
60. 6 George IV c 12 1825 *Statutes at Large* Vol. XXIV, p. 21.
61. SCHUMACHER, H. C. A comparison of the late Imperial Standard Troy Pound Weight with a Platina Copy of the Same *Phil. Trans. Roy. Soc.* (1836) Vol. CXXVI, p. 457. London. See p. 460. Professor Schumacher (1780 - 1850), German astronomer, was Director of the observatory at Altona near Hamburg and had an Imperial standard pound in platinum made for his own use. Capt. Nehus of the Royal Danish Engineers was one of his assistants.
62. KATER, Capt. Henry. An Account of the Construction and Verification of a Copy of the Imperial Standard Yard made for the Royal Society. *Phil. Trans. Roy. Soc.* (1831) p. 345. London.
63. SKINNER, F. G. The English Yard and Pound Weight. *Bulletin of the British Society for the History of Science* (1952) Vol. I, p. 186.
64. For example, Robert Brunton's book *A Compedium of Mechanics* 6th Edition, Glasgow: John Niven 1834, gives all those gallons, ale, wine, and Imperial equal prominence.
65. 4 & 5 William IV c 49 1834 *Statutes at Large* Vol. XXVII, p. 629.
66. 5 & 6 William IV c 63 1835 Ibid. Vol. XXVII, p. 977.
67. *The Times,* London 17 October 1834, p. 3 col. 3
68. Ibid. 18 October 1834, p. 5 col. 3.
    20 October 1834, p. 3 col 4.
69. MILLER, W. H. Loc. cit. p. 759.
70. Report of Commissioners appointed to consider the steps to be taken for Restoration of the Standards of Weights and Measures. 21 December 1841 *Parl. Papers* 1842 (356) xxv, p.263.
    From the list of recoveries, it appears that a remarkably large collection of standards had been immured. Besides the unrecovered standard Troy pound and the two yards of 1758 and 1760 there were also recovered the bed measures for these two yards, the 2 lb Troy weight of 1758 made by Harris for the Carysfort Committee together with his two 4 lb, two 8 lb weights, the 16 and 32 lb weights (see p. 248), together with a weight marked 'SF 1759', 17 lb 8 dwt Troy. 'Apparently a 14 lb stone Avoirdupois, allowing 7008 grains to the lb Avoirdupois'.
71. These matters were further discussed in 1862. See Minutes of Evidence taken

before the Select Committee appointed to consider the Practicability of Adopting a Simple and Uniform System of Weights and Measures. Report 15 July 1862. *Parl. Papers* 1862 vii, 187-, p. vi et seq. Also Minutes of Evidence 9 May 1862 p. 6 of the same report.

72. CHISHOLM, H. W. On the Science of Weighing and Measuring and the Standards of Weight and Measures (Pt. IV) *Nature* (1873) Vol. VIII, p. 367. London.
73. Papers relating to Standards of Weights and Measures 11 March 1864. *Parl. Papers* 1864 lviii, p. 621.
74. Report of the Commissioners appointed to superintend the Construction of the New Parliamentary Standards of Length and Weight, 28 March 1854. *Parl. Papers* xix, p. 933.
    This report and Refs. 24 and 54 are central to and contain much of what follows. Additional reference sources are given as we proceed which may be of easier access than the original report itself.
75. 18 and 19 Victoria c 72 1855 *Statutes at Large* Vol. XXXVI, p. 666.
76. Reference 72, p. 368.
77. AIRY, G. B. Loc. cit. p. 621 et seq.
78. It will be noted that the tubular yard of the Royal Astronomical Society does not appear here. Mr Bailey was of the opinion that its length had changed, so it was deemed unreliable for comparison purposes. It was later found that the yard was much better than thought for Mr Bailey's errors had been given the wrong sign (Reference 54, above).
79. Reference 72, p. 369.
80. McDONALD, Donald. *A History of Platinum* p. 147. London: Johnson, Matthey & Co. Ltd. 1960.
81. The mean of 300 weighings by Capt. Nehus in 1829 of Sp against the standard gave Sp = standard - 0.00857 grains and the mean of 140 weighings of RS gave
    RS = standard - 0.00205 grains, both in air.
    Whereas the volume of the lost standard had never been determined so that its weight could be corrected from a weighing in air to one *in vacuo*, it was assumed that its density was the same as the others made by Harris at the same time, namely 8.151. The density of Sp had been determined by weighing it in water as 21.1874.
    In a vacuum therefore Sp = standard - 0.52959 grains and
    RS = standard - 0.52441 grains.
82. For details see references 24, 72, and 74. CHISHOLM Ref. 72, p. 328 erroneously gives the shortage as 0.64266 grains.
83. Reference 74, Section 30.
84. MILLER, W. H. On the Imperial Standard Pound. *Phil. Mag.* (1856) Vol. XII, p. 550. London.
85. Reference 72, p. 328.
86. P.C. No. 4 was placed with Yard No. 4 in 'a stone case like one of the ancient Celtic Coffins' (Reference 73) and 'immured in the Cill of the Recess on the East Side of the lower Waiting Hall in the New Palace of Westminster' (Reference 75).
    The *Illustrated London News* for 9 April 1892 declared that the pound weight '... is

enclosed in a case of silver gilt, in another case of solid bronze, then placed in a mahogany box; this is put into a leaden case and the whole sealed up in an oaken box'.

87. Report from the Select Committee on Weights and Measures 15 July 1862. *Parl. Papers* 1862 vii, pp.187-478.
88. The standards were transferred to the newly created Standards Department of the Board of Trade in 1866, a department set up in response to Recommendation 2 of the Report of 1862 (Reference 71) which read:
    2. That a Department of Weights and Measures be established in connection with the Board of Trade. It would thus become subordinate to the Government and responsible to Parliament. To it should be intrusted the conversion and verification of the standards, the superintendence of Inspectors, and the general duties incident to such a department. It should also take such measures as may from time to time promote the use and extend the knowledge of the Metric system, in the departments of Government and among the people.

    In 1956 the name was changed to 'Standard Weights and Measures Department, Board of Trade'. In 1970 the whole organization changed its name and custody reposed in the 'Department of Trade and Industry, Weights and Measures Division'. In 1974 it became the Department of Prices and Consumer Protection; Standards; Weights and Measures Division'. Then in 1979 custody reposed in the 'Department of Trade, Legal Metrology Branch, Metrology, Quality Assurance and Standards Division', while in 1980 the standards were with the National Weights and Measures Laboratory, Department of Trade, Metrology, Quality Assurance, Safety and Standards Division (since 1981 called the Department of Trade and Industry). The address is 26 Chapter Street, London. (I am indebted to Mr Maurice Stevenson for this information).
89. 41 & 42 Victoria c 49 1878 *Public General Statutes* Vol. XIII, p. 308.

CHAPTER XIII

The Onset of Metrication: The Nineteenth Century*

The events in France towards the end of the eighteenth century had a profound influence on the British Government, even though Britain was not totally unaccustomed to rebellion and insurrection. The War of American Independence was not long over with the surrender of Yorktown in 1781 and the peace signed in Paris in 1783, while the Gordon riots in London of 1780 were still fresh in the memory, as were the Swiss and Dutch revolts of 1782 and 1783-7 respectively. The Government wished to keep anything and everything French at a distance lest the rebellious contagion spread. Anything appearing faintly seditious, far less treasonable, was vigorously eradicated as witnessed by a series of trials now infamous in legal history.[1]

It did not go unnoticed that the French had produced a new system of weights and measures, the primary units of which were the metre, kilogram, and litre, but neither did the tumult and the excesses of the revolution itself, which may explain the long delay experienced in Britain before this system of units received earnest consideration, far less sanction for use. Anything savouring of France at this time was an anathema to the Government. France meant revolution and the overthrow of established power. Had they not guillotined their king in 1793?

But slowly the wheels of change in England began to turn, though not without opposition. There were two main thrusts; one was to decimalize the pound sterling, the other was to abolish completely the existing system of weights and measures and to replace it with the metre-kilogram-litre system. The appeal of this latter system to its devotees was very great, for the metre defined the kilogram which in turn defined the litre. One single unit defined all, instead of the hodgepodge of units then current in England. Moreover the French coinage was defined in terms of grams and was decimalized.

These matters were raised in the House of Commons in 1816. Mr J. W. Crocker[2] stated that he

> wished to press on the House, that the present was a most favourable opportunity for giving all the parts of our circulating medium [i.e. coinage] a decimal relation to each other. Perhaps it might be well to follow the example set us by the French. The revolution had enabled them to make a change in their money. We now had a fair opportunity for doing the same.

On 15 March of the same year Mr Davies Gilbert introduced a motion into

*A brief outline of the origin and development of the metric system and the creation of its standard measures is to be found in Appendix C to which any reader unfamiliar with these units is referred.

the House for a comparison to be undertaken of the metre with the Imperial standard yard. The House agreed and the Royal Society undertook the task which was actually conducted by Capt. Kater. The metre was found to be 39.37079 inches but, as far as weights and measures were concerned, the actual introduction of a decimal scale was not being viewed with much favour. The Royal Commission appointed in 1816 'to consider how far it might be practicable or advisable to establish a more uniform system of weights and measures' reported in 1819[3] and recommended no change for 'the subdivisions . . . at present employed . . . appear to be far more convenient for practical purposes than the decimal scale'. The events outlined in the previous chapter which led to the Act of 1824 did nothing to further the cause of metrication, but the same day on which that Bill was introduced, Sir John Wrottesley (later Lord Wrottesley) introduced a motion into Parliament seeking an inquiry into the applicability of the decimal system to the coinage. He recommended the use of pounds, two-shilling pieces, and farthings, which latter were to be $\frac{1}{1000}$ of the pound rather than the current $\frac{1}{960}$. This was vigorously opposed on the ground of inconvenience by the Master of the Mint, Mr Wallace, but he 'did not deny the advantages of the plan'. The motion was withdrawn.

The destruction of the Imperial standards in 1834 did not advance the march of metrication either for, as we have seen, the Commission appointed to regain the standards switched from the Troy to the Avoirdupois pound as the primary standard of weight but not to the kilogram. Their report of 1841[5] also recommended no change in the standard of length, so the yard was re-established and they continued the 1824 definition of the gallon. They did, however, advocate the decimalization of the whole system of metrology and in particular the decimalization of the coinage,[6] but there was no action taken on these and other recommendations except that decimal bullion weights, being fractions and multiples of the Troy ounce, were permitted (16 & 17 Victoria c 29 1853) and it was enacted that a compulsory re-verification of local weight standards be undertaken every five years and of measures every ten years (22 and 23 Victoria c 56, 1859).

A comprehensive comparison of the kilogram with the Avoirdupois pound was undertaken by Professor Miller in 1844-46.[7] Having found that previous determinations of the kilogram varied among themselves by as much as six grains, that is by 4 parts in 10,000, a quite unacceptable situation, Miller went to Paris with Parliamentary Copies Nos. 1 and 2 of the British pound to measure the *Kilogramme des Archives**. As one of the auxiliary weights was somewhat in doubt, a platinum kilogram was made for him which was compared in detail with the archival standard of Paris. This platinum weight

*The kilogram and metre *des Archives* were the two initial standards of the Metric System deposited in the Archives in 1799. (See Appendix C).

was denoted by Miller by the Gothic letter '𝕮'. It was found to be 0.02412 grains (1.56 mg) lighter *in vacuo* than the standard of the Archives. The weight '𝕮' was later compared with the Imperial standard pound and all of the copies. Its weight so determined was 15432.32462 grains, of which the Avoirdupois pound contained 7000. The *Kilogramme des Archives* was therefore 15432.34874 grains *in vacuo*.

On 27 April 1847 a motion was brought forward in the House of Commons by Sir John Bowring to the effect that an Address be made to the Crown requesting a coinage and issue of silver coins of $\frac{1}{10}$ and $\frac{1}{100}$ of a pound. Sir Charles Wood (Chancellor of the Exchequer) stated he had no objection to the striking of a piece of $\frac{1}{10}$ of a pound, and so the motion was withdrawn on the understanding that this would take place. The florin, stamped 'one-tenth of a pound' and worth 2/-, was issued in 1849 and public reaction was awaited. There was an immediate but minor flurry when it was found that the inscription omitted the words '*Dei Gratia*' ('by the Grace of God') following the monarch's name. It was often called the 'Godless' or 'Graceless' florin but it was well accepted nonetheless. There was no further action relevant to the coinage at this time.

However on 26 March 1853 the Commission which had been appointed ten years earlier (1843) to superintend the construction of the new British standards communicated with the then Chancellor of the Exchequer (Mr Gladstone) to the effect that they were much impressed with the advantages of a decimal system for the coinage, recommending that copper coins of $\frac{1}{1000}$, $\frac{2}{1000}$, and $\frac{4}{1000}$ of a pound be produced.[8] The letter was signed by all members except Sir J. Herschel, the Master of the Mint and hence a government servant. The letter of 26 March 1853 is referred to in their Report of 1854. (p.264).

No doubt this was what led to a question in the House on 5 April 1853 as to whether it was the Government's intention to decimalize the coinage, and introduce the suggested copper coins. Mr Gladstone replied that there was no intention to change the copper coinage and that the Government, on the more general matter, was giving careful consideration and would support a motion to appoint a Select Committee to inquire into the subject. Just such a motion was introduced by Mr W. Brown on 12 April and a Committee of 25 people was nominated. Its report was presented to the Commons on 1 August 1853, which gives some indication of the speed with which it was working and the desire it had to see its recommendations implemented. These recommendations were substantially the same as those which had been made earlier with some additions, namely the retention of the pound sterling as the monetary unit, the introduction of copper coins of $\frac{1}{1000}$, $\frac{2}{1000}$, and $\frac{5}{1000}$ and new silver coins of $\frac{1}{100}$ and $\frac{2}{100}$ of a pound. The entire decimal system was strongly endorsed.

Widespread discussion followed the publication of the report. Numerous schemes were advanced for the subdivision of the pound or for the setting up of an entirely new coinage which eliminated the pound altogether. The Decimal Association, formed in 1854, sent a delegation to meet with Mr Gladstone, but he was of the opinion that the subject had not received sufficient airing to test public opinion. He felt that the people at large had not been consulted in depth nor had their reaction to such a change been ascertained. Mr Gladstone did not think the penny could be abandoned and the changes envisaged would in his view take years to obtain universal acceptance.

On 12 June 1855, on a motion by Mr W. Brown, it was agreed by the House that the introduction of the florin had been eminently successful and that a further extension of the system would be advantageous. A further part of the motion was withdrawn. This requested the issue of a silver coin to represent $\frac{1}{100}$ of a pound and a copper coin to represent $\frac{1}{1000}$, to be called the *cent* and the *mil* or such other names as might be appropriate. The issue of such coins was opposed by the Government as affecting too greatly the interests of the poorer people. Much more study was required in the view of the Government.

Following this a Commission was established to consider 'how far it may be practicable and advisable to introduce the principle of Decimal division into the coinage of the United Kingdom'. This Decimal Coinage Commission was doomed before it started for it consisted of only three people, Lord Monteagle, Lord Overstone, and a Mr T. G. Hubbard. The minutes of the meetings of the Commissioners and their correspondence leaves little doubt that Lord Monteagle (the Chairman) was very heavily committed elsewhere on government business, while Lord Overstone was steadfastly opposed to the introduction of a decimal coinage. However the Commission heard evidence from many opposed and many favourable to the introduction of a decimal currency and sought information from other countries which had so changed their currencies as to the advantages and disadvantages of the move. Lord Overstone prepared a list of questions described by himself as

> drawn up with a view of bringing under distinct notice and examination some of the advantages of the present system of coinage and some of the principal difficulties and objections which have been suggested with respect to the introduction of a system of decimal coinage.[9]

In their preliminary report of 4 April 1857 the Commission 'did not think it expedient . . . to express any opinion on the subject of this paper', but gave in appendices comparative tables of coins and a table of the eleven different schemes then under consideration for the reform of the coinage.

The record shows that, throughout the latter half of 1858 and the early part of 1859, Lord Overstone became increasingly irritated by the lack of meetings of the Commission. His eleven resolutions presented at the meeting of 1 March 1859 attended by Mr Hubbard and himself were entirely negative. Lord Monteagle withdrew from the Commission and the final report presented on 5 April 1859 consisted of Lord Overstone's earlier resolutions now extended to twelve in number and concluded[10]

> 12. That duly weighing the foregoing considerations it does not appear desirable under existing circumstances, while our weights and measures remain as at present, and so long as the principle on which their simplification ought to be founded is undetermined, to disturb the established habits of the people with regard to the coins now in use, by a partial attempt to introduce any new principle into the coinage alone.

Things were quiet for three years, then on 8 April 1862 a Select Committee on Weights and Measures was set up which reported on 15 July of that year.[11] It was to 'consider the practicability of adopting a simple and uniform system of weights and measures with a view not only to the benefit of our internal trade but to facilitate our trade and intercourse with foreign countries', and the membership numbered fifteen. The Chairman was Mr William Ewart.

The Report listed the anomalies of the existing system of weights and measures of which there were no lack. Insofar as the Report dealt with the generally chaotic system of Imperial and customary measures, that has been alluded to at the end of the previous chapter (p.270). The Report however had important things to say respecting the metric system. Its first recommendation was:[11] '1. That the use of the Metric System be rendered legal. No compulsory measures should be resorted to until they are sanctioned by the general conviction of the public'. The second recommendation requesting the establishment of a Department of Weights and Measures at the Board of Trade has already been referred to. It will be recalled that one of the recommended functions of the Department was to be the promotion of the Metric System. The third recommendation was that the Government should permit the use of metric units, together with the present ones, in levying customs duties. The fourth was that the metric system should form one of the subjects of examination for entry into and competitions within the Civil Service.

In all there were eleven recommendations, all referring directly or indirectly to the introduction of the metric system. They added that[12] '. . . a decimal system of money should as nearly as possible accompany a decimal system of weights and measures'. The final paragraph of this Report was prophetic in nature:

> Such is an outline of the course recommended by your Committee for introducing into this country a system which may tend to enlarge our foreign trade — hitherto

> imperfectly developed, if not neglected — with countries yearly becoming more and more mutually connected and mutually dependent, most of them composing the great European family of nations, and many of them near our own shores;

but a century was to go by before Britain joined with other nations, 'many of them near our own shores', in the European Economic Community, and an even longer period was to elapse before Britain declared for a decimal coinage and the metric system itself.

There was now some pressure to do something by way of recognition of the metric system. Thus far it was not a legal system of measures which could be used at all in the United Kingdom for trade.

Then the Metric (Weights and Measures) Act of 1864[13] was passed, but lest any might think that this was the big breakthrough it only legalized the use of metric terms in contracts but not the use of metric units for trading purposes. The metre, kilogram, and litre could be written about in commercial undertakings but metre bars could not be used to measure anything for sale, nor could kilogram weights be used to weigh anything in the market-place. They could be used for scientific purposes but not for business. The Act of 1864 did give metric equivalents in Imperial units, however:

1 m = 1 yd 0 ft 3.3708 in.
1 kg = 2 lb 3 oz 4.3830 drams (15432.3487 grains)
1 litre = 1.76077 pints
1 hectare = 2 acres 2280.3326 sq. yds.

These were substantially the length of the metre as determined by Kater (39.37079 inches), and the weight of the kilogram as determined by Miller (15432.34874 grains).

The Bill as introduced into the Commons would have made the metric system compulsory for certain things but the opposition of the Government led to this being deleted. As it finally emerged the entire Act occupies only three-quarters of a page of print.

In the First Report of the Commissioners appointed 'to enquire into the Condition of the Exchequer Standards' 1867-8[14] the cautious recommendation appears that a complete set of metric weights and measures be obtained.

In the Second Report, entitled 'On the Question of the Introduction of the Metric System of Weights and Measures into the United Kingdom', 1868-9,[15] the point is made that no complaints have been made respecting the present system, nor is there any movement for change by shopkeepers or workmen. The Commission was of the opinion that any enactment giving permission to use metric weights and measures for public sales must be accompanied by such provisions for their form as will make it impossible to mistake them for Imperial measures. Further the Commission called attention to the advantages of establishing a decimal coinage. Section 16 of the report concluded that 'the time had arrived when law . . . should provide for the introduction of metric weights and measures in the United Kingdom', with the French nomenclature being retained. But it was recommended that the introduction of the metric system should be permissive only and not compulsory, with no legislation respecting the metric system being brought in until the whole subject of weights and measures 'be brought to Parliament in one Bill'.

The recommendation of Section 16 was somewhat tempered by Appendix 1, written by G. B. Airy, the Commission Chairman and Astronomer Royal, in which he declared: 'My opinion is, therefore, distinct, that no step ought to be taken which can tend in any way to introduce material standards on the Metric or any other foreign system'. Attention was, however, called to the advantages of establishing a decimal coinage.

The Third Report, 'On the Abolition of Troy Weight',[16] made just such a recommendation as a means of improving the Imperial system.

But the metric system was rapidly taking hold in Europe and indeed throughout the world. In 1872 a new design for the metre was agreed to as well as a new kilogram standard. Instead of an end standard as heretofore, a line standard was proposed. The bar of winged section designed by Tresca was of a platinum-iridium alloy (90%–10%) rather than pure platinum and the distance between the parallel lines engraved in the neutral surface was declared to be the length of the *Mètre des Archives* 'in the state in which it was found'. In like manner the new kilogram was equal to the *Kilogramme des Archives* 'in the state in which it was found'.

The alloy for the new standards was cast in 1874, but the density did not meet the hoped-for specifications. While a number of metre bars and kilogram weights were produced from this enormous ingot (250 kg) the French Government decided to produce fresh material of the specified density and to make new bars and weights, but this task was not complete till 1889 as we shall see.

Meanwhile Europe was beginning to act as one at least in so far as metrology was concerned. Eighteen countries met in Paris in 1875 and signed the Metric Convention. Britain was not represented officially nor was she a signatory to the Treaty itself. The Convention authorized the setting up of the International Bureau of Weights and Measures at Sèvres near Paris. This body has performed magnificent work over the years since its inception and has a well deserved international reputation for its standards work, and is responsible to an international committee.

In 1871 another Bill in the British Parliament to make the use of the metric compulsory in certain areas was defeated by five votes. It was a fatal mistake to try to make compulsory the new system in any trade or industry. A gradual approach has always been the way of British governments.

As we have seen in the last chapter, the Weights and Measures Act of 1878 continued the Imperial system. In Schedule 3 Part I of the Act of 1878 a set of metric equivalents of which the principal values were the same as those of the 1864 Act and a list of metric standards now in the custody of the Board of Trade was given (Sch.3 Pt.II), but the Act repealed that of 1864. Nevertheless in a rather half-hearted fashion the British people were given the following authorization:

> Whereas the Board of Trade have obtained accurate copies of the metric standards and it is expedient to make provision for the verification of Metric Weights, the Board of Trade *may, if they think fit,* cause to be compared with the metric standards in their custody all metric weights and measures submitted for the purpose (our italics).

But even without the repeal of the Metric Act of 1864 it was an offence to use or possess a metre or kilogram *for the purposes of trade.*

This see-saw was discouraging and confusing, but the breakthrough was at hand, for Britain joined the Metric Convention in 1884 and finally in 1897 by the Weights and Measures (Metric System) Act[17] it was made lawful to use metric weights and measures *for trade.* The Board of Trade was now required to keep standard measures of the new system; metric measures were now to be certified in the same way that Imperial measures were and the Board acquired a copy of the international prototype metre and kilogram which had been authorized by the Convention in 1889 following the acceptance of the design in 1872 and the production of the denser alloy already mentioned.

Following the 1897 Act new equivalents for the metre and kilogram in the light of more recent comparisons were given, namely:

1 m = 39.370113 in★
1 kg = 2.2046223 lb Avoirdupois

Thus, as the nineteenth century closed, the Government of Britain had finally acknowledged that the metric system had some merit. It was now possible to use its measures in trading but in fact the vast majority of transactions were still being made in Imperial measures. Indeed, with few exceptions, the entire retail trade was being conducted, as before, in the old units.

Having regard to events that were to follow in the present century it might be well to summarize the reasons which led many in the United Kingdom in the nineteenth century to support the proposed decimalization of the coinage

★Confirmed by Order in Council No. 411, 1898.

coupled with a complete change in weights and measures. The minutes of evidence as laid before the various committees and commissions leave little doubt as to the motives.

Hard as it is to believe, far less understand, the truth is that none of the multitude of Acts passed in the British Parliament prior to 1878 was effective in regulating the use of weights and measures or of restricting usage to those measures which were contained in the most recent Act. The inspectorate, where it existed, was ineffective, with the result that local and customary measures flourished long after the Act of 1824, as we have seen, and not just in rural or remote areas either. The third quarter of the century was upon us before the British system of weights and measures had settled down to one recognizable system. The difficulties can be imagined, hence the grounds for discontent and the desire to move to a system apparently simpler even if of foreign provenance.

The Imperial system even in its purest form was not free from condemnation, as the 1862 Report shows. For example, we are told (p.iv)[11] that the Associated Chambers of Commerce of the United Kingdom in 1861 unanimously passed the following resolution: 'It is highly desirable to adopt the metric system which has been introduced into many European countries with great advantage to the saving of time in trading and other accounts'. On the same page the Report states: 'The superiority of a decimal system has long been acknowledged' while on p.vii one witness is quoted as saying: 'I think the difficulty of the English system is as great as it would be to make a calculation in the old Roman figures' and foreign students according to another witness were 'repelled and annoyed' by the English system of weights and measures. One educator observed: 'The waste of time to junior pupils in learning tables of weights and measures is immense', while Professor Miller stated that the metric system had been in use at least since 1836 for scientific procedures.

When the intricacies of the Imperial coinage are added to those of the weights and measures a very compelling argument could be advanced as that by Mr Dickson (Report p.viii) '. . . there is nothing so difficult to a man of imperfect education as to take an invoice of 10 tons 3 qrs or 7 cwts and 18 lbs at 25/11 per cwt'.

This was the nub of the nineteenth-century arguments. The labour in accounting was just too great. It called for abilities such as long division, multiplication at least as far as the twelve times table, and a certain aptitude with the basic arithmetical skills. Such skills were expected of those who had progressed through the full school programme, but as many had not, there were not a few 'men of imperfect education' who, notwithstanding this impediment, had found a niche in the commercial world of the day. To these and to their employers, a move to the metric system would be a great blessing, eliminating not only the labour but also much of the likelihood of error, especially if money was also to be in decimal notation. The pressure to go metric in the nineteenth century was aimed at facilitating computation and accounting procedures.

NOTES AND REFERENCES: CHAPTER XIII

1. See for example COCKBURN, H. *Examination of trials for sedition in Scotland* Edinburgh: David Douglas 1888 (2 vols.).
2. Hansard (Commons) 1816, Vol.XXXIV, 1024.
3. First Report of the Commissioners appointed to consider the Subject of Weights and Measures *Parl. Papers* 1819 (HC565) xi, p.307.
4. Preliminary Report of Decimal Coinage Commission 4 April 1857. *Parl. Papers* 1857 (Session 2) xix, p.1.
5. Report of Commissioners appointed to Consider the steps to be taken for the Restoration of the Standards of Weights and Measures. 21 December 1841. *Parl. Papers* 1842 [356], xxv, p.263.
6. Ibid, Section V.
7. CHISHOLM, H. W. On the Science of Weighing and Measuring and the Standards of Weight and Measure *Nature* (1873) Vol.VIII, p.490 et seq; MILLER, W. H. On the Construction of the New Imperial Standard Pound . . . and on the Comparison of the Imperial Standard Pound with the Kilogramme des Archives *Phil. Trans. Roy. Soc.* (1856) Vol.CXXXXVI, pp.753-946. London.
8. Reference 4, p.x; See also Report of Commissioners appointed to superintend the Construction of the New Parliamentary Standards of Length and Weight March 26th 1854. *Parl. Papers* 1854 xix p.933, Section 47.
9. Reference 4, pp.xiv-xv.
10. Final Report of the Decimal Coinage Commissioners; *Parl. Papers* 1859 (Sess.2) xi, p.1 et seq.
11. Report from the Select Committee on Weights and Measures 15 July 1862. *Parl. Papers* 1862, vii p.187 See p.ix.
12. Ibid, p.x.
13. The Metric Weights and Measures Act. 27 & 28 Victoria c 117 1864 *Statutes at Large* Vol.XL, p.575.
14. First Report of the Commissioners appointed to Enquire into the Condition of the Exchequer Standards *Parl. Papers* 1867-68 xxvii, p.1.
15. Second Report of the Commissioners appointed to Enquire into the Condition of the Exchequer (later Board of Trade) Standards, *Parl. Papers* 1868-69, xxiii, p.733.
16. Third Report of the Commisioners appointed to Enquire into the Condition of the Exchequer (later Board of Trade) Standards, *Parl. Papers* 1870, xxvii, p.81.
17. Weights and Measures (Metric System) Act 60 & 61 Victoria c 46, 1897 *The Statutes Revised* Vol.XIII, p.418.

CHAPTER XIV

Britain Goes Metric: The Twentieth Century

The Hodgson Report, 1950

Legislation on weights and measures was not lacking in the first half of the twentieth century. There were Acts in 1904, 1907, 1926, 1929, 1936, 1938, and 1950, all dealing with some aspect of the subject, but none was such as to change the direction of British metrology.*

As for the decimalization of the currency, the 1917 Report of the 'Committee on Commercial and Industrial Policy after War' dismissed the concept of a decimalized currency because a change in the value of the penny would cause too great an upheaval. Undaunted, a Royal Commission on Decimal Coinage was set up in 1918. Their 1920 Report feared that 'the abolition of the pound as a unit of value would endanger the position of London as the financial centre of the world'. The Commission considered and rejected two proposals for a new currency and concluded that the adoption of a decimal currency would be 'inadvisable'. These two reports of 1917 and 1920 between them laid low any such thoughts, or, for that matter, any thoughts of going metric with weights and measures, for a generation. Then in 1950, after half a century of relative inaction, a report of great significance was presented to the Minister on 13 December of that year and published in May 1951. It was the Report of the Committee on Weights and Measures Legislation[1] under the chairmanship of Sir Edward H. Hodgson with thirteen other members. The committee members had been appointed, some late in 1948 and others in early 1949. Their two years of work were most fruitful although at the time the Report caused hardly a ripple among the general public.[2]

A quick glance at the first chapter might lead one to believe that this was just another report going nowhere in particular, especially when section 6 is read:

> 6. The main conclusion to which our Enquiry has led us is that the existing principles of Weights and Measures are soundly based and stand in no need of fundamental revision. Its provisions appear to have provided, within the field they purport to cover, a reasonable protection to all engaged in trade without inflicting any undue burdensome requirements; and the machinery of administration seems to have worked well over the years. In the following Chapters we make many recommendations of varying scope and importance; but these represent in the

*All these Acts were administrative save those of 1936 and 1938 which introduced the cubic yard and its subdivisions as legal measures of volume.

> main additions to or simplifications of the present framework, rather than completely novel features.

If one may be allowed to disagree with these sentiments in the light of subsequent events, far from the Report being a recital of the status quo we may see in it the key to Britain's move toward metrication, for the arguments advanced in support of such a move in later chapters were, of themselves, quite compelling. Indeed, the second chapter (beginning on p.5) is entitled 'The Adoption of the Metric System' and points out that, whereas the metric system is a unity, the Imperial system is 'really a conglomeration of units which have in the past been found convenient for particular types of measurement . . .'. Moreover the Imperial systems of the British Commonwealth and of the USA do not comprise a single system of weights and measures, for the one relies on a standard yard bar and a platinum pound weight, whereas the other, that of the USA, has (since 1866)[3,4] defined its yard and pound in terms of the metre and kilogram: the gallons of the two countries differ by about 20 per cent and their two tons by about 10 per cent (2240 and 2000 lb respectively). Further there is no coordinating body for the Imperial system as there is for the metric (see Appendix C).

Again, the Federation of British Industry was aware of the tendency to use increasingly the metric system in preference to the Imperial but agreed the options available be continued. The Association of British Chambers of Commerce, on the other hand, wished to adopt completely the metric system.

The Committee was of the opinion that a change could only occur over a period of some twenty years and would not succeed if simply left to individuals and groups to decide voluntarily that such a move would be to their advantage. (It was to be a recurring theme of later government publications that the ground-swell to go metric came from business and industry and not from governmental persuasion, far less coercion.)

The Committee was of the unanimous opinion that the metric system was a 'better' system and that the Imperial system should eventually be abolished in favour of it and that the change should be preceded by:-

- (a) discussion with industry and commerce to determine the period of transition;
- (b) agreement with the Commonwealth and the USA for a simultaneous change on their parts;
- (c) a lengthy process of preparing the general public for the change;
- (d) the decimalization of the coinage;
- (e) the preparation of schemes for the compulsory change-over, trade by trade, during the period of transition, with provision for compensation wherever necessary (section 26, p.9).

As we shall see, of all their recommendations, this last one found very little favour with Government.

It was recognized that these recommendations might not be accepted, so in addition recommendations for improving the existing system were included. It was not difficult to see where improvements could be effected.

It had long been recognized that the yard measure in bronze did not match the quality of the metre bar of platinum-iridium in workmanship, engraving, or reproducibility. The accuracy of measurement using the Imperial unit is about 0.00002 inch, whereas with the metre it is about 0.000004 inch (that is a factor of five better); but more seriously, while the Imperial standard yard and its Parliamentary copies when checked one against another in 1912, 1922, and 1932 showed constancy to within 0.00005 inch, when compared with the metre it appeared that all the yards had shrunk progressively and in fifty years had decreased in length by two parts per million, probably due to the release of strains in the bronze metal. The metre, on the other hand, was found, by comparison with its copies and with the wavelength of light, to have remained remarkably constant. The additional Parliamentary copy (see Ch.XII) produced in 1879 in response to the 1878 Weights and Measures Act was made as closely as possible to the others. It has been carefully monitored since its first verification in 1886 and its shrinkage has been observed. If the others behaved in their early years as it has done, all the yards will have contracted by just over 0.0002 inches before reaching a steady stable configuration.[5]

The values of three determinations of the Imperial yard against the metre using a specially graduated nickel standard at the National Physical Laboratory,[6] Teddington, are:[7]

	Value of 1 m in inches	*Value of inch in mm*
1922	39.370147	25.399956
1932	39.370156	25.399950
1947	39.370186	25.399931

from which the shortening will be at once apparent. At the Laboratory the ratio 1 yd=0.91439841 m has been adopted, equivalent to the 1922 determination, and this has been the value accepted for *scientific purposes.*

The Imperial standard pound made in platinum in 1844 is likewise inferior to the standard kilogram. In relation to its copies, the Imperial standard 'diminished by 1 part in 3.5 million between 1846 and 1883 but remained constant to within about 1 part in 10 million between 1883 and 1933 . . . The decrease in mass was attributed to wear resulting from too frequent usage'.[8]

When compared to the standard kilogram (which had remained virtually unused up to the end of World War II) the story was worse. The Imperial standard had diminished by 1 part in 2 million from 1846 to 1883 and from

1883 to 1933 a further diminution of 1 part in 5 million. The values found were:[9]

Imperial Standard		
	lb in grams:	*Referred to:*
1846	453·592652	*Kilogramme des Archives*
1883	453·592428	*International Kilogram*
1922	453·592343	*International Kilogram*
1933	453·592338	*International Kilogram*

The kilogram, by numerous inter-comparisons, has remained essentially constant, the various weighings not differing among themselves by more than 2 parts in 100 million.

Nor was the British system helped by its own legislation. The yard was defined at a temperature specified as 'on Fahrenheit's thermometer' or '°F' but nowhere was the Fahrenheit scale defined legally. Further, Schedule 3 Part 1 of the 1878 Act[10] (still in force in 1950) stated:

1 m	=	39·3708 in
1 in	=	25·39954 mm
1 yd	=	0.91438 m

To begin with, the length given for the metre was that due to Kater in 1818. Later values were

39·370113 in	1895*
39·370147 in	1922, published in 1927
39·370138 in	1934

We cannot fault those writing in 1878 for not knowing that the 1818 value was a little large as there had been no intervening re-measurement, but, as was pointed out in the Hodgson Report, the equivalent for 1 inch is not 25·39954 mm. It is in fact 25·399535 mm so the figure has been rounded off. Equally 36 of these inches of 25·39954 mm amount to 0.91438344 m not 0.91438 m and if the metre is 39·3708 inches, the yard is not 0.91438 but is 0.9143832. Again rounding off has occurred. Following the 1897 repeal of Part 1 of Schedule 3 of the 1878 Act, (see Note 10), the subsequent Order in Council of 1898 (No. 411) while adhering to the 1895 value of the metre given above, further rounded off the value of the inch to 25.400 mm*. The 1898 values continued as the official equivalents until superseded by the Weights and Measures Act of 1963.

*See Ch.XIII p.286 and footnote thereto. The 1895 measurement was legalized in 1897 by the Weights and Measures (Metric System) Act, 60 & 61 Victoria c 46, and confirmed by Order in Council No.411 1898. The metre remained officially 39.370113 inches until 1963 when the relation, 1 yd = 0.9144 m was established i.e. 1 m = 39.37007874 . . . inches.

Moreover all the statutes to this date speak of weights. Strictly speaking these metal objects are not weights but masses. The weight of a body is the force of the earth's gravitational pull on the body. As gravity varies from point to point on the earth's surface the weight of a given body varies likewise. In the extreme case of a body in the depths of outer space, where the gravitational pull is essentially zero, its weight is zero regardless of its weight when measured on the earth's surface. Yet the amount of matter in the body or its inertia, which may be thought of as describing its mass, has not changed. In the last analysis, mass, like length and time, is undefinable, being one of the primary concepts on which the whole of physics is built. A spring balance, because it measures force, measures weight; a beam with scale pans compares unknown with known masses. In common parlance, the words mass and weight are used interchangeably though this is technically wrong.

So much for what had happened within our own borders. Across the Atlantic, however, the metre was declared to be 39.37 inches in 1866 and a later order declared that this value was to be interpreted as the relation defining the US yard. Hence the US inch was 25.400051 mm.

It may be said by some that these discrepancies are too trivial to be taken seriously, but this is not so. Standard reference gauges are today required with a precision of 1 part in a million or better. The difference between 25.400 and 25.39954 is 18 parts per million! To quote again from the Hodgson Report:[11]

> It is of the essence of a Standard that its definition shall be more precise than the most exacting practical demand made upon it; and from this point of view it is no mere scientific pedantry to say that Britain and the USA now have different standards of length.

As for the kilogram and the pound, the determination in 1844-46 by Miller of the pound in terms of the *Kilogramme des Archives* was

$$1 \text{ lb} = 453.5926525 \text{ grams}$$

The 1883 determination undertaken jointly by the Standards Department of the Board of Trade and the Bureau Internationale yielded 453.592428 grams. This was rounded off to 453.59243 by Order in Council dated May 1898.[12] The 1933 determination gave 453.592338 grams. This is to be compared with the declared US value of the Avoirdupois pound, namely:

$$1 \text{ lb} = 0.4535924227 \text{ (USA) kg}$$
$$1 \text{ lb} = 0.453592338 \text{ (British 1933) kg}$$

The difference between these two amounts to nearly 19 parts per hundred million.

The above considerations led the Hodgson Committee to *recommend:*

> . . . that the fundamental units of length and mass in the Imperial system should cease to be independently defined and that they should in future be defined in law as being specific fractions of the metre and kilogram respectively.

They further *recommended* that the inch be taken as 25.400 mm, yielding a yard of 0.9144 m, a value which involves no rounding off. They *recommended* the US Government be approached in order to see if a common definition of the pound was possible.

Notwithstanding their caution in the body of the Report in not giving a recommended equivalent for the pound, we find one in their list of 'recommended measures which alone should be lawful for trade', namely, 'pound = 0.45359237 kilogramme' (Report p.28). They also list the yard as 0.9144 metre as already stated.

As for the gallon, the Act of 1824 said it was equivalent in volume to 277.274 in^3 but a more recent actual determination of the volume of 10 lb of water gave a value of 277.42 in^3. The differences respecting dry and liquid measure in the USA and our own gallon and bushel left the Committee in a dilemma. They could see no hope of bringing the British and US standards into agreement and left to a subsequent Commission which they *recommended* be set up the problem of defining an acceptable gallon which had a definite relation to the litre.

For Apothecaries' measure, they *recommended* that after five years the scruple, drachm, Apothecaries' ounce, the minim, and the fluid drachm, and all current Apothecaries' weights and measures except grain weights and fluid ounces, together with the hieroglyphic symbols denoting these units, should be abolished and made no longer legal for trade.*

Grain weights and fluid ounces were regarded by the Committee as Imperial measure rather than Apothecaries' measure. It was further *recommended* that after five years the Troy ounce and pennyweight should be abolished and that the rod, pole, or perch be deleted from the list of legal measures and also the chaldron, quarter, bushel, and peck.

The setting up of a permanent Commission on Units and Standards of Measurement to advise the President of the Board of Trade was a further important *recommendation*. The scope of the Commission was to recommend on definitions, the nature and construction of primary standards, and on the custody of standards and the frequency of reverification and comparison of standards, etc. This recommended Commission was to be of the highest calibre, including in its membership the Astronomer Royal, the Director of the National Physical Laboratory (see Note 6), and the President of the Royal Society *ex officio*. The Minister responsible for the Department of Scientific

*See Ch.X pp.185-7 and Appendix D p.359.

and Industrial Research should nominate up to two members; the Lord Chancellor one member with a legal background. The President of the Board of Trade should nominate three more members and name the chairman.

All these recommendations were very far-reaching. They ranged from the adoption of the metric system as the sole system of metrology in the UK to the abandonment of many of the traditional units (which had been with us since mediaeval times and of which so much has been said in earlier pages), should the Government not accept the idea of total change. The proposal for a decimal coinage was made *en passant* on page 8 of the Report (which is 147 pages long), for the coinage was outside their terms of reference. But the point was made and was the stronger for being made *sotto voce*.

We must now turn our attention to the fate of these recommendations. It should be said at the outset that the Government of the day did not accept the suggestion of the replacement of the Imperial system by the metric system. They preferred to proceed slowly but surely along the path to complete metrication.

Nothing happened for a decade. Then in 1960 an independent report prepared jointly by committees of the British Association for the Advancement of Science and the Association of British Chambers of Commerce ('Decimal Coinage and the Metric System. Should Britain Change?') raised the whole matter again. It is not too much to say that these two reports of 1950 and 1960 released the brakes applied by the reports of 1917 and 1920. Metrication and a decimal currency became topics to which much thought was given yet again.

The first response of major significance was the announcement in the House of Commons by the Chancellor of the Exchequer on 19 December 1961 that a Committee of Inquiry would be set up

(a) to advise on the most convenient and practical form which a decimal currency might take including the major and minor units to be adopted.
(b) to advise on the timing and phasing of the changeover best calculated to minimize the cost.
(c) to estimate the probable amount and incidence of the cost to the economy of proposals based on (a) and (b).

The Committee consisted of the chairman, the Earl of Halsbury, with five others, all of whom were nominated to the Committee on 11 January 1962. The excellence of the work of this Committee is revealed in its report dated 19 July 1963 and placed before Parliament in the September of that year, but before the report was submitted to the Commons another major development occurred. A new Act, the Weights and Measures Act 1963, dated 31 July of that year, was passed, which Act we will now consider in detail before considering the subsequent recommendations of the Halsbury Committee.

Clearly the 1960s were to be years of decision for Britain.

The Weights and Measures Act, 1963[13]

The first section changed the foundations of British metrology forever. It said that the yard *or* metre shall be the unit of measurement for length and that the pound *or* kilogram shall be the unit of measurement of *mass* (not weight) and that

(a) the yard shall be 0.9144 metre exactly
(b) the pound shall be 0.45359237 kilogramme exactly,

that is, precisely the values recommended by the Hodgson Committee.

Here there is no rounding off. These are the precise fractions of the metre and kilogram which shall from now on represent the yard and pound. This means really that the yard has been increased in length by about $\frac{1}{10,000}$ inch, and the pound decreased by about $\frac{1}{1000}$ of a grain, amounts which in the normal course of trade would be insignificant. The metre in 1963 was no longer defined in terms of the international prototype metre but by the wavelength of light (see Appendix C), namely 'one metre is equal to 1650763.73 wavelengths in vacuum of the radiation corresponding to the transition between the levels $2p_{10}$ and $5d_5$ of the Krypton-86 atom' as contained in the Weights and Measures (International Definitions) Order 1963 (Statutory Instrument 1963 No. 1354).

Section 2 of the 1963 Act required the Board of Trade to maintain standards of these four units (that is, metal bars for yards and metres with platinum and other metal weights for the pound and kilogram), and until otherwise stated, the primary UK standard yard was to be our old friend the bronze bar of Baily's metal cast in 1845 by Troughton and Simms, the primary UK standard pound was to be the platinum pound of 1844, the UK metre was to be the bar of platinum-iridium marked No.16 (see Appendix C)*, and the UK kilogram was to be the British copy of the international prototype kilogram marked No.18. The authorized copies of the yard and the pound deposited at the Mint, the Royal Society, Royal Greenwich Observatory, and the Palace of Westminster and the additional yard and pound made in 1878 and 1879 respectively for the Standards Department of the Board of Trade, all appear in Schedule 2 of the Act.

Here we see the Government accepting the lesser of the recommendations, namely, that if the Imperial system was not to be abolished the units should be redefined metrically. The recommended length of the yard and weight of the

*The metre is now defined as 'the length of path travelled by light in vacuum during a time interval of $\frac{1}{299792458}$ of a second'. (Weights and Measures Act 1985 (30 October), and the equivalents given above were retained.)

pound was accepted. The value for the pound was very nearly the average of the value in the USA and the 1963 British determination. The original artifacts representing the standards were retained but were now redefined.

Section 7 of the Act set up the Commission which had been so strongly recommended in the Report with terms of reference closely those envisaged for it by the Committee. The members of the Commission were not to be the persons proposed by Hodgson's Committee but were to be individuals named by these distinguished people. Of the nine members, six were to be appointed by the Board of Trade upon recommendation, one by each of the Lord Chancellor, the Director of the National Physical Laboratory, the Astronomer Royal, and the President of the Royal Society, and two by the Minister for Science, with the chairman to be appointed from among the nine by the Board.

The metric carat[14] was restricted by the Act to the weighing of precious stones and pearls and the Troy ounce was restricted to the trade in precious metals. The Apothecaries' ounce, drachm, scruple, fluid drachm, and minim were continued for trade in drugs but were to be abolished not earlier than five years from the passage of the Act (that is 31 January 1969). The bushel, peck, and pennyweight were likewise to be abolished after five years. The fluid ounce continued to be a legal measure, being listed as an Imperial measure and not as one of the Apothecaries' measures. The Board of Trade was given the power to delete or add measures to the list approved as legal for trade 'but the Board shall not so exercise their powers under this subsection as to cause the exclusion from use for trade of Imperial in favour of metric units of measurement, weights, and measures' (Section 10.10). This power did not extend to abolishing the metric system either.[15]

Here we see the Government agreeing to a fair number of abolitions of ancient units, virtually the complete list sought. Some other measures were retained, others abolished, though they were not in the recommendations of the report of the Hodgson Committee, for example, the reputed quart was abolished (see Ch.X(h)) and the cran measure for herrings (see Ch.X(b)ii)[16] retained.

The Act concludes by listing ten schedules of which four are relevant to our inquiry. The first lists and defines the units which may lawfully be employed in trade. Comparing the list with the provisions of the 1878 Act we see that the chaldron and quarter have disappeared, although not specifically mentioned in the body of the Act. So have the rod, pole, or perch and the square rod. The gallon continues as heretofore. The litre equivalent is given as 1 gallon = 4.545964591 litres.[17] the quart, pint, and gill continue as before. The metric ton *(tonne)* is defined as 1000 kg and the *quintal* as 100 kg. The hectogram is given as $\frac{1}{10}$ kg and the metric carat as $\frac{1}{5}$ gram.

The second schedule lists the existing UK primary standards and the copies to which reference has already been made.

The third schedule lists measures and weights which are lawful for trade. These schedules need not detain us further, save to point out that the *cental* (100 lb), although defined in Schedule 1, does not appear here so might be presumed abolished[18] forthwith.[19] (See also p.305 for S.I.484 of 1978 and footnote p.263.)

Schedule 4 Part IV deals with bread, stating that any loaf exceeding 10 ounces shall be made for sale only if it is of a net weight of 14 ounces or a multiple thereof (see Ch.XI), while Part V dealing with milk declares that milk shall be pre-packed only if it is made up in one third or half a pint or a multiple of half a pint and the container marked to show the contents. These are two examples of many foodstuffs which by this Act may only be sold in Imperial measures.

We therefore see that, while all of the recommendations of the Hodgson Committee were not accepted and built into the 1963 Act, a very considerable number were. It may have been a disappointment to the members that the Government would not at once go completely metric but they could perhaps take comfort from the large measure of agreement they were able to find in government circles. Having got so far with weights and measures, what was happening to the coinage?

The Report of the Halsbury Committee, 1963

This Report[20] with its seventeen appendices ran to 253 pages which gives some idea of the detailed consideration given to the question in the two years of the Committee's existence. As the Committee was quick to point out, its terms of reference were not to consider yet again whether or not to decimalize the coinage but to advise the Crown how to achieve this end (see terms of reference above). The Government had now made up its mind to change, but, no doubt being mindful of Section 26 of the Hodgson Report, 'reserved to itself' the question as to whether any compensation was to be paid for losses or costs incurred in decimalization. No compensation was ever given. Any costs to the merchant in changing the currency were to be absorbed by the business.

The Committee, after receiving evidence both oral and written, decided for a system with only two units in it, and no intermediate unit of account, with the smaller unit defined as $\frac{1}{100}$ of the larger. But was the penny to be the smaller unit with the larger one of 100 pence, or was the pound to be the larger unit with a new denomination of $\frac{1}{100}$ of a pound for the smaller, or was 10 shillings to be taken as the larger unit, or should the smaller unit be the shilling with the larger unit worth £5? These are only four of the many possibilities. In fact the Committee boiled the number of practical schemes down to four in which the first three mentioned above figure together with a 5 shilling-cent scheme which showed merit.

A detailed review of the arguments for and against would be out of place here. The Report itself can be consulted for the pros and cons. Suffice it to say that the Committee recommended by a vote of four to two to adopt the pound-cent scheme, with the lowest coin to be ½ cent, which coin was thought 'tolerable, not desirable', it being of such low value it could rarely be needed in everyday transactions.* The scheme was known as the '£-cent-½' proposal. The two members opposed to this pound-cent-½ scheme submitted a minority report advocating the 10 shilling-cent scheme, which was one of the short list of four possibilities considered at length by the Committee as a whole. Indeed, Section 383 of the Report itself (p.89) states:

> From a purely domestic point of view the 10s-cent system without a fraction offers, on balance, the best system for the immediate future and the smoothest transition. . . .
> 384. The contention of those who have given evidence in support of the 'international' case for the £ is that risk is inseparable from adoption of any major unit other than the £ . . . we concluded that the disadvantages of the £-cent-½ system as opposed to a 10s-cent scheme were a price well worth paying, in the general interest to avoid risk to the international standing of sterling.

So the recommendation stood.

The Committee recommended ½, 1, and 2 cent coins in bronze; 5, 10, and 20 cent coins in cupro-nickel, all rounded at the edge; and a 50 cent bank note, the pound note being retained. The change-over was recommended to occur in the month of February with a three-year preparation time. As in the Hodgson Report (and the 1963 Act), an Executive Board was recommended to monitor the change-over, educate the public as to what was happening, and facilitate the introduction of the new coinage, but this Board was to see the transition through, unlike the Commission set up by the 1963 Weights and Measures Act which was more advisory in nature. The Report also gave detailed recommendations for the introduction of the new currency and the withdrawal of the old, save that the shilling and florin need not be withdrawn but could be recalled gradually, for the shilling was now worth five cents and the florin ten cents.

The Government did not wait thirteen years, as in the case of the Hodgson Report, before bringing in legislation to effect change. The Chancellor of the Exchequer reported to the House of Commons on 1 March 1966 that the change-over to decimal currency would take place in February 1971. Then a year later, the Decimal Currency Act 1967[21] provided for the creation of a new currency based on the pound and the 'new penny'†, the term 'cent'

*The lack of utility of the ½*p* coin was such that its production at the Mint ceased on 29 March, 1984, although it was to remain legal tender.
†After a decade of use, with everyone familiar with the appearance of this coinage, the word 'new' has been dropped. (See the 20*p* piece)

failing to gain approval. The new penny was to be $\frac{1}{100}$ £. Coins in bronze of the denominations of ½, 1, and 2 new pence were specified in size and weight as were coins of cupro-nickel of 5 and 10 new pence, these latter two being of the same size and weight as the coins they were to replace, that is the shilling and the florin. All coins described in the Act were to be round. There was no mention of the replacement for the ten shilling note — now 50 new pence — nor of the recommended 20p piece.*

A Decimal Currency Board was set up by the Act with much the same mandate as the proposed Executive Board and Decimalization day or D-day as it was popularly called (though termed in the Act the 'appointed day') was to be 'such day in 1971 as the Treasury may order', the month having already been specified.

The abbreviation for 'new pence' very rapidly became 'p' and was so used in government publications.

Then, in May 1968, the specification for the 50*p* unit to replace the 10 shilling note was released. It was to be a coin whose edge was a regular heptagon in cupro-nickel. The reasons for going to a coin were (a) that a 10 shilling note had an average life of 4-5 months whereas the proposed coin was expected to serve for half a century, and (b) there was a tendency in other countries to make their larger denominations coins rather than paper notes, no doubt having experienced something similar to that recorded in (a). The second Decimal Currency Act, that of 1969,[22] refers to this coin, which replaced the 10 shilling note on 14 October of that year. The public reacted unfavourably to this coin at first,[23] with an intensity which caught the Decimal Currency Board by surprise, but the reaction was short-lived.

*A characteristic of the old coinage had been its massive size and weight.

	Old Coinage	New Coinage
Farthing	43.75 grains (2.83495 grams)	½*p* 1.78200 grams
Halfpenny	87.5 grains (5.66990 grams)	1*p* 3.56400 grams
Penny	145.83 grains (9.44984 grams)	2*p* 7.12800 grams
Shilling	87.27 grains (5.65518 grams)	5*p* 5.65518 grams
Florin	174.55 grains (11.31036 grams)	10*p* 11.31036 grams

The new coinage is somewhat lighter but the 5 and 10*p* coins are the same as their counterparts. A 20*p* coin in cupro-nickel (84%-16%) made its appearance for the first time in June 1982. It was quite small, of diameter 21.40 mm, weighing 5.00 grams, the edge being a regular heptagon. The reduction in weight is a very welcome feature. A comparison with Canadian coins of approximately the same value shows the following:

2*p* 7.12800 grams	5¢ 4.6 grams
5*p* 5.65518 grams	10¢ 2.07 grams
10*p* 11.31036 grams	25¢ 5.05 grams

Now that currency is entirely token and the value is not represented by the weight of metal in the coin there is little merit in heavy coins, so it is interesting that the new 20*p* piece weighs less than its approximate Canadian counterpart, the 50¢ piece of 8.10 grams.

In April 1983 a £ coin of copper, nickel, and zinc (70, 5.5, and 24.5% respectively), was introduced, doubtless to remind one by its colour of the old gold sovereign. It weighs 9.5 grams. The pound note ceased to be issued on 31 December 1984 but still remains legal tender.

A schedule was prepared for the removal of the old coins. The farthing, which played so large a part in the mediaeval Assize of Bread, had already been withdrawn in 1960 and it was then followed by the halfpenny on 1 August 1969, neither coin any longer having any practical use. The half-crown (2*s* 6*d*) was to be demonetized on 1 January 1970. The 5 and 10*p* coins,having the same value, weight, and dimensions as the old shilling and florin, were brought into circulation early in 1968, well in advance of D-day. The old penny, three-penny bit, and sixpence were to continue for a while after D-day, but their usefulness rapidly declined, for 6*d* was now to be $2\frac{1}{2}p$, 3*d* was now $1\frac{1}{4}p$, and 1*d* was a mere $\frac{5}{12}p$, hardly units of convenience but as was pointed out,[24] a three-penny bit and three old pennies could be used for a $2\frac{1}{2}p$ purchase, small though that purchase might be.

On 5 January 1971 it was announced that D-day would be 15 February of that year. The next day the news was promulgated by the news media.[25]

Meanwhile the Mint had not been idle. A stockpile of 3400 million bronze coins of $\frac{1}{2}$, 1, and 2*p* denominations had been created for issue on D-day. With the issue of bronze the decimalization was complete. Looking back, its ready acceptance by the people and the relative ease with which it had been accomplished made one wonder why it had taken the better part of two centuries to reach that goal. In the meantime, the Government had at last made up its mind and had decided to go completely metric.

The Move to Metrication

After consulting its members, the Federation of British Industries (later the Confederation of British Industry) advised the Government in 1965 that a majority of its members favoured the adoption of the metric system, ultimately to be the only system in use in the kingdom, but at the beginning on a voluntary basis, with sector following sector as the advantages of such a move were made manifest. Specifically it was hoped that the Government would not impose a move to metrication but would do all it could to promote the transition on a gradual basis. The members of the British Standards Institution advised on similar lines.

The President of the Board of Trade stated on 24 May 1965 in the House of Commons:[26]

> The Government are impressed with the case which has been put to them by the representatives of industry for the wider use in British industry of the metric system of weights and measures. Countries using that system now take more than one-half of our exports; and the total proportion of world trade conducted in terms of metric units will no doubt continue to increase.Against that background the Government consider it desirable that British industries on a broadening front should adopt metric units, sector by sector, until that system can become in time the primary system of weights and measures for the country as a whole. . . .

> Practical difficulties attending the change-over will of course mean that the process must be gradual; but, the Government hope that within ten years the greater part of the country's industry will have effected the change.

In March 1966, a Standing Joint Committee on Metrication was established which presented a Report[27] in May 1968 which the Government accepted. The Report supported the ten-year transitional period, advised that a Metrication Board be established to advise and coordinate the plans for the transition, but, contrary to the proposal of the Hodgson Committee, there was to be no compensation for costs or losses incurred in the transition. New weights, measures, machinery, etc. were to be the responsibility of those acquiring them.

The Metrication Board

The Metrication Board was set up by the Minister of Technology in April and May of 1969 and its first meeting was held in May of that year. No time was to be lost! It was unfortunate that a delay of four years had occurred before the Board was established for it had only just commenced its work at what the Government hoped would be the half-way mark towards complete change-over or nearly so. Unlike the Decimal Currency Board's operation there was to be no legislated M-day, nor could there be with the myriad of individual items made, used, or consumed each day of modern living. The Metrication Board was not given the task of seeing metrication through to its conclusion. Its terms of reference were very general, enjoining it to 'facilitate' the transition and requiring it to 'examine in consultation with such organizations and persons as the Board consider appropriate the problems involved in the transition', 'to advise the responsible Minister', 'to make generally available information and advice', 'to furnish to any enquirer information . . .', 'to ensure that the relevant educational interests are kept fully . . . informed', etc, etc.

As the Board itself was to note, its terms of reference did not permit it to campaign in favour of the metric system.[28] It had no power to enforce any change.[29] It was purely advisory. It could only point out by precept and example that to delay might prove very costly. By 1970 about three quarters of British exports went to metric countries. To change to metric would be rather like the adoption of the Avoirdupois ounce from the city of Florence about the year 1300, a city with which England was then conducting an enormous wool trade, but many did not see it at all in that light. There was something of an affection or nostalgia for the Imperial units. Shoppers continued to ask for a pound of this or that and the retail trade obliged. Pre-packaged goods appearing in metric amounts appeared dearer to the consumer, not un-naturally, for a half kilogram of butter contained 10 per cent more than a pound did, but the packages looked alike. The fear of profiteering lingered in spite of measures to ensure this did not happen and surveys to show it had not happened.

The White Paper on Metrication, 1972, reflected the Government's continuing optimism that industry would be largely metric by 1975 as previously proposed (Section 5).[30] Then followed the statement (Section 6):

> Progress to metrication cannot be a haphazard affair, left to individual whim and decision. If that were to happen it could cause confusion throughout industry and would present untold difficulties to the consumer. It is in everybody's interests therefore to ensure that it takes place in a well ordered and properly regulated manner. To see to this is the job of the Metrication Board.

The Metrication Board's task was to coordinate in space and time the movement of the various sectors of the economy into the metric arena, but it had no power even to advocate the transition in one sector, the change in which would enable others to change too and in an orderly, logical fashion. The Government adhered to its policy of voluntary change. This may not have been individual whim but it was individual decision among the various sectors of trading and business life.

The White Paper indicated that speed limits and road signs were unlikely to be changed for a long time to come*, and that the Government had no wish to discourage the sale of draught beer or of milk by the pint. Nor would the suppliers. A pint is 568.3 ml and present-day milk bottles are so marked (568), retaining the pint measure but describing it metrically. Were litres alone to be used, the consumer of beer or milk would opt for the half litre, thus reducing the amount sold by 12 per cent, hardly a stimulus for the industry.

A plan to use metric bottles was shelved by the Government in 1979, and the EEC Commission accepted 'that British buyers will not accept a metric measure on the doorstep' (see *The Times*, 26 November 1979, 4f).

By April 1974, the Fifth Report of the Metrication Board[31] indicated that the metrication of consumer goods was running about two years behind schedule and could not be completed by 1975, but on the entire industrial and business front there was thought to be a good chance of substantial completion by that deadline date. It was important that furnishings and dress fabrics would be sold by retailers from 3 February 1975 by the metre and fractions thereof, to the nearest 10 cm (4 in). The traditional widths for cloth would be retained but would be expressed in centimetres rather than in inches.[32]

By this time there had been a good deal of looking over the shoulder to see

*They were still unchanged eleven years later (1986).

what was happening elsewhere. The Metric Study Group of the USA judged in their Report of 1971[33] that it would be inevitable that they should join the rest in adopting and using the metric system, and the Secretary for Commerce agreed with this conclusion when presenting the Report to Congress. He recommended a target date some ten years ahead, but nothing happened.

Not a few attributed the drive to metricate everything to Britain's entry into the Common Market on 1 January 1973. Britain had, however, applied for membership in 1961 but strong opposition from within EEC itself delayed matters so that negotiations were only concluded in 1971. Hence Britain's intent to join the EEC antedates the push to metricate by several years. However, now that Britain is a full member of EEC it has to abide by the rules. A series of directives have been issued, in particular one relating to uniformity of metric measures in many fields including the economic area.[34] It had been agreed that before 31 August 1976 a decision should be taken with respect to the category into which British Imperial weights and measures should be placed, but it has been agreed that it will be possible to extend that date in justifiable circumstances. To put it simply, there would be no problem with the EEC directives were Britain to be totally metric.

The deadline date of 1975 came and went with Britain's goal still far off. In the next year a new Weights and Measures Act was passed.[35] Entitled 'An Act to amend certain enactments relating to Weights and Measures', its Section 1 gave the Secretary of State power to add or remove units from Schedule 1 of the 1963 Act and the Act inserted at the end of this schedule a new list stating that the bushel, peck, fluid drachm, and minim were not to be used for trade, nor were the pennyweight (dwt), Apothecaries' ounce, drachm, scruple, nor (surprisingly) the metric ton (tonne) of 1000 kg. (Clearly the ton of 20 cwt each of 112 lb was to continue.) This Act also lifted some of the prohibitions for pre-packaged goods, many of which by the 1963 Act had to be in Imperial units. Powers were granted to enable dates to be set ending Imperial units in certain sectors of trade, but it could hardly be said that this Act was a major statute destined to move the nation rapidly into a state of complete metrication. The situation was no better when the Metrication Board's Report for 1977-78 appeared. The Report[36] begins:

> 1.1 The change to metric has moved forward in the 18 months covered by our Report. But we believe the advance has been insufficient . . .
>
> 1.2 . . . many people feel a strong attachment to Imperial weights and measures. There is a widespread, though far from universal, reluctance to make the adjustment to metric in the most familiar everyday activities.
>
> 1.4 The choice is not between Imperial or metric . . . the choice we have is between completing the change to metric within the next few years or settling for an indeterminate period during which the inefficiencies and inconveniences of using the two systems side by side will grow.

Using the powers granted under the 1976 Act, Orders were made phasing out Imperial packs of goods such as sugar, cocoa, cornflakes, etc., together with a variety of dried fruits and vegetables, flour and flour products. It would appear that virtually every article of normal retail grocery commerce required specific mention by name. Discussions were held in 1977 with the retail trade and consumers' organizations to see what could be done to set cut-off dates for the removal of Imperial measure for goods and foodstuffs weighed out or measured out by the retailer for the consumer. It was agreed to phase out the weighing by Imperial measure of such articles as meat, fish, and some poultry by 30 June 1981, other weighed-out food except fruit and vegetables by 30 September 1981 and fruit and vegetables by 31 December 1981, with petrol, diesel fuel, and paraffin being sold by the litre by 31 December 1981.[37] Draft Orders were prepared for all except petrol and paraffin but were not proceeded with because the Government 'was aware of the apprehension about the prospect of legal sanctions against retailers who did not apply metric units after the predescribed dates'.[38] The Report notes:

> Retailers' and consumers' organizations have always advised us that the best way of achieving an orderly change in the shops is to have legislation to phase out Imperial weights and measures in accordance with an agreed time-table. We are, therefore, disappointed that the Government has not felt able to continue with this course, but we intend to do all we can to achieve the change on the voluntary basis now proposed.

The chagrin of the Board can readily be imagined, especially as the Board knew full well that Commonwealth countries had to have recourse to legislation to effect the self-same changes. Indeed the Board had advised Government in 1974 that without legislation some retailers would not go metric.

In 1978, bowing to the EEC directive already mentioned, a number of Imperial units such as the chain, furlong, bushel, cental, etc. were no longer authorized for use by Statutory Instrument No.484 of 1978.

The Metrication Board, in its 'Report on the Change to Metric in the Retail Trade' to the Minister of State for Consumer Affairs in October 1979, pointed out that retailers of weighed out or measured out goods rejected the idea of voluntary metrication except for petrol. The next month, the Government announced it had no intention to issue further cut-off Orders. Throughout it had been the attitude of Government that metrication was to evolve on a voluntary basis. The Government made the decision to abolish the Metrication Board on 30 April 1980 when the terms of office of all the members and the chairman expired. The Board therefore submitted its 'Final Report July 1978 - April 1980' in 1980.

The Final Report, while recognizing that many sectors had gone metric, declared 'Nevertheless, taken as a whole, Britain is far from being wholly metric'.[39] Factually and fairly, the Board recorded the problems it had encountered; the voluntary approach by Government, with the Board in an advisory capacity only, the problems with the retail trade, the lack of metrication on road signs and speed limits, and surveys which showed that a majority of people were still not in favour of the transition to the metric system.

After a meeting of the Confederation of British Industry and the Metrication Board in 1979, it was agreed among others that the engineering industry had not yet attained the level of 75 per cent metric which had been the goal for 1975 and that there was a clear need for a more positive commitment to metrication by Government.[40] A survey of 800 firms showed that 50 per cent of the production was still in Imperial units.[40]

The Report drew attention to a new EEC Directive on Units of Measurement by which a decision on the future of the 'inch, foot, yard, mile, ounce, pound, pint, and gallon, does not have to be taken until 31 December 1989. The Directive thus reflects the Government's view that the rate of change to the use of SI (metric) units in the United Kingdom should not be determined by the Community'.[41]

Presumably someone feels there is yet time. Even now, in 1986, one does not sense much movement towards metrication. In the first prosecution of its kind (October 1980) a shopkeeper was fined for selling sugar and sea-salt in Imperial measure, a procedure made illegal in 1978, but the fine was only £1 and three hundred people petitioned in support of the shopkeeper, who pleaded not guilty. Moreover on 24 October 1980 it was reported in the media that the UK would not include the yard and square yard in the 'harmonizing' plan with the EEC, so presumably they will continue in use for some time.

The Weights and Measures Act, 1985

This Act, while being extensively regulatory and procedural, is of importance to our inquiry in that it set up a National Metrological Co-ordinating Unit to review the operation of the Act, to make available information, and to advise local weights and measures authorities, etc. (s.55), while the Schedules clearly indicated a move away from the traditional units.

Schedule 1 Part 1 defined the yard as 0.9144 m as before, but the metre is now defined as the distance light travels in a vacuum in $\frac{1}{299792458}$ of a second. Part III declares that 1 litre = 1 cubic decimetre, Part IV gives the gallon as 4.54609 cubic decimetres, and Part V says that 1 lb = 0.45359237 kg, and the Troy ounce = $\frac{12}{175}$ of a pound (N.B. no grains mentioned, see below). Part VI lists a large number of units which may not be used for trade, namely the furlong, chain, square mile, rood, square inch, cubic yard, cubic foot, cubic inch, bushel, peck, fluid drachm, minim, ton, hundredweight, cental, quarter, stone, dram, grain, pennyweight, apothecaries' ounce, drachm, scruple, metric ton = 1000 kg (but the tonne or metric tonne = 1000 kg shall be allowed for trade), and the quintal of 100 kg.

Schedule 2 Part I defines Baily's yard of 1845. Part II defines the platinum lb, 'PS 1844', Parts III and IV define the British copies of the prototype metre (No. B16) and prototype kilogram (No. 18). Part V describes the authorized copies of the Standards: the yards of 1845 and the pounds of 1844 in platinum deposited at the Mint, the Royal Society, the Royal Observatory, Greenwich, and the immured standards in the Houses of Parliament, as well as the yard cast in 1878 and the platinum-iridium pound 'PC5, 1879' at the National Weights and Measures Laboratory of the Department of Trade and Industry.

Schedule V deals with solid fuel which is to be sold by weight in units of 7, 14, 28, 56, 112, or 140 lb or 25 or 50 kg or any multiple of 50 kg, with the Secretary of State being empowered to exclude areas of Scotland in which regions it may be sold by volume in quantities of 0.2 cubic metres or a multiple of 0.2 cubic metres.

Some of the units now disallowed had been excluded in previous Acts but other traditional units have been added to the list, and while the grain has gone after who knows how many centuries, the pound, gallon, and yard are still with us. The Troy ounce is now given as a fraction of the Avoirdupois pound as grains have disappeared.

It would not take a great deal now to complete the move to the metric system, apart from public opposition and the reluctance of some areas of industry to change.

A Personal Comment

Talking to people in the UK in 1982, 1983, and 1985, the impression was gained that somehow the movement to metricate completely had simply run out of steam, and that while some continuing progress was being made it was so slight that little headway was being observed. Yet petrol moved to being sold by the litre in 1982 although the price per gallon was also being displayed. Many grocery items were now in metric packs. Close examination revealed many items to be imports from various European countries which, being pre-packaged in the country of origin, naturally came in metric sizes.

It is, on the face of it, anomalous that Britain, whose major trading partners use metric units, should be hesitating so long before taking the final plunge while Canada, whose major trading partner is the largely non-metric USA, should have rushed ahead in an endeavour to go metric on all fronts with change being mandatory. This resulted in Canada being now farther ahead than Britain though it started later. Road signs are all in kilometres and gasoline (petrol) is sold by the litre. By August 1982 it was exceedingly rare to find a notice giving the price per gallon. Automatic scales frequently gave the weight in kilograms and in pounds so that a pound of something could still be bought, though the sales slip recorded the weight in grams.

But then a reaction set in.

Some Conservative Members of Parliament acquired a petrol station and began selling by the gallon, defying the then Liberal Government to prosecute. The Government announced in 1984 its intention to dissolve the Metric Commission of Canada and this was carried out by the new Conservative Government, thus removing the pressure to go metric. The result is a frozen immobility which is now producing the strangest anomalies. A colleague in November 1985 wished to build an electric fence. He was asked by the supplier how many *chains* of wire he required. His property is a rectangular area, one side of which is bounded by a provincial road which has been surveyed in metres, the other side is bounded by a municipal road which is still using the old survey in yards. His survey report advised him that he owned so many *meter-yards* of land. Presumably this hybrid unit of area represents a rectangle whose adjacent sides are respectively one metre and one yard long.

School-children in both Britain and Canada are being taught in metric notation. They are going to be confused when they encounter a dual system with little logic to it. The Canadian scene will be particularly unfortunate, having gone so far and stopped so close to completion. A recent (November 1985) letter to the editor of the *Winnipeg Free Press* lists the following anomalies among others; prices per pound in supermarkets, lawn mowers with fuel tank capacities in pints, carpeting advertised as so much per square meter but sold by the square yard. Clearly this situation cannot be tolerated for long.

One cannot but wonder if it will be the EEC directives which finally push the UK totally into the metric camp. The deadline of 31 December 1989 will be awaited with interest if no further definitive moves to metricate occur in the interval. One can but recall a statement from the Final Report of the British Metrication Board 1980, fifteen years after the decision to go metric, that 'a majority of people are not in favour of changing' (Section 1.21).

But there is what might be called the moral issue involved. If it is true, as we are told, that there is no turning back, why not bite the bullet and get it over with? The Imperial system was an honourable system, infinitely better than that which went before. It deserves an honourable demise. It should not simply be bled to death over a long period of time, to be finally dumped when some pressure, internal or external, becomes too great. Its obsequies should be attended with proper ceremony.

And what if she had seen those glories fade,
Those titles vanish and that strength decay;
Yet shall some tribute of regret be paid
When her long life hath reach'd its final day:
Men are we, and must grieve when even the shade
Of that which once was great is pass'd away.

William Wordsworth:
'On the Extinction of the Venetian Republic.'

NOTES AND REFERENCES: CHAPTER XIV

1. *Report of the Committee on Weights and Measures Legislation,* (the Hodgson Report) *Parl. Papers* 1950–51 (Cmd 8219), xx, p.913.
2. The Report goes unmentioned in the *Encyclopaedia Britannica* Year Books for 1950–53.
3. Sanction was given by Act of Congress 1866 for the relation
   1 m = 39.370000 inch
   and by the later Mendenhall Order this ratio is now used to define the American inch in terms of the metre and not vice versa, i.e. the American inch is 25.400051 mm.
4. See also SEARS, J. E. The Standards of Length *Proc. Roy. Soc.* (1946) Vol.CLXXXVI, pp.152–64. See p.159.
5. Ibid, p.157.
6. The National Physical Laboratory (N.P.L.) was set up in 1900, to serve much the same function on a national scale that the Bureau International de Poids et Mesures at Sèvres does on an international scale. The N.P.L. enjoys a well deserved reputation for excellence throughout the entire range of standards work.
7. *Units and Standards of Measurement employed at the National Physical Laboratory I — Length, Mass, Time, Volume, Density and Specific Gravity, Force and Pressure:* p.4. London: Dept. of Scientific and Industrial Research 1951.
8. GOULD, F. A. Standards of Mass *Proc. Roy. Soc.* (1946) Vol.CLXXXVI, pp.171–179. See p.174.
9. Ibid, p.175.
10. Repealed by the Weights and Measures (Metric System) Act 1897 (60 and 61 Victoria c 46 s 2(2)) which left the table of equivalents to Order in Council.
11. Reference 1, p.13.
12. Statutory Rules and Orders, 1898 No.411.
13. The Weights and Measures Act 1963, Elizabeth II c 31 1963. *Public General Acts and Measures, 1963,* p.500, London: HMSO 1963.
    An invaluable commentary on this Act is given by O'KEEFE, John Alfred *The Law of Weights and Measures* London: Butterworth 1966.
14. The ancient carat was equated with 4 grains and is so defined in Johnson's *Dictionary* of 1755. The more modern carat used for weighing diamonds and precious stones was somewhat variable but in 1877 merchants in London, Paris, and Amsterdam settled on a weight of 205 mg. The metric carat by which all international trading in precious stones is carried out today is exactly 200 mg. This use of the word should not be confused with that for describing the fineness of gold. Pure gold is considered to be 24 carats fine. Eighteen carat gold is $\frac{18}{24}$, or $\frac{3}{4}$ gold and $\frac{1}{4}$ alloy.
15. O'KEEFE, John Alfred. *The Law of Weights and Measures* p.135. London: Butterworth 1966.
16. Reference 13, Sections 59 and 60 respectively.
17. O'KEEFE, John Alfred. Loc. cit. p.324.
18. Ibid, p.332.

19. Yet it does not appear to have been abolished, in spite of the interpretation of Reference 18, for it is given in the Schedule of lawful measures in the 1976 Weights and Measures Act. (See p.305 for S.I. No. 484 of 1978.)
20. *Report of the Committee of Inquiry on Decimal Currency* (The Halsbury Committee) *Parl. Papers* 1962-63 (Cmnd 2145) xi, p.195.
21. Elizabeth II c 47. London: HMSO 1967 *Public General Acts and Measures of 1967* Pt.1, p.1019. The Decimal Currency Act 1967.
22. Elizabeth II c 19. London: HMSO 1969. *Public General Acts and Measures of 1969* Pt.1, p.151. Decimal Currency Act 1969.
23. *Decimal Currency Board Third Annual Report 1969/70* pp.2-3 and 10-11. London: HMSO.
24. *Decimal Currency Board. Decimal Currency: Three Years to Go* London: HMSO 1968.
25. E.g. *The Times* of London, 6 January 1971.
26. *Hansard, House of Commons* Vol.CCCXIII, Col.32-33 (W) 1965.
27. *Change to the Metric System in the United Kingdom.* Report by the Standing Joint Committee on Metrication. London: HMSO 1968.
28. *Final Report of the Metrication Board* London: HMSO 1980.
29. *White Paper on Metrication* February 1972. *Parl. Papers* 1971-72 (Cmnd 4880) xxii, p.817 (see Section 92 p.23).
30. Ibid, Section 5 p.1.
31. *Going Metric — The Next Phase.* The Fifth Report of the Metrication Board. London: HMSO 1974.
32. *Going Metric No.13* October 1974. London: HMSO. The Metrication Board.
33. *A Metric America. A decision whose time has come.* US Government Printing Office Washington, D.C. July 1971
34. *E.E. Council directive on Units of Measurement.* Directive 71/354/EEC of 18.10.71 (Journal Officiel L 243 of 29.10.71 p.29). This directive is reproduced in Annex 1 to the White Paper on Metrication (Reference 29 above, p.29).
35. Elizabeth II c 77. London: HMSO 1976 *Public General Acts and Measures of 1976.* Part II, p.1895. The Weights and Measures Act 1976.
36. *Going Metric; progress in 1977/78. Report of the Metrication Board for the period January 1977 to June 1978.* p.7. London: HMSO.
37. Ibid, p.7.
38. Ibid, p.8.
39. *Final Report of the Metrication Board* p.7. London: HMSO 1980.
40. Ibid, p.18.
41. Ibid, pp.19-20.

APPENDIX A

(a) *Assisa Panis et Cervisie;* Assize of Bread and Ale; Statutes of the Realm I pp.199-200.
(b) *Judicium Pillorie;* The Judgment of the Pillory; Ibid pp.201-2.
(c) *Statutum de Pistoribus;* Statute concerning Bakers; Ibid pp.202-4.
(d) *Tractatus de Ponderibus et Mensuris;* The Assize of Weights and Measures; Ibid pp.204-5.
(e) *Statutum de Admensuratione Terre;* Statute for the Measuring of Land; Ibid pp.206-7.

(a) *Assisa Panis et Cervisie;* Assize of Bread and Ale; Attributed to 1266.

When a Quarter of Wheat is sold for xii*d.* then Wastel Bread [made with fine flour] of a Farthing shall weigh vi *l.* and xvj*s.* But Bread Cocket [inferior bread] of a Farthing of the same Corn and Bultel [sieve], shall weigh more than Wastel by ii*s.* And Cocket Bread made of Corn of lower Price, shall weigh more than Wastel by v*s.* Bread made into a Simnel [boiled and then baked] shall weigh ii*s.* less than Wastel. Bread made of the whole Wheat shall weigh a Cocket and a half, so that a Cocket shall weigh more than a Wastel by v*s.* Bread of Treet [coarse, cheap bread] shall weigh ii Wastels. And Bread of common Wheat shall weigh Two great Cockets.

When a Quarter of Wheat is sold for xviii*d.* then Wastel Bread of a Farthing white and well baked shall weigh iv *l.* x*s.* viii*d*

When for ii*s.*	iii *li.* viii*s.*
When for ii*s.* vi*d.*	liv*s.* iv*d.* *ob.* *q.* [*ob.* = $\frac{1}{2}d$, $q = \frac{1}{4}d$]
When for iii*s.*	xlviii*s*
When for xix*s.*	7*s* 2*d*
When for xix*s.*vi*d.*	6*s* 11*d.*$\frac{1}{2}$
When for xx*s.*	6*s* 9*d.*$\frac{3}{4}$

And it is to be known, that then a Baker in every Quarter of Wheat, as it is proved by the King's Bakers, may gain iv*d.* and the Bran, and Two Loaves for Advantage, for Three Servants 1*d.* *ob.* for Two Lads *ob.*, in Salt *ob.*, for Kneading *ob.*, for Candle *q.*, for Wood ii*d.*, for his Bultel *ob.*

When a Quarter of Wheat is sold for iii*s.* or iii*s.* iv*d.* and a Quarter of Barley for xx*d.* or ii*s.* and a Quarter of Oats for xvi*d.* then Brewers in Cities ought and may well afford to sell Two Gallons of Beer or Ale for a Penny, and out of Cities to sell iii or iv Gallons for a Penny. And when in a Town iii Gallons are sold for a Penny, out of a Town they ought and may sell four; and this Assise ought to holden throughout all England.

⋆And if a Baker or Brewer be convicted that they have not kept the foresaid Assises, the First, Second, and Third Time they shall be amerced [fined], according to the Quantity of their Offence; and that as often as a Baker shall offend in the Weight of a Farthing Loaf of Bread not above ij*s.* Weight, that then he be amerced as before is said; but if he exceeds ij*s.* then is he to be set upon the Pillory without any Redemption of Money.

In like Manner shall it be done if he offend oftentimes and will not amend, then he shall suffer the Judgment of the Body, that is to say, the Pillory if he offend in the Weight of a farthing loaf under two shillings weight as is aforesaid. Likewise the Woman Brewer shall be punished by the Tumbrell, Trebuchit [ducking stools], or Castigatorie [instrument of chastisement], if she offend divers Times and will not amend.*

*These sentences are not in *Lib. Horn;* nor in *Rot. Pat.* 2 Ric. II, the latter of which concludes in the following Manner:

> The Assize of Bread (as it is contained in a Writing of the Marshalsey [court of a knight-marshal] of our Lord the King delivered unto them) may be holden, according to the Price of Wheat, that is to say, as well Wastel, as other Bread of the better, second, or third sort, shall be weighed, as is aforesaid, by the middle Price of Wheat; and the Assise or Weight of Bread shall not be changed but by Sixpence increasing or decreasing in the Sale of a Quarter.
>
> By the Consent of the whole Realm of England, the Measure of our Lord the King was made; that is to say, That an English Penny, called a Sterling, round and without any clipping, shall weigh xxxii Wheat Corns in the midst of the Ear; and xx*d* do make an ounce; and xii Ounces one Pound; and viii Pounds do make a Gallon of wine; and viii Gallons of Wine do make a London Bushel; which is the Eighth Part of a Quarter.
>
> Forasmuch as in our Parliament holden at Westminster, in the First Year of our Reign, we have granted that all good Statutes and Ordinances made in the Times of our Progenitors aforesaid, and not revoked, shall be still held, we have caused at the Request of the Bakers of our Town of Coventry, that the Ordinances aforesaid, by tenor of these Presents, shall be exemplified. In Witness whereof, &c. Witness the King at Westminster, the xxii Day of March.

(b) *Judicium Pillorie; The Judgment of the Pillory; Attributed to 1266.*

If a Baker or a Brewer be convict, because he hath not observed the Assise of Bread and Ale, the First, Second, and Third Times, he shall be amerced according to his Offence, if it be not over grievous; but if the Offence be grievous and often, and will not be corrected, then he shall suffer Punishment of the Body, that is to wit, a Baker to the Pillory, and a Brewer to the Tumbrel, or some other Correction.

First, Six lawful Men shall be sworn truly to gather all Measures of the Town, that is to wit, Bushels, Half and Quarter Bushels, Gallons, Pottles [half gallons], and Quarts, as well of Taverns as of other Places; Measures and Weights, that is to wit, Pounds, Half Pounds, and other little Weights, wherewith Bread of the Town or of the Court is weighed; that is to say, one Loaf of every sort of Bread. And upon every Measure, Bushel, Weight, and also upon every Loaf, the Name of the Owner distinctly written; and likewise they shall gather the Measures of Mills. After which Thing done, Twelve lawful Men shall swear to make true Answer to all such Things as shall be demanded of them in the King's Behalf upon Articles here following; and such Things as be secret, they shall utter secretly, and answer privately. And the Bailiffs shall be commanded to bring in all the Bakers and Brewers with their Measures, and all Things under written.

First, they shall inquire the Price of Wheat, that is to wit, how a Quarter of the best Wheat was sold the last Market Day, and how the Second Wheat, and how the Third; and how a Quarter of Barley and Oats.

After, how the Baker's Bread in the Court doth agree, that is to wit Wastel and other Bread after Wheat of the best, or of the Second, or of the Third Price.

Also upon how much Increase or Decrease in the Price of Wheat a Baker ought to change the Assise and Weight of his Bread.

Also how much the Wastel of a Farthing ought to weigh, and all other Manner of Bread, after the Price of a Quarter of Wheat that they present.

And for default in the Weight of the Bread, a Baker ought to be amerced, or to be judged unto the Pillory, according to the Law and Custom of the Court.

Also, if any Steward or Bailiff, for any Bribe, doth release Punishment of the Pillory and Tumbrel, being already judged, or to be judged of Right.

Also if they have in the Town a Pillory of convenient Strength, as appertaineth to the Liberty of their Market, which they may use, if need be, without bodily Peril either of Man or Woman.

After, they shall inquire of the Assise and Price of Wine, after the Departure of the Justices in Eyre [in circuit], or of them that were last in Office of the Market of the Town; that is to say, of the Vintners Names, and how they sell a Gallon of Wine: And if any corrupted Wine be in the Town, or such as is not wholesome for Man's Body.

Also of the Assise of Ale in the Court of the Town how it is, and whether it be observed; and if not, how much Brewers have sold contrary to the Assise; and they shall present their Names distinctly and openly, and that they be amerced for every Default or to be judged to the Tumbrel, if they sell contrary to the Assise.

Also if there be any that sell by one Measure, and buy by another. Also if any do use false Ells, (or 'false yards'; '*falsis ulnis*'), Weights or Measures.

And if any Butcher do sell contagious [noxious] Flesh, or that died of the Murrein [plague]. Also they shall inquire of Cooks that seethe Flesh or Fish with Bread or Water, or any otherwise, that is not wholesome for Man's Body, or after that they have kept it so long that it loseth its natural Wholesomeness, and then seethe it again, and sell it: or if any do buy Flesh of Jews, and then sell it to Christians.

And also Forestallers [those who intercept goods on their way to market], that buy any Thing afore the due and accustomed Hour, against the good State and Weal of the Town and Market, or that pass out of the Town to meet such Things as come to the Market, being out of the Town, to the Intent that they may sell the same in the Town more dear to Regrators [those who 'corner' a market], that utter it more dear than they would that brought it, in case they had come to the Town or Market.

When a Quarter of Barley is sold for Two Shillings, then Four Quarts of Ale shall be sold for a Penny; when for Two Shillings Sixpence, then Seven Quarts of Ale shall be sold for Two-pence; when for Three Shillings, then Three Quarts for One Penny; when for Three Shillings Sixpence, then Five Quarts for Two-pence; when it is sold for Four Shillings, then Two Quarts at One Penny. And so from henceforth the Prices shall increase and decrease after the Rate of Sixpence.

(c) *Statutum de Pistoribus; Statute concerning Bakers; Attributed to 1266. Here beginneth the Rule for punishment of the Infringers of the Assise of Bread and Ale, Forestallers, Cooks, &c,*

The Assize of Bread shall be kept, according as it is contained in the Writing of the Marshalsey [court of a knight-marshal] of our Lord the King, delivered unto them, after the Sale of Wheat, that is to wit, the better, the worse, and the worst. And as well Wastel Bread, as other of what sort soever they be, shall be weighed, according as it is said, of the Sale of the meaner Wheat: Neither shall the Assise or Weight of Wheat be changed more than Six-pence increasing or decreasing, as it is in the Sale of the Quarter.

A Baker, if his Bread be founden a Farthing Weight lacking in Two Shillings Six-pence, or under, shall be amerced; and if it pass the same Number he shall suffer Punishment of the Pillory, which shall not be remitted to the Offender either for Gold or Silver. And every Baker shall have a mark of his own for his Bread.

Every Pillory or Stretchneck must be made of convenient Strength; so that Execution may be done upon Offenders without Peril of their Bodies.

The Toll of a Mill shall be taken, according to the Custom of the Land and, according to the Strength of the Water-course, either to the Twentieth or Four and twentieth Corn. And the Measure whereby the Toll must be taken shall be agreeable to the King's Measure, and Toll shall be taken, by the Rase [smoothed level], and not by the Heap or Cantel [mound]. And in case that the Fermors [ecclesiastical landowners] find [procure from] the Millers their Necessaires [their dues], they shall take nothing besides their due Toll; and if they do otherwise, they shall be grievously punished.

The Assise of Wine shall be kept, that is a Sextertium [gallon] at Twelve Pence; and if the Taverners exceed the same Assise, their Doors shall be shut up.

The Assise of Ale shall be assessed, proclaimed, and kept according to the Price of the Corn whereof the Malt is made. And the Brewer shall not increase more in a Gallon, but according to the Rate of Sixpence rising in a Quarter of Malt. And if he break the Assise the First, Second, and Third Time, he shall be amerced; but the Fourth Time he shall suffer Judgement of the Pillory without Redemption.

A Butcher that selleth Swines Flesh meazled [infected], or Flesh dead of the Murrain, or that buyeth Flesh of Jews, and selleth the same unto Christians, after he shall be convict thereof, for the first Time, he shall be grievously amerced; the Second Time he shall suffer Judgement of the Pillory; and the Third Time he shall be imprisoned and make Fine; and the Fourth Time he shall forswear the Town. And in this Manner shall it be done of all that offend in like Case.

The Standard of Bushels, Gallons, and Ells, [or yards, 'ulne'], shall be sealed with an Iron Seal of our Lord the King, and safe kept, under the Pain of a Hundred Pound. And no measure shall be in any Town unless it do agree with the King's Measure, and marked with the Seal of the Shire Town. If any do sell or buy by Measures unsealed, and not examined by the Mayor or Bailiffs, he shall be grievously amerced. And all the Measures of every Town, both great and small, shall be viewed and examined twice in the Year. If any be convict for a Double Measure, that is to wit, a greater for to buy with, and a smaller to sell with, he shall be imprisoned for his Falshood, and shall be grievously punished.

The Standard, Bushels and Ells [or yards, 'ulne'], shall be in the Custody of the Mayor and Bailiffs, and of Six lawful Persons of the same Town being sworn, before whom all Measures shall be sealed.

No Manner of Grain shall be sold by the Heap or Cantle, except it be Oats, Malt, and Meal.

But especially be it commanded on the Behalf of our Lord the King, that no Forestaller be suffered to dwell in any Town, which is an open Oppressor of Poor People and of all the Commonalty, and an Enemy of the whole Shire and Country, which for Greediness of his private Gain doth prevent others in buying Grain, Fish, Herring, or any other Thing to be sold coming by Land or Water, oppressing the Poor, and deceiving the Rich, which carrieth away such Things, intending to sell them more dear; the which come to Merchants Strangers [foreign] that bring Merchandize, offering them to buy, and informing them that their Goods might be dearer sold than they intended to sell; and an whole Town or a Country is deceived by such Craft and Subtilty: He that is convict thereof, the first Time shall be amerced, shall lose the Thing so bought, and that according to the Custom and Ordinance of the Town; he that is convict the Second Time shall have Judgement of the Pillory; at the Third Time he shall be imprisoned and make Fine; the Fourth Time he shall abjure the Town. And this Judgement shall be given upon all Manner of Forestallers; and likewise upon them that have given them Counsel, Help, or Favour.

And if any presume to sell the Meal of Oats adulterated, or in any other deceitful manner, for the first offence he shall be grievously punished; for the second he shall lose all his Meal; for the third he shall undergo the judgement of the Pillory; and for the fourth he shall abjure the Town.

All the Things before written shall be observed by command of the King, so that if any, great or small, shall presume to contravene the before written Statutes, in any Thing by Word, Counsel, Help, or Favour he shall be apprehended as a contemner of the King's Commandments, and imprisoned, and shall not be delivered out of Prison, until he be delivered by the command and the express Writ of the King. And the present Schedule shall be delivered to the Mayor and Bailiffs and Six lawful Men of the Town sworn, together with the Standard Bushel, Gallon, Yard, and Stone, to be observed; and when need shall be, they may be certified by the same Schedule.

(d) *Tractatus de Ponderibus et Mensuris; The Assise of Weights and Measures; Attributed to 1302/3.*

By Consent of the whole Realm the King's Measure was made, so that an English Penny, which is called the Sterling, round without clipping, shall weigh Thirty-two Grains of Wheat dry in the midst of the Ear; Twenty-pence make an Ounce; and Twelve Ounces make a Pound and Eight Pounds make a Gallon of Wine; and Eight Gallons of Wine make a Bushel of London; which is the Eighth Part of a Quarter.

A Sack of Wool ought to weigh Twenty-eight Stone, that is Three hundred and fifty Pounds, and in some Parts Thirty Stone, that is Three hundred and seventy-five Pounds, and they are the same according to the greater or lesser Pound; Six times Twenty Stone, that is fifteen hundred Pound, make a Load of Lead, to wit the great Load of London, but the Load of the Peak is much less.

The Load of Lead doth consist of Thirty Formels [fotmals], and every Formel containeth Six Stone, except Two Pound; and every Stone doth consist of Twelve Pound, and every Pound consisteth of the Weight of Twenty-five Shillings, whereby the Sum in the Formel is Seventy Pound. But the Sum of the Stones in the Load is Eight Times Twenty and Fifteen, and it is proved by Six Times Thirty which is Nine Times Twenty. But of every Formel there are abated Two Pound in the aforesaid Multiplication, which are Sixty, which make Five Stone. And so there are in the Load Eight Times Twenty and Fifteen as is aforesaid. According to some other, it consisteth of Twelve Weights,* and this is after Troy Weight. And the Sum of Stones in the Load is Eight Times Twenty and Eight Stones, and it is proved by Twelve Times Fourteen. There is a Weight,* as well of Lead as of Wool, Tallow, and Cheese, and weigheth Fourteen Stone. And Two Weights* of Wool make a Sack, and Twelve Sacks make a Last. But a Last of Herrings containeth ten thousand, and every Thousand containeth Ten hundred, and every Hundred six score. A Last of Leather doth consist of Twenty Diker, and every Diker consisteth of Ten Skins. And a Diker of Gloves consisteth of Ten Pair of Gloves. Item a Diker of Horse-shoes doth consist of Ten Shoes. Item a Dozen of Gloves, Parchment, and Vellum in their Kinds contain Twelve Skins, and Twelve Pair of Gloves. Item a Hundred of Wax, Sugar, Pepper, Cinnamon, Nutmegs, and Allum, containeth Thirteen Stone and a Half, and every Stone Eight Pound. The Sum of Pounds in a Hundred One hundred and eight

*or Weighs

Pounds; and the Hundred consisteth of Five Times Twenty, and every Pound of Twenty-five Shillings. Item it is to be known, that the Pound of Pence, Spices, Confections, as of Electuaries, consisteth in weight of Twenty Shillings. But the Pound of all other Things weigheth Twenty-five Shillings. Item of Electuaries and Confections the Pound containeth Twelve Ounces, and an Ounce hereof is of the Weight of Twenty-pence. Item a Hundred of Canvass, and Linen Cloth consisteth of One hundred Ells [or yards, 'C ulnis'], and every hundred containeth Six Score. But the hundred of Iron and Shillings consisteth but of Five Score. The Seeme of Glass containeth Twenty-four Stone, and every Stone Five Pound. And so the Seeme containeth Six score Pound. The Dozen of Iron consisteth of Six Pieces. A Bind of Eels consisteth of Ten Stikes, and every Stike Twenty-five Eels. But the Bind of Skins consisteth of Thirty-three Skins. A Timber of Coney-Skins and Grayes consisteth of Forty Skins. A Chef of Fustian consisteth of Fourteen Ells [or yards, 'XIIII ulnis']. A Chef of Sindon containeth Ten Ells [or yards]. A Hundred of Garlike consisteth of fifteen Ropes, and every Rope containeth fifteen Heads. A Hundred of Hard Fish is Eight Score.

(e) *Statutum de Admensuratione Terre;* Statute for the Measuring of Land; Attributed to 1305.

The Manner of admeasuring Land, Meadow, Wood, Pasture, Heath Land, and the like.

Hereunder mention is made of the measuring of Land, and of the Yard of our Lord the King.

When an Acre of Land containeth Ten Perches in Length, then it shall be in Breadth Sixteen Perches. When Eleven in Length, then Fourteen Perches and a half, and one Foot and five Inches. When Twelve in Length, then Thirteen Perches, one quarter of a Perch, one Foot, and one Inch and a half.

[Then follow corresponding entries from 13 perches in steps of one perch to 80 perches in length.]

When Eighty in Length, then in Breadth Two Perches.

And Be it Remembered, That the Iron Yard of our Lord the King, containeth three Feet and no more. And a Foot ought to contain Twelve Inches, by the right measure of this Yard measured; to wit, the Thirty-sixth Part of this Yard rightly measured maketh one Inch, neither more nor less. And Five Yards and a half make one Perch, that is Sixteen Feet and a half, measured by the aforesaid Iron yard of Our Lord the King.

[The following short Article, intitled, 'Compositio Ulnarum & Perticarum', is inserted in MS. Cott. Claudius D. II. fo. 241 b. and in some of the Old Printed Copies: The translation is from Rastall's Collection; 1603.]

It is Ordained that three Grains of Barley dry and round do make an Inch; Twelve Inches make a Foot; Three Feet make a Yard; Five Yards and a Half make a Perch; and Forty Perches in Length and Four in Breadth make an Acre.

APPENDIX B

Regulation & Enforcement[1]

We know that, from the days of Edgar, the king maintained standards at his capital by which all others were supposed to be sized. Where a municipal standard was so sized, it was carried off to its new home in town or city where replicas were made. In mediaeval England these were often far from accurate and in turn were used to make even more copies of even more dubious validity. Thus the role of the Crown was restricted to the provision of what we would now call the primary standards with local authorities entrusted with the propagation of these standards. Over and over again we hear the call that all measures should be sealed with the king's seal, which tells us unequivocally that a considerable number of measures were in use which had no seal at all. These were measures of doubtful parentage, and not everything that carried a seal was an accurate copy of the measure whose name it bore.

Some of the assizes or inspections were to be conducted by 'lawful men' (*'legales homines'** as the document *Judicium Pillorie*[2] (attributed to 1266) and Fleta[3] (1290) called them), others by the magistrates, still others by groups such as the master and wardens of a livery company of London. The 'lawful men' belonged to town or city as did the magistrates, and the livery companies were part of the life of the City of London. There were no rural representatives even though the rural population exceeded the urban in the Middle Ages and, while there were urban regulations and assizes to be satisfied, the country districts were free of them until the jurisdiction of the lord of the manor was encountered. A large part of the difficulty in achieving uniformity was the number of statutory exclusions to any Bill intended to promulgate the size of a measure or proclaim the applicability of such and such a measure. The 'rents and ferms of the lords' were frequently excluded from any such regulation, as were the dues owed to the king, all of which were to continue using the old customary measures, and as the measures of each manor could be and often were different from one another, the number of such measures was legion. Moreover, in mediaeval England, monasteries and ecclesiastical institutions could and frequently did have their own weights and measures. For example, in the *Tractatus de Ponderibus et Mensuris,*[4] attributed

*Inspection by *legales homines,* or annoyance or nuisance juries as they were called, continued in four London parishes (Paddington, St. Marylebone, St. Pancras, and St. Mary's, Islington) until surrendered under Section 55 of the Weights and Measures Act 1878. They are specifically mentioned in the Fourth Report of the Commissioners appointed to Enquire into the Condition of the Exchequer Standards, 1870 (*Parl. Papers* 1870, xxvii, p.249).

to 1302, we read of the 'weigh' (wey) of cheese being 14 stone. At 13 lb to the stone this is 182 lb. Other contemporary documents[5] say 14 stone each of 12½ lb, that is 175 lb. In Register B of Canterbury Priory[6] of the early fourteenth century we read:

> The weigh of cheese according to the king's weight contains 26 great pounds and each great pound contains 7 small pounds. And the small pound contains 25 shillings sterling. The weigh of cheese according to Lanfranc's weight contains 32 great lb.

Here the small pound is the *libra mercatoria* of 15 oz Troy (25/-): the great pound is the 7 lb clove. By the king's weight (Exchequer) the weigh is 182 *librae mercatoria*. By Lanfranc's[7] weight it is 224. Then we find a century later the king being petitioned,[8] as cheese had always been weighed by the auncel and that had been abolished, to declare what the weigh of cheese was to be by the balance and the king assented to the proposed weight, namely 32 cloves each of 7 lb or 224 lb, just the weight of the weigh according to Lanfranc. The weigh of 224 lb continued, being so mentioned at the end of the sixteenth century[9] and again at the beginning of the nineteenth.[10]

On occasion a manor belonging to a church would pay the farm in kind either by customary or by king's measure as for St. Paul's Cathedral, 1283, where the unit was to be 18½ quarters at 7 bushels to the quarter (an anomalous quarter) or 16 quarters at 8 bushels of the king's measure.[11] The two levies are almost the same, the one being 129½ bushels, the other 128. The two descriptions were no doubt intended to represent the same amount of grain, nearly enough for the purposes of the time.

The Leet Courts

Because weights and measures were so local an affair, it is not surprising to find the earliest recorded inspections being those carried out by juries under the auspices of the leet courts, the manorial courts presided over by the lord of the manor, or, more frequently, by his steward. The authority to hold such courts was usually given by royal charter and these courts were empowered to maintain the king's peace. Every resident male over twelve years of age was required to swear allegiance to the king before the court and had to find sureties (pledges) for the correctness of statements made in evidence to that court. For those reasons the leet court was sometimes called the View of Frankpledge *(leta visus francii plegii)*. In the Statute Book we find an enactment so entitled dated 1325[12] which gives to these courts jurisdiction over;

> 24. Of the Assize of Bread and Ale broken.
> 25. Of false measures, as of bushels, gallons, yards, and ells.

26. Of false balances and weights.
27. Of such as have double measure, to buy by the great and sell by the less,[13]

among many other things ranging from housebreaking to rape. The leet courts continued into the nineteenth century with much the same powers. The Astronomer Royal (Mr Airy) in a written submission to the Select Committee on Weights and Measures in the mid-nineteenth century declared[14] 'The Government now takes no part either in supplying the standards or in compelling local bodies to supply themselves'. Mr Airy went on to point out that, while the Government takes much care over the coinage, 'a want of precision in the measure of commodities is exactly equivalent to a want of precision in the measure of gold. Who would remit the power of coining to a leet court?'.

The jurisdiction of the lord of the manor and his leet court, while lasting into the nineteenth century, did not continue into the twentieth. Indeed, after 1835 manorial jurisdiction progressively declined and the last manor held by normal succession, that of Wakefield, was dissolved in 1892. (But see p.49 for the Manor of Laxton.) The manor had been held originally by the Dukes of York. When Richard, Duke of York, was defeated and slain at the battle of Wakefield in 1460 during the Wars of the Roses, the title reverted to his son Edward who assumed the kingship as Edward IV following the Yorkist victory in 1461. Thereafter the manor was in the hands of the Crown until, enabled by the Act 55 and 56 Victoria c 18,[15] the County Council of the West Riding of Yorkshire purchased the franchise of weights and measures in 1892.

*The Clerk of the Market**

While local officials were very much on part-time short-term appointments, there was one officer whose position continued. He was the Clerk of the Market, an office which was in existence for several centuries. The principal clerk was that of the King's Household but each shire and city had its own. The Clerk of the King's Household maintained the royal standard measures and was to compare those in use throughout the kingdom with his own royal standards. Fleta (1290) tells us:[16]

> Of the Office of Clerk of the Marshalcy
> There is committed to the charge and custody of a clerk or layman the king's measures, which are taken as the standard and pattern measures of the realm, that is to say, of yards†, gallons, weights, bushels, and the like, and he should be

*For a description of the role of Clerk of the Market of the Royal Household in Middlesex from the thirteenth to the sixteenth century, see WILLIAMS, N. *Transactions of the London and Middlesex Archaeological Society*, Vol.XIX, Pt.2. 1957 London.

†The Latin reads '*ulnarum*'. The word *ulna* was used for both yards and ells. The editors translate the word as 'ells', but at this time 'yard' is more likely to be correct.

> experienced in (the administration) of the assizes of bread, wine, measures, and ale so that he may fully understand them . . .
> Furthermore the clerk of the marshalcy ought to know the profit a baker can make on every quarter of wheat, and the particulars of the assize of ale.
> Furthermore the clerk of the marshalcy should understand the origin and characteristics of weights and measures so that he can well and truly know the amount of corn contained in a gallon-measure and in a bushel-measure.

Thus in 1321 the 'keeper of the City of London' and the aldermen and sheriffs of the same were required to show to the Clerk of the King's Market the standards of London, the measures for wine, ale, and corn, so that the 'king may cause other measures to be made throughout the realm' based on the assayed standards. The king had 'learned from frequent complaints that divers merchants and others use false measures'.[17]

Just as the clerk could examine municipal and city weights and measures, so the local Clerk of the Market could be subject to scrutiny by the king's commissioners. A visitation of 1281 is mentioned in the Domesday Book of Kent (L. B. Larking, (ed. & tr.) 1869) while we find on 3 November 1324[18] a major scrutiny being authorized in no fewer than forty counties and ridings. To each, two (or three) commissioners were named, their mandate being to survey all measures of wine, ale, and corn, to burn false measures, and to punish those using such measures. Although the warrant ended with the king expressing his unwillingness that the work of the Clerk of the Market should be impeded, there is little doubt that the arrival of the visitors was met with unveiled resentment by the clerk and the local officials. Until shown otherwise, the implication was that the measures of the clerk were faulty or, if correct, merchants were using their own fraudulent measures, or both. Not surprisingly, objections were made to the king concerning these visitations. For example, in 1326[19] the citizens of Norwich expressed their displeasure in that (1) the city's charter gave the ordering of the city's weights and measures to the bailiffs unless they default, (2) the city possessed standard measures, and (3) the bailiffs had punished transgressors. Now two commissioners, Robert Baynard and Simon de Hedersete, intended entering the city to examine weights and measures, to enquire the names of those using false measures, and to punish them 'to the prejudice of the citizens and contrary to the charter'. Clearly a case of double jeopardy was feared, but the king (Edward II) was adamant. Robert and Simon were ordered to the city and, if things were found in order, with correct measures and the guilty punished, 'they are then not to intermeddle further with the measures', but if they find the bailiffs to be negligent 'they [Robert and Simon] are to certify to the king of what they shall find so that he may cause to be done herein what he shall see to be just and reasonable'.

In 1340 steps were taken by statute[20] to improve the availability of measures sized to the king's standard. After reciting that

> none shall sell by the bushel if it be not marked by the King's seal, and that it be according to the King's Standard; and also it is ordained that he which shall be attainted for having double measure that is to say, one greater to buy and another less to sell, shall be imprisoned as false, and grievously punished,

the Treasurer was to have standard bushels, gallons, and weights made in brass and sent into every county where such standards had not in the past been sent. Further, two people (more if needed) were to be appointed to survey measures and weights with power to punish those making or using defective measures. The sheriff was to imprison wrong-doers at the command of the appointees. The Act ends with a repetition of the declaration that 'it is not the King's Mind, but that the Clerk of the Market shall do his Office where he will, according as he was wont to do in times past; nor the Lords of Franchises shall not be ousted of the Franchises by the occasion of this Ordinance'.* Of any fines levied, the two 'visitors' were to retain one quarter with the king receiving three quarters. This 'payment by conviction' would only ensure that many, guilty and innocent alike, would be arraigned, found guilty, and fined. It could be a quick way to make money but the resentment caused would be very great. Indeed it was so great that four years later another Act[21] stated 'Commissioners to assay Weights and Measures shall be repealed and none such granted'. But as the Carysfort Committee was to observe in its second report[22] (1759):

> . . . experience soon testified that the ordinary jurisdiction in the country was insufficient to enforce the Law of Weights and Measures, and that, where either the Crown had no immediate interest or no particular persons were appointed to watch over that special set of laws, obedience to regulations of weights and measures could not be expected, therefore the legislature was forced again[23] to have recourse to persons specially appointed by the Crown, to hear, determine, and punish offences respecting weights and measures, with such directions for exercising their jurisdiction as seemed proper to avoid the objections to which the first Commission of this kind was liable.
>
> This power in the Crown, to assign special Commissioners, was thought by Parliament of such much consequence that it was continued expressly even after Justices of the Peace were appointed,[24] after Magistrates of cities and boroughs were directed to execute those laws, and stands hitherto unrepealed, though it does not appear that such Commissions were afterwards issued.

We do find that from the mid-fourteenth century, the role played by

*On 6 June 1341, Richard Spynes, Edmund de Lacy of Folkton, and Nicholas Halden of Snayton were appointed surveyors of weights and measurers for the County of York 'without prejudice to the Clerk of the Market in his Office or the Lords of Liberties within the County', according to the Calendar of Patent Rolls, Edward III 1340-43, p.318.

mayors, bailiffs, clerks of the market, and other local officials* in the area of enforcement increased, with magistrates dispensing summary justice with the aid of the sheriff. Serious cases could be referred to the Justices of Oyer and Terminer.[25]

Meanwhile the office of Clerk of the Market continued and flourished. Towards the end of the fourteenth century his role was restated by statute,[26,27] but if he failed to perform his duty or did so fraudulently he was liable to a fine of 100 shillings for the first offence, £10 for the second, and £20 for the third. He was to have his own sets of weights and measures stamped with the Exchequer insignia (a checker board) to demonstrate they were sized in accordance with those at the Exchequer, and was to carry them around the country as he went on his perambulations 'making assay'. A further Act of 1389[28] provided 'imprisonment of half a year' for using measures other than the standard, with double restitution to the aggrieved party, 'except it be in the county of Lancaster because in that county it hath always been used to have greater measure than in any other part of the realm'.

The abortive Act of Henry VII of a century later (1494)[29] has been referred to elsewhere (Ch.XII). His praiseworthy attempt to provide standards in brass to forty-two shires ended in disaster when the standards so issued proved to be defective (they were made good in 1496-7). The Act of 1494 provided for an examination twice a year (if the city or town officials thought it necessary) of all weights and measures. Any defective items were to be destroyed at once and the owner fined, 6*s* 8*d* for the first offence, 13*s* 4*d* for the second, and 20*s* for the third with a period in the pillory thrown in for good measure, 'to the ensample of others'. Other towns and cities were not slow in acquiring their own sets of standards, as did Windsor in 1499 which had a 'brasyn' bushel, 2 gallons, and 1 quart of 'brasse', a 'pair of balance and a troy weight to weigh bread withall'.[30]

The duties of the clerk were reviewed in 1640.[31] The Act, 16 Charles I c 19, 1640, found that the Clerk of the Market had on occasion exceeded his duty, with 'the subjects much troubled by unnecessary summons', and observed that the office of clerk had been farmed out 'for great sums of money'. Now the role of the clerk was to be restricted to 'within the verge of the King's court where it shall reside for the time being' and not nationwide. Mayors and other local officers were to have plenary powers, but nonetheless the office of clerk continued for another two centuries and more.

*The officers regulating the market and those regulating weights and measures tended to be the same people and this combination of offices continued in some areas up to the time of the local government reorganization of 1974, with the greater part of their duties relating to the regulation of the market, rather than the superintendence of weights and measures.

The Livery Companies

The Coopers

Once established, the City of London's standards began to be enforced to a limited extent by the Livery Companies, either by Act of Parliament or by charter. We have already alluded to the role of the Merchant Taylors' Company and the Drapers' Company with respect to the linear measures for cloth (Ch.XII). Henry VIII in 1531[32] established an assize for the making and verifying of beer, ale, and soap casks, with the Wardens of the Coopers' Company being empowered to search, view, and gauge within the City and suburbs, and two miles compass thereof, all barrels and other vessels for these three commodities, provided they took with them an officer of the Mayor of the City. Any untrue measure was to be burned and its owner fined 12*d* for each defective vessel. In other cities the mayors, bailiffs, sheriffs, constables, etc. were to have these powers of search also.

Correct barrels were to be marked with St. Anthony's Cross[33] and one farthing was to be paid for the examination and marking of each cask. If anyone reduced the capacity of a barrel, the false measure was to be burned and a fine of 3*s* 4*d* paid, which sum was to be shared between king and complainant. A like penalty was to be levied on any cooper who made a deficient barrel. The smallness of the penalties and the need for court procedures to extract the fine which itself had to be shared gave little incentive for action. It was concluded in 1758 that no effect could reasonably be expected from this Act.[34] The practice of marking casks continued in theory to that date but had in reality almost ceased, for the Excise authorities had established their own procedures for marking casks and there was little point in repeating the process. Moreover the farthing did not meet the expenses of checking and marking the casks. The last formal circuit of inspection by the Coopers' Company was on 24 May 1744 when 48 faulty vessels were found, but no one was prosecuted, nor was any penalty levied.[35]

The Plumbers

The Plumbers' Company, by charter granted 12 April 1611,[36] was authorized 'to search, correct, reform, amend, assay, and try' all weights of lead within the City of London and within seven miles of the City, with powers of entry and seizure if any were found defective. All correct lead weights were to be stamped with the Company's seal, an image of St. Michael the Archangel (patron saint of the Company), holding a balance in the left hand and a sword in the right. Lead weights appear to have been very common throughout England from the Middle Ages, and the Company had been sealing such weights at least since 1488. Early in the seventeenth century we find the Plumbers passing over payment for the task of sealing lead weights to 'the

Officer of the Guildhall'. However it was thought appropriate that the Company should seal iron weights and be paid '6*d* for every hundred (weight)'.[37] This was the same rate of remuneration as was allowed the Company under its charter for sealing lead weights. As iron is almost impossible to stamp with die and hammer, lead would be run into the hollow surrounding the staple holding the ring of the weight. This served a double purpose. It could receive a stamp and also, because it was soft, it could be the medium for adjusting the weight into conformity with the standard. Iron weights became very popular; however, weights and measures of brass or bronze dominated the field by the end of the seventeenth century.

Weights of lead were very easily tampered with, for the metal was soft and it was easy to remove a portion without it being apparent. Their use was finally made illegal by the Weights and Measures Act 1834, Section IV.[38]

The Founders

In the sixteenth century relatively few brass or bronze weights were in use. The Founders' Company in 1591 applied to the Corporation of London, requesting that all weights under 2 lb should be of brass and not of lead. The Corporation passed along the request to the Privy Council on 26 April of that year, the Corporation not recommending the change. The Corporation considered the arguments of the Founders insubstantial, for brass weights could be tampered with just as readily as lead weights, the burden on small shopkeepers would be too great, and, if the material be changed from lead to brass, the plumbers would complain that their livelihood was being removed.

However, on 18 September 1614 the Founders' Company obtained their charter from James I.[39] This empowered them to size and mark all brass weights made in the City or within three miles thereof. The now customary rights of entry and search, destruction of false weights, and fees for sizing and sealing were given. The fee was to be that anciently (that is prior to the date of the charter) received by the Company, namely 3*d* for 12 weights marked or $\frac{1}{4}$*d* per weight. The right to stamp weights of brass sold or used in the City of London remained with this Company* till 1889,[40,41] though not exclusively, for we find the officers of the Guildhall also engaged in stamping weights and sealing measures from the sixteenth to eighteenth century, and they were in possession of a very complete set of brass weights ranging from 56 lb to $\frac{1}{2}$ oz and a variety of brass measures ranging from a coal bushel down to one gill. During the examination of the deputy Sealer of Weights and Measures by the Carysfort Committee in 1758, it transpired that this individual was also the Beadle of the Founders' Hall.[42] The fees were still as before, namely $\frac{1}{4}$*d* per weight stamped, be it of brass, lead, or iron, and for 56 lb weights the fee was

*The Weights and Measures Acts of 1834 and 1878 specifically reserve the rights of the Founders.

20*d* per ton, that is $\frac{1}{2}d$ for each 56 lb weight. The sealing of a measure was more lucrative: a coal bushel 18*d;* a half bushel 15*d,* a peck 10*d,* a half peck, 8*d;* a vat 2 shillings; a corn bushel $4\frac{1}{2}d$; half a corn bushel $2\frac{1}{2}d$; and so on. Measures were sealed only at the Guildhall; brass weights were sealed and stamped at the Founders' Hall by the deputy sealer, and the weights were stamped with a crowned letter 'G' and a dagger, the symbol of the City of London and also with an ewer, the symbol of the Founders' Company and the letter 'A' to denote Avoirdupois weight. This continued until 1889, when, by the Weights and Measures Act of that year, only a single stamping was required, and the role of the Founders' Company ceased in this respect. Iron and lead weights were sized at the Guildhall using the standards of reference kept there. The deputy sealer had a salary paid him by the Sealer of the Guildhall and he also received payment from the Founders' Company. For the sealing of lead weights he was employed by the Plumbers' Company to impress their seal upon approved lead weights. Although thus employed he had no appointment and so presumably no salary from the Plumbers' Company.*

The Fruiterers
Although seldom noted, the measures of the members of the Fruiterers' Company could be sealed by officers of the Company. Attention has already been drawn to this Company being exempted from the provision of the Act of 1 Anne Stat. 1, c 15 s 2, 1701[43]† which prescribed the dimensions and form of water measure (See Ch.IX and Ch.X.)

The Inspectorate

We therefore see by joint appointments a certain measure of continuity and uniformity in the sizing and sealing of weights and measures in London in the mid-eighteenth century, but there was no systematic system of inspection throughout the land. This had to wait until 1795, when the first 'examiners' of weights and measures were required to be appointed by the justices at Quarter Sessions in accordance with the Acts 35 George III c 102 1795[44] and 37 George III c 143, 1797.[45] Shops of all kinds were to be inspected at monthly intervals and the necessary salaries and other costs were to be borne by the county rates (taxes). The 'Clerk of the Peace' or other proper person was to be supplied with a set of standard weights, also at the cost of the county rates, and this issue of standards, together with scales, was made to the various local authorities.[46] The Act was not to extend to or lessen the powers of the leet courts but did

*With a single individual representing City, Founders, and Plumbers as sealer we can see why the Plumbers would pass payment to the 'officer at the Guildhall', though at other times there could have been two sealers involved. From Elizabethan to Victorian times we find in London the stamps of the Guildhall and the Plumbers on all lead weights and for brass weights those of the Guildhall and the Founders, as all surviving weights show.
†Given in C.T.S. as 1 Anne c 9 s 2 1702.

empower authorities to appoint the parish constable as the examiner, should they so wish. In 1797[45] it was enacted that the appointment of examiners was to be made at the petty sessions rather than the quarter sessions and the vestries of the various parishes were to recommend suitable persons for the office of examiner. These enactments were extended to corporate towns and cities by the Act 55 George III c 43 of 1815.[47] Following these enactments, most county Weights and Measures Departments were established using police officers as Examiners or Inspectors. These were the very first full-time officers of the local authorities charged with the enforcement of weights and measures legislation and they were joined by 'Inspectors' in accordance with the Acts 4 and 5 William IV c 49, 1834[38] and 5 and 6 William IV c 63 1835.[48] The Act of 1834 enabled magistrates at the quarter sessions to procure for the use of inspectors 'good and sufficient stamps'. The inspectors were to be bonded in the amount of £100, an enormous sum for a working man in those days. As might be expected, this Act was not to infringe on the rights and privileges of the Founders' Company (Section 28). Both Acts directed that weights of lead or pewter were not to be stamped, and that of 1835 directed that a sufficient number of inspectors were to be appointed by the Justices of the Peace 'in general or quarter sessions assembled'. As there was no weight greater than 56 lb in the Exchequer the 1835 Act directed that no weight in excess of 56 lb could be stamped or inspected. In 1859 and 1861 the right of such appointments was extended to certain town councils and borough councils.

Although steps were being taken to endeavour to regulate and control British metrology it cannot be said that even by the mid-nineteenth century uniformity had been achieved. Indeed the Report of 1862 referred to earlier contains the following *cri de coeur:*[49]

> The silent influence of usage has baffled the decrees of legislation; and we are still far distant from the uniformity at which we have so often, yet so vainly, aimed. Omitting many specific anomalies, we have no less than ten different systems of Weights and Measures, most of them established by law. Our neighbours the French, and many other nations, have only one, founded on the metre, which is a near approximation to the English yard. We find in our country the following different systems:
>
> 1. Grains, computed decimally for scientific purposes.
> 2. Troy weight under 5 George IV c 74, and 18 and 19 Victoria c 72.
> 3. Troy ounce, with decimal multiples and divisions called bullion weights under 16 and 17 Victoria c 29.
> 4. Bankers' weights to weigh 10, 20, 30, 50, 100, and 200 sovereigns.
> 5. Apothecaries' weight.
> 6. Diamond weights and pearl weights, including carats.
> 7. Avoirdupois weight, under 5 George IV c 74, and 18 and 19 Victoria 1 c 72.
> 8. Weights for hay and straw*.

*One truss of straw = 36 lb, of old hay = 56 lb, of new hay = 60 lb; and in the same ratio, one load of straw = 11 cwt 2 qr 8 lb, of old hay = 18 cwt, of new hay = 19 cwt 1 qr 4 lb.[50]

9. Wool weight, using as factors 2, 3, 7, 13, and their multiples.
10. Coal weights, decimal under 1 and 2 William IV c 76, and 8 and 9 Victoria c 101; Nos. 1, 0.5, 0.2, 0.1, 0.05, 0.025.

The Report then proceeded to list measures of a given name but of different magnitudes at different places or for different commodities at the same place showing that in reality matters had not improved significantly since the 1820 Report whose Appendix A devoted 32 pages to just this point.

The Government set up, by Royal Warrant of 9 May 1867 (reconstituted 4 May 1868), a Commission 'to enquire into the condition of the Exchequer standards constructed under 29 & 30 Victoria c 82', under the chairmanship of the Astronomer Royal, Mr G. B. Airy. Five reports were issued in the period 1867/8-1871[51] which were to be influential in moulding the important Weights and Measures Act of 1878, itself destined to be the principal Act till 1963. The Fourth Report, 'On the Inspection of Weights and Measures', of 1870 is relevant to our present inquiry.

Prior to 1835, the Report noted, the inspection of weights and measures in shops was still being carried out by 'Nuisance Juries' so named in the Act 29 George II c 25 of 1756 (see p.323) and the system of inspection introduced in 1835 was intended to supplement, not to replace, the earlier system, but in fact, very quickly, the old local inspectorate was replaced by the new, appointed under the Acts of 1834 and 1835. Inspection by leet courts and nuisance juries was discontinued almost completely but the Clerk of the Market still continued to wield exclusive jurisdiction at markets.

Section XIV of the Report pointed out that of 731 inspectors in Scotland, England, and Wales in 1870 some 476 were police officers whose remuneration was nominal and only served to cover expenses. There was no fee charged for verifying and stamping weights except in Ireland. It was felt that the costs of a salaried inspectorate could be met from revenues were appropriate charges to be made. Moreover the system lacked central control as its operations were being conducted exclusively at the local level.

The Commissioners discovered a practice which in certain trades was fairly widespread, of tradesmen 'adjusting' their scales to negate the demand by the public that true weight was obtained with the balance pans tipped towards the goods (See Ch.VIII):

> Section XXV . . . It appears now to be the general custom in certain trades for the tradesmen to have their counter scales thus adjusted and it is alleged that they are fully entitled to such draught in order to counter balance the turn of the scale in favour of the buyer which the public always demanded. A demand is thus made that one unjust practice should be met by the sanction of another equally unjust. We think that both practices are wrong and that no definite allowance for draft in scales should be legally recognized.

It was recommended that the authority over verification and inspection of weights and measures be withdrawn from city corporations, town councils, 'Courts of Burgessee', Universities, lords of manors and leet courts, ward inquests, parish vestries, etc.* and be transferred to the county magistrates except in such large towns and cities as Parliament might decide upon, and in these places the authority was to go to the local magistrates. It was further recommended that there be separate officers for verification and inspection, who were to be authorized and instructed by the Board of Trade which itself was to receive the necessary powers. The inspectors were to be special police officers with appointments approved by the Board of Trade.

In the City of London the Report stated that weights and measures are verified by the Stamper at the Guildhall. Brass weights are verified by the Founders' Company under their James I charter.

The 1878 Weights and Measures Act which followed adopted many of the recommendations contained in these five reports.

s.20	All goods were to be sold by Avoirdupois weight except bullion which was to be sold by the troy ounce.
s.25	It is an offence if balances . . . are unjust.
s.28	All weights are to be stamped.
s.33	The Board of Trade is to be in charge of the inspectorate.
s.40	Local authorities are to provide local standards.
s.43	Every local authority shall appoint a sufficient number of inspectors. They may appoint different people for verification and for inspection. The bond is to be £200.
s.55	'In the Metropolis . . . vestry commissioners may resolve it expedient that their powers in relation to the appointment of Weights and Measures inspectors should cease and after notice to the Court of Quarter sessions, the appointment of inspector or examiner appointed under such local Act, charter or otherwise shall cease . . .'
ss.67–8	contained saving clauses for the rights of the Founders' Company and for the City of London.

Surprisingly, having regard to the wording of s.55, we read in s.69 that nothing is to diminish the authority of a person appointed by a leet court.

While authority was being granted to local authorities to appoint examiners and inspectors there was no provision for the training and education of those appointed or about to be appointed. This was remedied in part by the Weights and Measures Act of 1889[52] which prescribed that inspectors should qualify by passing an examination in practical knowledge of weights and measures. Surprisingly the inspectors had to be appointed by the local authority *before* sitting the examination. No instruction being offered,

*Some of these changes were long in coming. Compare the jurisdiction of the Lord of the Manor of Wakefield, p.325, above.

private study was the means of success, but some help was at hand. The 1889 Act, Section 9, empowered local authorities to make, with the approval of the Board of Trade, general regulations with respect to 'the verification and stamping of weights, measures, and weighing and measuring instruments . . .'. As a guide to local authorities, the Board made up a booklet of regulations, dated 1 January 1890, which, if made by local authorities, the Board was prepared to approve. The result was that these Model Regulations became the rule book without significant local alteration.

The Model Regulations[53] ran to 77 sections and were detailed in every respect. Here, for the first time, was a manual of operation for the inspection and verification of weights and measures, with the duties, responsibilities, and powers clearly defined together with practical directions as to what to do. Moreover, Section IV gave information as to the subjects of examination for prospective inspectors, namely reading, writing, arithmetic including decimals, elementary mechanics, elementary physics, and practical verification of weights, measures, and weighing and measuring instruments. To begin with, the examinations were to be held in London, Bristol, Manchester, and Edinburgh. The examination fee was £1, a fairly large sum for those days. The Regulations were reissued in 1901 identical in substance to those of 1890 and continued in force until 1 October 1927, although new regulations were brought into service in 1907 which necessitated a new publication[54] giving the new and old in parallel columns.

The Survival of the Eight-pound Stone

A good example of the tenacity of custom is afforded by the long history of the stone of eight pounds. It is mentioned in the White Book of Peterborough Abbey of about 1253[55] as the stone for wax, sugar, pepper, cumin, almonds, and alum. Fleta of 1290[56] concurs but he omits the alum and substitutes wormwood. The *Tractatus* of 1302 tells the same story but keeps the alum, omitting the wormwood. In 1496 it appears again,[57] '. . .viii lb . . . of old tyme called the stone of London', while eight-pound weights are mentioned twice in the report of Elizabeth's jury holding the inquisition into standard weights in 1574.[58] It crops up again about 1600,[59] 'waxe and spyce, sugare, peper, cinamond, nuttmegs . . . and every stonne 8 lb . . .', and again in 1708[60] for the selling of meat. The oft-mentioned Report of 1820[61] says the stone of meat is eight pounds and in 1834[62] we read:

> And whereas by local customs in the markets, towns, and other places throughout the United Kingdom, the denomination of the stone varies being in the country generally deemed to contain 14 lbs Avoirdupois and in London commonly 8 of such lbs or otherwise (be it) enacted . . . after 1 January 1835 a stone shall in all cases consist of 14 standard lbs Avoirdupois and that the weight denominated a

hundredweight shall consist of 8 such stones.

A century later the eight-pound unit is still alive and well, for we read in the Memorandum to Inspectors of Weights and Measures from the Standards Department of the Board of Trade, dated December 1934:[63]

> Weights and Measures Regulations 1907. Regulation 18 and Instruction 18. Graduation in units of 8 lbs.
> . . .butcher's steelyards and other types of weighing machines graduated in 8 lb units (the old London or butcher's stone), the attention of inspectors of weights and measures is directed to the following arrangements which have now been made with local authorities of the areas principally affected.
> New machines graduated in 8 lb units will be accepted for verification and stamping until 31 January 1935 it being understood,
> (1) that after that date no new machines so graduated are to be stamped
> (2) that such machines as may then be in use will be restamped from time to time after repair and adjustment until 31 December 1939
> (3) that after this latter date all machines submitted for verification or re-verification may be graduated in *legal* units only [author's italics]
> (4) that during the transition period machines may be graduated both in 8 lb units and in legal units or fitted with an attachment of the nature of a conversion table in any manner which may be regarded as satisfactory by the Board of Trade. . .

It boggles the imagination how the eight-pound stone could have continued right up to well within living memory in spite of the host of declarations, Parliamentary Acts, regulations, and decrees that the stone for all items should be 14 pounds, and when a Statute of 150 years ago so declares and we find precisely 100 years later an official document speaking of the eight-pound unit on the one hand and legal units on the other, we have a situation fit to make the angels weep.

The Modern Period

We therefore see the nineteenth century as being a turning point in the annals of enforcement. Responsible and ethical attitudes on the part of traders, coupled with the activities of the examiners and inspectors (who were predominately police officers until the qualifying examination was established by the Act of 1889*), all acted together to produce a new climate for buying and selling. The Acts mentioned above through to 1889 strengthened

*In some county jurisdictions, such as Lancashire and Leicestershire County Councils, the police controlled the administration of weights and measures until the mid-twentieth century.

the entire system.* Heavier penalties removed any incentive to try to make small gains by giving short weight in individual purchases, so that by the twentieth century true and just weights and measures were to be had generally. The emphasis changed at the appropriate department of the Board of Trade from weights and measures only to the very broad field of consumer protection, information, and education. The nation is currently well served, with everything from prepackaged goods to petrol pumps and gas meters being under constant examination and review. But it has taken a long, long time to achieve this goal.

The National Weights and Measures Laboratory (NWML)[64]

Whereas policy relating to weights and measures is and has been the province of the central government, enforcement is and has been the province of the local authorities of which there are currently 89. The National Weights and Measures Laboratory as part of the Metrology, Quality Assurance, Safety, and Standards Division of the Department of Trade and Industry, is a part of the mechanism of central government today and is a development of the old Standards Department of the Board of Trade. It has responsibility for (1) maintaining the national standards of length, mass, and capacity as used in trade, (2) specifying measurement standards and testing equipment needed to ensure appropriate and uniform levels of accuracy in local government weights and measures offices, (3) ensuring that patterns of weighing and measuring equipment are suitable for trade, and (4) formulating and representing the national viewpoint on legal metrology both in the European Community and in the Organisation Internationale de Metrologie Legale (O.I.M.L.). This latter organization of 48 states and 2 corresponding members was set up in 1955 and has a general secretariat in Paris. It is inter-governmental in scope and seeks to coordinate at that level 'the administrative and technical regulations on measurements and measuring instruments issued in various countries'.

The NWML is the legal custodian of the primary standards of the realm and now handles a broad spectrum of consumer-related functions, from the statutary five-year inspection and verification of the traditional standards used by local authorities to assessing the suitability of electronic equipment associated with new weighing and packaging machines and the weighing of commodities like coal while in continuous motion on a conveyor belt.

The Laboratory is equipped to check the accuracy of all devices used in

*But not so much as to totally exclude unacceptable scales like the wooden butter scales of the nineteenth century shown in Figure 55. It is 18 inches high and the beam is 18 inches long. It is very similar to the balance shown on a bas-relief from a tomb at Gizeh, Egypt dating from about 2500 BC. Being thoroughly insensitive and unadjustable for correct weighing, the nineteenth-century inspectors seized them whenever found.

weighing and measuring for trade, be they standards of mass, length, or volume or machines designed to replace such standards. The work ranges from the calibration of petrol pumps to the gauging of fishing nets.

This Laboratory serves the nation well, though it is seldom in the public eye. It is at present located at 26 Chapter Street, London, but there are current discussions underway relating to its co-location with the National Physical Laboratory (NPL) on the Teddington site. All standards used by NWML are traceable to the NPL which, as has been mentioned earlier, is the research centre for English metrology.

The Institute of Trading Standards Administration[65]

Finally a word should be said about the very important role played in all this by the Institute of Trading Standards Administration, which began its life as the British Association of Inspectors of Weights and Measures when Mr Tom Wimhurst, Manchester's Chief Inspector, called together some twenty inspectors and a number of scale-makers for an inaugural meeting and exhibition of equipment on 8 and 9 May 1881. The inspectorate, having been re-established by the Act of 1878, wisely saw an opportunity to influence the march of events which they seized and in time were enabled to press for modifications and changes to the laws of weights and measures. Such modifications were urgently required, for prior to the introduction of the Board of Trade's Model Regulations (1890) divergent modes of operation were prevalent among the different local authorities. Not the least of the activities of the Association was the education and upgrading in expertise of the appointed inspectors, and the exchanging of ideas and information without which an inspectorate of the type now in place could not have discharged its mandate adequately.

The role to be played by members of industry, makers of balances, weights, yardsticks, and measuring devices of all kinds, was unclear and at a time when anything savouring of trade as opposed to the exercise of a profession was looked down upon, it was not surprising that some inspectors felt that they alone should form the Association. Matters reached such a pass that in 1892 a new association of inspectors called the Society of Inspectors of Weights and Measures was formed in London as a break-away group from the Association. To their credit one month later the members of the new Society established a journal — the *Monthly Review* — which has continued to display a high degree of technical and literary excellence ever since. The Society was incorporated in 1893 and the next year the Association and the Society were reunited in one group under the name, the Incorporated Society of Inspectors of Weights and Measures.

It is not too much to say that the Weights and Measures Act of 1904 stemmed from the Society. Originally presented as a Private Member's Bill, it was taken over by the Government, watered down until almost unrecognizable by the removal of 'controversial' clauses, but finally enacted.

With the Board of Trade empowered under the Act to make mandatory rather than permissive regulations respecting the testing, verification, and stamping of weights and measures and measuring devices, the inspectors worked in committee to produce the 1907 regulations referred to earlier.

It is surprising to us in the present day to find that, of all things, tea was being sold by weight in the first decade of this century and later, a process which included the weight of the wrapping. When lead foil was used as wrapping material, it could account for no less than 25 per cent of the gross weight of small (4 oz) packages. The work of the Society against such anomalies was finally successful, but not until the Sale of Tea Act of 1922 which required that commodity to be sold by net weight.

The Society made important representations to the Hodgson Committee of 1948–50, but as we have seen there were long delays before any government action occurred.

In 1949 the Society became the Institute of Weights and Measures Administration. Continual reports from inspectors of anomalies in the operation of the existing law, together with the Hodgson Report, culminated in the 1963 Weights and Measures Act, but many of the recommendations of the Hodgson Report were not addressed in the 1963 Act. The Institute and other groups had continued to press the Government who set up a Committee on Consumer Protection — the Molony Committee — as a direct result of which the Consumer Protection Act of 1961, among other pieces of legislation, was enacted, all dealing with safety in the home and workplace and with regulations applicable to a wide spectrum of hazardous products and procedures.

In 1962 the Final Report of the Molony Committee[66] contained no fewer than 214 recommendations impinging on every aspect of interest to the consumer.

In particular, the recommendation that any legislation stemming from the Report prescribing mandatory powers of enforcement should be given to Weights and Measures Authorities was accepted. This meant that, with the passage of the comprehensive Trade Descriptions Act 1968,[67] the role of the Standards Department of the Board of Trade underwent a complete revision. This Act of 1968 deals with the description of goods offered for sale, the marking of such goods, oral and written declarations made in the course of trading, trade marks, and false or misleading labelling of goods, and it prescribes as an offence making statements known to be false in the course of trade or business. Section 26(1) states that it is the duty of every local Weights and Measures Authority to enforce this Act and the Weights and Measures Act 1963, while Sections 27 and 28 give these Authorities and their officers power to make test purchases, to enter premises, inspect and seize goods, and inspect documents.

Previous legislation dealing with some of these topics had failed to designate an agency of government whose duty it was to ensure enforcement. This was particularly true of the Merchandise Mark Acts of 1887, 1891, 1894, 1911, 1926, and 1953. Now there was no doubt as to whose responsibility it was in these matters. This greatly extended the powers of the department beyond simply matters relating to weights and measures, and the entire organization changed in 1970 (See Ch.XII Note 84). (For commentary on the Trade Descriptions Act 1968 see Current Law Statutes Annotated, 1968).[68]

Nor was the work of the Institute entirely devoted to this field of endeavour, important as it may be. A library was established in 1955 and located at the University of Sussex in 1967 with Mr Maurice Stevenson as Honorary Librarian. This is a unique collection of material on the entire subject of weights and measures to which the writer, through the kindness of the Honorary Librarian, has been given access.

It is beyond the scope of this work to endeavour to do justice to the entire range of activities of the Institute — Britain's entry into the European Economic Community, the Local Authority Co-ordinating Committee on Trading Standards, the presentation of a public image for the Institute and its work, the development of a Diploma in Consumer Affairs — to name but a few.

Suffice it to say that the new name of the Institute, the Institute of Trading Standards Administration, has been chosen to reflect more appropriately the role discharged by this body. Their in-house educational programmes which up-grade those already appointed as inspectors, enabling them to meet the ever-changing complexity of modern trading practices, deserves great commendation.

This body has passed its first centenary. Its service to the people of the United Kingdom has been of cardinal importance. As a British local government professional body it will doubtless continue to perform in the future as significantly as it has in the past.

NOTES AND REFERENCES: APPENDIX B

1. For further details see:
   (1) CHISHOLM, H. W. *Seventh Annual Report of the Warden of the Standards* Appendix VII, p. 41. London: HMSO 1873.
   (2) CHANEY, H. J. *Our Weights and Measures* London: Eyre & Spottiswoode, 1897.

(3) O'KEEFE, John Alfred. *The Law of Weights and Measures* p. 52 et seq London: Butterworth 1966.

For a more extensive treatment see:

ZUPKO, Ronald Edward. *British Weights and Measures.* Madison, Wisconsin: University of Wisconsin Press 1977.

2. *Judicium Pillorie* 1266 *Statutes* Vol. I p. 201.
3. RICHARDSON, H. L. and SAYLES, G. O. (eds. & trs.) *Fleta* Vol. II, Book II, p. 120. Selden Society Publication Vol. 72 of 1953, London: Selden Society 1955.
4. *Tractatus de Ponderibus et Mensuris* 1302 *Statutes* Vol. I, p. 204.
5. Reports from the Committee appointed to inquire into the Original Standards of Weights and Measures in this Kingdom (Lord Carysfort's Committee) 26 May 1758 *Reports from Committees of the House of Commons* Vol. II, (1737-65) pp. 419-20.
6. Register B. Canterbury Priory f 423 v.
I am indebted to the Librarian of Canterbury Cathedral for supplying me with a photocopy of several pages of this valuable document, parts of which are to be found printed in HALL, Hubert and NICHOLAS, Frieda J. *Select Tracts and Table Books relating to English Weights and Measures (1100-1742)* p. 31. Camden Miscellany Vol. XV, London: Camden Society 1929.
7. LANFRANC. Archbishop of Canterbury 1070-89.
8. In 1430 the weigh of cheese was confirmed as 224 lb (See Ch. VIII p. 139 and STRACHEY, J. (ed.) *Rotuli Parliamentorum* 1430-31, 9 Henry VI). There was thus a decided increase in the size of this weigh in the period of a century, Lanfranc's figure being adopted.
9. HALL, Hubert and NICHOLAS, Frieda J. Loc. cit. p. 23. 'the wey of cheese is 32 cloves conteyning 224 pounds waight haberdepois' (*c*1600).
10. Second Report of Commissioners to consider the Subject of Weights and Measures 13 July 1820. *Parl. Papers* 1820 (HC314) vii, p. 473. Appendix A p. 37, viz.
'Weigh or Wey
Of cheese 2 cwt; but in Essex 256 lbs, otherwise 416 but in Suffolk 3 cwt'.
11. HALE, W. H. (ed.) *The Domesday of St. Pauls 1222;* Compotus bracini of St. Paul's 1283, p. 164*, London: Camden Society 1858.
12. Statute for View of Frank Pledge; 18 Edward II 1325 *Statutes at Large* Vol. I, pp. 184-5.
13. Compare the biblical injunction against having in the house a great and a small measure - Deuteronomy 25, 13-15.
14. Report of the Select Committee on Weights and Measures 15 July 1862. *Parl. Papers* 1862 vii, 187-, see pp. viii-ix.
15. Weights and Measures (Purchase) Act; 55 & 56 Victoria c 18 1892 *Public General Statutes* Vol. XXIX, 1892, pp. 136-7.
16. RICHARDSON, H. L. and SAYLES, G. O. Loc. cit. Ch. 8 p. 117; Chapters 9 and 12 pp. 118-9.
17. *Calendar of Close Rolls* Edward II 1318-1323 p. 362; 14 Edward II 1321, March 2nd. London: HMSO 1895.
18. *Calender of Fine Rolls* Edward II 1319-27 pp. 314-5; 17 Edward II 1324, November 3rd. London: HMSO 1912.

19. *Calendar of Close Rolls:* Edward II 1323-27 p. 532-3; 19 Edward II 1326, January 20th. London: HMSO 1898.
20. 14 Edward III Stat 1 c 12 1340 *Statutes at Large* Vol. I, p. 225.
21. 18 Edward III Stat 2 c 4 1344 Ibid. p. 236.
22. Second Report from the Committee appointed to Inquire into the Original Standards of Weights and Measures in this Kingdom (The Carysfort Committee) 11 April 1759 *Reports from Committees of the House of Commons* 1737-65, Vol. II, p. 460.
23. 34 Edward III 1360 *Statutes at Large* Vol. I, p. 291.
24. Justices of the Peace were first appointed in England in 1360. The appointees were usually local squires and other gentry. Of the other local officials, the office of the coroner who recorded crimes, inquired into homicide, took care of royal property, and had jurisdiction over treasure trove dates from the late twelfth century, while the Office of Sheriff as shire-reeve dates from before the Conquest.
25. The Justices of the Assize courts. The Assizes were abolished in 1972 with the institution of Crown courts.
26. 13 Richard II c 4 1389 *Statutes at Large* Vol. I, pp. 357-8.
27. 16 Richard II c 3 1392 Ibid. p. 377.
28. 13 Richard II c 9 1389 Ibid p. 360.
29. 11 Henry VII c 4 1494 Ibid. Vol. II, p. 72.
30. Reference 1 (1) p. 48 footnote.
31. 16 Charles I c 19 1640 *Statutes at Large* Vol. III, pp. 139-40.
32. 23 Henry VIII c 4 1531 Ibid. Vol. II pp. 151-3.
33. The Tau cross, T, which soon degenerated into a simple 'X'.
34. Reference 5, p. 425.
35. Ibid, p. 427.
36. See STEVENSON, Maurice. Old English Lead Weights *Libra* (1962) Vol. I, pp. 2, 10, 18; *Libra* (1966) Vol. V, p.23 and *Libra* (1969) Vol. VIII, p. 32 for a discussion of lead weights.
37. STEVENSON, Maurice. Loc. cit. Vol. I, p. 10.
38. Weights and Measures Act 1834; 4 & 5 William IV c 49, 1834 *Statutes at Large* Vol. XXVII, p. 629.
39. Reference 5, p. 428.
40. ZUPKO, Ronald Edward. Loc. cit. p. 85.
41. CHANEY, H. C. Loc. cit. p. 56.
42. Reference 5, p. 429.
43. 1 Anne Stat 1 c 15 s 2 1701 (Given in C.T.S. as 1 Anne c 9 s 2 1702) The Water Measure of Fruit; *Statutes at Large* Vol. IV p. 97.
44. Ibid. Vol. XIII, pp. 182-3; 35 George III c 102 1795.
45. Ibid. Vol. XIII, pp. 677-9; 37 George III c 143 1797.
46. I am indebted to Mr Maurice Stevenson for enabling me to examine the issue of metal scales and weights of this date inscribed 'Hundred of East Bourne' (Eastbourne, East Sussex).
47. 55 George III c 43 1815 *Statutes at Large* Vol. XX, pp. 76-8.
48. 5 & 6 William IV c 63 1835 Ibid. Vol. XXVII, pp. 977-986.
49. Reference 14, pp. iii-iv.
50. Reference 14, Minutes of Evidence p. 29.

51. These were (1) The First Report of Commissioners appointed to Enquire into the Condition of the Exchequer Standards; 24 July 1868. *Parl. Papers* 1867-68 xxvii, pp. 1-8.
(2) The Second Report of Commissioners appointed to Enquire into the Condition of the Exchequer (later Board of Trade) Standards; 'On the Question of the Introduction of the Metric System of Weights and Measures into the United Kingdom'; 3 April 1869. *Parl. Papers* 1868-69 xxiii, pp. 733-865.
(3) The Third Report of Commissioners appointed to Enquire into the Condition of the Exchequer Standards; 'On the Abolition of Troy Weight'; 1 Feb. 1870. *Parl. Papers* 1870 xxvii, pp.81-247.
(4) The Fourth Report of Commissioners appointed to Enquire into the Condition of the Exchequer Standards; 'On the Inspection of Weights and Measures'; 21 May 1870. Ibid pp. 249-739.
(5) The Fifth Report of Commissioners appointed to Enquire into the Condition of the Exchequer Standards; 'On the Business of the Standards Department . . .'; 3 August 1870. *Parl. Papers* 1871, xxiv. p. 647-1048.
Index to these five reports; *Parl. Papers* 1873 xxxviii. p. 447.
52. Weights and Measures Act 1889; 52 & 53 Victoria c 21 1889. *Public General Statutes* Vol. XXVI, p. 58.
53. *Weights and Measures. Inspectors and Inspection. Model Regulations* London: HMSO 1890.
54. CUNLIFFE, Howard. *Parallel Weights and Measures Regulations.* London: Shaw & Sons 1908.
55. HALL, Hubert and NICHOLAS, Frieda J. Loc. cit. p. 11.
56. RICHARDSON, H. L. and SAYLES, G. O. Loc. cit. Ch. 12 p. 119.
57. Harleian Mss 698 f 64-5. Printed in CHISHOLM, H. W. Loc. cit. p. 29.
58. Ibid. p. 11 (Sec 7), and p. 15.
59. HALL, Hubert and NICHOLAS, Frieda J. Loc. cit. p. 25.
60. CHAMBERLAYNE, John. *Magnae Britanniae Notitia* p. 207 London: Godwin Timothy 1708.
61. Reference 10, Appendix A, p. 33.
62. Reference 38, Section XII.
63. I am indebted to Mr Maurice Stevenson for making this Memorandum available to me.
64. See SAMUELS, F. L. N. Maintaining Measurements in the Market Place. *Electronics and Power,* February 1982, pp. 150-154. London.
65. See JOHNSON, David and STEVENSON, Maurice. *The Centennial Review - A history of the first 100 years of the Institute.* London: The Institute of Trading Standards Administration 1981.
66. Final Report of the Committee on Consumer Protection (The Molony Committee). *Parl. Papers* 1961-62 (Cmnd 1781) xii, pp. 271-609.
67. The Trade Descriptions Act; Elizabeth II, 1968 c 29.
68. Current Law Statutes Annotated, 1968 c 29.

APPENDIX C

A Brief Account of the Development of the Metric System[1]

Incredible as it may seem to those who have read the main text of this book, English metrology of the eighteenth century was an orderly uniform system when compared to that of France, which can only be described as utterly chaotic. The cause can in large measure be laid at the door of the long duration of the feudal system in France, with each seigneur having the right of establishing and enforcing his own system of weights and measures throughout his territories. The variety of units can scarcely be imagined. For example we are told the *canne* for cloth measure had at least six different lengths in the Department of Haute-Garonne and twelve in l'Ariège.[2] Such disparities ensured fraudulent dealings, with the burden falling most heavily on the peasantry and the poor of the cities.

Even before the eighteenth century there had been suggestions for a reform of weights and measures with new units being established but, at the time, they came to nothing. In 1670 Gabriel Mouton, vicar of St. Paul's Church, Lyons,[3] proposed a decimalized metrology based on the length of one minute of arc of a terrestrial great circle, the length of the degree having been determined earlier that year by the astronomer Jean Picard. Then Picard and fellow astronomer Ole Römer proposed a unit of length based on the length of the seconds pendulum, a length which was also favoured in Britain at one time, as we have seen. As the length of the pendulum varies from place to place with the local value of the acceleration of gravity, it clearly would need to have a place or at least a latitude specified. Hence the pendulum at Paris or at 45°N latitude or at the equator was suggested, but, as the equator cuts through relatively few countries and certainly very few countries of commercial or industrial importance in the seventeenth and eighteenth centuries, this latter proposal was dropped.

In 1739-40 Lacaille and Cassini[4] undertook the survey of a terrestrial arc from Dunkirk, through Paris, to the Spanish border. On this triangulation survey the map of France was based and the length of one degree of latitude at 45°N was determined, together with the length of the seconds pendulum at Paris, but nothing happened to reform the system of weights and measures until after the Revolution, which began with the storming of the Bastille on 14 July 1789. In the following year the Constituent Assembly received a report from Talleyrand, Bishop of Autun, on the position of French metrology, in which a new system of weights and measures was proposed based on a standard derived from the length of the pendulum at 45°N.

Talleyrand also proposed a collaboration between the Academy of Sciences at Paris and the Royal Society of London in order to determine authoritatively the pendulum's length with the greatest accuracy. Indeed on 8 May 1790 the Assembly passed a decree asking Louis XVI to write to His Britannic Majesty

> requesting that Parliament should meet with the National Assembly for the fixation of national units of weights and measures so that commissioners of the French Academy could meet with an equal number from the Royal Society in the most convenient place to determine at 45° latitude or at any other preferred latitude the length of the [seconds] pendulum and produce an invariable model for all weights and measures.

The king concurred on 22 August, but whether Louis wrote or not is uncertain. There is no trace of such a letter in the British Royal Archives* and, bearing in mind the prevailing political climate, it is more than likely that the French monarch was occupied with matters which he felt to be of greater moment.

Be that as it may, the Academy did not wait for a reply but before the end of 1790 set up a commission consisting of five of the greatest scientists and mathematicians France has ever produced, Lagrange, Laplace, Borda, Monge, and Condorcet (a sixth, Lalande was added shortly afterwards), to enable the preliminary work to get underway. Their report, presented to the Academy on 19 March 1791, initiated the metric system.

They recommended that the seconds pendulum should not be used because of the arbitrary nature of the second of time, but one ten-millionth part of the arc of the earth's quadrant, pole to equator, be taken as the standard of length. The alternative of a fraction of the equatorial quadrant was rejected as only few countries are equatorial but all countries lie under polar meridians. For the unit of weight they recommended a declared volume of distilled water *in vacuo* at the freezing point. To this end a number of undertakings were required:[5]

1. To determine the difference of latitude from Dunkirk to Barcelona (sea-level to sea-level).
2. To remeasure the ancient bases which had served for the measurement of a degree at the latitude of Paris and for making the map of France (survey of 1739-40).
3. To verify by new observations the series of triangles employed for measuring the meridian survey of 1739-40 and to prolong them as far as Barcelona.
4. To make observations in latitude 45° for determining the number of vibrations in a day, and in a vacuum at sea-level, of a simple pendulum equal in length when at the temperature of melting ice, to the ten-millionth part of the meridian quadrant with a view to the possibility of restoring the length of the

*BERRIMAN, A. E. *Historical Metrology* p.142. London: J. M. Dent & Sons Ltd. 1953.

new standard unit, at any future time, by pendulum observations.

5. To verify carefully and by new experiments the weight in a vacuum of a given volume of distilled water at the temperature of melting ice.
6. To draw up tables of existing measures of length, surface, and capacity and of the different weights in use in order to ascertain their equivalents in the measures and weights of the new system as soon as they should be determined.

Legislation authorizing the construction of this proposed system was passed on 26 March 1791. Five commissions were set up to supervise the various aspects of the project with a central directorate of Borda, Condorcet, Lagrange, and Lavoisier.[6]

The Metre

Clearly the new unit of length (subsequently to be called the *metre*) would have to be expressed in terms of an existing standard. In the seventeenth and eighteenth centuries there had been two standards of length which had received considerable acceptance and stood out from among the plethora of local and customary measures with which the kingdom had been plagued. These were (1) the *Toise du Grand Chatelet,* made in 1668 but which had been supplanted in 1766 by (2) the *Toise de l'Académie,* better known as the *Toise de Perou,* this having been the standard used by La Condamine, Godin, and Bouguer in their survey of a meridian arc in Peru in order to determine the length of the equatorial seconds pendulum and the length of a meridian degree at the equator. This toise (1.949 m) had also been used in the 1739-40 survey. In Paris measure, the toise consisted of 6 feet each of 12 inches and each inch consisted of 12 lines *(lignes),* a total of 864 lignes, and this was the unit to be used in the new survey. The standard measure of the toise had been made in 1735 by La Condamine in iron. It was and is an end measure about 4 cm broad and 1 cm thick. The bar is cut to half its breadth at each end with the measure of 6 Paris feet being the distance between the end faces exposed by the portions cut away.[6]

The new survey was entrusted to Delambre and Méchain who commenced work in 1792, and it lasted till 1798. There were two major reasons for the survey to take so long. One was the political turmoil of the country with the Revolution in full swing. The surveyors were in possession of a royal warrant calling on municipalities to afford them every facility. This was a red rag to a bull. With monotonous regularity they were arrested and their instruments impounded, but their story was so utterly implausible to the rural people that it just had to be true. No one in reality a royalist spy could dream up such a ridiculous story as being on a national survey in such times. The work was allowed to proceed, but at a snail's pace. The second difficulty was with the survey points themselves. It had been the intention to use the same triangulation points as had been used in the 1739-40 survey. Alas, some no longer existed, while others were obscured by new taller buildings. In the previous half century much had been demolished and much built. What happened was that the survey had to be made completely afresh with new bases and new reference points, and was quite independent of the earlier survey.

Two base lines were measured off; the one at Melun was 6075.90 toises in length, the other at Perpignan measured 6006.249 toises. These bases were measured using four rulers of platinum, each 2 toises long and about $\frac{1}{2}$ inch broad and $\frac{1}{12}$ inch thick. Along each ruler a brass bar $11\frac{1}{2}$ feet long was placed, the two being lined up and fixed at one end, with the other ends free. Together this constituted a bi-metallic thermometer. The difference in length of the two rods could be measured by a vernier device, the temperature obtained, and any necessary correction made to the length of the platinum ruler due to expansion. Experiments to determine the coefficients of expansion of the two metals were carried out in the garden of Lavoisier's house between 25 May and 5 June 1793.

Considering all things it became clear to the directorate that the survey would not be speedy. Borda therefore proposed (29 May 1793), that the new (metric) system be brought into being at once on a provisional basis using the data of the 1739-40 survey. This was agreed to on 1 August. Provisionally, then, the metre was stated to be 443.44 lignes* of which the Toise de Perou contained 864. Then the greatest hold-up of all occurred. On 8 August the Academy itself was suddenly suppressed. Lavoisier was arrested on 28 November and on 23 December, Borda, Lavoisier, Laplace, Coulomb, Brisson, and Delambre were removed from membership of the Academy. It will be remembered that the Reign of Terror did not end till 1794. These events brought matters to a standstill until, on 7 August 1795, the Republican Convention just as suddenly ordered the work to recommence and all the members of the Academy were reinstated except Lavoisier who had been guillotined in 1794. The Academy was now called L'Institut National, the name 'Academy' savouring too greatly of royal patronage.

Lenoir, an instrument-maker, constructed a number of brass metre bars each differing slightly from the next. One, designated 'No.2', was closest to the desired length being 443.4519 lignes at 12.96° on Réaumur's scale (16.2°C), so it was selected as the provisional standard.

The field work on the survey was finished at last in 1798. The Report is dated 30 April 1799. The findings as they apply to the metre were as follows:

(1) The arc Dunkirk to Barcelona amounted to 9° 40′ 45″ and measured 551,584.72 toise.

*Specifically given equivalently as 2 m = 6 feet 1 inch $10\frac{22}{25}$ lignes. See Ref.1(j) p.40.

(2) Assuming the ellipticity of the earth to be $\frac{1}{334}$*, the meridian quadrant is 5,130,740 toise.

(3) The new unit of length (metre) being 10^{-7} of the quadrant is therefore equal to 0.5130740740 toise when all the necessary corrections are made, that is 443.296 lignes of the Toise de Perou at 13°R† (16.25°C).

(4) The length of seconds pendulum at Paris at 0°C in vacuum at sea level is 0.99385 m.‡ (This last is equivalent to the period of a pendulum of length 1 m being 2.00618 seconds at Paris, latitude 48°52′.)

Some idea of the remarkable accuracy achieved by Delambre and Méchain can be obtained by comparing the measured length of the base line at Perpignan with the value computed from the Melun base through a series of 53 triangles:

Measured = 6006.249 toise
Computed = 6006.089 toise.

The difference is only 3 parts in 100,000. As for the actual length of the meridian, the ellipticity of the earth was not well known at the end of the eighteenth century, and this in all probability was the major source of error in the determination of the quadrant. Modern work shows the quadrant to be about 1 part in 5000 longer than that given in the 1799 Report. So the metre of 1799, and indeed the metre of today, is in fact about one fifth of a millimetre too short if it is to represent truly the actual 10^{-7} of the quadrant. But as we shall see this point was attended to later by a change in definition. The final result is really remarkable when it is further recalled that the quadrant is 90° of arc whereas the survey Dunkirk to Barcelona was only a little over $9\frac{1}{2}$°. This makes the computed length of the quadrant quite sensitive to the figure chosen for the ellipticity.

Meanwhile, in 1795, the Government of France, though often on shaky ground and beset with food shortages and rampant inflation, had the good sense to take time by the forelock, bringing the court jeweller, Marc Etienne Janety (Janetti), back to Paris from Marseilles, to which city he had judiciously withdrawn when the troubles started. This was not for his skill as a jeweller but for his skill as a worker in platinum. Late in 1795 he began to make metre bars (and kilogram weights) in this refractory§ metal. One of his metre bars, an end

*Ellipticity of the earth is given by $\frac{\text{equatorial} - \text{polar radius}}{\text{equatorial radius}}$

A more modern value would be $\frac{1}{298 \cdot 3}$ as determined from the motion of artificial satellites. The Report (Ref. 1(c)) gives the length of the quadrant for each of a series of assumed values of the ellipticity.

†The Réaumur Scale of temperature had 0°R as the freezing point and 80°R as the boiling point of water and was introduced by the French physicist R.A.F. de Réaumur (1683-1757).

‡This result had been obtained by Borda and Cassini in 1792.

§In every sense of the word.

standard, 25.3 mm broad and 4 mm thick with plane parallel end faces,* was adopted as being at 0°C as near as possible to 443.296 lignes.[7] This measure, known as the *Mètre des Archives,* was deposited in the National Archives on 22 June 1799.

Although legalized on 10 December 1799, the metric system was slow to be generally adopted. Contrary to common belief, Napoleon was not a devotee of the system. It is recorded[8] that he once declared 'What are metres? Speak to me in toises'. It was not until 1 January 1840 that the system became obligatory in France. In the interval the old names were given to the new units; thus the *toise* was now 2 metres, the foot *(pied)* was $\frac{1}{3}$ metre and the inch *(pouce)* was $\frac{1}{36}$ metre and the ligne $\frac{1}{432}$ of a metre. This kept the population in a contented frame of mind insofar as measures were concerned but did little to establish the system. However, once made obligatory strenuous efforts were made to encourage adoption of the system by other countries. These moves proved to be very successful.†

Such was the extent of the interest on the international front that the French Government called two conferences, one in 1870, the other in 1872, under the name Commission Internationale du Mètre to promote discussion and to consider suggestions for the design, construction, and distribution of new metric standards. It was decided that the new metre bars should be made not of platinum but of a platinum-iridium alloy (90%-10%)‡, that the standards should be line and not end measures, that the distance between the defining lines should be the length of the Mètre des Archives 'in the state in which it is found', thus abandoning any claim that the length of the metre bar was 10^{-7} of the earth's quadrant, and that the form of the bar itself should be the winged design attributed to Tresca§ which had the merit of possessing the greatest rigidity for a given quantity of metal (see Fig.56). The engraved lines were to be on the central flat surface which was the neutral plane and was therefore exposed throughout its entire length. Work commenced at once and twenty countries participated in the next conference, held in 1875. Of the twenty, eighteen subscribed to a treaty, the Convention du Mètre (Metric Convention), by which an international organization on weights and measures was set up, with offices and a laboratory to be known as the Bureau International des Poids et Mésures. This organization was to be funded by the participating countries and responsible for the custody and verification of metric standards,

*The end faces are now neither plane nor parallel.

†By 1900 thirty-five countries had gone metric, by 1960 some eighty, and by 1980 one hundred and eleven had gone metric with a further forty-two in process of change.

‡This alloy was chosen for its 'high density, high melting point, great resistance to humidity and air, fine grain, perfect polish, great hardness, and full malleability'.[9]

§Henri Tresca, Professor of Mechanics at the Conservatoire des Arts et Métiers and secretary to the French Section of the Commission Internationale. (For the dimensions of the winged form see Fig.56.)

comparing the copies sent to the several nations, etc. The Bureau was indeed set up and still exists today, being located at Sèvres, near Paris, but it was quite another matter to create an ingot of the alloy which met all the specifications, especially as it was felt desirable that all the measures should be produced from a single ingot. The metallurgy of platinum and iridium was far from being completely known. Indeed, really pure platinum had not yet been produced, while the situation with iridium was much worse, in that the greatest degree of purity attained to that date was only about 50%.[10] The melting points of platinum and iridium are 1769 and 2443°C respectively, so furnaces capable of producing these extremely high temperatures were necessary. While a metre bar of the 90%-10% alloy had been produced in 1870 with a rectangular cross section it was clearly going to be a different matter with an ingot weighing some 250 kg being forged and cold-worked into the winged form.

The melting took place on 13 May 1874 at the Conservatoire des Arts et Métiers. The final ingot weighed 236 kg. It was forged when red-hot into a bar 5 × 5 × 450 cm. This was then forged into a bar 25 mm square and cut into nine pieces totalling 16.4 m in length. It had been hoped to produce 65 metre bars but this was not to be. Some bars cracked during further extrusion but 27 metre bars of the winged section were produced weighing 90 kg. Then it was discovered that the density prescribed, 21.385-21.455 g/cm^3, had not been attained. The mean density was only 21.1. Some were for accepting the alloy as it was, others were for rejecting the whole production, and while a few metres (and kilograms) were retained from this alloy, known as the 'Conservatoire alloy of 1874' the decision was made in 1877 to reject it.

Two improvements in technique were discovered which made the eventual completion of the task in hand possible. These were (1) to cast the preliminary ingot three times and not just once, so as to ensure the rejection of impurities, and (2) to plane the winged form from a rectangular bar rather than draw the metal through a steel die, which process had apparently introduced iron into the bars in the past.

It was decided to move to the production of a provisional standard metre and then to the standard itself. The firm of Johnson, Matthey & Co. of London was commissioned to make the alloy and to fabricate the bars. This was done with consummate success in 1878-9 by George Matthey. One of the metres so produced became the provisional standard in 1881, notwithstanding it was 0.006 mm longer than the Mètre des Archives.

The French Government then in 1882 ordered thirty metre bars of the winged section (and forty kilogram weights in the form of cylinders) from Johnson, Matthey & Co. The production of the metal was a triumph, with all specifications being met. Of the thirty metres so produced, one, No.6, became the standard. It was 0.006 mm shorter than the provisional standard so was exactly the length of the original Mètre des Archives. The metres were finished by Brunner, Collot, and Laurent of Paris and engraved by Tresca. By the time (1889) the first General Conference on Weights and Measures convened in Paris the new standards were all ready for distribution. The primary standard was named the International Prototype Mètre (bar No.6) and, with this metre as the unit, the lengths of all the other metres were determined. All have remained remarkably constant in length since their manufacture. The British copy is No.16. In 1889 it was stated to be 0.59 μ m* shorter than the Prototype at 0°C with an accuracy of ±0.2 μ m. The metre was originally defined at 0°C, which is not always the best temperature at which to carry out inter-comparisons, so the British metre, after being compared with the International Prototype Metre in 1956, had its defining lines removed and was re-ruled with the appropriate lengths at 0°C and also at 20°C. It was re-certified in 1956 as being 1 m less 0.84 μm at 0°C and 1 m less 0.54 μ m at 20°C. A comparison in 1957 showed the new markings on bar 16 to be equal to the old to an accuracy of within ± 0.1 μ m. In recent years other national standards have been similarly re-engraved.

The International Prototype Métre was formally accepted at the first General Conference on Weights and Measures in 1901 and remained the primary reference for the metric system until 1960 when the metre, defined as the distance between two lines engraved on the prototype bar at 0°C, was replaced by the metre defined in terms of the wavelength of light.

The Optical Metre

The idea of using the wavelength of light as a fundamental and natural standard of length was first mooted by Babinet in 1829. During the nineteenth century optical interferometers of several designs were developed which could determine changes in position of a fraction of the wavelength being used. The first direct measurement of the metre optically was that carried out by Michelson and Benoît in 1892-3 and reported in 1895.[11] The radiation used was the red line in the cadmium spectrum and the experiment was repeated by Benoît and others in 1905-6 with improved equipment. The value so obtained, λ_R† $= 6438.4696 \times 10^{-10}$m, was adopted in 1907 by the International Solar Union (now the International Astronomical Union) as the reference standard for all wavelength measurements. This value of the wavelength was accepted provisionally by the 7th General Conference on Weights and Measures in 1927 as a supplementary definition of the metre, the measurement being conducted in air under specified conditions of temperature and pressure and carbon dioxide content.‡ It was not until pure

*$1\ \mu m = 10^{-6}\ m = 10^{-3}\ mm$

†The Greek letter λ denotes wavelength.

‡Dry air at 15°C under a pressure of 760 mm of mercury at 0°C, containing 0.03% by volume of CO_2.

isotopes of a single even mass number with even nuclear charge and therefore zero nuclear spin became available that optical measurements exceeded in precision measurements made with line standards of first quality. The insistence on zero nuclear spin meant no coupling between it and the electron spins and hence no hyperfine structure to the line under examination. The removal of the attendant hyperfine lines greatly sharpened up the principal line and meant ever finer interferometric precision. While natural cadmium is a mixture of eight isotopes, mercury of mass number 198 and charge 80 can be obtained in separated form from the other isotopes of mercury present in the natural element and likewise krypton mass number 86 and charge 36 can also be obtained in a pure state. The 9th General Conference (1948) saw the possibility of achieving an optical definition of the metre in terms of the radiation from an even-even isotope and later it was proposed that it be the wavelength *in vacuo*, not in air, that be used. The vacuum value of λ_R is $6440.2490 \times 10^{-10}$ m. Tests with krypton − 86, mercury − 198, and cadmium − 114 showed the first to be superior and capable of the greatest accuracy, and this led the 11th General Conference to pass the following resolution (1960):

1. The metre is the length equal to 1,650,763.73 wavelengths in vacuum of the radiation corresponding to the transition between levels $2p_{10}$ and $5d_5$ of the atom of krypton − 86.
2. The definition of the metre in use since 1889 based on the International Prototype Metre of platinum-iridium is repealed.
3. The International Prototype Metre authorized by the First General Conference on Weights and Measures in 1889 will be preserved by the International Bureau of Weights and Measures under the same conditions as those laid down in 1889.

The form and conditions of operation of the krypton light source were also specified and tests have shown the wavelength to be determinable in reproducible fashion to one part in one hundred million. More recent work with lasers has shown that these coherent light sources are capable of length determinations at least two and probably three orders of magnitude better than the krypton apparatus, but to go further into this is beyond the scope of this book.

As to the constancy of the lengths of the various metre bars over time, this has been shown to be extremely satisfactory using standard methods of comparison as well as wavelength measurements. Any changes that have occurred are virtually at the limits of error of the measuring equipment. For the British copy, bar No.16 has been verified six times between 1889 and 1956 when it was re-ruled. The mean of these measurements shows no changes at all in the length over this interval of time.[12]

The Kilogram

The essence of the metric system is that the unit of length defines the units of mass (weight) and capacity. The 1791 Report★ (p.345) recommended the unit of mass as a standard volume of distilled water at its freezing point. Later, after the ground-work of Lavoisier and Haüy, it was decided to select 4°C, the temperature at which water has its maximum density, and the chosen volume was to be a cubic decimetre, called initially the *grave,* but after 1795 the *kilogramme.* In 1795 Lefèvre-Gineau was entrusted with the determination of this new standard of mass, the kilogram, the gram, originally called the *gravet,* being the mass of 1 cm^3 of water under the same conditions.

A decision had to be made between two avenues of approach: either (1) to try to construct a vessel whose capacity was a cubic decimetre which, when weighed empty and filled with water, would yield the desired volume of the kilogram, or (2) to construct a body whose volume could be accurately determined by external measurement which could be weighed in air, then, when immersed in water, the apparent loss of weight, the upthrust, being by Archimedes' Principle the weight of the body's volume of water, could be found. The second alternative was chosen. There is no doubt as to the wisdom of the choice, for the accuracy attainable by this method is much greater. A cylinder of height equal to the diameter of the base was made of brass. It was hollow so that it just sank when immersed in water. Special rulers were prepared for measuring the cylinder's height and diameter. The end result of numerous repeated measurements gave the mean height as 2.437672 decimetres and the mean diameter as 2.428368 decimetres at 17.6°C. Using the known coefficient of expansion of the brass, the computed volume of the cylinder at 0°C was almost exactly 11.28 cubic decimetres.[13] Special weights were created for the weighing, being of the same density as the brass of the cylinder so as to eliminate buoyancy correction when weighed in air†, and the weighings were made in terms of a provisional kilogram of brass which had been constructed closely approximating the expected value for the kilogram. This had, of course, to be evaluated in terms of the old, then existing, French weights, the so-called Poids de Marc of the Pile de Charlemagne. The *livre* (lb) was 9216 French grains, the *once* (oz) was $\frac{1}{16}$ lb and was divided into 8 *gros* (drachms). The special weights were carefully compared with the old standards and the final conclusion, involving the work of Fortin and Fabbroni as well as Lefèvre-Gineau, was that a cubic decimetre of distilled water at its maximum density weighed 0.9992072 of the provisional kilogram and was equal to 18827.15 grains of the old system.[14] The provisional kilogram was therefore slightly too heavy.

★And Ref.1(j), the *Instruction* of 1793-4 pp.46 and 52.

†The interior of the cylinder communicated with the external air via a fine tube screwed into the top of the cylinder. The end of the tube was out of the water when the cylinder was being weighed immersed and allowance was made for the submerged portion of the tube and the weight of the air inside the cylinder.

It is known that Janety was making kilogram weights in platinum in 1796[14] and a little later Fortin supervised the construction of one kilogram in platinum and one in brass, and these were adjusted to the value of the newly determined standard (18827.15 French grains). The platinum weight became the Kilogramme des Archives. It is a cylinder about 39.4 mm in diameter and 39.7 mm in height with slightly rounded edges. It was legalized with the metre in 1799.[15]

As was mentioned in the preceding section on the metre, the decision to have the primary standards in platinum-iridium alloy applied to the kilogram also, so a new standard, the International Prototype Kilogramme, was prepared with the same care as was lavished on the metre. On its acceptance in 1901 it displaced the Kilogramme des Archives and has continued as the primary standard of mass to the present. Subsequent measurements to those of 1795 showed the kilogram was not precisely the weight of a cubic decimetre of water under the stated conditions but was the weight at 4°C of 1.000028 dm^3. Thus the kilogram is 28 parts per million too large for the original definition. It was decided to retain the kilogram mass as represented by the prototype cylinder and to abandon any reference to the cubic decimetre of water (see the following section on the litre).

The British copy of the kilogram is that designated No.18. Its volume was found to be 46.414 ml at 0°C giving a density of 21.5454 g/cm^3. It is 0.070 milligrams heavier than the prototype, with an error of $\pm$0.002 milligrams. The standard kilogram is shown in Figure 57.

Unlike the metre there is no external agency like the wavelength of light to which a given mass can be compared. One must rely on repeated inter-comparisons with as many of the certified copies as possible. These comparisons have been undertaken from time to time and only four copies showed a change exceeding 0.02 mg which is two parts in one hundred million. The three measurements on the British copy of 1889, 1924, and 1933 do not differ among themselves by more than this amount.

There can therefore be little doubt as to the stability of the kilogram and the metre. Indeed one copy of the kilogram was kept in a stream of air at 100°C for a year without producing a detectable change in mass,[16] all of which is a tribute to the magnificent work of George Matthey in producing so perfect an alloy in the first place.

The Litre

The unit of volume was originally defined as a cubic decimetre and called a *cadil*, later a *litre*, which volume of water at its maximum density (4°C) under standard atmospheric pressure weighed a *grave* (kilogram). With the discovery that the volume of a kilogram of water under these conditions was 1.000028 dm^3, a redefinition was called for and accepted in 1901, namely that

1 litre (l) is the volume of a kilogram mass of pure water at its maximum density under standard atmospheric pressure. Thus 1 litre = 1.000028 dm^3, though for many practical purposes the difference can be ignored except for work of the highest precision.[17] A photograph of a cadil (1 litre) made in Year II of the Republic (1793-94) is shown in Figure 58.

The litre is however under a cloud insofar as scientific work is concerned, which is ironic for it was the scientific community which decided its graduated glassware should show 'ml' rather than 'cm^3', not that the divisions were accurate to 28 parts per million but because this was the proper unit to use. In October 1964 the General Conference on Weights and Measures decided to abolish the litre as the *scientific* unit of volume, the reason being that 'water' is not just 'water' but even in the purest state is a mixture of light water (H_2O) and heavy water (D_2O) where D represents the atom of deuterium (heavy hydrogen). This mixture is not constant throughout the world. Some sources contain more heavy water, others less, so that the precise volume of a given weight of water varies — not greatly but enough to cause problems where the most precise work is undertaken.

Even for purposes of trade we can detect a cautionary note in the British Weights and Measures Act 1963 where in Schedule 1 Part IV we find:

> litre — shall have the meaning from time to time assigned by order of the Board (of Trade) being the meaning appearing to the Board to reproduce in English the international definition of the litre in force at the date of making the order.

An identical statement appears in Part V for the kilogram.

The current regulations of the Board of Trade contained in the Weights and Measures (International Definitions) Order 1963 (Statutory Instrument 1963 No.1354) read:

> *Litre*
> The litre is the volume occupied by a mass of 1 kilogramme of pure water at its maximum density and under standard atmospheric pressure.
>
> *Kilogramme*
> The kilogramme is the unit of mass represented by the mass of the international prototype kilogramme.

These are the definitions adopted by the first General Conference on Weights and Measures at Paris in 1901.

This does not mean that petrol pumps in Britain which have only recently switched from gallons to litres will soon have to experience another change. For purposes of trade errors of 28 parts per million or even smaller errors occasioned by isotopic variations hold no significance. But in the ever-increasing search for accuracy it would be foolhardy to think that there will be no further change in the units of the metric system or their definitions.

Indeed the General Conference on Weights and Measures in October 1983 adopted the proposal that the metre should be defined as the distance a light beam would travel in vacuum in $\frac{1}{299792458}$ of a second.[18] As the speed of light is now known to such high accuracy, this definition would be more precise than any of the previous ones. So much for rejections based on the 'arbitrary nature of the second' (see p.345). One cannot but recall 'The stone that the builders rejected . . .'.[19] This definition of the metre is to be found in the Weights and Measures Act 1985 (30 October) c 72.

NOTES AND REFERENCES: APPENDIX C

1. The principal sources drawn upon for the material of this Appendix are:
   - (a) BARRELL, H. The Metre. *Contemporary Physics* (1962) Vol.III, p.415. London.
   - (b) CHANEY, H. J. *Our Weights and Measures* London: Eyre & Spottiswoode 1897.
   - (c) CHISHOLM, H. W. On the Science of Weighing and Measuring and the Standards of Weight and Measure, *Nature* (1873) Vol.VIII, p.386 et seq. London.
   - (d) DARWIN, Sir Charles et al. A Discussion on Units and Standards *Proc. Roy. Soc.* (1946) Vol.CLXXXVI, pp.149-152. London.
     SEARS, J. E. The Standards of Length. Ibid pp.152-164.
     BARRELL, H. The Standards of Length in Wavelengths of Light. Ibid pp.164-170.
     GOULD, F. A. The Standards of Mass. Ibid pp.171-179.
   - (e) MECHAIN, P. F. A. and DELAMBRE, J. B. J. *Base du System Metrique Decimal* Paris: Baudouin 1806-10.
   - (f) HALLOCK, William and WADE, Herbert T. *Outlines of the Evolution of Weights and Measures and the Metric System* London & New York: Macmillan 1906.
   - (g) McDONALD, Donald. *A History of Platinum* London: Johnson, Matthey & Co. 1960.
   - (h) O'KEEFE, John Alfred. *The Law of Weights and Measures* London: Butterworth 1966.
   - (i) TATON, René. L'Evolution des Units de Longueur, Surface, Volume, et Poids. *La Nature* Dec. 1949. Supplement to issue 3176. pp.387-393. Paris.
     PAPIN, Maurice Denis-, Les Systemes d'Unites de Mesure. Ibid pp.393-395.
     MOREAU, Henri. Unites et Etalons Principaux. Ibid pp.396-412.

(j) ANON, *Instruction sur les Mesures* Poitiers: La Commission temporaire des Poids et Mesures republicaines, Michael-Vincent An.II (1793-4).

2. ZUPKO, Ronald Edward. *French Weights and Measures before the Revolution* p.34, Bloomington & London: Indiana University Press 1978.
3. MOUTON, Gabriel. *Observationes diametrorum solis et lunae apparentium.* Lugduni (Lyons): Liberal 1670.
4. Reference 1(e).
5. Reference 1(e) and 1(c) p.387.
6. Reference 1(a) p.417.
7. Reference 1(a) p.420.
8. Report from the Select Committee on Weights and Measures 15 July 1862. *Reports from Committees* (2) Vol.VII, p.59. 1862. (Minutes of Evidence M. Michael Chevalier).
9. Reference 1(g) p.193.
10. Reference 1(g) p.208.
11. MICHELSON, A. A. and BENOIT, J. R. *Travaux Bureau International des Poids et Mesures* (1895) Vol.XI, p.3. Paris.
12. Reference 1(a) p.425.
13. Reference 1(c) p.389.
14. Reference 1(g) p.127.
15. Reference 1(c) p.489.
16. Reference 1(d) GOULD, F. A. p.174-5.
17. The litre as a volume of water at 4°C weighing 1 kg was never a particularly good choice of definition, for it can only be arrived at by computation. The decimetre cube would depend on the metre which is true at 0°C. We would therefore have the anomalous situation of a vessel at 0°C containing a kilogram of water at a temperature of 4°C.
18. BALIBAR, François. Le mètre au fil du temps, *La Recherche* (1984) Vol.XV, No.152, p.263. Paris.
19. MATTHEW, 21:42.

APPENDIX D

Tables of Pre-metric British Measures

Linear Measure

12 inches	= 1 foot
3 feet	= 1 yard
$5\frac{1}{2}$ yards	= 1 rod, pole or perch
40 rods	= 1 furlong
8 furlongs	= 1 mile
(22 yards	= 1 chain
10 chains	= 1 furlong)

Superficial Measure

144 sq inches	= 1 sq foot
9 sq feet	= 1 sq yard
$30\frac{1}{4}$ sq yards	= 1 sq rod
40 sq rods	= 1 rood
4 roods	= 1 acre

Cubic Measure

1728 in^3	= 1 ft^3
27 ft^3	= 1 $yard^3$

Troy Weight

24 Troy grains	= 1 pennyweight
20 pennyweights	= 1 Troy oz (480 grains)
12 Troy ounces	= 1 Troy pound (5760 grains)

Avoirdupois Weight:

16 drams	= 1 oz (437.5 grains)
16 ounces	= 1 lb (7000 Troy grains)
14 lb	= 1 stone
2 stones	= 1 quarter
4 quarters	= 1 hundredweight (112 lb)
20 cwt	= 1 ton* (2240 lb)

Apothecaries' Weight:

20 grains (Troy)	= 1 scruple (℈)
3 scruples	= 1 drachm (ʒ)
8 drachms	= 1 ounce (℥)
12 ounces	= 1 lb (5760 grai

Capacity

4 gills	= 1 pint
2 pints	= 1 quart
4 quarts	= 1 gallon
2 gallons	= 1 peck
4 pecks	= 1 bushel
8 bushels	= 1 quarter
1 chaldron	= 32 bushels, originally, later 36 bushels heaped

Apothecaries' Measure

20 minims	= 1 fl. scruple
3 fl scruples	= 1 fl. drachm
8 fl drachms	= 1 fl. oz

*The word *ton* in maritime circles generally refers to a volume and not to a mass or weight. For example, for cargo whose volume exceeds 40 ft^3 per ton, the *freight tonnage* is calculated at 40 ft^3 = 1 ton and is charged accordingly. Likewise the *measurement ton* or *shipping ton* is a unit of volume equal to 40 ft^3. Prior to 1982 the *registered tonnage* of a vessel was calculated at 100 ft^3 per ton. Today a more complex formula is used. The *displacement ton* measures a ship's displacement and is taken as 35 ft^3 of sea-water or 2240 lb. The *Thames Measurement Tonnage* of a small vessel or yacht is given by the formula $\frac{(L-B) \times B \times \frac{1}{2}B}{94}$ where L is the length and B the breadth of the deck.

The 1963 Weights and Measures Act defined the yard as 0.9144 metres and the Avoirdupois pound as 0.45359237 kg. The gallon was given as 4.545964591 litres. The Act abolished the bushel, peck, and pennyweight with effect from 31 January 1969. It also abolished our old friend the rod, pole, or perch and the square rod. The Apothecaries' system was to cease to have effect on a date not sooner than 31 January 1969.

The 1976 Weights and Measures Act abolished the bushel, peck, fluid drachm, minim, pennyweight, Apothecaries' ounce, drachm, and scruple. The 1985 Weights and Measures Act abolished for purposes of trade the furlong, chain, square mile, rood, square inch, cubic yard, cubic foot, cubic inch, bushel, peck, fluid drachm, minim, ton, hundredweight, cental, quarter, stone, dram, grain, pennyweight, Apothecaries' ounce, drachm, scruple, metric ton, and quintal. But the square foot remains, as do the gallon and its fractions, Avoirdupois weight from 56 lb to $\frac{1}{32}$ oz., Troy weight from 500 oz to 0.001 oz, the yard, and the pound defined as in 1963. While the hundredweight has gone we may still buy 112 lb of coal or other solid fuel, though it has to be called 112 lb and not 1 cwt.

Metric Equivalents

Linear		*Superficial*		*Cubic*	
1 inch	= 2.54 cm	1 in^2	= 6.45 cm^2	1 in^3	= 16.387 cm^3
1 foot	= 30.48 cm	1 ft^2	= 929.0 cm^2	1 ft^3	= 28.317 dm^3
1 yard	= 91.44 cm	1 yd^2	= 0.83618 m^2	1 yd^3	= 0.7646 m^3
1 rod	= 5.03 m	1 sq rod	= 25.29 m^2	For most purposes	
1 furlong	= 201.17 m	1 rood	= 1011.71 m^2	1 dm^3 (1000 cm^3) =	
1 mile	= 1.609 km	1 acre	= 4046.86 m^2	1 litre	
		1 hectare (ha)	= 10,000 m^2		

Troy Weight		*Avoirdupois Weight*		*Apothecaries' Weight*	
1 Troy grain	= 64.8 mg	1 oz	= 28.35 g	1 scruple	= 1.296 g
1 dwt	= 1.555 g	1 lb	= 453.59 g	1 drachm	= 3.888 g
1 Troy oz	= 31.104 g	1 stone	= 6.350 kg	1 oz	= 31.104 g
1 Troy lb	= 373.248 g	1 quarter	= 12.70 kg	1 lb	= 373.248 g
1 g	= 15.43 grains	1 cwt	= 50.80 kg		
		1 ton	= 1016.0 kg		

(One tonne is the mass of 1000 kg)

Capacity	*British*	*US Liquid*	*US Dry*
1 gill	0.142 litres	0.11829 litres	–
1 pint	0.568 litres	0.47317 litres	0.55060 litres
1 quart	1.136 litres	0.94633 litres	1.10120 litres
1 gallon	4.54596 litres	3.78533 litres	–
1 peck	9.092 litres		8.8096 litres
1 bushel	0.3636 hl		0.35238 hl
1 quarter	2.909 hl		
1 hectolitre (hl)	= 100 litres		

Cubic Equivalents:

	British	*US Liquid*	*US Dry*
1 gill	8.6694 in^3	7.21875 in^3	–
1 pint	34.6775 in^3	28.875 in^3	33.60 in^3
1 quart	69.355 in^3	57.75 in^3	67.20 in^3
1 gallon	277.42 in^3	231.0 in^3	–
1 peck	554.84 in^3		537.6 in^3
1 bushel	1.284 ft^3		1.2445 ft^3
1 quarter	10.275 ft^3		–

1 in^3	= 0.57674 fl. oz* (British)	= 0.5541 fl. oz* (US)
1 ft^3	= 6.2288 gallons (British)	= 7.480 gallons (US)
1 fl. oz	= 1.7339 in^3 (British)	= 1.80469 in^3 (US)

Apothecaries' Measure:

1 minim	= 0.00361 in^3	= 0.05915 ml
1 fl. scuple	= 0.0722 in^3	= 1.183 ml
1 fl. drachm	= 0.2167 in^3	= 3.55 ml
1 fl. oz*	= 1.7339 in^3	= 28.41 ml

* The British fluid ounce is defined as the volume at 16.7°C (62°F) of 1 Avoirdupois ounce of distilled water, i.e. $\frac{1}{160}$ of the gallon of 277.42 in^3, i.e. 1.7339 in^3. So 1 gallon (British) = 160 fl oz.

The US fluid ounce is defined as $\frac{1}{16}$ of the liquid pint which is $\frac{1}{8}$ of the US gallon of 231 in^3, i.e. the US fluid ounce is 1.80469 in^3. Thus 24 US fluid ounces = 25 British fluid ounces to better than 1 part in 1000.

Glossary of Unit Terms

Readers are referred to Appendix D (page 358-360) for exact definitions of all modern units.

Acre: The land ploughable in a day by a team of oxen: a rectangular area 40 rods long (1 furlong) by 4 rods broad: 160 square rods: 4 roods.

Actus: A Roman measure of length of 120 feet.

Aeginetan Measure: One of two great scales of measurement in the Greek world. (The other was the Attic-Euboic System). Introduced from Egypt to Greece in the seventh century BC and in use thereafter.

Amber: A unit of capacity: in Saxon times, probably six gallons: in the thirteenth century, 4 bushels.

Amphora: eight *congii* (see *congius):* a Roman measure of capacity about 24 – 28 l: a *quadrantal* (q.v.): the cube of a Roman foot.

Apothecaries' Measure: The system of measures used by apothecaries in dispensing medicines. In the early seventeenth century the capacity table was regulated by weight of water of the various volumes. Thus (1618):

A tablespoon weighed 3 *drachms*
A wine glassful weighed 1 ½ oz (Troy)
A *cotyla* weighed 9 oz
A pound weighed 12 oz
A *sextarius* weighed 18 oz
A gallon weighed 108 oz. (Volume 269.4 in^3, close to the gallons of Henry VII and Elizabeth I).

In the mid-nineteenth century (post-Imperial), fluid measures were used. Thus:

60 *minima* = 1 fluid *drachma*
8 fluid *drachmas* = 1 fluid oz
20 fluid oz. = 1 *octarius*
8 *octarius* = 1 *congius*

But increasingly Avoirdupois weight was used for drugs. The 1878 Weights and Measures Act (Schedule 2) made small changes to the table, thus:

20 *minims* = 1 fluid *scruple*
3 fluid *scruples* = 1 fluid *drachm*
8 fluid *drachms* = 1 fluid oz
160 fluid oz = 1 Imperial gallon

This shows the *congius* to be the Imperial gallon, hence the *octarius* is the Imperial pint.

Today, metric measures apply in pharmacy.

Apothecaries' Weight: The system of weights used by apothecaries in dispensing medicines, based on the Apothecaries' pound (the Troy pound) of 5760 grains, or 12 oz, 96 *drachms* or 288 *scruples*. Today, metric measures are used for pharmaceutical weighings.

As (plural **Aes**): One pound of bronze or copper used as a weight and currency in Roman times.

Attic-Euboic Measure: A system of weights associated with Solon's reforms of 594 BC. See *Talent, Mina, Drachma,* and *Obol.*

Aureus: The gold coin of the Roman Empire, valued at 25 silver *denarii* (see *Denarius).*

Avoirdupois Weight: Weight scale based on a pound of 7000 grains, introduced into England *c*1300, used for commercial weighing.

Baker's Dozen: Thirteen, to avoid the penalties for giving bread in short weight.

Barley Corns: Grains of barley traditionally weighing one twenty-fourth of a Troy ounce and measuring in length one-third of an inch; the weight unit of all English weights, e.g. 1 lb Troy = 5760 grains, 1 lb Avoirdupois = 7000 grains, 1 lb Tower = 5400 grains, 1 *libra mercatoria* = 7200 grains.

Barrel: A measure of capacity; for beer usually 36 gallons; for wine $\frac{1}{8}$ tun (q.v.) or 31 ½ gallons. Various other volumes for different commodities (See Ch. X. (b)).

Bovate: One-eighth of a *hide* (q.v.): term used in East Anglia. The ploughland of one ox.

Bullion Weights: The weights for gold and silver, that is Troy weight (q.v.); the weights used by goldsmiths and silversmiths.

Bushel: A measure of capacity of eight gallons (q.v.), one eighth of a quarter (q.v.).

Butt: A measure for wine of 126 gallons, also called a *pipe.*

Cade: A cask for herring holding 720 fish.

Carat: Originally the seed of the Carob plant *(Ceratonia Siliqua);* in the early Middle Ages it equalled the weight of 3 barley grains or 4 wheat grains, later equated with a weight of 4 grains. In the nineteenth century the carat equalled 205 mg. The metric carat of today is 200 mg.

Carucate: In Danelaw, the same as a *hide* (q.v.).

Cental: A weight of 100 lb Avoirdupois.

Chain: A measure of length introduced by Gunter in the seventeenth century to decimalize the acre: a length of four rods or 66 ft; the surveyor's chain.

Chaldron: A measure of capacity, originally 32 bushels, later 36 bushels. In 1695 the chaldron of coal = 53 cwt = 3 wagon loads = 6 cart loads. (See Ch. VI (e)).

Clove: One sixteenth of a hundredweight, for wool. Variously 6 ¼, 6 ½, 6 ¾, and 7 lb, according to whether the hundredweight was 100, 104, 108, or 112 lb.

Congius: A Roman measure, one-eighth of an *amphora,* approx. 3.5 l.

Coomb: A measure of capacity of the Middle Ages equal to 4 bushels in the thirteenth century; half a *seam* (q.v.).

Cotyla: A measure of capacity in early seventeenth-century pharmacy; the volume of 9 oz Troy of distilled water.

Cran: A measure of capacity, being a box or basket, for fish, especially herring; 45 English wine gallons or 37 ½ Imperial gallons in capacity.

Cubit: An ancient measure of length being the distance from elbow to fingertips, approximately 18 inches (45 cm). The Samian cubit equalled the Roman Egyptian cubit of 52.3 cm.

Cyanthus: A small Roman measure of capacity, one twelfth of a *sextarius* (q.v.) or one five hundred and seventy-sixth part of an *amphora* (q.v.); approximately 48 ml.

Denarius: A Roman silver coin originally, as its name implies, worth 10 *aes.* (q.v.), first minted in 211 BC. weighing 4 scruples (q.v.) or $\frac{1}{6}$ of a Roman ounce (4.55 g) at a purity of 96% silver.

Denier: The eighth-century Frankish penny minted at 240 to the pound, each of weight 20 Troy grains (1.3 g), later increased to about 1.70 g.

Didrachma: A silver coin of the third century BC. of southern Italy weighing 6 scruples (6.75 – 6.8 g).

Digit: An early body measure of length, the breadth of a finger. In Roman times 16 digits made a Roman foot.

Drachm: An apothecaries' weight, one ninety-sixth of an apothecaries' (Troy) lb, and one eighth of a Troy ounce.

Drachma: A Greek weight, one hundredth of a *mina* (q.v.) equal to six *obols* (q.v.); also called a *drax*. Also a silver coin of weight about 6 g.
A Roman weight, one ninety-sixth of the pound *(libra)* or one-eighth of a Roman ounce.
Fluid apothecaries' measure, one eighth of a fluid ounce.

Dragma: see Roman drachma.

Dram: A unit of weight, one-sixteenth of an ounce Avoirdupois, 27⅓ Troy grains (1.77 g).

Ell: A measure of length especially for cloth. In Saxon times its length is unknown. The English ell was 45 inches, the Scottish 37 inches, the Flemish ell 27 English inches.

Faat: An unofficial measure of capacity, comprising a 'quarter' of nine bushels instead of eight, used in London, early fifteenth century.

Fathom: A body measure of length, the distance between fingertips with the arms outstretched, approximately one's height. The Greek fathom was four *cubits* (q.v.) or six feet in length. In modern times the fathom equalled six feet English measure.

Firkin: *Wine:* a measure of capacity of 84 gallons, i.e. one third of a Tun (q.v.)
Ale or beer: a quarter barrel originally 9 ale gallons, in Imperial measure 9 gallons.
Butter: A cask of butter weighing 64 lb gross, 56 lb butter with the cask weighing 8 lb.

Florin: An English silver coin worth two shillings, originally stamped 'One tenth of a pound', first issued in 1849.

Foot: A measure of length: *Drusian:* a foot of eighteen digits or 13.11 inches (333 mm), the foot of the Tungri mentioned by the Roman General Drusius.
Natural: the length of a human foot; in the Middle Ages 9.7 – 9.9 inches (246 – 251 mm), frequently taken as 9.9 inches.
Olympic: A Greek foot of about 12.14 English inches (308 mm).
Roman: The length of 16 digits, 11.65 English inches (295.9 mm).
Welsh: a foot of three palm breadths each of three inches, so a nine inch foot (228.6 mm).

Forpit: A measure of capacity, one fourth of a peck often measured by weight and not by volume as 3½lb Avoirdupois, in use in Scotland.

Fotmal: A weight for lead (thirteenth-century) equal to 70 lb.

Furlong: The length of the long side of a rectangular acre measuring 40 rods, originally the length of a furrow in ploughing an acre. Later one-eighth of a statute mile.

Gallon: A measure of capacity based on the volume of eight lb of wheat, also of eight lb of wine.

Ale: a measure of volume 282 in^3.

Corn: a measure of 268.43 in^3 later raised to 272 or 272¼ in^3 and called Winchester measure.

Guildhall: a measure of volume 224 in^3. A Tudor standard deposited in the Guildhall, London.

Wine: a measure of 231 in^3 used for centuries in determining duty on imported wine without apparent authority. Legalized 1707.

Gradus: A step (Latin), a unit of length of 2½ Roman feet.

Grain: The fundamental weight in theory in English metrology, being the supposed average weight of a barley corn from the middle of the ear, equal to 64.8 mg, and the basis of Avoirdupois weight (1lb = 7000 grains), Apothecaries' and Troy weight (1 lb = 5760 grains), Tower weight (1 lb = 5400 grains), and the *libra mercatoria* (1 lb = 7200 grains).

Paris: The supposed weight of a wheat grain, 53 mg, thirty-two of which was the weight of Charlemagne's new penny of AD 793/4 (see Grain – wheat, below).

Troy: The supposed weight of a barley grain, 64.8 mg (see above).

Wheat: A weight unit of 48.6 mg supposedly the average weight of a grain of wheat, thirty-two of which made the weight of an English penny or sterling.

Gyrd: An Anglo-Saxon unit of length, now known to be equal to the later land rod of 16½ English feet.

Hand: A unit of length, being a body unit, the breadth of the whole hand, taken as four English inches.

Hide: A unit of assessment of land, originally based on productivity, being the land required for the support of a family throughout the year. It was the ploughland of a team of eight oxen and almost always referred to arable land, found variously as 64 to 240 acres but usually taken as 120 acres, this being a long hundred (see below).

Hogshead: A measure of capacity for wine equal to 63 wine gallons (q.v.) or 52½ Imperial gallons.

Hundred: In the Middle Ages, six score or 120, sometimes called the *long* hundred. The Saxon and Norse hundred was 120. In England this was the customary meaning of the word till the sixteenth century, this mode of reckoning continuing for some commodities into the nineteenth century.

Hundredweight: A unit of weight, variously 100, 104, 108, or 112 lb Avoirdupois depending on commodity and period, used for heavy merchandise.

Imperial Measure: The system of English measurement introduced by the Act 5 George IV c. 74 of 1824 which defined the yard, the gallon (as the volume of 10 lb of water), and the Troy pound of 5760 grains as the basis of British measurement.

Inch: In English metrology, one twelfth of a foot; in the Middle Ages commonly taken as the thumb's breadth at the root of the nail.

Ingot: Axe-head, a slab of cast silver hammered out at both ends to resemble a double-edged axe, often inscribed, weighing approximately one Roman pound, and associated with the mintage of Roman coins at so many to the pound.

Jugerum: A Roman measure of area equal to two square *actus* (q.v.).
Jugum: A measure of land area, being one-quarter of a *sulung* (q.v.).

Kilderkin: For ale or beer, half a barrel (q.v.). For butter, a half barrel, weighing 132 lb gross, of which 112 lb was butter, the cask itself weighing 20 lb.
Kilogram: The unit of weight (mass) on the Metric System, originally intended as the mass of a cubic decimetre of distilled water at its maximum density (4°C). The International Prototype Kilogram is a cylinder of platinum-iridium alloy but it is not precisely the mass of a cubic decimetre of water at 4°C but is the mass of 1.000028 dm^3 at 4°C. The kilogram is no longer defined in terms of water but is the mass of this platinum cylinder.
Knight's Fee: A unit of land assessment. From each such unit held of the king one knight was to be provided to give annually 40 days' and nights' service in the field. Later this personal service was commuted to a money payment for the hiring of mercenaries, called *scutage* or shield money. As a land area it was very variable. If land was assessed as one fee it was one fee, liable for this service, regardless of its actual area, but the most frequently encountered areas were 480 and 640 acres, that is 4 or 5⅓ *hides* (q.v.).

Last: *Herrings:* A measure of quantity or capacity defined (thirteenth century) as ten thousand fish, each thousand of ten hundred and each hundred (q.v.) of six score. Thus the last was 12,000 fish.
Leather: 200 skins
Wool: 12 sacks (q.v.).
League: *English:* 12 furlongs (q.v.) in thirteenth, fourteenth century i.e. 1½ miles; 2 miles in late fourteenth century and 3 miles in fifteenth century and later.
Gallic: 16 furlongs, later (nineteenth century) defined as $\frac{1}{25}$ of a degree of latitude or 2.76 miles.
Nautical: On the globe the length of the arc of $\frac{1}{20}$ of a degree of latitude. In the seventeeth century the degree was wrongly taken as 60 miles, giving 3 miles for the league. In the eighteenth century the degree was found to be 69.1 English miles giving 3.456 miles for a nautical league.
Libbra: The pound of 12 oz of the city of Florence (*c*340 g), also of 16 oz (*c*454 g).
Libra: The Roman pound of 5050 grains (327.45 g); the original weight of the *as*(q.v.). The *solidus* (gold) of Constantine weighed $\frac{1}{72}$ of this pound.
Libra Mercatoria: A pound weight of the Middle Ages, the weight of 25 shillings in pence (15 Troy oz), so called by Fleta (1290). This was used for weighing everything except pharmaceuticals, bullion, and bread.
Litre: The unit capacity on the Metric System, originally the volume of a cubic decimetre. The weight of such a volume of distilled water was intended to be the kilogram (q.v.) but as it was found that 1kg of water at 4°C occupied 1.000028 cm^3 the litre was redefined (1901) as the volume of a kilogram mass of water at 4°C.
Livre: The old French pound of 9216 French grains (488.4 g) each grain of 53 mg (see Grain, Paris). The livre was divided into 16 oz or 128 *gros*.
Load: A weight of 30 *fotmals* (q.v.), 2100 lb of lead.

Mancus: An old English gold coin, also a weight of gold equal to that of the coin. The coin was worth 30 silver pennies.

Manupedes: See *Pes Manualis*.

Mark: *Weight:* Eight ounces, two thirds of a pound, or half a 16 ounce pound. *Money:* Two thirds of a pound sterling, or 13 shillings and 4 pence.

Metre: The unit of length in the Metric System. Originally intended to be one ten-millionth part of the earth's quadrant, pole to equator through Paris. The actual metre bar made according to this definition turned out to be about one fifth of a millimetre too short were it to truly represent this fraction of the actual quadrant. The metre was redefined as the actual length of the metre bar as it was. It is now defined (1983) as the distance travelled by a light beam in a vacuum in 1/299792458th of a second. (The speed of light is now accepted as 299792458 metres per second).

Metric System: The system of measurement based on the metre, kilogram, and litre (qq.v.), legalized 10 December 1799 in France, later adopted by many countries.

Mil: (plural **Mila**): The Anglo-Saxon mile of 5000 natural feet (q.v.).

Mile: A measure of length. *English:* The statute mile is 8 furlongs or 5280 feet.
Nautical: one third of a nautical league (q.v.) i.e. 1.152 statute miles.
Old English: A customary mile of pre-Elizabethan England measuring 1.3 – 1.5 statute miles.
Roman: The distance of 1000 paces each of 5 Roman Feet (q.v.); also, 8 *stades* (q.v.), equal to 4854 English feet (1479.5 m).

Millarium: A Roman measure of distance, 1000 paces. The Roman mile (q.v.).

Mina: A Greek weight, about 435 g in the Attic-Euboic scale or 600 g on the Aeginetan.

Minim: Apothecaries' measure, one sixtieth of a fluid drachma, often referred to as a 'drop'.

Mitta: An Anglo-Saxon unit of capacity, equal to two *ambers*(q.v.).

Nail: A measure of length, being one-sixteenth part of a yard, 2¼ inches.

New Pence: The new currency introduced on 15 February 1971. The pound previously was worth 240 pennies, now it was worth 100 'new pence', thus breaking a tradition that had lasted since the end of the eighth century but introducing decimalization into the coinage. The word 'new' is now seldom used as everyone is accustomed to the change. The abbreviation is 'p' and there are ½p, 1p, 2p, 5p, 10p, 20p, 50p, and, since 1983, £1 coins.

Obol: A Greek unit of weight and currency. As a weight, the obol was one-sixth of a *drachma*. The name derives from *obelos* = a spit, and the early currency consisted not of circular coins but long iron spits, six of which made a handful (*drachma* or *drax*). With the mintage of silver the coin displaced the iron spit and the name was transferred to the coin which weighed about 1 g.

Octarius: One eighth of a gallon, Apothecaries' measure.

Ora or **Ore:** A weight and a unit of monetary account of Danish origin used in Saxon and Norman times. Originally and most commonly there were 15 ores to the pound and 1 ore equalled 16 pence, but in the revenues from the royal estates the dues in ores were listed as 20 pence and are so listed in Domesday Book and elsewhere. As a weight, Domesday Book considers it to be one ounce.

Ounce: A unit of weight:
Apothecaries': see Troy ounce.
Avoirdupois: one-sixteenth of an Avoirdupois lb, 437.5 grains (28.35 g).
Roman: one-twelfth of a Roman lb, 420.8 grains (27.27 g).
Troy: one-twelfth of a Troy lb., 480 grains (31.1 g).
A unit of volume, fluid oz: Originally Apothecaries' measure equal to 8 fluid drachmas, later regarded as Imperial measure, viz the volume of 1 Avoirdupois ounce of water (1.733 in^3). In the U.S.A. the fluid ounce is one-sixteenth of their fluid pint and has a volume of 1.80 in^3 and, as water, weighs 1.04 oz.

Ox: The standard of value in the Homeric world, equivalent to one *talent* of gold (8.5 – 8.75 g).

Oxgang: One eighth part of a *hide* (q.v.). In East Anglia also called a *bovate* (q.v.).

Palm: A body-measure of length representing the breadth of the four fingers of the hand.

Peck: A measure of capacity usually for dry goods equal to two gallons or a quarter of a bushel.

Penny: A silver coin first minted in England in the late eighth century. Its theoretical weight, given in later documents, was 32 wheat grains which would have been 24 Troy grains. In fact its weight was quite variable until Norman times when it stabilized at 22½ Troy grains; also called a *sterling* (q.v.). The quality of the silver was usually good, 90-95% pure. Abbreviation, *d* (for *denarius*). See also *New Pence*.

Pennyweight: A unit of weight on the Troy Scale, equal to 24 Troy grains or one-twentieth of a Troy ounce. Only exceptionally did a silver penny weigh a pennyweight, a fact which led to much confusion. Abbreviation, dwt.

Perch: A measure of length dating back to Saxon times, also called a *rod* or a *pole*, equal to 16.5 English feet (5.029 m), which defined the *acre* (q.v.). The Saxon *gyrd* (q.v.) has been shown to be identical to this unit. In the Middle Ages the length of the perch could be quite variable depending on the land's productivity.

Pes: A measure of length, one foot (q.v.). The Roman *pes* measured 11.65 inches (295.9 mm), also called the *Pes Monetalis* as the Roman Standards were kept in the Temple of Juno Moneta in the Capitol.

Pes ad manus: see *Pes Manualis*.

Pes Manualis: A body-measure of length, a 'foot' measured by the hands. Grasp a stick with both hands with the thumbs extended and touching: a *pes manualis* is the distance between the extremities of the hands. This was approximately 13.1 inches (333 mm), one half of which was a *shaftment* (q.v.).

Pied Manuel: see *Pes Manualis*.

Pint: A measure of capacity being one-eighth of a gallon. In the Imperial System, 34.66 in^3. In the U.S.A., 28.875 in^3.

Pipe: see Butt.

Pole: see Perch.
Pot: A measure of capacity for butter weighing 20 lb containing 14 lb butter, the vessel weighing 6 lb.
Pottle: A measure of capacity equal to half a gallon.
Pound: A unit of weight on all systems of English Measurement. (see also *Libra Mercatoria*).
Apothecaries': see pound Troy.
Avoirdupois: A pound of 7000 Troy grains, of 16 oz each ounce of 16 drams, used for commercial purposes.
Celtic: Two pounds used in ancient Britain, one said to have weighed 309 g, the other 619 g.
Roman: A pound of 5050 Troy grains (327.453 g).
Tower: The pound of the English moneyers, the weight of 240 silver pennies each nominally of 22½ Troy grains, hence 5400 Troy grains: The weight of the 'pound of pence'.
Troy: A pound of 5760 Troy grains, of 12 oz or 240 pennyweights. The weight for bullion and in the Middle Ages for bread and generating the unit of volume ('eight pounds of wheat do make a gallon').
Money: The basis of the English monetary system; One pound worth twenty shillings or 240 pence. Since metrication the same pound is worth 100 new pence (q.v.).
Puncheon: see Firkin.

Quadrans (Quadrantes): In Roman times, one quarter of the monetary unit, the *as* (q.v.).
Quadrantal: A Roman measure of capacity, the same as *amphora* (q.v.), being the volume equal to the cube of a Roman foot.
Quart: A measure of capacity, being one-quarter of a gallon.
Ale: 70.5 in^3.
Imperial: 69.36 in^3.
Reputed: 46.24 in^3, the volume of the standard wine or spirit bottle, in fact only $\frac{2}{3}$ of an Imperial quart.
Whisky: see Reputed Quart (above).
Winchester: For liquid chemicals, (twentieth century) containing two Imperial quarts (*c*2¼ litres). Today 'standardized' at 2½ litres.
Wine: 57.75 in^3.
U.S.A: for liquids: 57.75 in^3.
for dry goods: 67.20 in^3.
Quartarius: A Roman measure of capacity being the one hundred and ninety-second part of an *amphora (quadrantal)*, or one quarter of a *sextarius* which was about 567 ml, so very close to the Imperial pint. The *quartarius* contained therefore about 141.75 ml.
Quarter: A measure of capacity being legally eight bushels or sixty-four gallons. In earlier times also called a *seam*. In Imperial measure the volume of 640 lb of pure water, i.e. 10.275 ft^3. (2.909 hectolitres)
Quintal: A weight of 100 lb; sometimes used in modern times to describe 100kg.

Rast: An itinerary measure of length of two Gallic leagues or three *miliaria* (qq.v.).
Rod: see Perch.
Rood: A measure of area, one quarter of an acre; a strip of land, 40 perches long and one perch broad.
Rundlet: A measure of capacity for wine of 18 ½ gallons or, since 1700, 18 gallons, being one-fourteenth of a tun (q.v.). In Imperial measure, 15 Imperial gallons.

Sack: A unit of weight for wool, originally 350 lb, later 364 lb.
Sceatta: Originally one Troy grain of gold, one-twentieth of a gold shilling (seventh century AD). In the late seventh and eighth centuries the word was used to describe the first (uninscribed) silver coins of England weighing 20 grains when the gold currency was displaced.
Scruple: A Roman weight, one twenty-fourth of the Roman ounce or 1.14 g. In Apothecaries' measure, a unit of weight equal to one twenty-fourth of a Troy (Apothecaries') ounce. One scruple was thus 20 grains (1.3 g).
Seam: see Quarter.
Semis: A Roman unit of monetary weight being half an *as* (q.v.).
Semimodius: A Roman measure of capacity; half a *modius* and one-sixth of an *amphora* (q.v.).
Sester: A unit of capacity in Anglo-Saxon times, also mentioned in the Domesday Book; close to the volume of a Roman *sextarius* (q.v.), being about a pint, but for some goods given as fifty per cent more. By the thirteenth century the unit was much larger. For wine it was then stated to be four gallons.
Sestertius, (sestercius): A unit of Roman monetary weight and a coin, originally worth 2.5 *asses,* later in Augustan times, worth 4 *asses.*
Sesuncia: A unit of Roman monetary weight, one eighth of an *as* (q.v.).
Sextans: A unit of Roman monetary weight, one sixth of an *as* (q.v.).
Sextarius: A Roman measure of capacity, being one-sixth of a *congius* (q.v.), approximately 567 ml, almost one Imperial pint.
Sextula: A unit of Roman monetary weight, one seventy-second part of an *as* (q.v.).
Shaftment: A unit of length, body measure; half of a *Pes Manualis* (q.v.), the breadth of a fist with thumb extended.
Shilling: A unit of account and a coin. In the seventh century a gold coin weighing 20 Troy grains, displaced at the end of that century by a coinage of silver, called *sceattas,* of weight 20 Troy grains. The shilling of Mercia and of Wessex were worth four and five pence respectively. At the end of the eighth century the Frankish shilling was worth 12 pence (duodecim denarios) and as early as the mid-century there were 240 pence to the Frankish pound. The Norman reckoning was likewise 12 pence to the shilling. At this time there was no coin of this value. It was simply a unit of account. The silver coin worth 12 pence (originally called a *testoon,* later a shilling) was issued for the first time in England in 1504.
Sicilus: A Roman unit of monetary weight, being one quarter of a Roman ounce or one forty-eighth of the *as* (q.v.).
Siliqua: An ancient unit of weight, that of the seed of the carob plant *(Ceratonia Siliqua)* also called the *carat* or *keration*, weighing 4 wheat grains or 3 barley grains. In Roman times, reckoned as one seventeen hundred and twenty-eighth part of the Roman pound. From the known weight of the Roman pound this gives

almost three Troy grains i.e. barley grains as stated in earlier times.

Solidus: A Roman unit of weight and a coin. The gold solidus of Constantine was minted at 72 to the Roman pound in AD 309, i.e. the weight was one-sixth of a Roman ounce, approximately 4.55 g, 70 Troy grains or 24 carats. The third of the solidus, a coin weighing 1.5 g, was called the *tremissis.*

Sovereign: A gold coin worth one pound or twenty shillings first issued in England in October 1489 by Henry VII.

Stade: A Greek and Roman measure of length, both of the same length.
Greek: A distance of 600 Olympic feet (607 English feet, 185.0 m).
Roman: A distance of 625 Roman feet or 125 paces.

Stater: The name of a weight and several gold coins e.g. that of Philip II of Macedon (359 – 336 BC) weighing about 8.75 g (135 grains) or that of Solon (*c*600 BC) of 196 grains (12.7 g).

Sterling: *Coin:* A silver penny weighing on average, theoretically, 22.5 grains (1.45 g).
Silver: Silver bullion of the quality used in minting the silver penny being 92.5% pure silver, the rest base metal alloy.

Stone: A unit of weight whose magnitude depended on the commodity being weighed and the period in history, ranging from 5 to 12, 12½, 13, and 14 lb, e.g. 1302/3, Stone for glass = 5 lb, stone for cheese, wool, etc. = 12½ lb, for lead = 12 lb. In 1281 the stone for wool was 13 lb. By 1340 the stone for wool was 14 lb. The 14 lb stone was in common use for most commodities before metrication.

Sulung: A unit of land assessment in Kent equal to the *hide* (q.v.).

Tablespoon (ful): A rough measure of capacity, supposedly 3 *drachms* (q.v.) when measuring water.

Talent: An ancient unit of weight. A small weight of gold (8.5 – 8.75 g) worth an ox in Homeric times, alternatively a large weight for general commerce ranging from 25 to 37 kg, depending on its place of origin and period. One sixtieth of a talent was called a *mina* (q.v.).

Tertian: see Firkin.

Testoon: see Shilling.

Thumb: A body-measure of length. The breadth of the thumb at the root of the nail was commonly called an inch in the Middle Ages.

Thurdendel: A drinking vessel, seventeenth-century, for malt liquors, somewhat larger than the requisite capacity so that a full measure of liquid may be obtained with the froth on top, similar in intent to the modern 'line' beer glass.

Tierce: A measure of capacity for wine, 42 gallons, being one-sixth of a *tun* (q.v.).

Tod: A unit of weight for wool of 28 lb.

Toise: An old French measure of length equal to six French feet. The *Toise de Perou* (1766) consisted of 6 French feet, each of 12 inches, each of 12 'lines' for a total of 864 lignes. Provisionally the metre was declared to be 443.44 'lines' (1793), later amended (1799) to 443.296 'lines'.

Tremissis: A coin and weight in Roman times being one third of a *solidus* (q.v.).

Triens: A Roman monetary weight, one third of an *as* (q.v.).

Troy Weight: Believed to be the most ancient weight system of the kingdom and to have entered England via the city of Troyes, in France, hence its name. Troy weight and Apothecaries' weight have the same pound, ounce, and grain. They only differ in their other subdivisions for reasons of expediency and are doubtless of common origin. For centuries this was the standard weight system at the Exchequer. Apothecaries' weight (q.v.) especially and Troy weight to a lesser extent show the influence of the Roman mode of subdivision. The Troy pound consists of 12 ounces each of 20 pennyweights, each of 24 grains. Today only the Troy ounce of 480 grains (31.1 g) is in use solely for weighing precious metals.

Ton: A unit of weight on the Avoirdupois system, being 2240 lb, subdivided into 20 hundredweights each 112 lb, and 160 stones each of 14 lb.

Tun: A measure of capacity for wine and ale, originally 256 gallons for both, with the ale barrel for centuries being one-eighth of this. For wine the tun lost four gallons and dropped to 252 gallons, some time prior to the fourteenth century, at which figure it remained for centuries. With the change to the Imperial gallon the tun became equivalent to 210 Imperial gallons.

Ulna: A measure of length, a Latin word signifying either a yard or an ell as the same word is used for both indiscriminately, although the yard is quite different from the ell (qq.v.).

Uncia: A Roman unit of length of one twelfth of a Roman foot: also a Roman unit of monetary weight being one-twelfth of a Roman pound.

Unit: (a) A measure of capacity for herring adopted on Britain's entry into the European Economic Community, amounting to 100 kg of fish, subdivided into 4 'boxes' of 25 kg each.

(b) Of liquor bottles, one unit is one bottle containing not less than 23 nor more than 28 fluid ounces. One case consists of 12 such 'units' (Weights and Measures Act 1963 s.59).

Virgate: A measure of area or of assessment. The term is used for both as a quarter of an acre (q.v.) and a quarter of a *hide* (q.v.).

Vaat: see Faat.

Wey: A unit of weight for wool, tallow, and cheese, being first defined in 1302/3 as 14 stone each of 12½ lb. Two weys of wool became the sack of 350 lb, later raised to 364 lb. In 1430 the wey of cheese was declared to be 224 lb at which figure it continued into the nineteenth century.

Wist: A unit of area or assessment mentioned in the Chronicle of Battle Abbey which gives 1 wist as 4 *virgates*. A *virgate* (q.v.) was a quarter *hide*, so a *wist* is a *hide* normally. (The Abbey at this time was given beneficial hidage so that in assessing the dues per hide the Abbey could count 8 *virgates* to the assessable *hide*).

Winchester Measure: The measures deriving from those of Henry VII and Elizabeth I for dry goods, seeds, etc., so mentioned in 1670 and defined 8 and 9 William III c 22, s. 9 and s. 45 of 1696-7. This made the Winchester bushel a little larger than that of Elizabeth, that is 2150.42 in^3. The gallon should have been 268.8 in^3 but subsequent statutes refer to 272.25 in^3 which is close to the gallon of Henry and Elizabeth but not one-eighth of the bushel. These measures were widely used into the nineteenth century.

Wineglassful: A rough measure of capacity for Apothecaries' use. The volume of a wine glass without size specified but taken as approximately 1 ½oz for water.

Yard: The primary measure of length prior to metrication, introduced about 1100 by Henry I. Iron bars representing the yard were issued following Richard I Assize of Measures of 1196. The yard consists of 3 feet or 36 inches. Currently the yard is defined in terms of the metre (1 yard = 0.9144 m) (Weights and Measures Act 1963. s. 1).

Yoke: A measure of area or assessment being one quarter of a *sulung* in Kent, i.e. one quarter of a *hide* (q.v.). The ploughland of two oxen.

Bibliography

ADDYMAN, P. V. and HILL, D. H. Saxon Southampton: A Review of the Evidence: Pt.1 History, Location, Date, and Character of the Town. *Proceedings of the Hampshire Field Club* (1968) Vol.XXV pp.61-93 Winchester.

AGRICOLA, Georgius See HOOVER, Herbert Clark (ed. & tr.).

AIRY, G. B. An Account of the Construction of the New National Standard of Length and of its Principal Copies. *Phil. Trans. Roy. Soc.* (1857) Vol.CXXXX-VII, p.621 et seq. London.

ALLEN, Derek. *An Introduction to Celtic Coins* London: British Museum Publications Ltd 1978.

ALLEN, Derek *Belgic Dynasties of Britain and their Coins* London: Society of Antiquaries 1944.

ALLEN, Derek. Iron Currency Bars in Britain. *Proc. Prehist. Soc.* (1967) Vol.XXXIII, p.307. London.

Anglo-Saxon Chronicle Garmonsway, G. N. (tr.) London: J. M. Dent and Son 1962.

ANON. *A Metric America: A decision whose time has come* July 1971. Washington, D. C.: US Government Printing Office 1971.

ANON. An Account of a Comparison lately made by some Gentlemen of the Royal Society of the Standard of a Yard ... *Phil. Trans. Roy. Soc.* (1742-3) Vol.XXXXII, pp.541-556. London.

ANON. An Account of the Proportions of the English and French Measures and Weights from the Standards of the Same, kept at the Royal Society. *Phil. Trans. Roy. Soc.* (1742-3) Vol.XXXXII, pp.185-188. London.

ANON. *Units and Standards of Measurement employed at the National Physical Laboratory I. Length, Mass, Time, Volume, Density and Specific Gravity, Force, and Pressure.* London: Dept. of Scientific and Industrial Research 1951.

ANON. *Instruction sur les Mesures.* Poitiers: La Commission temporaire des Poids et Mesures republicaines, Michael-Vincent An.II (1793-4).

ANON. Weights and Measures in History ... *Chemist and Druggist* (1969) Vol.CX, p.816 et seq. London.

ANON. *Weights and Measures. Inspectors and Inspection. Model Regulations* London: HMSO 1890.

ARNOLD, Richard. *The Customs of London, commonly called Arnold's Chronicle* London: F. C. and J. Rivington 1811 (Reprint of 1st ed. *c* 1502).

ASSER. *Life of King Alfred.* STEVENSON, William Henry (ed.) Oxford: Clarendon Press 1904, reprinted 1959.

ATKINSON, Tom. *Elizabethan Winchester.* London: Faber and Faber 1963.

ATTENBOROUGH, F. L. (ed.) *The Laws of the Earliest English Kings.* Cambridge: C.U.P. 1922.

BAILEY, F. Report of the New Standard Scale of this Society. *Memoires of the Royal Astronomical Society* (1836) Vol.IX, pp.35-184. London.

BALIBAR, François. Le mètre au fil du temps. *La Recherche* (1984) Vol.XV, No.152, p.263. Paris.

BARRELL, H. The Metre. *Contemporary Physics* (1962) Vol.III, p.415-434.

BARRELL, H. The Standards of Length in Wavelengths of Light. *Proc. Roy. Soc.* (1946) Vol.CLXXXVI, pp.164-170. London.

BASS, G. Cape Geldgonya: A Bronze Age Shipwreck. *Trans. Am. Philos. Soc.* (1967) N.S. Vol.LVII, Pt.8. Philadelphia.

BATESON, Mary (ed.) *Records of the Borough of Leicester* London and Cambridge: C. J. Clay and Sons 1899.

BEDE. *A History of the English Church and People.* PRICE, Leo Sherley (tr.) Harmondsworth: Penguin Books 1972.

BENTON, W. A. The Soham and Kings Lynn Steelyards. *Newcomen Society Transactions* (1938-9) Vol.XIX, p.241. London.

BERESFORD, M. and HURST, J. G. (Hurst, J. G. (ed.)) *Deserted Mediaeval Villages.* Woking: Lutterworth Press 1971.

BERRIMAN, A. E. *Historical Metrology.* London: J. M. Dent and Sons 1953.

BIDDLE, Birthe Kjolbe-. *Preliminary Report. The Old Minster at Winchester in the 7th Century.* Privately circulated, February 1974.

BIDDLE, Martin. Excavations at Winchester 1962-3. Second Interim Report. *Antiq. J.* (1964) Vol.XLIV, pp.188-219. Oxford.

BIDDLE, Martin. Excavations at Winchester 1969. Eighth Interim Report. *Antiq. J.* (1970) Vol.L, Pt.II, pp.277-326. Oxford.

BIDDLE, Martin. Excavations at Winchester 1970. Ninth Interim Report. *Antiq. J.* (1972) Vol.LII, pp.93-131. Oxford.

BIDDLE, Martin (ed.). *Winchester in the Early Middle Ages. Winchester Studies* Vol.I. Oxford: Clarendon Press 1976.

BIDDLE, Martin. Winchester, the Development of an Early Capital. *Vor-und Frühformen der europäischen Stadt im Mittelalter. Symposium in Reinhausen bei Gottingen 18-24 April 1972* Teil 1 pp. 229-261 Gottingen.

BIRCH, Walter de Gray. *Cartularium Saxonicum* (3 vols.) London: Whiting 1885-93.

BLAIR, Peter Hunter. *Roman Britain and Early England 55 BC — AD 871.* London: Nelson 1963.

BOYS, W. *Collections for a History of Sandwich in Kent.* The Customal of Sandwich. Vol.II. Canterbury 1892.

BRAUNFELS, W. (ed.). *Karl der Grosse* Dusseldorf: L. Schwann 1965.

BROOKE, George Cyril. *English Coins from the Seventh Century to the Present Day* 3rd ed. London: Methuen 1950.

BROWNBILL, J. The Tribal Hidage. *Eng. Hist. Rev.* (1925) Vol.XXXX, pp.497-503.

BRUNTON, Robert. *A Compendium of Mechanics* (6th Ed.) Glasgow: John Niven 1834.

BURRELL, L. The Standards of Scotland. *The Monthly Review, The Journal of the Institute of Weights and Measures Administration* (March 1961) Vol.LXIX, pp.49-62. London.

BUTLIN, R. A. Some Terms used in Agrarian History. *Agricultural History Review* (1961) Vol.IX, pp.98-104. London.

Calendar of Charter Rolls Edward I 1257-1300, Vol.II. London: HMSO 1906.
Edward III 1327-1341, Vol.IV. London: HMSO 1912.

Calendar of Close Rolls Edward II 1318-1322, London: HMSO 1895
Edward II 1323-1327, London: HMSO 1898.

CARSON, R. A. G. (ed.). *Mints, Dies and Currency. Essays dedicated to the memory of Albert Baldwin* London: Methuen 1971.

CHADWICK, H. Monro. *Studies in Anglo-Saxon Institutions* Cambridge: C.U.P. 1905.

CHADWICK, H. Monro. *The Origin of the English Nation* Cambridge: C.U.P. 1907.

CHAMBERLAYNE, John. *Magnae Britanniae Notitia* London: Goodwin Timothy 1708.

CHANEY, H. J. *Our Weights and Measures* London: Eyre and Spottiswoode 1897.

CHISHOLM, H. W. On the Science of Weighing and the Standard of Weight and Measures. *Nature* (1873) Vol.VIII, pp.268-270, 307-309, 327-329, 367-370, 386-389, 489-491, 552-555; *Nature* (1874) Vol.IX, pp. 47-49, 87-89, London.

CHISHOLM, H. W. *Seventh Annual Report of the Warden of the Standards* London: HMSO 1873.

CLODE, C. M. *Early History of the Guild of Merchant Taylors of the Fraternity of St. John the Baptist.* London: Privately printed 1888.

CLOSE, Col. Sir Charles. The Old English Mile. *Geog. J.* (1930) Vol.LXXVI, p.338. London.

CONANT, Kenneth J. Mediaeval Academy (of America) Excavations at Cluny IX: Systematic dimensions of the buildings. *Speculum* (1963) Vol.XXXVIII, p.1. Cambridge, Mass.

CONWAY, R. S. (ed.). *Melandra Castle. Being the Report of the Manchester and District Branch of the Classical Association for 1905* Manchester: Manchester University Press 1906.

COOK, G. H. *Old St. Paul's Cathedral.* London: Phoenix Press 1955.

COOPER, C. H. The Foot of St. Paul. *Gentleman's Magazine* (1852) N.S. Vol.XXXVIII, p.57. London.

COOPER, Major E. R. The Steelyard at Woodbridge. *Newcomen Society Transactions* (1938-9) Vol.XIX, p.185. London.

COPE, L. H. Roman Imperial Silver Coinage Alloy Standards. *Numis. Chron.* (1967) Vol.VII, pp.107-127. London.

COURBIN, Paul. Valeur comparée du Fer et de l'Argent lors de l'introduction du monnayage. *Annales: Economies, Societes, Civilisations* (1959) Vol.XIV, pp.209-233. Paris.

COWELL, J. *The Interpreter* Cambridge: John Legate 1607.

CRAWFORD, M. H. *Roman Republican Coinage* (2 vols.) Cambridge: C.U.P. 1974.

CROCKER, J. W. *Hansard (Commons)* 1816 Vol.XXXIV 1024.

CRUMMY, Philip. The System of Measurement used in Town Planning from the Ninth to the Thirteenth Centuries. *British Archaeological Reports* No.72 1979 No.8 pp.149-155 Oxford.

CUNLIFFE, Barry. *Winchester Excavations 1949-60* Winchester: City of Winchester Museum and Library Committee 1964.

CUNLIFFE, Howard. *Parallel Weights and Measures Regulations* London: Shaw & Sons. 1908.

CUNNINGHAM, W. *Growth of English Industry and Commerce — Early Middle Ages* Cambridge: C.U.P. 1890.

DARWIN, Sir Charles. A Discussion on Units and Standards. *Proc. Roy. Soc.* (1946)

Vol.CLXXXVI pp.149-152 London.

DAVIS, Norman. *Paston Letters and Papers of the 15th Century* Pt.I Oxford: Clarendon Press 1971.

DENT, Major Herbert C. Bronze Wool Weights of England. *Apollo* 1929 (July) pp.25-33 London.

DENT, Major Herbert C. *Old English Wool Weights* Norwich: H. W. Hunt 1927.

DILLON, Myles and CHADWICK, Nora K. *The Celtic Realms* (2nd ed.) London: Weidenfeld & Nicolson 1972.

DOLLEY, Michael. *Anglo-Saxon Pennies* London: British Museum 1970.

DOLLEY, R. H. M. (ed.). *Anglo-Saxon Coins* London: Methuen 1961.

DOLLEY, Reginald Hugh Michael (ed.) *The Norman Conquest and the English Coinage* London: Spink 1966.

DOWNER, L. J. (ed. & tr.). *Leges Henrici Primi* Oxford: Clarendon Press 1972.

DU CAGNE, Charles du Fresne. *Glossarium mediae et infirmae Latinatis* (10 vols) Niort: L. Favre 1883-87.

DUEMMLER, Ernst (ed.). *Monumenta Germaniae Historica, Epistolarum* Tomus IV Karolini Aevi II Berlin: Weidmannsche Verlagsbuchhandlung 1895.

DUGDALE, Sir William. *Monasticon Anglicanum* (J. Caley, H. Ellis and B. Bandinel, eds.) (6 vol. in 8) London: Longmans et al 1817-30.

DUGDALE, Sir William. See SPELMAN, Henry.

ELLIS, Sir Henry (ed.) *Chronica Johannis de Oxenedes* London: (Rolls Series) Longmans et al 1859.

ELLIS, Sir Henry (ed.) *Registrum Vulgariter Nuncupatum, The Record of Caernarvon* London: Record Commission 1838.

EVANS, Allan (ed.). See PEGOLOTTI, Francesco Balducci.

FERNIE, Eric. The Greek Metrological Relief in Oxford. *Antiq. J.* (1981) Vol.LXI pp.255-263 Oxford.

FERNIE, Eric. Observations on the Norman Plan of Ely Cathedral. *British Archaeological Association Conference, Transactions for the Year 1976*. DRAPER, Peter and COLDSTREAM, Nicola (eds.) Pt.II Mediaeval Art and Architecture at Ely Cathedral 1979.

FINBERG, H.P.R. *Early Charters of Wessex* Leicester: Leicester University Press 1964.

FLEETWOOD, Bishop. *Chronicon Preciosum* 2nd Ed. London: T. Osborne 1745.

FORBES, J. S. and DALLADAY, D. B. Composition of English Silver Coins (870-1300) *Brit. Numis. J.* (1960) Vol.XXXI pp.82-7 London.

FOWLER, W. On the Ancient Terms Applicable to the Measurement of Land. *Royal Institute of Chartered Surveyors' Transactions* (1884) Vol.XVI pp.275-316 London.

FREEMAN, E. A. *The Norman Conquest* (6 vols.) Oxford: Clarendon Press 1876.

FREEMAN, E. A. *The Reign of William Rufus* (2 vols) Oxford: Clarendon Press 1882.

FRY, Sir Frederic Morris and TEWSON, Ronald Stewart. *An Illustrated Catalogue of Silver Plate of the Worshipful Company of Merchant Taylors* London: Privately printed 1929.

GARMONSWAY, G. N. (tr.) See *Anglo-Saxon Chronicle*.

GIBBS, Marion (ed.) *Early Charters of the Cathedral Church of St. Paul* Camden 3rd Series Vol.LVIII London: Camden Society 1939.

GILES, J. R. (ed.) *William of Malmesbury Chronicle* London: Henry G. Bohn 1847.

GLAZEBROOK, Sir Richard. Standards of Measurement: Their History and Development. Supplement to *Nature* July 4, 1931 Vol.CXXVII p.17 et seq. London.

GOODMAN, Arthur Worthington. *The Manor of Goodbegot in the City of Winchester* Winchester: Warren and Son 1923.

GOULD, F. A. The Standards of Mass *Proc. Roy. Soc.* (1946) Vol.CLXXXVI pp.171-179 London.

GRAHAM, J. T. *Weights and Measures* (new edition revised by STEVENSON, M.) Princes Risborough: Shire Publications 1987.

GREAVES, J. *A Discourse on the Romane Foot and Denarius* London: William Lee 1647.

GRIERSON, Philip. *English Linear Measures. The Stenton Lecture 1971* Reading: The University of Reading 1972.

GRIERSON, Philip. La Fonction Sociale de la Monnaie en Angleterre aux VIIe-VIIIe Siècles. *Settimane di Studio del Centro italiano di studi sull 'alto medioevo* (1961) pp.341-385 Spoleto.

GRIERSON, Philip. Presidential Address. Coin Wear and the Frequency Table. *Numis. Chron.* (1963) Vol. III pp. i-xvi London.

GRIERSON, Philip. Presidential Address. Weight and Coinage. *Numis. Chron.* (1964) Vol.IV pp.iii-xvii London.

GRIERSON, Philip. *The Origins of Money. The Creighton Lecture in History 1970* London: The Athlone Press, University of London 1977.

GRIERSON, Philip. See also BRAUNFELS, W.

GRUNDY, G. B. The Old English Mile and Gallic League *Geog. J.* (1938) Vol.XCI p.251 London.

HALE, William Hale (ed.) *The Domesday of St. Paul's of the Year MCCXXII.* Camden Society Publications Vol.LXIX. London: Camden Society 1858.

HALL, Hubert (ed.) *The Red Book of the Exchequer* (3 vols) London: Eyre and Spottiswoode 1896.

HALL, Hubert and NICHOLAS, Frieda J. (eds.) *Select Tracts and Table Books Relating to English Weights and Measures (1100-1742)* Camden Miscellany Vol.XV London: Camden Society 1929.

HALLOCK, William and WADE, Herbert T. *Outlines of the Evolution of Weights and Measures and the Metric System* London and New York: Macmillan 1906.

HARMER, Florence Elizabeth (ed. & tr.) *Select English Historical Documents of the Ninth and Tenth Centuries* Cambridge: C.U.P. 1916.

HARRIS, Mary Donner (ed.) *Coventry Leet Book* London: EETS 1907-13.

HART, William Henry (ed.) *Historia et Cartularium Monasterii Sancti Petri Gloucestriae* (3 vols) London (Rolls Series): Longman et al 1863-67.

HARVEY, John H. (ed.) *William Worcestre Itineraries* Oxford: Clarendon Press 1969.

HARVEY, S. Royal Revenue and Domesday Terminology. *Econ. Hist. Rev.* (1967) 2nd Ser. Vol.XX No.2 pp.221-8 London.

HASSALL, W. O. (ed.) *Cartulary of St. Mary Clerkenwell* London: Roy. Hist.Soc. 1949.

HEAD, B. V. *Historia Numorum* 2nd Ed. Oxford: Clarendon Press 1911.

HERLIHY, D. *History of Feudalism*. London & Basingstoke: Macmillan 1970.

HERODOTUS, *History*. See MACAULAY, G. C. (ed.).

HILL, David. The Burghal Hidage — Southampton. *Proceedings of the Hampshire Field Club for the Year 1967* (1969) pp.59-61 Winchester.

HOLLINGS, Marjory (ed.) *The Red Book of Worcester* (4 vols) London: Worcestershire Historial Society 1934-50.

HOOVER, Herbert Clark and HOOVER, Lou Henry (eds. & trs.) AGRICOLA, Georgius *De Re Metallica* New York: Dover Publications Inc. 1950.

HORN, W. and BORN, E. *The Plan of St. Gall* (3 vols) Berkeley: University of California Press 1979.

HORWOOD, Alfred J. (ed. & tr.) *Year Books of the Reign of Edward 1st Years XX and XXI* London: HMSO and Longmans, Green, Reader & Dyer 1866.

HOVEDEN (HOWDEN) Roger de. See STUBBS, William (ed.).

HUDSON, Revd William and TINGEY, John Cottingham. *The Records of the City of Norwich* (2 vols) Norwich and London: Jarrold and Son 1906.

HUGHES, Bernard. Old English Bronze Wool Weights of the Period Edward IV to George II *The Connoisseur* 1969 (March) pp.153-9 London.

HULTSCH, Friedrich Otto. *Griechische und Romische Metrologie* Berlin: Weidmann 1862.

HULTSCH, Friedrich Otto. *Metrologicorum Scriptorum Reliquiae* (2 vols) Stuttgart: Teubner 1971 (Reprint of edition of 1864-66)

HUSSEY, Robert. *Ancient Weights and Money* Oxford: J. H. Parker 1836.

HYDE, J. Wilson. *The Early History of the Post Office* London: Adam & Charles Black 1894.

ISIDORE, Hispalensis Episcopi. *Etymologiarum sive originum libri XX* LINDSAY, W. M. (ed.) Oxford: Clarendon Press 1911.

JOHNSON, Revd A. H. *History of the Worshipful Company of Drapers of London* (5 vols) Oxford: Clarendon Press 1914-22.

JOHNSON, Charles (ed. & tr.). *The 'De Moneta' of Nicholas Oresme and English Mint Documents* London and New York: Nelson 1956.

JOHNSON, Charles (ed.) *Dialogus de Scaccario* London & New York: Nelson 1950.

JOHNSON, David and STEVENSON, Maurice. *The Centennial Review. A History of the First Hundred Years of the Institute*. London: The Institute of Trading Standards Administration 1981.

JONES, R. P. Duncan. Length Units in Roman Town Planning. The Pes Monetalis and the Pes Drusianus. *Britannia* (1980) Vol.XI pp.27-33 London.

JOSSET, Christopher Robert. *Money in Great Britain and Ireland* Newton Abbot: David and Charles 1971.

KARSLAKE, Lt. Col. J. B. P. Silchester and its Relation to the pre-Roman Civilization of Gaul. *Proc. Soc. Antiq.* (1920) Vol.XXXII p.198 London.

KARSLAKE, Lt. Col. J. B. P. Further Notes on the Old English Mile. *Geog. J.* (1931) Vol. LXXVII pp.358-360 London.

KATER, Capt. Henry. An Account of the Comparisons of Various British Standards of Linear Measure. *Phil. Trans. Roy. Soc.* (1821) Vol.CXI Pt.1 p.75 London.

KATER, Capt. Henry. An Account of the Construction and Adjustment of the new Standards of Weights and Measures of the United Kingdom and Ireland. *Phil. Trans. Roy. Soc.* (1826) Vol.CXVI Pt.II p.1 London.

KATER, Capt. Henry. An Account of the Construction and Verification of a Copy of the Imperial Standard Yard made for the Royal Society. *Phil. Trans. Roy. Soc.* (1831) Vol.CXXI p.345 London.

KATER, Capt. Henry. On the Error in Standards of Linear Measure arising from the thickness of the Bar on which they are traced. *Phil. Trans. Roy. Soc.* (1830) Vol.CXX p.359 London.

KEARY, C. F. *A Catalogue of English Coins in the British Museum: Anglo-Saxon Series* Vol.I London: British Museum Trustees 1887.

KELLY, Patrick. *Metrology* London: Privately printed 1816.

KELLY, Patrick. *The Universal Cambist* (2 vols) London: Privately printed 1821.

KEMBLE, J. M. *Codex Diplomaticus aevi Saxonici* (6 vols) London: Eng. Hist. Soc. 1839-48.

KING, Frank Alfred. *Beer has a History* London & New York: Hutchinson's Scientific and Technical Publications 1947.

KINGDON, J. A. *The Strife of the Scales* London: Rixon and Arnold 1905.

KISCH, Bruno. *Scales and Weights* Newhaven and London: Yale University Press 1965.

KOBEL, Jakob. *Geometrei* Frankfurt am Meyn: Christian Egenolffs Erben 1535.

KRAAY, Colin M. *Archaic and Classical Greek Coins* Berkeley: University of California Press 1976.

LARKING, L. B. (ed.) *The Domesday Book of Kent* London: J. Toovey 1869.

LENNARD, R. *Rural England 1086-1135* Oxford: Clarendon Press 1959.

LIEBERMANN, F. *Die Gesetze der Angelsachsen* (3 vols) Halle: a.s. M. Niemeyer 1903-16.

LINDSAY, W. M. (ed.) See ISIDORE, Hispalensis Episcopi.

LLOYD, T. H. *English Wool Trade in the Middle Ages.* Cambridge: C.U.P. 1977.

LUARD, Henry Richards (ed.) *Annales Monastici* (5 vols) London: (Rolls Series): Longmans et al 1864-91.

LUARD, Henry Richards (ed.) *Bartholomaei de Cotton, Historia Anglicana* London (Rolls Series): Longman et al. 1859.

LUARD, Henry Richards (ed.) *Matthaei Parisiensis Chronica Majora* (7 vols) London (Rolls Series): Longmans et al 1872-83.

LYELL, L. and WATNEY, D. *Acts of the Court of the Mercers' Company 1453-1527* Cambridge: C.U.P. 1936.

LYON, Stewart. Historical problems of the Anglo-Saxon Coinage (3) Denominations and Weights *Brit. Numis J.* (1969) Vol.XXXVI pp.204-222 London.

MACAULAY, G. C. (ed. & tr.) *The History of Herodotus* (2 vols) London: Macmillan & Co. 1904.

MACCANCE, Robert Alexander and WIDDOWSON, Elsie May. *Breads White and Brown* London: Pitman Medical Publishing Co. 1956.

MACHABEY, A. *La Mètrologie Dans les Musées de Province* Paris: Revue de

Mètrologie Pratique et Legale et Centre National de la Recherche Scientifique 1962.

MACK, R. P. *The Coinage of Ancient Britain* 2nd ed. London: Spink & Son, B. A. Seaby 1964.

MACK, R. P. *Sylloge of Coins of the British Isles* Vol.XX London: British Academy; O.U.P. 1973.

MACKENZIE, N.P.B. The Sale of Herring at Mallaig Fishing Port. *The Monthly Review* Dec. (1979) Vol.LXXXVII No.12 p.214.

MCLAUGHLIN, Revd Thomas (ed. & Tr.) Dean of Lismore's Book. Edinburgh: Edmonson and Douglas 1862.

MADDEN, Sir Frederic (ed.). *Matthaei Parisiensis Historia Anglorum* (3 vols) London (Rolls Series): Longman et al 1866.

MADOX, Thomas. *History and Antiquities of the Exchequer* London: John Matthews 1711.

MAGNUS, Olaus. *Historia de Gentibus Septentrionalibus* Rome: Joannes Miriam de Viottis Parmensem in aedibus Brigittae 1555.

MAITLAND, Frederic William. *The Constitutional History of England* Cambridge: C.U.P. 1926.

MAITLAND, Frederic William. *Domesday Book and Beyond* Cambridge: C.U.P. 1897.

MATTHAEI PARISIENSIS *Chronica Majora* See LUARD, Henry Richards (ed.)

MATTHAEI PARISIENSIS *Historia Anglorum* See MADDEN, Sir Frederic (ed.)

MATTHEWS, L. G. *History of Pharmacy in Britain* Edinburgh and London: E. & S. Livingstone 1962.

McDONALD, Donald. *A History of Platinum* London: Johnson, Matthey & Co. Ltd. 1960.

MCKECHNIE, William Sharp. *Magna Carta* Glasgow: James Maclehose & Sons 1905.

MECHAINE, P.F.A. and DELAMBRE, J.B.J. *Base du Systeme Metrique Decimal.* Paris: Baudouin 1806-10.

METCALF, D. M. and MERRICK, J. M. Composition of Early Mediaeval Coins. *Numis. Chron.* (1967) Vol.VII p.167 London.

METRICATION BOARD, *Going Metric – The Next Phase* The Fifth Report of the Metrication Board. London: HMSO 1974

Going Metric: Progress in 1977/78. Report of the Metrication Board for the period January 1977 to June 1978. London: HMSO 1978.

Final Report of the Metrication Board London: HMSO 1980.

Going Metric No 11 April 1974 London: HMSO 1974.

Going Metric No 13 October 1974. London: HMSO 1974.

MEYERS, A. R. (ed.). *English Historical Documents 1327-1485* Vol.IV London: Eyre & Spottiswoode 1969.

MICHELSON, A. A. and BENOIT, J. R. *Travaux Bureau International des Poids et Mesures* (1895) Vol.XI p.3 Paris.

MILES, J. Observations on Some Antiquities found in the Tower of London in the year 1777. *Archaeologia* (1778) Vol.V London & Oxford. (Vol. dated 1779).

MILLER, W. H. On the Construction of the New Imperial Standard Pound and its copies of Platinum . . . *Phil. Trans. Roy. Soc.* (1856) Vol.CXLVI p.753 et seq. London.

MILLER, W. H. On the Imperial Standard Pound. *Phil. Mag.* (1856) Vol.XII p.550 London.

MISKIMIN, Harry A. Two Reforms of Charlemagne? Weights and Measures in the Middle Ages. *Econ. Hist. Rev.* 1967 Second Series XX No.1 pp.35-52.

MITFORD, R. Bruce-. *The Sutton Hoo Ship Burial* Vol.I London: British Museum Trustees 1975.

MOLL, P. D. The Winchester Quart. *The Industrial Chemist* (September 1954) London.

MOODY, B. E. The Origin of the 'Reputed Quart' and other Measures. *Glass Technology* (April 1960) Vol.I No.2 pp.55-68 Sheffield.

MOREAU, Henri. Unités et Etalons Principaux. *La Nature* (Dec. 1949) Supplement to issue 3176 pp.396-412 Paris.

MORGAN, A. de. 'Mile', *The Penny Cyclopaedia* Vol.XV pp.210-13 and 'League' *Ibid* Vol.XIII pp.375-6 London (1839).

MORGAN, J. F. *England under the Norman Occupation* London: Williams and Norgate 1858.

MORGAN, Morris Hickey (tr.) see VITRUVIUS.

MORRISON, K. F. The Monetary Reform of Charlemagne *Speculum* (1963) Vol. XXXVIII pp.403-432, Cambridge, Mass.

MOUTON, Gabriel *Observationes diametrorum solis et lunae apparentium* Lugduni (Lyons): Liberal 1670.

MOWAT, John Lancaster Gough. *Sixteen Old Maps of Properties in Oxfordshire . . .* Oxford: Clarendon Press 1888.

NEF, J. U. *The Rise of the British Coal Industry* Vol.II London: G. Routledge & Sons 1932.

NEVILLE, William S. Sessions of the Clerk of the Market of the Household in Middlesex *Trans. of the London and Middlesex Archaeological Society* (1957) Vol.XIX Pt.2 pp.76-89 London.

NICHOLS, Francis Morgan (ed. & tr.) *Britton* (2 vols) Oxford: Clarendon Press 1865.

NICHOLS, J. G. (ed.) *Chronicle of the Grey Friars of London* Camden Society Publications No.53 London: The Camden Society 1852.

NICHOLS, J. G. The Foot of St Pauls. *Gentleman's Magazine* (1852) N.S. Vol.XXXVIII Pt.2 pp.276-7 London.

NICHOLS, John F. The Extent of Lawling in the Custody of Essex AD 1310. *Trans. Essex Arch. Soc.* (1933) Vol.XX Pt.2 pp.173-98. Colchester.

NICHOLSON, Edward *Men and Measures* London: Smith Elder & Co. 1912.

NORDEN, John. *The Surveyor's Dialogue* London: John Astley 1607.

NORRIS, H. Ancient Weights and Measures prior to Henry VII *Phil. Trans. Roy. Soc.* (1775) Vol.LXV p.48 London.

NORRIS, H. On Greaves' Weights and Measures *Archaeologia* (1781) (vol. dated 1782) Vol.VI p.221 London and Oxford.

O'KEEFE, John Alfred. *The Law of Weights and Measures* 1st ed. London: Butterworth 1966; 2nd ed. London: Butterworth 1978.

OSCHINSKY, Dorothea. *Walter of Henley and other Treatises* Oxford: Clarendon Press 1971.

OWEN, George Alfred (Revised by POOLE, A. W.) *The Law Relating to Weights and Measures* 2nd ed. London: C. Griffin 1947.

PAINTER, K. S. A late Roman silver ingot from Kent. *Antiq. J.* (1972) Vol.LII pp.84-92 Oxford.

PAPIN, Maurice Denis-. Les Systemes d'Unites de Mesure *La Nature* (Dec. 1949.) Supplement to issue 3176 pp.393-95 Paris.

PARLIAMENTARY PAPERS Papers Related to Standards of Weights and Measures No 115, 11 March 1864. *Parl. Papers* 1864, lviii, p.621.

Report from the Committee Appointed to Inquire into the Original Standards of Weights and Measures in the Kingdom (Lord Carysfort's Committee) 26 May 1758 *Reports from Committees of the House of Commons* (1737-65) Vol. II, pp.411-51.

Report from the Committee Appointed on the First day of December 1758 to Inquire into the Original Standards of Weights and Measures in the Kingdom 11 April 1759 (Lord Carysfort's Second Report) *Reports from Committees of the House of Commons* (1737-65), Vol. II, pp.453-463.

Report of the Committee on Weights and Measures 1 July 1814. *Parl. Papers* 1813-14 (HC290), iii, pp.131-153.

Report of the Select Committee on the Laws Relating to the Manufacture, Sale and Assize of Bread 6 June 1815. Ibid 1814-15 (HC186) v, pp.1341-1490.

Report (First) of Commissioners to Consider the Subject of Weights and Measures 7 July 1819. Ibid 1819 (HC565) xi, pp.307-323.

Report (Second) of Commissioners to Consider the Subject of Weights and Measures 13 July 1820. Ibid 1820 (HC314) vii, pp.473-512.

Report (Third) of Commissioners to Consider the Subject of Weights and Measures 31 March 1821. Ibid 1821 (HC383) iv, pp297-302.

Report of the Select Committee on Weights and Measures 23 May 1821. Ibid 1821 (HC571) iv, pp.289-295.

Report of the Commissioners Appointed to Consider the Steps to be taken for the Restoration of the Standards of Weights and Measures 21 December 1841. Ibid 1842, xxv pp.263-375.

Report of the Select Committee on Poor Rate Returns 10 July 1821. Ibid 1821 (HC748) iv pp.269-286.

Report of Commissioners . . . on the Royal Mint (undated). Ibid 1849 (1026) xxviii pp.347-665.

Report of Commissioners Appointed to Superintend the Construction of the New Parliamentary Standards of Length and Weight 28 March 1854. Ibid 1854 xix, pp.933-955.

Report (Preliminary) of the Decimal Coinage Commissioners 4 April 1857. Ibid 1857 (Session 2) xix, p.ix. (This is also contained in the Final Report.)

Report (Final) of the Decimal Coinage Commissioners 5 April 1859. Ibid 1859 (Session 2) xi, pp.1-115. Appendix: Ibid 1860, xxx, pp.387-634.

Report of the Select Committee Appointed to Consider the Practicality of Adopting a Simple and Uniform System of Weights and Measures 15 July 1862. Ibid 1862, vii, pp.187-478.

Report (First) of Commissioners Appointed to Inquire into the Condition of the Exchequer Standards 24 July 1869. *Parl. Papers* 1867/8 xxvii, pp.1-8.

Report (Second) of Commissioners Appointed to Inquire into the Condition of the

Exchequer Standards 3 April 1869. Ibid 1868/9 xxiii, pp.733-865.
Report (Third) of Commissioners Appointed to Inquire into the Condition of the Exchequer Standards 1 Feb. 1870. Ibid 1870 xxvii, pp.81-247.
Report (Fourth) of Commissioners Appointed to Inquire into the Condition of the Exchequer Standards 21 May 1870. Ibid 1870 xxvii, pp.249-739.
Report (Fifth) of Commissioners Appointed to Inquire into the Condition of the Exchequer Standards 3 August 1870. Ibid 1871 xxiv, pp.647-1048.
Report of the Committee on Weights and Measures Legislation (The Hodgson Report). Ibid 1950-51 (Cmd 8219) xx, pp.913.
Report of the Committee of Inquiry on the Decimal Currency (The Halsbury Committee). Ibid 1962-63 (Cmnd 2145), xi, p.195.
Report (Final) of the Committee on Consumer Protection (The Molony Report). July 1962. Ibid 1961-62 (Cmnd 1781), xii, pp.271-609.

PEGOLOTTI, Francesco Balducci. *La Pratica della Mercatura.* EVANS, Allan (ed.) Cambridge, Mass.: Mediaeval Academy of America (Publication No.24) 1936.

PELHAM, R. A. Exportation of Wool from Winchelsea and Pevensey in 1288-9 *Sussex Notes & Queries* (1935) Vol.V No.7 pp.205-6 Lewes.

PERCY, Thomas (ed.) Regulations for the Establishment of the Household of H. A. Percy, 5th Earl of Northumberland 1512-25. London: W. Pickering 1770.

PERTZ, G. H. (ed.) *Monumenta Germaniae Historica, Leges* Vol.IV Stuttgart: Kraus Reprint 1965.

PETRIE, W. M. Flinders. The Old English Mile. *Proc. Roy. Soc. Edin.* (1882-4) Vol.XII pp.254-66 Edinburgh.

PETRIE, W. M. Flinders. 'Weights and Measures' *Encyclopaedia Britannica* 11th ed. Vol.28 pp.477-488 Cambridge: C.U.P. 1911.

PIGGOTT, Stuart. *Ancient Europe* Chicago and Edinburgh: Aldine Publishing Co. and Edinburgh University Press 1970.

Pipe Roll 8 Henry II 1162 Vol.V London: Pipe Roll Society 1885.

Pipe Roll 9 Richard I 1197 Vol.XLVI London: Pipe Roll Society 1931.

PLOT, Robert. *The Natural History of Oxfordshire* 2nd ed. Oxford: L. Lichfield 1705.

POLLOCK, Sir Frederick and MAITLAND, Frederic William. *The History of English Law* (2 vols) Cambridge: C.U.P. 1895.

POOLE, Austin Lane. *From Domesday Book to Magna Carta* 2nd ed. Oxford: Clarendon Press 1955.

POOLE, Stanley Lane-. (POOLE, Reginald S. ed.) *Catalogue of Arabic Glass Weights in the British Museum* London: Trustees of the British Museum 1891.

POWELL, J. *Assize of Bread* London: R. Scot 1684.

PRELL, Heinrich. *Bemerkungen zur Geschichte der Englischen Langenmass-Systeme* Berlin: Bericht äber die Verhandlungen der Sächsischen Akademie der Wissenschaften zu Leipzig: Mathematische-Naturwissenschaftliche Klasse, Band 104 Heft 4 1962.

PRIOR, W. H. Notes on the Weights and Measures of Mediaeval England. *Bulletin du Cange* (1924) Vol.I pp.77-97 and pp.140-170 Paris.

RADFORD, C. A. Ralegh. The Later Pre-Conquest Boroughs and their Defences. *Mediaeval Archaeology* (1970) Vol.XIV pp.83-103 Reading.

RAMSAY, Sir James H. *A History of the Revenue of the Kings of England 1066-1399* (2 vols) Oxford: Clarendon Press 1925.

RAPER, Matthew. An Inquiry into the Measure of the Roman Foot *Phil. Trans. Roy. Soc.* (1760) Vol.LI (2) p.774 London.

REECE, Richard. *Roman Coins* London: Benn 1970.

REYNARDSON, Samuel. A state of English Weights and Measures of Capacity . . . *Phil. Trans. Roy. Soc.* (1749-50) Vol.XLVI pp.54-71 London.

RICHARDSON, H. G. and SAYLES, G. O. (eds. & trs.) *Fleta* Book II Selden Society Publications Vol.72 of 1953. London: The Selden Society 1955.

RIDGEWAY, W. *The Origin of Metal Currency and Weight Standards* Cambridge: C.U.P. 1892.

RILEY, Henry Thomas (ed.) *Munimenta Gildhallae Londoniensis. Liber Albus, Liber Custumarum, Liber Horn* (3 vols in 4) London: Rolls Series: Longmans et al. 1859-62.

ROBERTSON, Agnes Jane. *Anglo-Saxon Charters* 2nd ed. Cambridge: C.U.P. 1956.

ROBERTSON, A. J. *Laws of the Kings of England from Edmund to Henry I.* Cambridge: C.U.P. 1925.

ROBERTSON, E. William. *Historical Essays* Edinburgh: Edmonston and Douglas 1872.

ROBERTSON, Revd W. A. Scott. The Assize of Bread. *Archaeologia Cantiana* (1878) Vol.XII p.321 Maidstone and London.

ROBINSON, G. J. A. The Duties of the Ale Taster. *Libra* (1968) Vol.VII p.10 Eastbourne.

ROBINSON, G. J. A. Some Malpractices of Innkeepers and Brewers. *Libra* (1968) Vol.VII p.21 Eastbourne.

Rotuli Parliamentorum. See STRACHEY, J. (ed.)

ROUND, J. H. *Feudal England* London: George Allen & Unwin 1964.

ROUSE, Rowland. The Weight of an English Penny at Various Periods. *Gentleman's Magazine* (1797) Vol.LXVII Pt.1 p.394.

ROWE, John Frank. *The Bread Acts* London: Hampton and Co. 1894.

ROY, Major-General William. An Account of the Measurement of a Base on Hounslow Heath. *Phil. Trans. Roy. Soc.* (1785) Vol.LXXV pp.385-480 London.

RUDING, Rogers. *Annals of the Coinage of Great Britain and its Dependencies* 3rd ed. London: J. Hearne 1840.

SALTMAN, A. (ed.). *Cartulary of Dale Abbey* London: Historical Manuscripts Commission J.P. 11 1967.

SALTMAN, A. (ed.). *Cartulary of Tutbury Priory* London: Historical Manuscripts Commission J.P. 2 1962.

SALTZMAN, L. F. *English Trade in the Middle Ages* Oxford: Clarendon Press 1931.

SAMUELS, F. L. N. Maintaining Measurements in the Market Place. *Electronics and Power* (1982) Vol.XXVIII pp.150-4 London.

SCARGILL-BIRD, S. R. (ed.). *The Custumals of Battle Abbey in the Reigns of Edward I and Edward II 1283-1312* Camden Society N.S. Vol.41 London: Camden Society 1887.

SCHUMACHER, H. C. A Comparison of the late Imperial Standard Troy Pound Weight with a Platina Copy of the Same. *Phil. Trans. Roy. Soc.* (1836) Vol.CXXVI p.457 et seq. London.

SCHWARTZ, G. T. Gallo-Romische Gewichte in Aventicum, *Schweizer Munzblatt* (1964) Vol.13/14 pp.150-7.

SEARLE, Eleanor (ed. & tr.). *The Chronicle of Battle Abbey* Oxford: Clarendon Press 1980.

SEARLE, Eleanor. Hides, Virgates, and Tenant Settlement at Battle Abbey. *Eng. Hist. Rev.* (1963) 2nd Ser. Vol.XVI pp.290-300.

SEARS, J. E. The Standards of Length. *Proc. Roy. Soc.* (1946) Vol.CLXXXVI pp.152-164.

SEEBOHM, Frederic. *Customary Acres and their Historical Importance* New York: Longmans, Green 1914.

SELLERS, Maud (ed.). *York Memorandum Book, Part 1 1376-1419* Durham: Surtees Society (Publication 120) 1912.

SELTMAN, Charles Theodore. *Greek Coins* 2nd ed. London: Methuen 1955.

SHARPE, R. R. *London and the Kingdom* (3 vols) London: Longmans & Co., 1894-5.

SHEPPARD, J. B. (ed.). Certa Mensura Cantuariensis. Second Report on Historical Manuscripts belonging to the Dean and Chapter of Canterbury, in the *8th Report of the Royal Commission on Historical Manuscripts* London: HMSO 1881.

SHEPPARD, T. and MUSHAM, J. F. *Money, Scales and Weights* London: Spink & Co. 1923.

SHUCKBURGH, Sir George. An Account of Some Endeavours to Ascertain a Standard of Weights and Measures. *Phil. Trans. Roy. Soc.* (1798) Vol.LXXXVIII pp.133-182 London.

SIMPSON, W. Sparrow- (ed.). *Documents Illustrating the History of St. Paul's Cathedral* Camden Society Publications N.S. Vol.XXVI London (Westminster): Camden Society 1880.

SKINNER, F. G. The English Yard and Pound Weight. *Bulletin of the British Society for the History of Science* (1952) Vol.I p.184 London.

SKINNER, F. G. *Weights and Measures* London: The Science Museum and HMSO 1967.

SMITH, Reginald A. Early Anglo-Saxon Weights. *Antiq. J.* (1923) Vol.III pp.122-9 Oxford.

SMITH, R. A. Currency Bars and Weights. *Proc. Soc. Antiq. London* (1903-5) (Pub. 1905) Vol.XX pp.179-195.

SMITH, William. *The Particular Description of England 1588* WHEATLEY, Henry B. and ASHBEE, Edmund W. (eds.) London: Privately printed 1879.

SNEYD, Charlotte Augusta (tr.). *A Relation . . . of the Island of England* Camden Society Publications Vol.XXXVII, London: Camden Society 1847.

SPELMAN, Henry *Glossarium Archaiologicum* DUGDALE, Sir W. (ed.) London: Alicia Warren 1664.

SPRATLING, Mansel. Iron Age Settlement of Gussage All Saints Pt.II *Antiquity* (1973) Vol.XLVII pp.117-130 Cambridge.

STAPLETON, T. (ed.) *De Antiquis Legibus Liber* London: Camden Society 1846.

STATUTES Statutes of the Realm (11 vols) London: HMSO 1810-28.

Statutes at Large RUFFHEAD, Owen and RUNNINGTON, Charles (eds.) (41 vols) London: HMSO 1780-1865.

Acts and Ordinances of the Interregnum 1642-1660 (3 vols) London: HMSO 1911.

The Law Reports (Statutes), Public General Statutes (87 vols.) London: HMSO 1861–1951.

Public General Acts and Measures (annual) London: HMSO 1949–date.

Statutes in Force (88 vols) London: HMSO; various dates to the present.

Public General Statutes, see *The Law Reports (Statutes)* above.

STENTON, Doris Mary (ed.). *Preparatory to Anglo-Saxon England: being the collected papers of Frank Merry Stenton* Oxford: Clarendon Press 1970.

STENTON, Sir Frank M. *Anglo-Saxon England* 3rd ed. Oxford: Clarendon Press 1971.

STEVENSON, Revd Joseph (ed.). *Chronicon Monasterii de Abingdon* (2 vols) London: (Rolls Series) Longmans et al 1858.

STEVENSON, Maurice. A Note on the Sandwich Steelyard Weight. *Libra* (1963) Vol.II No.2 pp.15–16 Eastbourne.

STEVENSON, Maurice. The Size of Liquid Measures in the 17th and 18th Centuries. *Libra* (1964) Vol.III No.2 (3) pp.9–12 and Vol.III No.3 (4) pp.16–19 Eastbourne.

STEVENSON, Maurice. The Assize of Bread. *Libra* (1966) Vol.V pp.3, 10, 18, 26 Eastbourne.

STEVENSON, Maurice. The Assize of Ale (3) *Libra* (1969) Vol.VIII p.20 and (4) *Libra* 1970 Vol.IX p.54 Eastbourne.

STEVENSON, Maurice. Old English Lead Weights *Libra* (1962) Vol.I pp.2, 10, 18; 1966 Vol.V p.23; 1969 Vol.VIII p.32 Eastbourne.

STEVENSON, Maurice, see JOHNSON, David.

STRACHEY, J. (ed.). *Rotuli Parliamentorum* (6 vols) London: HMSO 1767–77.

STUBBS, William (ed.). *Chronica Magistri Rogeri de Houedene* (4 vols) London: Record Commission (Rolls Series): Longmans et al 1868–71.

STUBBS, William (ed.). *Willelmi Malmesbiriensis Monachi De Gestis Regum Anglorum* (2 vols) London: (Rolls Series) Longmans et al. 1887–9.

SUTHERLAND, Carol Humphrey Vivian *Anglo-Saxon Gold Coinage in the Light of the Crondall Hoard* Oxford: O.U.P. 1948.

TATON, René. L'Evolution des Unites de Longueur, Surface, Volume et Poids. *La Nature* (Dec. 1949) Supplement to issue 3176 pp.387–393 Paris.

TAYLOR, E. W. and WILSON, J. S. *At the Sign of the Orrery* London: Cooke, Troughton & Simms Ltd. (undated, post 1950).

THOMSON, T. and INNES, C. (eds.) *Acts of Parliament of Scotland* (12 vols) London: Record Commission 1844 et seq.

THORPE, Benjamin. *Diplomatarium Anglicum aevi Saxonici* (3 vols) London: Macmillan 1865.

TODD, Michael. Romano-British mintages of Antoninus Pius. *Numis. Chron.* (1966) Vol.VI pp.147–153 London.

TUDOR, T. L. The Lead Miners' Dish or Measure. *Derbyshire Archaeological and Natural History Society Journal* (1937) Part 1 pp.95–106 (1938) Part 2 pp.101–116 London.

TYLECOTE, R. F. *Metallurgy in Archaeology*. London: E. Arnold 1962.

URDANG, George (ed.) *Parmacopoeia Londinensis of 1618* Madison, Wisconsin: State Historical Society of Wisconsin 1944.

VITRUVIUS *The Ten Books on Architecture* MORGAN, Morris Hicky (tr.) New York: Dover Publications Inc. 1960.

WATNEY, J. *Beer is Best — A History of Beer* London: Peter Owen 1974.

WATSON, Sir Charles M. *British Weights and Measures* London: John Murray 1910.

WELBORN, Mary Catherine. The De Ponderibus et Mensuris of Dino de Garbo. *Isis* (1935) Vol.XXIV pp.15-36 Brussels.

WHITELOCK, Dorothy. *English Historical Documents ca 500-1042 AD* Vol.I London: Eyre & Spottiswoode 1955.

White Paper on Metrication February 1972 Cmnd. 4880 London: HMSO 1972.

WILDE, Edith E. Weights and Measures of the City of Winchester. *Papers and Proceedings of the Hampshire Field Club* (1931) Vol.X Pt.3 pp.237-248 Winchester.

WILLIAMS, Howard R. and MEYERS, Charles J. *Manual of Oil and Gas Terms* 5th ed. New York and San Francisco: Matthew Bender 1981.

Year Books of the Reign of King Edward the First Years XX and XXI See HORWOOD, Alfred J. (ed. & tr.)

ZUPKO, Ronald Edward. *A Dictionary of English Weights and Measures* Madison, Wisconsin: University of Wisconsin Press 1968.

ZUPKO, Ronald Edward. *British Weights and Measures* Madison, Wisconsin: University of Wisconsin Press 1977.

ZUPKO, Ronald Edward. *French Weights and Measures before the Revolution* Bloomington and London: Indiana University Press 1978.

Index

Main treatments are denoted by page-references in **bold** type. Foreign expressions, and English terms not normally found in dictionaries, are printed in *italic*. A page-number followed by *n* indicates a footnote-reference. Illustration-references (e.g. *fig 1*) follow the page-references.

NB The indexing of names of persons, places and books is selective only; further information will be found in the chapter references, and in the Glossary and Appendices preceding this index.

account, money/unit of 103*n*, 106*n*
acre **36-8**, **48-9**, 89-90, 96, *Fig 22*
  as area 38, 48-9
  'decimalization' 38
  and hide 57
  as length-measure 37-8
  'nominal' 60-61
  as rectangle 36-8, 89, *Fig 18*
  as taxable area 61
    scutage 63, 64
  variability 45-9
  as work-unit 37-8
'acre's-breadth' *see* chain
Acts *see* Parliament; *and individual monarchs*
actus 12
Aegina/Aeginetan 6, 7
Aelgyfa (Emma of Normandy) (Lady of England) 41, 42-3
aes grave 14
aes rude 13
aes signatum 14
Aethelbert I, king of Kent 32, 100*n*, 101, 102
  currency 100*n*, 101-2, 103
Aethelbert II *see* Ethelbert
Aethelflaed (Lady of Mercia) 37
Aethelred I, king
  acre 36
  currency 106
Aethelred II the Unready, king 41, 42
  currency 105*n*, 121, 125*n*
  hundred 59
  'Troy' weight 120
Aethelred, earl of Mercia 37
Aethelstan, king 32, 100
  ell 80
  foot 27
  shilling 102
  Southampton 40
Aethelwulf, king 82
Agenda (Science Council of Canada) 220
aid (levy) 62
Airy, G.B. (Astronomer Royal) 259, 285, 325
  Commissions (1838/43) 261, 267, 269
Alaric I, king of the Visigoths 31
alcohol *see* ale; wine *etc*
ale **193**, **220-27**
  assize system 220-27
    inspectors/conners 222-3
  and beer *defined* 221, 223
  bride-ale *etc* 220
  duty 158, 188, **226**
  ingredients 223-4
  'off-licence' sales 223
  penalties for dishonest trading 221, 222, **223**
    ale-conners 222
  standards **243**, 329
    Imperial 256
  weight 154

see also beer
ale-boon 30
ale-conners/founders/tasters 222-3
Alexandria 119
Alfred the Great, king 6, 32, 54*n*, 108
ale-testers 222
Burghal Hidage 39
cattle as fine 6-7
currency 102, 108-9
Orosius 68
Algar's foot (foot of St Paul's) 85-6
almonds 127, 136
alnage/alnager 87, 93, 236
alum
adulteration of bread 209, 212
weighing 127, 130*n*, 135, 136, 320
amber (capacity measure) 149
amerce (fine) 198
amphora 12-13, 117, 149
Anglo-Saxon Chronicle
Anglo-Saxon invasion/settlement 31
Britain, size of 68
Danish invasions 32
Edward the Confessor 43*n*
ring-giving 100
Roman withdrawal 31
wergeld 104*n*
Anglo-Saxons **31-3**
capacity 149
currency **100-110**, 125
foot 29
king's peace 29
laws 105*n*, 105-6
mile 68
shaftment 29
weights **119-21**
see also individual monarchs etc
Anne, queen
bread 207, 209
capacity measures 163-4, 172, 179
coal bushel 182
and US measures 188
wine gallon 246, 272, *Fig 37*
wine quart 187
cloth measures 93
inch 80
salt 184
Antoninianus 15
Antoninus Pius, emperor 20
Apothecaries' measure **185-6**, 297
Hodgson Committee 294
weight equivalents 185
see also individual units
Apothecaries' weight **117-19**, 126, 252
obsolescent 118, 186, 294, 297
volume equivalents 185
see also Troy weight; *and individual units*
aqua-vitae/vite *see* brandy
Arabic civilization
glass weights 3
and Troy weight 119
arc, degree of *see* latitude; meridian
architecture *see* building
area/superficial measure 36-8, **48-9**, **54-64**
hectare permitted 284
jugerum 12
see also individual units
Arnold's Chronicle 69-70, 75, 95, 113, 197, 202
as 13-16, *Fig 9*
libral *as* 13-14
Assisa Panis . . . see Assize of Bread and Ale
assize
Clerk of Market 326-8
foot of 86
loaf of 209-14
see also bread; ale
Assize of Bread and Ale (Assisa Panis et Cervisie) (?1266) **313-14**
ale 220
bread **193-6**, **203-5**
baker's allowances/penalties **198**, 200, 204
renounced 207, 209
capacity 151
Compositio 123
Tractatus 123
Assize of Measures (1196) 89, **90-91**, 95-6, 149-50, 234
Assize of Weights and Measures (?1302/3) *see Tractatus*

Association of British Chambers of Commerce/*formerly* Associated Chambers of Commerce of the United Kingdom 287, 290, 295
Astronomer Royal 253, 259, 294, 297
  see also Airy
Athelstan *see* Aethelstan
Athenian measure 9
Attic-Euboic (Solonic) system 5-6
Augustine, saint (archbishop of Canterbury) 32
Augustus, emperor 15
aulne (yard) 92
auncel/*scheft*/*pounder* **22**, **133-5**, 137
  Chicheley's intervention 139
  obsolescent 138-40
  weights for 233, 234
  see also steelyard
aureus 15
Avoirdupois weight (haber de peyse; haber-de-pois; haberty-poie; habur de peyse) 123, **127-43**, 233
  and Apothecaries' 186
  bread 207-19, 298
  Carysfort Committee 247, 248
  established 262-4
  origin of term 128
  recognized in law 129
  standards 122, **252-3**, **262-3**, **267-70**
  Troy, confused with 127-8, 129, 131, 146 (ref 55), 154, 242
  see also weights; *and individual units*
axe-head ingot 20-21, *Fig 11*

Babinet, Jacques 351
Baily, F.
  Commissions (1838/43) 261
  'metal no. 4' **265**, 266, 296
  yard (Imperial Standard) **264-5**
baker's dozen 198-9
balance/beam scale **21**, **134-9**, *Fig 30*
  earliest 3
  Kater's 258
  king's beam 134, 136
  'tipped' 135-6, 137
  tron 118, 135, 138-9, **234**
Banks, J. 253
Barker, Thomas 163
barley
  bread 194, 209, 210, 212, 215
  brewing 220-21
barleycorn 35, **124**
  as linear unit 3, 4-5, 35, 79-80, 89, *Figs 4, 17*
    and foot 28, 81
  as weight unit (*see also* grain: Troy) 2, **3-4**, 35, 101, **118**, **124**
barm 209*n*
barrel **172-7**
  ale/beer 170-71, **172-3**, 224, **225**, 226
    distinguished 225
  butter 176-7
  Coopers' standards 329
  half-barrel 176
  herring 173-5
  honey 176
  oil 177
  soap 176-7
  weight (of barrel) 176-7
  wine **171**, 172-3
  see also cask
Barrow (instrument-maker) 268
barter 100
Barth (Emden) hoard 103
batch bread 216, 217
Bate (instrument-maker) 257-8, 272
beam scales *see* balance
beans 161
  for bread 199*n*, 201, 209, 210, 212, 215
  horseload/seam 149
Bedcan Ford, battle of 32
beef 174
beer (malt liquor) 193, **220-27**
  defined 221, 223
  duty 64, 188, **226**
  early 225
  measures 172-3, **224-6**
    barrel **172-3**, 225-6
    Imperial 256
    post-metrication 303
  see also ale
beer/bier (yarn-carrier) 88
Belgae 19, 75

Belgium *Fig 30*
Benoît, J.R. 351
bequa system *Fig 5*
Bethune, J.E.D. (Commissions 1838/43) 261
Bible xxiv, 5
bier/beer (yarn-carrier) 88
bind (of eels *etc*) 321
Bird (instrument-maker) **246-9**, 250, 252-5, 259-61
Birmingham 3, 22
bismar 21-2, 134, 140, *Figs 14, 15*
  see also auncel
blanching 121, 122
blanc peyne (white bread) 195*n*
Blois, bishop Henry de 42
body measures **1-2**, 35
  kings *etc* 83
Borda, Jean Charles de 345-7, 348*n*
bottle (as liquor measure) 187-8
  milk 303
bovate (oxgang) 55
Bowring, Sir John, MP 281
box (herring-measure) 175
brandy/aqua-vite/aqua-vitae 64, 226
brass
  Roman coins 15
  weights 330-31
bread **193-220**, *Fig 41*
  assize systems **193-216**, **313-17**
    disadvantages **203-5**, 214
    'juries' 199-202
    late 205-14
    London 199, **201-2**, 212-15
    mediaeval 194-205
    obsolescent 214-16
  baker's dozen 198-9
  Cocket Office 212
  expenses/profits 205, 207-10, **210-11**, 212-13
    medieval 197-200, 203-5
  import/export regulation 214
  ingredients 194-7, 209, 212-13, 215, 217-19
    additives permitted 219-20
  loaves of assize 209-14
  penalties for dishonest trading 206-7, 313-14
    fines 198-200, 206, 209, 215
    medieval 196-202, *Fig 41*
  prices prescribed 205-14, 218-19
    Frankfort Capitulary 194
    medieval 194-6, 200, 201-5
  prized bread 209-14, 218
  rationed 219
  shape prescribed 218
  varieties 205-11, 215-16
    fancy/rolls **215-16**, 217-19
    horsebread **199**, 201, 207
    medieval 194-6, **199**, 203, 204
    spicebread 207
  weights prescribed/allowed
    Avoirdupois 207-19, 298
    'currency' weights 194-200, 204
    metric 219-20
    quarter, peck *etc* 185, 209-16
    scales/weights 213-15, 217, *Figs 42, 44*
    Troy 125-7, **197**, 200, 202-9; *obsolescent* 125-7, 207-9
  World War II 217-19
Bremen (weights) 119
Bridport (gallon) 161, 243*n*
Bristol (capacity measures) 161-2
Britain (Celts) 23
  Anglo-Saxon invasion/settlement 31-3
  currency **16-21**, 100
  Roman withdrawal 31; *see also* Rome
  weights 16, **18-19**, 23
  see otherwise England *etc*
'Britannia' (depicted on coins) 20
British Association for the Advancement of Science 295
British Association of Inspectors of Weights and Measures 338
British Pharmacopeia (1864) 186
British Standards Institution 301
bronze 10*n*
  Baily's metal 265
  speculum 19
  see also aes
Brown, W, MP 281, 282

Brunner, Collot & Laurent company 351
buckwheat 212, 215
building measures **39-42**
body measures 1-2
chain 38-9
foot 81, 82
Drusian 29, 81
Roman 22
rod **39-42**, 44, **46-8**
bulk weights/goods *see* Avoirdupois; *and individual commodities*
bullion
as currency standard 5
early 5-8
weighing of 123, 125, 135, **252**, 263
decimal element 280
ounce as primary 271
see also gold *etc*
Bureau international des poids et mésures/ International Bureau of Weights and Measures 285, 293, **349-50**, 352
Burghal Hidage 36, **39-41**, 59, 60
burh (fortified town) 38, **39-41**
Burton Monastery 194
bushel 151-5, **155-66**, **252**, **270**, *Figs 34-5*
abolished 304, 305, 307
and amber 149
and chaldron 180-81
coal bushel **182**, 272, *Fig 40*
Edward III 150
half-bushel 150, 154
imputed **183-5**
king's 232
London 124, 151, 157-8, **232**
obsolescent 166, 294, 297
and quarter 150, 156-8, 166, 221-2
and sack 182-3, 212*n*, 257
and seam 149-50
standards 232-3, 238, 239, 246
Exchequer 155-6, **160**, 232-3
Imperial 256-8, *Fig 53*
statute of 1824 166
stricken/heaped 156-8, 256-7
US 188
water-measure 178-80
weight of 154
as weight-unit 183-5
wheat for bread 207-13
Winchester **164-6**, 181-4, 252, 272
butt/pipe
salmon 173
wine 163, **170-71**, 172, 177, **257**
butter 176, 206
amber 149
barrel 174, **176-7**
kilderkin 176-7
scales 337*n* *Fig 55*

cade 173*n*
cadil (litre) 354-5 *Fig 58*
Cambridge
ale privileges 223, 224
weights 240, 242
wine pottle 155
Canada 220, 300*n*, **307-8**
candles
and inches 79
as time-measures 30
canne 344
cannel 182
Canterbury 21, 109, 324
cantle, by (heaped) *see* capacity: heaped measure
capacity/volume measure **149-66**, **170-89**, 193-255
cubic yard *etc* legalized 289*n*
dry/liquid systems 150, 151-2, 158-9, 161, 164-5, 170-71
early measures 1, 10, **12-13**
Founders' standards 330-31
heaped/stricken measure **149-50**, 151, 152, **156-8**, 165
abolished 182, 260, 272
coal 181, 182
statute (1824) 179-80, 256-7
water-measure 178-80
metric permitted 284
origin of terms 151
'oversized' 13
Saxon 149
standards **232-3**, 235, **239**, 243

Richard I 90
statutes (1706) 163-4; (1824) 165-6, 179-80, 256-7; (1963) 297
US 177, 187, **188-9,** 290
water-measure 178-80
weights of capacity units
established 254
notional 183-5
carat
as carob seed 2-3
metric 279
and solidus 16
and tremissis 16
see also carob seed
carato (carat) 117
Carey/Cary (instrument-maker) 255
Carloman, ruler of Franks 107
carob seed (*keration; siliqua*) **3-4**, 16, 35, 101, 117
see also carat
Carolingians
Carloman 107
Charlemagne *see* Charlemagne
score 43
see also Franks
cart-load 181
carucate/*carucata* **54-5**, 58
see also hide
Cary, William 255, 259
Carysfort Committee (1758-60) 160-5, 171, 174, **246-9**, 251, 327
barrel (of fish) 174
bushel 161, 232-3, 246, 249
Founders' weights 330
gallon 153, 160, **162-3**, **165**, 246, 249
Tractatus 123*n*
tun 171
case
tobacco 141
wine *etc* 188
cask **329**
tobacco 141
wine 257
cassatura/cassatus see hide
Cassini, Giovanni Domenico (Jean Dominique) 76, 344, 348*n*
cattle (as units of value) 5, 6-7, 13, 35
Celts *see* Britain; Gaul
'cent' (UK) 282, 298-300
see also penny, new
cental 263, 298, 305, 307
'centner' 263
Ceratonia siliqua (carob plant) 3
Cerdic (*later* king of Wessex) 31
chain (acre's breadth) **38-9**, 86, 263, 270
abolished 305, 306
Canada 308
chaldron 170, **180-82**, 257
and coal bushels 182
obsolescent 294, 297
Chaney, H.J., 178, 235
Founder's weights 330
lead-dish 178
Troy weight 118
yard 235
char (load) 130
Charlemagne, emperor **107**
bread 194
currency 104, 107-8
modius 194
Paris grain 2
pied du roi 83
pile of 353
Troyes 119
wool 132
Charles I 183, 328
Charles II
capacity 164, 179, 183
cloth measures 93
coal 181
duty on drinks 226
knight's service 64, 226
wool 142
cheese 130, 139, 206, 320, 324
chef 321
chemicals 166
see also Apothecaries' measures/weights
chest (of tobacco) 141
Childeric III, king of Franks 107
Chisholm, H.W. 260
chocolate 64, 226

church-*mitta* 149
cider 64, 226
cinnamon 130*n*, 135, 320
Claudius, emperor 15, 20
Clerk, Sir George 253, 255
Clerk of the Market/Marshalcy 239-40, 241, **325-8**, 333
  Powell, J. 199, 205-7, 212, 224-5.
cloffe/cloff/tret 136-8, 140, 141
  'inverse' 140-41
cloth
  duty 131, 133
  measures **87-9, 90-96, 234-7**
    broadcloth 91-4
    foot 82-3
    hundred 58, 94, 321
    post-metrication 303
  'searching' 235, **236-7,** *Fig 45*
  trade 132-33, 142
clove (weight-unit) 129-30, **135-40,** 142, 233, *Figs 29, 31*
Cnut the Great, king of England and Denmark 42, 106 125*n*
coal **180-83**
  bushel 182-3, 256, 272, *Fig 40*
  chaldron 170, **180-82**
  heaped measure 179-80
  solid fuel regulations (1985) 307
  weight of 181, 182, **183**
    conveyor-belt weighing 337
cocket (bread) **195-6**, 203, 206
Cocket Office 212
cocoa 141, 305
Coenwulf, king of Mercia 56, 61
coffee 64, 141, 226
coinage *see* currency
Cole, Humfrey 35, *Fig 17*
Cologne 117, 123
Commission Internationale du Mètre (1870/72) 349
Commission (permanent) on Units and Standards of Measurement 294-5, **297**
Commissions, Royal/Parliamentary *see* Parliament
Common Market *see* European Economic Community
Commonwealth, British 290, 305
  Canada 220, 300*n,* **307-8**
Compositio Mensurarum (Compositio Monete) 123-5
Condorcet, Marie J.A.N. Caritat, marquis de 345-6
confections 125-6
Confederation of British Industry/ *formerly* Federation of British Industries 290, 301, 306
congius 12, **13**, 185, 186
Constantine the Great, emperor 15, 119
consumer protection **313-40**
  Molony Committee 339
Convention du Mètre/Metric Convention 285, 286, **349**
coomb 149
Coopers' Company 329
copper
  currency 113
    Roman 13-14, 15
  oxhide ingots 6
  weighing of 127, 136
  see also cupro-nickel
coppice pole (land measure) 48
corbus (corbis) (basket) 194
corn 157, 160, 163, **164-6**, 183
  bushel, standard 233, 239
  gallon, standard 239
  'measure' of 150
  see also wheat *etc*
corn, Indian (maize) 212, 215
cornflakes 305
corn laws 214
'corn measure' (Roman) 13
corn-rents 166
coterie (furlong) 75
cotyla 185
Court of Wards and Liveries and Tenures 226
Coventry 127, 197, 199, 223
cow *see* cattle
cran **174-5**, 297
  quarter-cran 175
Croker, J.W., MP 279
Croesus, king of Lydia 7

Crondall (Hants) hoard 100-103
cubic measure *see* capacity; *and individual units*
cubit **1-2**, 8, 9, *Fig 3*
  and yard 83-4
culm 179, 182, 256-7
cumin 127
cupful 232
cupro(copper)-nickel 113, 300-301
currency **5-8, 100-113, 298-301**
  Belgic 19
  blanched 121
  British **16-20**, 100
  coins-notes (pre-decimalization) 300-301, 300*n*
  decimal **298-301**
    Acts (1967/69) 299-300
    BAAS/ABCC Report 295
    coins/notes 299, 300-301
    costs of change 298
    Decimal Currency Board 300
    Executive Board 299
    Halsbury Committee 295, 298-9
    Hodgson Report 290, 295
    primary unit 298-300
    proposed 263-4, 279-80, **281-3**, 285, **286-8**, 289
  Greek 5-8
  hundred 58-9
  manipulated 113
  Pyx, Trials of 112-13
  Roman **13-16**, 101
  Romano-British 18-20, **20-21**, 100
  Saxon **100-110**, 125
  score 43-4
  see also individual units
currency bars/spits 6, 7-8, **16-18**, *Fig 10*
customary measures abolished 182, 272
customs duties *see* duty
cwt *see* hundredweight
cyathus 13, 117, 232
Cynegils, king of Wessex 47
Cyprus 6*n*, *Fig 6*

'd' *(denarius) see* penny
Daily Telegraph 219
Danegeld 56, 125*n*
Danelaw 32, **54**
Danes 32-3
David I, king of Scotland 44, 79-80
'D(decimalization)-day' 300-301
decempeda **11-12**, 30
Decimal Association 282
Decimal Coinage and the Metric System: Should Britain Change 295
Decimal Currency Board 300
decimalization 38, 262, 263
  currency *see* currency: decimal
  see also metric system
decimetre (dm), cubic 306, 353-5
Delambre, Jean Joseph 346-8
denarius 12, **14-15**, *Fig 9*
  and Apothecaries' weight 119
  'd' abbreviation 103
  post-Roman 103, 105*n; see also* denier
denaro (denier) 117
denier *(denarius)* 107-8
Derby 1
  lead-dishes 177-8
  pint 160, *Fig 36*
  weight 1
Derbyshire (lead-measures) 177-8, *Fig 39*
Dickson (1862 Committee) 287
didrachm 14
die (for coin-striking) 8
diesel oil 305
digit/finger 1, **2**, 11
  and foot 27, 81
  Roman 8, 11
  Saxon 81
diker (dicker) 320
Dio Cassius 23
Diocletian, emperor 11, 15
distance *see* linear measure
'dm' *see* decimetre
Dollond (instrument-maker) 257, 258-9, 260
Domesday Surveys (1086)
    area for taxation **59-61**
    currency 120-21

hide 54, 57
hundred 58
sester 149
12 cent 42
13 cent 58
of Kent 122, 326
dozen
baker's 198-9
and case (liquor) 188
cloth 87
iron 321
drachm (weight unit) **118**, 186, 294, 297
abolished 304, 307
and spoonful 160
drachm, fluid (dram; drachma) **186**, 294, 297, 304, 307
drachma *(dragma; drax)* (weight and currency unit) 6, 7-8, 16, 102, 117
fluid *see* drachm, fluid
dram (weight unit) 129, 207, 253, 263, 307
see also drachm
Drapers' Company 237
drax see drachma
drop (capacity measure) 186*n*
drugs *see* Apothecaries' weights/measures
Drusus, Nero Claudius 28
dry goods *see* capacity; *and individual units*
Dublin 258, 259, 267, 269
dupondius 15
Durotriges 19
duty, customs/excise 64, 140-41, 226, 283
ale/beer 64, 158, 188, **226**
brandy/aqua-vite 64, 226
casks, standard 329
chocolate 226
cider 226
cloth 131, 133
coal 180, 181
cocoa 141
coffee 141, 226
glass 141
grain 214
herring 174
lead 178
malt 164-5, 226, 252
mead 226
measures, privileged 323
metheglin 226
paper 141
perry 226
salt 184
sherbet 226
spirits 188
tea 226
tobacco 140-41
wine 153, 158, 171
gallon 162-4, 165, 172
wool 131, 133, **135**
dwt *see* pennyweight

Eadred, king 58
Earth (and 'natural' measures) 35
quadrant and metre 344-8
Ecgberht, king of Kent 107
Edgar, king of Mercia and England 56, 96, 323
currency 109, 125*n*
weights 132
wool 132
Edinburgh 87, 307
Imperial standards 258, 259, 267, 269
Edward I
bread 195, 199
capacity measures 150, 156, 221
currency 111
land measures 90
Magna Carta (1279) 150
scutage 63, 64
Tractatus 124
tron 135
weights 132, 136-7, 138
clove 129, 233, *Fig 29*
Edward II 46, 201
Edward III
Avoirdupois 128, 131
balance 137-8, 139
capacity 150-51, 156, 171
cloth 94
currency 113

Troy 121
weights 129, 142, **233-4**, 248, *Fig 28*
Woolsack 131
Edward IV 199, 236
Edward VI 22, 93
Edward the Confessor, king 39, 42-3
currency 120, 125*n*
mints 109
hidages 59-60, 61
Edward the Elder, king 32, 39
capacity 149
land measure 82
shilling 102
Edward the Martyr, king 125*n*
Edwin, king of Northumbria 32
EEC *see* European Economic Community
eels 173, 174, 321
Egbert (of Kent) *see* Ecgberht
Egbert, king of Wessex and England 32
Egypt
measuring devices 2, 9, *Figs 2-3*
scales/weights 337*n*, *Figs 5, 55*
talent 119
electrum 7
electuaries *see* Apothecaries' weights/ measures
Elizabeth I **240-43**, 272
capacity **159-61**, 164, 165, 174, **243**, *Fig 35*
bushel 160-61, 246
gallon 185-6
pint 13, 252
quart 159
cloth measures 87, 93, **240-41**
ell **94-5**, 244-5
yard 238, 244-5, *Figs 46-7*
currency 113
Hanse 22
mile 70
post 73*n*
weight 127-8, 129, 155, 241-2, *Figs 49-51*
Elizabeth II 112*n*
ell 82-3, 91, **94-6**, 260
Flemish 95
and rod 44
rules/standards 87-8, 235, 236, 240-41
Saxon 80
Scotland 44, 80, **87-8**, 95, **260**
as yard 83
elne (ell) 80
Ely Cathedral 47
Emden (Barth) hoard 103
Emma of Normandy (Aelgyfa, Lady of England) 41, 42-3
England
Anglo-Saxon invasion/settlement 31-2, 107
Danish invasion/settlement 32-3
early *see* Britain
hidage 59-60
inflation (18 cent) 214
Roman withdrawal 31
World War II 217-19
Ethel- (names beginning with) *see also* Aethel-
Ethelbert II, king of Kent 36
Euboic-Attic (Solonic) system 5-6
Europe 284, **285**
see also individual countries
European Economic Community (EEC) 175, **304**, 305, 306, 308
NWML 337
Ewart, William 283
Exchequer standards **121-3**, 138, **163-4**, **237-8**, 239-71
bread 213-14, *Fig 42*
bushel 155-6, 160, 165-6, 232, 239, *Figs 34-5*
coal bushel 182, *Fig 40*
Carysfort Committee 246-9
comparative tests 244-5
ell 94-6, 240-41, 245
Exchequer Standards Commission (1871) 185
gallon 13, 154-66, 173-5, 225, 239, *Figs 34-5, 37*
Imperial 257-71
insignia 328
museum collections 272
pint 13, 252, 254, *Figs 35-6, 38*

quart 163, 172, 243, *Figs 35, 38*
standards moved to Board of Trade 271
weights
Edward III 129, 234
Elizabeth I 127-8, 155, 242, **245-6**, *Fig 49*
Imperial 268-9
Tower pound 110
Trial of Pyx 112
yards 84, 94, 235, **239-40**, 267, *Figs 46-7*
comparative tests 244-5, 250-51, 266
excise/export duty *see* duty

faat (fat) 157
Fabbroni, Giovanni Valentino Mattia 353
familia see hide
farms/ferms (commissions) **122**
on coinage 110*n*, 111
paid in kind 151, 156
privileged 156, 323-4
farthing 195-6, 300*n*, 301
bread 194-206, 317
'decimalized' 263, 280
fathom **2**, 82-3, **270**
Greece 8-9
and 'old English mile' 73
Federation of British Industries/*later* Confederation of British Industry 290, 301, 306
ferlingate 63
fermors 317
ferms *see* farms
fifth (liquor measure) 187
fines 238, 323*ff*
bakers 198, 199, 200, 206, 209, 215
brewers 221, **223**, 224, 225
ale-conners 222
Clerk of the Market 328
paid in livestock 6-7, 13
finger *see* digit
firkin/puncheon/tertian
ale/beer **172**, 225
butter 176-7
soap 176
wine **171**, 172, 177
fish 58, **173-5**, 184, 320-21
heaped measure 180, 256-7
metrication (weight) 305
salmon 173, 174, 177
Flamsteed, John 152
flax 127
flesh *see* meat
Fleta
acre 57
assizes of bread/ale 194-5, 200, 221, 323
Clerk of the Market 325-6
Compositio 124
hundred 58
sester 149
weights 126-7, 130, 131, 138, 335
yard 83
Florence 117, 131-2, 138
florin **281**, 282, 300*n*
'decimalized' 299, 300
'Godless'/'Graceless' 281
proposed 263, 280
flour
bread 207, 209-15; *see also* bread: varieties
additives 219-20
National 217-19
metrication 305
millers 317
sack 210-14, 216*n*, 217
see also wheat *etc*
fluid measure *see* drachm, fluid; ounce, fluid *etc*
food 298, 305, 307
see also bread *etc*
foot **27-9**, **80-83**, 89, 91, 96
Algar's (St. Paul's) 85-6
of assize 86
body measure 1, 2
cubic 306
Drusian/Great Northern **28-9**, 30, 81
half-foot *see* shaftment
and mile 73
and rod 43-7, 82
France 244, 346, 349

Great Northern *see* Drusian *above*
Greece 8-10
and land measure 85
and mile 68, 69
'natural' **27-8**, 68
and palm 80
and rod 43-7, 82, *Fig 21*
Roman 8, 9, **10-11**, 22, 28, 82
and amphora 12
and mile 68
on rule 241, 243, 244, 247
rules (Roman) 10, *Fig 8*
St. Paul's (Algar's) 85-6
standards 84-7, 244
and yard 83-4
forestallers 316, 318
forest perch 45-6, 48
formel see fotmal
'forpit' 185
Fortin, Jean Nicolas 353, 354
fotmal/*formel* 130, 320
Founders' Company 245-6, **330-31**, 332
France
Académie des Sciences/Academy of Sciences 244, 345, 347
Avoirdupois pound 132*n*
canne 344
Carolingians *see* Franks
Cluny Abbey 22
foot 244
Drusian 30*n*
pied du roi 83
inch 346, 349
Institut National 347
league, Gallic 73, 74-6
map surveys 344-8
Merovingians *see* Franks
metrication/decimalization 279, **344-56**
mile 73, 74
mille de Paris 73
Paris weight 117, 244, 353
Revolution 279, 344-9
standards 244, 280-81, 285, 344-56
toise 346-9
Troyes 118-19
see also Gaul; Franks
frankpledge, view of (leet court) 324
Franks **107-8**
bread 194
Carolingians 43, 107
coinage 100-101, 104, 105*n*, 107
Merovingians 100-101, **107**
see also Charlemagne
French bread 215, 216
Frisia 107
fruit 178-80, 256-7, 305
Fruiterers' Company 179, 331
fuel *see* coal *etc*
furlong 35, **36-8**, 96, *Fig 18*
abolished 305, 306
and league 75
and mile 69-74
and stade 9, 10, **69-70**

Gallic league 73, 74-6
gallon **151-66**, **248-9**, **251-6**, 294, *Figs 34, 52*
ale/beer **172-3**, 233
assize 220-26
Carysfort Committee 252, 254
City 243
as sole gallon 248, 249
Apothecaries' 185-6
bread 185, 209, 214, *Fig 43*
Carysfort Committee 162-6, 252, 254
and congius 186
corn 233, 239, 246, 249, 254
Edward III 150
Exchequer *see* Exchequer
excise 226
fish 173-5
Imperial **165-6**, 174, 177, 256
metric equivalent 297, **306**
and seam 149-50
and sester 149
stricken/heaped 156-7
unified **256**
US 177, 187, **188-9**, 290
as weight unit 123, 185, 209, 214
Winchester 160, 165
wine 170-73, 246, 249, 252, 254, 272

garbol (garble) commodities 206
garlic 321
Gaul **19**, 20
  Belgae 19, 75
  league 73, 74-6
geld *see* taxation; Danegeld
General Conferences on Weights and Measures 351-2, 355, 356
'geographical' league 76
Geometrie (Kobel) *Fig 21*
geometry, origins of 2
George I 184, *Fig 33*
George II
  bread 209, 213
  coal bushel 182, 272, *Fig 40*
George III
  bread 211-14
  capacity 174, 187, 225
  cloth measures 94, 95
  inspectors 331-2
  salt 184
  weight 140-41, 142, *Fig 33*
  wool 143
George IV **255**
  bread 215
  capacity 150, 165-6, 187, *Fig 52*
  ell 95
  mile 70
  yard 95
Germany
  Altona Observatory (*formerly* in Denmark) 269
  Hanse 22
  rast 74, 76
  rod 44, *Fig 21*
  score 43
  Tungri 28, 43
  weights 117, 119, 123
  see also Franks
Gilbert, Davies, MP 119, 279-80
  Commissions 253, 261, 264
ginger 136
Gladstone, W.E., MP 281-2
glass 130*n*, 135, 141
gloves 320
Godbegot *see* Winchester
gold 3
  currency 15, 19, 100-103
  mancus 58-9
  see also bullion
Goldsmiths' Company 112*n*, 239, 241, 247-8
Goodbegot *see* Winchester
Government
  Dept of Education and Science: Minister for Science 297
  Dept of Scientific and Industrial Research 294-5
  Ministry of Technology and Power 302
  Board of Trade *see* Trade
  Dept of Trade and Industry 307, 337
  see also Parliament
gradus 11
Graham, George **244-5**, 246, 249, 266
grain (as dry goods) 90, 149, 156, 263*n*
  see also corn *etc*
grain (as weight unit) **2**, 117, **124-5**, **245**, **270**
  abolished 307
  Avoirdupois 127-8, 253, 256, **263**, 268-70
  Hodgson Committee 294
  Paris 2, 107, 117
  Troy **118**, 119
    as barleycorn 2, 4, 35, **118**
    and penny 108-13, 151
    sceatta 102
    as siliqua 101
    and tremissis 16
gram 353
'grave' 353, 354
'gravet' 353
Great Charter *see* Magna Carta
Great Northern foot *see* foot, Drusian
Great Wishford (Wilts) 214, *Fig 43*
Greece
  capacity 10
  currency 5-8
  linear measures 8-10
  weights 3
Grocers' Company 134
grocery commodities *see* food; spices
'gros' 353

Guildhall *see* London
guilds/livery companies **329-31**
Clockmakers 243, 244-5
Coopers 329
Drapers 237
Founders 245-6, **330-31**, 332
Fruiterers 179, 331
Goldsmiths 112*n*, 239, 241, 247-8
Grocers 134
Mercers 236-7
Merchant Taylors 234-6, 237, *Fig 45*
Pepperers 134
Plumbers 329-30
guinea 103*n*
gunmetal 247
gunpowder 59
Gunter, Edmund 38, 86
Guthrum (*later* king of East Anglia) 32, 54*n*
Gyges, king of Lydia 7
gyrd 39-42

haber de peyse/haber-de-pois/haberty-poie/haburdepeyse see Avoirdupois
half-crown 301
halfpenny/ha'penny (½d) 113, 300*n*, 301
bread 201, 202, 203, 206
halfpenny, decimal (½p) 299*n*
Halley, Edmund 152
Halsbury Report 295, 298-9
hand
as linear unit 1, 2
and Drusian foot 28-9
shaftment 29
and yard 85
'yard and a' 87
selling/weighing by 1, 134
see also palm *etc*
Hanse (Hanseatic League) 22
Harold I Harefoot 42
Harris, Joseph **246-8**, 251, 254-6, 267-8
Harrowby Committee 119
Harthacnut, king of England and Denmark 33, 42
Hastings (capacity measures) 161-2
Haüy, René Just 353
haverdepois *see* Avoirdupois
hay 270, 332
Heaberht, king of Kent 107
heaped measure *see* capacity: heaped measure
'heavy goods' *see* Avoirdupois
hectare 61, 284
hectogram 297
hemina 13
hemp 127
Hengist (*later* king of Kent) 31
Henry I
currency 105, 106-7
foot 85
inch 79
king's peace 29-30
Laws 105*n*
scutage 62, 63
yard 83
Henry II
Assize of Bread 194-6, 198
knight's pay 64
rod 46
scutage 63
taxation 56
Henry III 22
bread *see* Assize of Bread and Ale
capacity 150, 156
league 75
Magna Carta 150
perch 45
Statutum de Pistoribus 199
Trial of the Pyx 112
weights 130*n*
yard 88
Henry IV 92, 125
Henry V 157, 180, 222
Henry VI 139, 171
cloth measures 92, 235
Henry VII 237-40, 272, 328, *Fig 47*
bread 203, 205
capacity 153, **155-60**, 164, 174
bushel 232-3, 239, *Fig 34*
gallon 13, 185-6, 239, *Fig 34*
water measure 178-9
sovereign 106*n*
weights **239-40**, 241, *Figs 31, 48*

yard 84, 87, 235-6, **238-9**, **244-5**, *Figs 46-7*
Henry VIII
acre 38*n*
capacity 171*n*, 225
casks 329
cloth measures 93
meat-weighing 1
pound 110
Winchester 43
Henry, Matthew 193
Henry of Huntingdon 54
Heptarchy (Anglo-Saxon kingdoms) 32
Herodotus 2, 7, 8, 9
herring **173-5**, 320
cran measure 174-5, 297
weight 174, 175, 297
Herschel, Sir John (Master of Mint) 281
Commissions (1838/43) 261, 269
hide/hidage 12, 39-40, **54-6**, **57-61**
and burhs 36, 39-41, 59, 60
and league 75*n*
names, alternative 54-5
and sulung 56-7
as unit of assessment 54, 55-6, **59-61**
knight's fee 62-4
variability 55-6, 57
as virgate 61
Hodgson Report 187, **289-95**, 296-8, 339
hogshead 177
cider/perry 226
tobacco 141
wine 163, **171**, 172, **257**
Holinshed, Raphael 48
Homer 5
honey 149, 171, 173, 176, 177
barrel 176
weight 176
Honorius, emperor 20-21, 31, *Fig 11*
hops 221, **223-4**
Horsa (Saxon war-leader) 31
horsebread **199**, 201, 207
horse-load 90, 149
horse-shoes 320
Hounslow Heath (Mddx) 249, 254
'household' bread 195*n*, 205-13
Hubbard, T.C. 282-3
hundred 55*n*, **58**, 94, 135, 320-21
cloth 58, 94
fish 173, **320-21**
half-hundred 154
'long' **58**
see also hundredweight
hundredweight (cwt) **59**, **135-8**, 140, 270
abolished 307
coal 181, 183
quarter 140, 172
and quarter (capacity) 154-5
spicers' 136
hurdle (penalty) 201-2, *Fig 41*

Imperial measures/weights 254, **256-70**
barrels 172-177
bushel 183, *Fig 53*
gallon 166, 172, 188-9, 254
ale/beer 225
Apothecaries' 186
herring 174, 175
Hodgson Report 291-5
post-metrication 302-8
pound *Fig 54*
Troy 256
quart 187
redefined (1963) 296-8
sack 183
and US measures 188-9, 290
yard 256, *Fig 54*
import duty *see* duty
'imputed' bushel 183-5
'in' *see* inch
inch ('in') **79-80**, 96
as three barleycorns 3, 28, 79-80, 89, *Figs 4, 17*
cubic 306
French 346, 349
metric equivalent 294
Roman 8
on rules/yardsticks 239, 241, 243, 244, 247, *Fig 47*
square 306

'yard and an' 87-8, 92-4
see also thumb
Incorporated Society of Inspectors of Weights and Measures 338-9
Indian corn 212, 215
Ine, king of Wessex
capacity 149
coinage 102, 103, 104*n*
digit 81
Inspectorate 287, **331-5**, 338-40
associations 338-40
Clerks of the Market 325-8
King's commissioners 326-8
leet courts 324-5
Model Regulations 335
Institute of Trading Standards Administration **338-40**
Institute of Weights and Measures Administration 339
International Astronomical Union 351
International Bureau of Weights and Measures/*Bureau international des poids et mésures* 285, 293, **349-50**, 352
International Solar Union 351
international standards 285, 293, 337, 349-50
kilogram 296, 307, *Fig 57*
metre 285, 296*n*, 306, *Fig 56*
nautical mile 76
OIML 337
Ireland
foot 27
Northern 218
Scots 31
standards 258, 259, 267, 269
iron
currency bars 16-18, *Fig 10*
hundred of 58, 321
obols 7-8
weighing of 127, 136
weights 330, 331
Italy
Drusian foot 30*n*
weights 117, 131-2, 138
itinerary measures 70-74

James I (James VI of Scotland) 73, 330
weight 140, *Fig 32*
James II (James VII of Scotland) 70
Janety (Janetti), Marc Étienne 348-9, 354
jewellery (as medium of exchange) 100, 102
jewels, weighing of *see* bullion; stones, precious
John, king
capacity 150
cloth measures 88, 91
scutage 64
weights 130*n*
Johnson, Samuel *(Dictionary)* 75-6, 84
Johnson & Cock/Johnson, Matthey & Co 268, 350, 354
Judicium Pillorie (Judgment of the Pillory) 196, 199-200, 221, **315-16**
jugerum 12
jugum (yoke) 57
Julius Caesar 20
Julius Hyginus 11
Jutes 31, 43
see otherwise Anglo-Saxons

Kater, Captain Henry 251, 253, **257-60**, 262, 272
balance 258
scale 258-9, 262, 272
support surfaces 259, 265, 267
yard comparisons 255
yard/metre comparisons **280**, 284, 292
weights 257-8, 268
keel (coal-ship) 180
Kendal (Cumbria) 128
Kent (sulung) 56-7, 63
Kenwalh, king of Wessex 47
keration see carob
kg *see* kilogram
kilderkin
ale/beer 172, **225**
butter 176-7
kilogram (kg) **352-4**
France 279, **280-81**, 285, 352-4
origination 345-6

Imperial equivalent 284, 286
International Prototype **354**
Kilogramme des Archives 280-81, 285, 292, 293, **354**
king's beam **134**, 136
king's bushel 232
King's Lynn (Norfolk) 135, 138
king's peace 29-30, 69
king's perch 44, 45, 46, 69, 89*n*
king's tron 135
king's yard 86, 88-9, 80-91
kings *etc* (and standard measures) 83
knight's fee/service **61-4**, 226
daily payment 64
Kobel, Jakob 44, *Fig 21*
krypton 352

£/'L'/'l' *(libra) see* pound
'l' *see also* litre
Lacaille, Nicolas Louis de 76, 344
laces 127
La Condamine, Charles Marie de 346
ladleful 232
Lasgrange, Joseph L., comte de 345-6
Lalande, Joseph J. le F. de 345
Lambton scale 255
Lancashire (perch) 48
land measure **35-9, 54-64, 81-2, 89-90**
king's peace 29-30
Laxton 49
Statutum de Admensuratione Terre 322
survey measures **249-50**, 254, 255
Canada 308
France 344-8
rule 35, *Fig 17*
toise de Perou 346
variability 30, **44-9**
yards 85
'with the inch' 88
see also area; linear measure
Laplace, Pierre S., marquis de 345, 347
lasers 352
last (weight/quantity unit) 58, 173, 320
latitude (computed in degrees of arc) 76, 344-5
Lavoisier, Antoine Laurent 346-7, 353
lawful men *see* nuisance juries
Laxton (Notts) 49
'lb' (libra) *see* pound (weight)
'£-cent-½' scheme 299
lead
levies on 178
measures for 177-8, *Fig 39*
weighing of 127, 130, 130*n*, 177, 329-31, 332
load 320
league 38, 73-4, **74-7**
Gallic/French 73, **74-6**
and hide 75*n*
nautical/geographical 76
leather/skins *etc* 320, 321
leet court 49, **324-5**, 332, 333, 334
Lefevre, J.S. (Commissions 1838/43) 261
Lefèvre-Gineau, 353
legales homines see nuisance juries
length *see* linear measure
Lenoir (instrument-maker) 347
Leonardo da Vinci 2, *Fig 1*
leuca/leuga (Gallic league) 74
leuga (ley land) (Measure of land extent) 55
leuva/leweke (Gallic league) 74
Lewes (Sussex) (weight) 19
'li' *see* pound
libbra (pound) 117, **131-2**, 138
Liber Albus 195, 198-9, 202*n*, *Fig 41*
libra see pound: Roman
libral *as* 13-14
libra mercatoria (commercial/mercantile pound) **125-7**, 128-32
confused with Avoirdupois 242
and capacity 158, 159
libra sutil ('light pound') 117
light, speed of 35, 396*n*, **306**, 356
ligne (linear unit) **346**, 347, 349
lime 179-80, 256-7
Lindisfarne Gospels 102
linear measure 27-30, **35-49, 68-96**
building measures *see* building
early **1-2**, 4-5, **8-12**
'itinerary' measure 70
land measures *see* land
metric permitted 284

'natural' 35
parallel systems 81
standards *see* standards
see also individual units
line/brim measures 224-5
liquid measures *see* capacity
litre ('l') 279, **306**, **354-6**
Imperial equivalent 284
livery companies *see* guilds
livre 353
load (weight unit) 320, 332*n*
cart-load 181
char 130
horse-load 90
wagon-load 170, 181
loaf, farthing 194-6, 200, 201, 203-4, 207, 313-16 *see also* quartern loaf
London
acre's-breadth 38
ale-conners 223
bakers 199, 201-2, 212-15
as capital 39
City weights 130, 241-2, 334
see also Guildhall *below*
cloth ordinances 235-7
Edgar's standards 96
Exchequer *see* Exchequer
Great Fire (1666) 248
Greenwich 267, 269, 296
Guildhall (standard measures) 86-7, 181, 245, 330, 334
gallons 152-5, 158-9, 161, 163
Imperial standards 258, 259
quart, pint *etc* 161
Hanse 22
House of Commons *see* Parliament
livery companies *see* guilds
measure of
bushel 124, 151, 157-8, **232**
quarter 150
see also Guildhall *above*
Mint *see* Tower *below*
museums *see* museums
nuisance juries 323*n*
Palace of Westminster *see* Parliament
pound of *see libra mercatoria*
rules, Roman 10
St. Bartholomew's Fair 235
St. Paul's Cathedral 27-8, 55-6, 58, 85, 109
Steelyard 22
Tower
Board of Ordnance *see* Ordnance
Mint 20, **109-13**, 241
pint 160
pound of **109-10**, 110-13, **122-5**, 126; *and* gallon 151-3, 158
Trafalgar Square 87
as 'Troy' 119
weights of 130, 241-2
Westminster 39, 109
London Pharmacopeia (1618) 185
(1851 and 1864) 186
Louis XVI, king of France 345
Lubbock, J.W. (Commissions (1838/43)) 261

'm' *see* metre
Macclesfield, Lord (of Royal Society) 245
mace (spice) 136
madder 127
Maecianus, L. Volusius 117
Magna Carta
(1215) **64**, 91, 130*n*, 150
later 13 cent 91, 150
Magnus the Good, king of Norway 42
Malaysia 22, *Fig 15*
malt 156, 157, 164, 222
brewing 221-4, 225
ale-yields 223
duty on 164-5, 226, 252
mancus 58-9, 105, 120
manetium/*mansatus*/*mansura see* hide
manupes see foot, Drusian
maps ('old English miles') 71-4
mark/*marc* (currency unit) 63, 131
mark/*marc* (weight) **117**, 120, **123**, 353, *Fig 27*
and London pound 126
and Paris ounce 244
markets 316
Bartholomew Fair 235
see also Clerk of the Market

Marshalcy, Clerk of *see* Clerk of the Market
Mary I, queen (Philip and) 93
mass
  defined 293
  established as standard 296
  measures of *see* weight
mastin 210
Matthey, George 350
Maundy money 113
mead 64, 226
meal 149, 156
measure (unmodified) *see* capacity
measuring rods *see* rules
meat
  assize 316, 317-18
  auncel weight 134
  Avoirdupois 206
  barrel of 174
  butcher's stone 335-6
  metrication 305
  weighed 'by hand' 1
Méchain, P.F.A. 346-8
medicine
  'hieroglyphs' 118
  weighing/measuring of *see* Apothecaries'
mercantile pound *see libra mercatoria*
merce (fine) 199
Mercers' Company 236-7
Merchant Taylors' Company 234-6, 237, *Fig 45*
meridian 344-8
Merovingians 100-101, **107**
  see also Franks
Meter, Common (measures-official) 236
'meter-yard' 308
metheglin 64, 226
metre ('m') 296*n*, **306**, 344-6, **346-52**
  France 279, 285, 346-52
  Imperial equivalent 284, 286, **292**
    scientific standard 291
  International Prototype **351-2**, 356
  optical 351-2, 356; *see also* light
  rules/standards *see* rules
Mètre des Archives 280*n*, 285, 349
Metrication Board **302-6**, 308
Metric Convention/*Convention du Mètre* 285, 286, **349**
metric system/metrication 279-88, **289-98**, **301-7**, 344-56
  adopted internationally **349**
  advantages summarized 286-8, 290
  areas of non-application 302-3, 304, **306-8**
  Canada 307-8
  costs 302
  currency *see* decimalization
  fish 175
  foreshadowed 38, 262, 263
  Hodgson Report **289-95**, 296-8
  international/trade considerations 283-4, **285**, 286-7, 290, **301-2**, 307
    EEC **304**, 305, 306, 308
  Italy 132*n*
  legalized for trade 286
  limited uses sanctioned (1864) 284-6
  Metrication Board **302-6**, 308
  'natural' basis 35
  origination 279, **344-56**
  pre-packaged goods 304, 305, 307
  profiteering fears 302-3
  recommended (1862) 283-4
  'single-standard' basis 279
  standards redefined 296-7
  statute (1985) 306-7
  US 290
  'voluntary change' concept 285-6, 303, 305-6
Michelson, Albert Abraham 351
micrometer microscope calibration 251, 264, 265
mil (mile) 68, 71
'mil' (proposed UK currency unit) 282
mile **68-74**, 263, 270
  French 75
    mille de Paris 73
  Greek 9-10
  and league 74-7
  nautical 76
  'old English' **70-74**, 76-7, *Fig 23*
  Roman 8, **11-12**, 68

square 306
mileage, early estimates of 68, 70-74
milestones 68, 73
miliare/milliarius 68, 74
miliarium/milliarium 11-12, 68, 69, 71, 76
and league 74, 75
milion 9-10
milk 298, 303
millarium 11
mille de Paris 73
mille passuum (Roman mile) 8, 68
Miller, Professor W.H. 264, 287
pound **264-70**, 293
yard/metre 280-81, 284
millers 317
milliarium see miliarium
milliarius/miliare 68, 74
millilitre ('ml') 355
millimetre ('mm') 292
'milyard' 263
mina 6
minim 186, 294, 297, 304, 307
mints 109-10, **110-13**, 122-5, 241
Roman 20
weights 245-6, 258, 296
yard 267, 296
mitta 149
'ml' *see* millilitre
'mm' *see* millimetre
Model Regulations 335
modius **13**, 194
Molony Committee 339
money *see* currency
moneyer *see* mint
Monge, Gaspard 345
mons Badonicus (Badon Hill), battle of 31-2
Monteagle, Lord (Commission, 1855) 282-3
morgen 37
Mouton, Gabriel 344
museums **272**
Birmingham: Avery 3, 22
Cambridge: Archaeology and Ethnology 242
Canterbury 21
Derby 1
Derby: Industrial 177-8
Lewes: Barbican 19
London: British
axe-head ingot 20-21
currency bars 17
ox-hide ingot 6*n*, *Fig 6*
rules, Roman 10*n*
weights, Roman 132*n*
London: Museum of London
ale gallon 160, **243**
bread weights 214, *Fig 42*
Kater's balance 258*n*
rules, Roman 10*n*
London: Science **272**
capacity measures 155-6, 159, 160, 239, *Fig 34*
ell 240-41
lead-dish 178
Roy's scale 250
weights 129-30, 136-7, 233, 240, 241-2, *Figs 29, 48-51*
yard 239, 240-41, 243, *Fig 47*
London: World of Beer 153
Oxford: Ashmolean 9, *Fig 7*
Reading 120
Winchester: City
currency bars 17
weights 129, 142, 233-4, *Fig 28*
yard 238, 239, *Fig 46*
Yale: Streeter Collection 120

nail (linear measure) **84**, 235, 240
'yard and a' 93
nail (weight unit) 140
Napoleon I, emperor 349
National flour 217-19
National Metrological Co-ordinating Unit 306
National Physical Laboratory (NPL) 291, 294, 297, 338
National Weights and Measures Laboratory (NWML) 307, **337-8**, *Fig 54*
'Nature'
as origin of measures 35
pendulum 253

see also light; wavelength
nautical distances 76
Nef, J.U. 181
Nehus, Captain 259, 268
Nero, emperor 15
Newcastle-upon-Tyne (capacity measures) 161-2
Newton, Sir Isaac 76
Normans
league 75
mile 74
see also France; William I *etc*
Norse hundred 58
Norway
bismar 21, *Fig 14*
Viking invasions 32
Norwich (weights) 239-40
NPL *see* National Physical Laboratory
nuisance juries *(legales homines)* **315-16**, 319, 323, 333
records of 146 (ref 55), 156, 202, 326
see also assize
numero, accounting 122
nutmeg 130*n,* 135, 320
nuts 136
almonds 127, 136
NWML *see* National Weights and Measures Laboratory

oats 156, 157
bread 194, 196-7, 209, 210, 212, 215
brewing 220-21
obol/*obolus* 6, **7**, 16, 117
octarius 186
Offa, king of Mercia 32, 107, 108-9, 132
oil 171, 173, **177**
OIML *see Organisation Internationale etc*
'old English' mile 70-74, 76-7, *Fig 23*
Olympic measure 9
once 353
ora/*ore* 59, 120-21, 122
Ordinacio facta de modo ponderandi per balanciam 136, 138
Ordnance, Board of/Ordnance Survey
A1/A2 yards 264-6
Roy's scale 249-50, 254
Tower/Rowley's yard 243-5, 250
Organisation Internationale de Métrologie Légale (OIML) 337
orgyia 9
orichalcum 10, 10*n,* 15
Orkney (steelyard) 22
ounce ('oz')
Apothecaries'
abolished 304, 307
obsolescent 297
see otherwise Troy *below*
Commissions *etc* 249, 253, 263
fifteen to pound 126
Florentine 131-2
fluid 186, 188-9, 358, 360
and mark 117*n,* 244
of Paris 244
and penny 123, **125-6**, 151
and pint 154
Roman 8, 13, 102
sixteen to pound 127
Troy/Apothecaries' **110**, **117-18**, 119, 294
and Avoirdupois confused 127-8
and Avoirdupois distinguished 155
metric equivalent 306
as primary 271
redefined (1985) 307
restricted 186, 263, 297
twelve to pound 151
uncia 8
ounce, fluid 186, 294, 297
and reputed quart 187
US 188-9
Overstone, Lord (Commission, 1855) 282-3
ox
carucate 54
sulung 56
as unit of value 5, 6, 13, 35
see also ploughing
Oxford
ale privileges 223, 224
Arundel relief 9, *Fig 7*
Corpus Christi College 48-9, *Fig 22*
oxgang/bovate 55

'ox-hide' ingot 6, *Fig 6*
'oz' *see* ounce

'p' *see* penny, new
pace 11, 68, 75
palm 1, **2**, 11, 80
  and foot 28
pan loaf 216
paper 141
paraffin 305
Paris measures/weights
  foot 244
  grain 2, 108, 117
  mark 117, 244
  mile *(mille de Paris)* 73
  ounce 244
  see also France
Parliament/Palace of Westminster
  Bread Act (1836) 216
  Bread Orders/Instruments (1917; 1940-46; 1976) 217-19
  Carysfort Committee *see* Carysfort
  Commercial and Industrial Policy Committee (1917) 289
  Consumer Protection Act (1961) 339
  Cran Measures Act (1908) 175
  Customs and Excise Act (1952) 188
  Decimal Coinage Commission (1855) 282-3
  Decimal Coinage Commission (1920) 289
  Decimal Currency Acts (1967/69) 299-300
  Decimal Currency Committee (1962-3) 295, **298-9**
  Exchequer Standards Commission (1871) 185, 333-4
  fire (1834) 261
  Herring Fishery (Scotland) Act (1815) 174-5
  Imperial standards deposited **247-8**, 251, 252-3, **256**; fire 260-61; restored
  267, 269; post-metrication 296
  Laxton 49
  Manufacture . . . of Bread Committee (1815) 211
  Medical Act (1858) 186
  Merchandise Mark Acts (1887/91/94; 1911/26/53) 340
  Metric Act (1864) 284, 286
  Metrication Committee (1966-8) 302
  Metrication White Paper (1972) 303
  Sale of Food (Weights and Measures) Act (1926) 216, 217
  Sale of Tea Act (1922) 339
  Salt Duties Act (1702) 184
  Standards Act (1855) 264-70
  Standards Commission (1867-8) 284-5, 333-4
  Standards Committee (1813-14) 251-3
  Standards Restoration Commissions (1841-54) **261-70**, 281
  Trade Descriptions Act (1968) 339-40
  Union, Act of (1706) 165
  Weights and Measures Act (1824) 165-6, 179-80, **255-60**
  Weights and Measures Acts (1834/35) 182, 272, 330
  Weights and Measures Act (1878) 180, 182, 186, **271-2**, 286, 330
  Weights and Measures Act (1889) 331, 334-5
  Weights and Measures Act (1904) 338-9
  Weights and Measures Act (1963) **296-8**
    Apothecaries' measures 186-7
    bread 217, 219
    coal 182
    cran 175
    Imperial/metric equivalence 292*n*, 296
    litre/kilogram 355
    reputed quart 188
  Weights and Measures Act (1970) 186-7
  Weights and Measures (amendment) *Act (1976)* 304
  Weights and Measures Act (1985) 296*n*, **306-7**, 356
  Weights and Measures Commission (1819-21) **253-4**

bread 215
cloth measures 94
decimalization 280
league 76
rod 48
Winchester bushel 165
Weights and Measures Committee (1821) 254-5
Weights and Measures Committee (1862) 270, **283**, 287, 332-3
Weights and Measures Legislation Committee (1948-51) 289-95
Weights and Measures (Metric System) Act (1897) 286
see also individual monarchs
Peacock, Revd G. (Commissions, 1838/43) 261
peas 150, 161
for bread 199*n*, 209, 212, 215
peck 178-9
abolished 166, 304, 307
bread 209-14
and modius 194
obsolescent 294, 297
quarter-peck (forpit) *etc* 185
salt 185
as weight unit 185, 209-14
Pegolotti, Francesco Balducci
cloth measures 94
hundredweight 136, 138
silver-weights 123
wool-weights 140
Peiresc, N.C.F. de 11
pendulum 344-5, 348
and Imperial yard 256, 262
as natural measure 253-4
penny ('d') **103**, 104, **107-13**, 125*n*, *Figs 25-6*
abbreviations for 103, 194*n*
bread-loaves 202-11, 214
copper 113, 300*n*
cross/crux coins 125*n*
as decimal primary 298
decimal value 301
half-penny *see* halfpenny
metric 104
and ora 120
quarter-penny *see* farthing
shear-accuracy 111
and shilling 104-6
silver
fineness 110-13
weight in wheat-grains 4, 123, 151
sixpenny bit (coin)/sixpence 264, 301
threepenny bit (coin)/threepence 301
and Troy pound 120-21
weight of **109-13**, 123, **124-5**, 151
bread 194-200, 204
for Troy weight-unit *see* pennyweight
penny, 'new' ('p') **299-301**
fractions/multiples 300, 300*n*, 301
see also cent
pennyweight (dwt)/penny **110**, 113, **118**, **125**, 126, 194*n*
abolished 304, 307
barleycorns 35
bread 194-6, 203
obsolescent 118*n*, 263, 294, 297
for weight of penny *see* penny
wheatgrains 197
pensum, ad accounting 122
Pepin the Short, king of the Franks 107, 108
pepper 127, 130*n*, 135, 320
Pepperers' Guild 134
perch (land measure) *see* rod
perry 64, 226
pertica (Roman perch) 30
pes see foot, Roman
pes ad manus/manualis see foot, Drusian
pes monetalis 11
pes naturalis 27
peso sottile ('light' weight) 117
pes palme see foot, Drusian
petrol 305, 307-8
pewter weights 332
peyem system *Fig 5*
pharmaceuticals *see* Apothecaries' measure/weight
Pheidon, king of Argos 6, 7-8
Philip II (king of Spain; husband of Mary I) *see* Mary

Philip II, king of Macedon 5
Philippus (coin) *see stater*
Picard, Jean 76, 344
Picts 31
pied see France: foot
pied manuel see foot, Drusian
pied du Roi 83
pillory 238, 317, *Fig 41*
  bakers 197-202, 206, *Fig 41*
  brewers 221
  see also Judicium Pillorie
pint 166, **252**
  ale **223**, 224
  dry 188
  half-/quarter pint 161-2, 252
  half-thurdendel 224-5
  hooped 224-5
  Imperial 258
  metric equivalent 303
  and octarius 186
  'oversized' 160-61
  post-metrication 303
  and sextarius/sester 12, 149
  standards 153, 160, **161-2**, 165, 223, *Figs 35-6, 38*
    Exchequer 13, **160**, 246
  US 188
  weight of 154, 252
pipe/butt (of wine) 163, **170-71**, 172, 177, **257**
platinum/iridium alloy 349-50
Playfair, Professor (of Edinburgh) 251
plethron 9
Pliny 13
ploughing 37-8, 54-5, 56, 57, 60
Plumbers' Company 329-30
poids de marc 353
pole (land measure) *see* rod
pollards (flour mixture) 213
pondre (pounder) *see* auncel
pork 174
porter (yarn-carrier) 88
postal service 73
pot (capacity measure)
  butter 177
  Roman 13
potatoes 179-80, 185, 256-7
  for bread 212, 215, 218
potin coins 20
pottle 150, **154**, 155, 162, 221
pouce (inch) 349
pound ('l'/'lb'/'li')
  Apothecaries' **118**
  Avoirdupois, origins of 117, **131-2**, 138
  as capacity measure 185
  Celtic 16, **18-19**
  currency/monetary (£/'l'/'li') **106-7**, **298-300**
    coin 300*n*
    decimal, proposed 263, 280, **281-2**
    as decimal primary 298-300
    'one-hundredth' unit proposed 263, 281-2
    'one-tenth' coin *see* florin
    and pence 104-5, 108-13
    and shillings 105-6, 107, 194*n*
    as unit of account 106*n*
  European 130-32
  and gallon 123, 151-5
  'great' 324
  Imperial (Avoirdupois) established 262-4
  libra mercatoria/London pound *see libra mercatoria*
  metric equivalent 293-4, **296-7**, 306
  and pint 153
  Roman 8, **13**, **15-16**, 108-9, 117
    axe-head ingots 21, *Fig 11*
    and volume measures 12
  and sack 140
  and shilling 107, 321
  standards *see* weights
  Tower pound *see* Tower pound
  Troy **110-13**, **118-26**
    abolished 271
    Imperial established 256
  Troy/Avoirdupois recommendations 248-9, 252-3
  water-displacement 252
  weights *see* weights
pounder (auncel) 133
Powell, John 199, 205-7, 224-5
Praetextatus (prefect) 23

pressure, barometric (and liquid volume) 166
Prisciani Libro de Figuris Numerorum 3-4
prized bread 209-14, 218
puncheon *see* firkin
Pyx, Trials of 112-13

qedet system *Fig 5*
quadrans 15-16
quadrantal 12-13, 117
quart 150, 154, 160, 166, 252, *Fig 35*
  ale/beer 187, 221-2, 224-5, 243
  and coal bushel 183
  hooped 224-5
  Imperial 258, *Fig 52*
  'oversized' 160-61
  'reputed' 187-8, 297
  standards 161-2, 165, 223, 252, *Fig 52*
    ale 172, 246, *frontispiece*
  thurdendel 224-5
  US 188
  weight of 252
  whisky 187
  of Winchester 166
  wine 187, 272
  wine/ale/spirit bottles 187-8
quartarius 13
quartentine 37
quarter (as capacity unit) 124, 150, 151-4, **156-8**, 166
  abolished 307
  and chaldron 180-81
  dry **170**
  faat 157
  grain for brewing 220-23
  liquid **170-72**
  of London 150
  and mitta 149
  obsolescent 294, 297
  and seam 149-50
  stricken/heaped 156-8
  weight of 154
  wheat for bread 194-6, 198, 200, 202-7, 211-13
quarter (of cloth) *see* yard; ell
quarter (as weight unit) 172
quartern loaf 185, 209, 211, 216
  half-loaf 216
  standard weight 213-14, *Fig 42*
quintal 297

Ramsden (Jesse) scale 255, 266
rast 74, 76
rati seeds 3
Read, Samuel 244, 255
Reading (Berks) (weights) 120
regrators 316
Renolds *see* Reynolds, John
rents
  and hide 54, 62-4
  paid in kind 151, 156
    corn-rents 166
  paid in service 30, 54, 61-4
    socage 120
  privileged 156, 323
'reputed' quart 187-8, 297
Reynolds/Renolds, John
  wine measures 153, 155, 159
  yard 243, 272
rice 136
  for bread 212, 215
Richard I
  army 62
  Assize of Measures 89, **90-91**, 95-6, 149-50, 234
  capacity 149-50
  hundred 58
  perch 44, 45
  scutage 63, 64
    ransom levy 62
  standards 150
  yard 89
Richard II 92, 157, 201
Richard III 93, 171*n*
ring-giving 100
road-signs 303, 306, 307
rod/pole/perch (linear measure) 35-49, 69, 89, 96, 322, *Fig 18*
  'acre's breadth' *see* chain
  coppice 48
  fencing 48
  and foot 81-3, *Fig 21*
  forest *see* woodland *below*

German 44, *Fig 21*
king's 37, 44, 69
and mile 69-70
obsolescent 294, 297
Roman 11-12, **30**
square 297
urban 44, **46-8**, 82
variability **44-9**
woodland/forest **45-6**, 48
and yard 83
romana see steelyard
Rome (ancient)/Romano-British 8, 23, 31, 117
area measure
jugerum 12
capacity measure **12-13**, 117
currency **13-16**, 20-21, 100, 101, *Figs 9, 11*
laws 106
linear measure **10-12**, 22
rules 9, 10-11, *Fig 8*
Porchester 41
Southampton 40
standards 11
weight measure **13-16**, 117, *Fig 11*
pound of Offa 108-9
scales/weights **21-2**, 132*n*, *Figs 12-13*
seeds 3-4
Winchester 40
Rome (early modern) 117
Römev, Ole 344
rood **37**, 306, *Fig 18*
rope (as measuring-device) 2, 44, 236, *Fig 2*
rope (as quantity-unit) 321
ropes (weighing of) 127
Rosse, earl of (Commission, 1843) 264
Rowley, John **243**, 244-5
Roy, Maj-Gen William **249-50**, 253-5, 266, 272
Royal Astronomical Society 264
Royal Observatory (Greenwich) 267, 269, 296
Royal Society
Measurement Commission 294, 297
pendulum 262, 345
rules **244-5**, 246-7, 249-50, 253, 267
Kater's yard 260
museum collections 272
standards deposited 296
weights 245, 259, 264
yard/metre test 280
rules/measuring rods/foot rules/yardsticks **234-7**, **243-5**, **249-72**
artificer's (16 cent) 80
barleycorns 35, *Figs. 4, 17*
Carysfort Committee 246-9
Drusian foot 30*n*
Egyptian 2, 9, *Fig 3*
ell 87-8, **94**, 235, 236, **240-41**
inches 239, 241, 243, 244, 247, *Fig 47*
Lambton's scale 255
metre 296, 307, 351-2, *Fig 56*
Mètre des Archives 280*n*, 285, 349
micrometer calibration 250-51, 264, 265
museum collections 272
nail 84
Ramsden scale 255, 266
Roman 9, **10-11**, *Fig 8*
Royal Astronomical Society scale 264
Royal Society 244-5; *see also* yard *below*
Roy's scale **249-50**, 253-5, 266, 272
Shuckburgh scale **250-51**, 255, 258-9, 262, 264-6, 272
support-surfaces **259**, 264*n*, 265, 267, **271**
temperature 266, 267, 271
Troughton's scale 250, 265, 266
yard **234-7**, **243-72**, 307
'A1'/'A2' 264-6
Baily's *see* Imperial *below*
Clockmakers' 243, **244-5**
Exchequer 94, **239**, **240-41**, *Figs 46-7*
Henry VII **84**
Imperial **256-62**, 264-7, 271-2, 296, 307, *Fig 54;* inaccuracy 291
iron *see* king's *below*
Kater's **257-60**, 262, 272
king's (iron) **86**, **88-91**, 322;

Richard I 89, 91, 96, 150
Merchant Taylors' 86, **234-6**, *Fig 45*
Ordnance/Tower/Rowley's 243-5, 246, 250, 266, 272
regulation 316, 318-19, 323-40
Royal Society's 244-7, 249-50, 255, 264-6, 272; Kater's 260
standards: Bird (1758/60) 247-55, 258-61, 266; compared 244-5, 255
deposited Palace of Westminster 259, **260-61**; Imperial yard 256
Tower *see* Ordnance *above*
Winchester 238, 239, *Fig 46*
'yard with an inch' 87, 88
rundlet 171
rye (for bread) 209, 210, 215

's' (*semis;* half-drachm) 118
's' *(solidus) see* shilling
sack (capacity unit) **182-3**, 257
sack (weight unit) **130-31**, 132, 137-8, 140, 234, 320
and bushel 212*n*
flour 210-14, 216*n*, 217
half-sack 234
quarter-sack 129, 142, 233-4, *Fig 28*
saffron 136
St. Bartholomew's Fair 235
St. John's Bread (carob plant) 3
St. Paul's, foot of 85-6
salmon 173, 174, 177
salt **183-5**, 306
Samian measure 9
Sandwich (Kent)
ale 221
bread 200-201
weights 233, 234
Saxon hundred 58
Saxons *see* Anglo-Saxons
scalam, ad, accounting 122
scales **21-2, 133-41**, 234
auncel *see* auncel
balance *see* balance
beam *see* balance
bismar **21-2**, 134, 140, *Figs 14, 15*
bread 215, **217**, *Fig 44*
butter 337*n*, *Fig 55*
early 3, *Fig 55*
electronic 337
money-balances 122-3
public/official 1, 150, **233-4**
king's beam 134, 136
tron 118, 135, 138-9, **234**
Roman 21-2, *Fig 12*
steelyard *see* steelyard
tipped balance/turn of the scale 135-6, 137, 140-41, 333
weights *see* weights
'weights lying'/'*couchants*' 139
Scandinavia
jewellery as currency 102
score 43, 44
see also individual countries
sceaftmund/scaeftmund see shaftment
sceatta 101-5, 107, 110*n*, *Fig 25*
and score 43
scheft/shaft (auncel) 133
Schumacher, Professor N. C. 268, 269
Science, Minister for 297
Scientific and Industrial Research, Dept of 294-5
scilling (shilling) 101-2
score **43-4**, 58, 171
Scotland
bread 216, 217, 218
Edinburgh (standards) 258, 259, 267, 269
public 87
ell 44, 80, **87-8**, 95
Herring Fishery Act (1815) 174-5
inch 80
mile 70, 71
Picts 31
rod 44
solid fuel 307
Stirling jug 252
Act of Union (1706) 165
Scots (tribe) 31
scripulum (scruple) 14, 117
scruple/*scripulum*
currency unit 14

and denarius 119
weight unit **118**, 253, 294, 297
abolished 304, 307
scruple, fluid 186
scutage 62-4
seam **149**
capacity measure **149-50**, 170
weight unit 135, 321
seeds (as dry goods) 161, 164
seeds (as weights) 2-5
see also grain
seignorage (royalty payment) 111
semis 15-16, 118
semodius 13
semuncia 16
sester 149
sestertius 15
sesuncia 16
Severus, L. Septimius, emperor 11
sextans 16
sextarius 12, 13, 149, 185
sextertium 317
sextula 16
shaft (scales) *see* auncel
shaftment 29-30
Shakespeare, John 222
sheep (as unit of value) 13
Sheepshanks, Revd R. (Commissions, 1838/43) 261, **265-7**
shekel 5
sherbet 64, 226
sheriff 112, 121, 122
shield weights 128, 129, 136-7, **141-2**, 233
shilling ('s' *or* '/') **101-6**, 194*n*, 300*n*
'decimalized' 264, **299**, 300
five shillings as primary 298
as primary 298
ten shillings 'decimalized' 300
ten shillings as primary 298-9
hundred of 321
of Mercia and Wessex 104-6
metric equivalent 103*n*
and score 43-4, 321
two-shilling coin *see* florin
as unit of account 103*n*
variant values 105
as weight unit 127, 194-200
wergeld 101-2
Shrewsbury, earl of (Wirksworth Dish) 177
Shuckburgh, Sir George **250-51**, 255, 258-9, 262, 264-6, 272
sicilicus 16
siliqua see carob
silks 127, 134
silver
axe-head ingots 20-21, *Fig 11*
currency 15, 103
fineness 110-13, 120
obsolescent 113
hundred 59
weighing of 3, 123
see also bullion
simnel bread 195-6, 203, 207, 313
Sissons, Jonathan 244, 249
SI units *see individual metric units*
slack (coal) 182
snuff 140-41
soap 174, **176-7**, 329
socage 120*n*
Society of Inspectors of Weights and Measures 338
sokeman 120
solidus **15-16**, 21
Constantine's pound xxiv, 119
Frankish 101, 105*n*
solidus (oblique stroke) 194*n*
Solonic (Attic-Euboic) system 5-6
sovereign (pound coin) ***106n*,** 300*n*
Spain (wool trade) 131, 135, 138
Valencia 117
span 2
speculum (alloy) 19
speed-limits 303, 306
spices 125-6, 127, 134, 136, 140, 154, 320
cinnamon 130*n*, 135, 320
nutmeg 130*n*, 135, 320
spice-bread 207
spicers' hundredweight 136
spillage 161, 162
spirits (distilled liquor) **187-8**, 256
see also brandy

spits (currency units) 6, **7-8**
see also obol
spoonful 160
ladleful 232
tablespoon 185
spring-balance 217
square measure *see* area; *and individual units*
stade *(stadion; stadium)*
and furlong **69-70**, 75
Greek 9-10
Roman 10, 11, 68, 69
standards (artefacts)/standardization 48, **232-72**
Act of 1824 255-60
Assize of Measures (1196) **90-91**
Board of Trade 271
Carysfort Committee **246-9**
Commission (1841) 261-4
Commission (1854) 264-70
comparative tests
French/British 244
pound 244, 245-6, 268
pound/kilogram **280-81**, 284, 291-2, 293
yard 244-5, 249-51, 255, 264-6
yard/half-toise 244
yard/metre **280**, 284, 291
Edgar 48
Exchequer *see* Exchequer
France 244, 280-81, 285
Hodgson Report 291-5
Imperial **254-72**
destroyed 260-61
restated (1963/85) 296-7, 307
restored after fire 261-71
International Kilogram 292, 296, *Fig 57*
International Metre 291
linear measure: survey scales 249-50
metric 286, 296
Middle Ages **232-7**, 323-31
museum collections 272
'natural origin' 253
nautical distance 76
19 cent **251-72**
NWML **337-8**
regulation/enforcement 313-22, **323-40**
sealing of 318
17/18 cent **243-51**
toise **346**
Tudor **237-43**
verification procedures 271, 280, 334, 337
weight: Troy-Avoirdupois vacillation 252-62
see also rules; weights; *and individual units*
Standerd, Edwarde 49, *Fig 22*
stater *(Philippus)* **5**, 15
Celtic 19-20
statera see steelyard
Statute of Northampton (1328) 91, 92, 94
Statutum de Admensuratione Terre (Statute for the Measuring of Land) (?1305) 322
Statutum de Pistoribus (Statute for Bakers and Brewers) (13 cent) **317-19**
bread 196-7, **199**
brewing 221
stricken/heaped measure 156
steel 127
steelyard *(romana; statera)* 21-2, 134, 137, 140, *Fig 15*
bread 217, *Fig 44*
Roman 21, *Fig 13*
weight for 233, 234, *Fig 16*
see also auncel
step (linear measure) 11
and fathom 2
see also pace; *passus*
Stephen, king 63
sterling (penny) 110*n*, 123, **124**
sterling (as standard of silver-purity) 110, 120
stike 321
stone (weights made of) 3, *fig 5*
stone (weight unit) **270**, 307
eight-pound 130*n*, 135, 320-21, **335-6**
fourteen-pound 131, 136, 137, 140
wool-weight 142, *Fig 33*

and sack 140
various weights 130-31, 135, 136
stones, precious 3
see also bullion
straw 270, 332
stricken measure *see* capacity
strip-farming 37-8, 49
sugar 127, 130*n*, 135, 136, 305, 320
sulung **56-7**, 63
see also hide
Sunday Times 216*n*
superficial measure *see* area
survey measures *see* land measure
Sutton Hoo (Suffolk) 100
Sweden (bismar) 21, *Fig 14*
Swift, Jonathan 193

tablespoon (capacity measure) 185
Tabula Codicis Bernensis 3, 118
Tabula Codicis Mutinensis Prioris 118
Tabulae Oribasianae 3
tale, by (accounting method) 5, 122
talent (weight/currency unit) 5-6
'greater' 119
Talleyrand (Charles Maurice de Talleyrand-Périgord) 344-5
tallow 127, 130, 206
Tarrage Survey 42, *Fig 19*
taxation
and hide 54, 55-6, 57, 59-61
scutage 62
see also Burghal Hidage
Taylors *see* Merchant Taylors
tea 64, **226**, 339
Technology and Power, Ministry of 302
temperature
and kilogram 353
and liquid-volume 165-6
and metre-establishment 347, 348, 351
and yard-accuracy 266, 267, 292
tertian *see* firkin
testoon 103*n*
Teutonic hundred 58
thegns 61
thermometer
bi-metallic 347
Réaumur 348*n*
Sheepshanks' 266
see also temperature
thousand 58
'long' 320
thread 127
thumb 1, 2, **79-80**
distinguished from inch 94
Roman 8, 11
as three barleycorns 3
see also inch
thurdendel 224-5, 230 (ref 71)
tierce 171
timber (as quantity-unit) 321
time-measures
candles 30, 79
'day' 79*n*
Times, The 303
tin 127, 136
hundred 59
speculum 19
'tin currency' 19-20
Titus, emperor 11
tobacco 140-41
tod 140, *Fig 32*
toise 86, **346**, 349
half-toise 244, 245
de Perou 346-8
ton (Avoirdupois) 139, 304, 307
coal 180*n*, 181
and tun 172
ton, metric *see* tonne
ton (of wine etc) *see* tun
tonne 297, 304, 307
tourte (bread) 195*n*
Tower pound **109-10**, 110-13, **122-5**, 126
and gallon 151-3, 158
Tower yard **243-5**, 246
town-planning 38-42
Tractatus de Ponderibus et Mensuris (Assize of Weights and Measures) (?1302/3) **123-5**, 129-130, **320-21**
capacity 151-2, 158
fish 173
hundredweight 135

pound 125-6, 129
variable standards 323-4
wool *etc* 130, 137-8
Trade, Board of 283, 335, 337, **339-40**
metrication 286, 355
Model Regulations 335
permanent Commission 294-5, 297
pound/kilogram comparison 293
Standards Department 337, **339-40**
standards deposited 269*n*, **271**, 296, *Fig 54*
Trade and Industry, Dept of 307, 337
treet (bread) 195-6
tremissis **15-16**, 101
Tresca, Henri 285, 349-51, *Fig 56*
tret/*cloffe*/cloff 136-8, 140, 141
'inverse' 140-41
Tribal Hidage 59, 61
triens 16
tron/*trone* 118, 138-9, **234**
king's 135
tronager 234
Troughton (instrument-maker) 257
scale 250, 265, 266
weights 251
Troughton & Simms company (founders) 267, 296
Troyes (and Troy weight) 118-19
Troy weight 4, **117-31**, 206
and Avoirdupois, confused 127-8, 129, 131, 154, 242
bread 125-7, **197**, 200, 202-9
and capacity measures 151-5, 158, 160
Apothecaries' measures 185-6
Carysfort Committee 247-8
Committee (1814) 252
currency 101, 108-13
decimal element 280
grain **118**, 119
as barleycorn 2, 4
obsolescent 207, 262, 271, 285
origin of name 118-19
Paris/London compared 244
pound abolished 271
standards *see* weights
and wine gallon 187
see also individual units
truss 332*n*
tun/ton (of wine) 163, **170-72**, 177, **257**
and ton weight 172
Tungri **28**, 43
Tuscany (weights) 132*n*
see also Florence

ulna (ell *or* yard) 80, 91
uncia (ounce *or* inch) 8, 16, 117
'unit' (for fish) 175
(for liquor) 188
United States of America (US) **188-9**, 290, 293-4
bushel 188
gallon 177, 187, **188-9**, 290
inch 293
kilogram 293
liquor measures 187
metrication 304
nautical mile 76
oil barrel 177
ounce, fluid 188-9
pint, dry/liquid 188
pound 293-4, 297
quart 188
ton 290
yard 293
urna 13
US/USA *see* United States
Utrecht, treaty of 22

Valencia (weights) 117
vat 181
see also faat
vegetables 305
Venice (weights) 117
Verus, C. Julius 20
Vespasian, emperor 12, 13
Vienna loaf 217, 218, 219
Vikings *see* Danes
vinegar 172
vinegar-beer (beer for vinegar production) 226
virgate *(virga; virgata)* 37, 55, 63
as hide 61

scutage 63
'Vitruvian Man' 2, *Fig 1*
volume *see* capacity
Vortigern (prince of south-east Britain) 31

wagon-load
coal 181
and wey 170
Wakefield Manor 325
Wales
amber 149
foot 28
Wallace (Master of Mint) 280
Warden of the Standards 260, 271
wastel (bread) **194-6**, 198, 199, 200, 204, 207, 313
water
density determined 251
and volumes 165-6, 179, 186, **188-9**, 354-5
established by weight 254-6, **258**, 271, 345-6
and weights 256, 262, 353-4
water (maritime) measure 178-80
wavelengths (and linear measure) 35, 291, **296**, **351-2**
wax 127, 130*n*, 135, 136, 320
weigh (weight unit) *see* wey
weight/mass measures **117-43**, **248-9**, **270**
Assize of Measures (1196) **90-91**
Avoirdupois/Troy vacillation 252-62
Commission (1841) 263
early 1, 2-5
mass/weight defined 293
metric permitted 284
'natural' 35
parallel systems (1862) 270
Roman 13-16
scales *see* scales
standards/weights (artefacts) *see* weights
see also Avoirdupois; Troy; *and individual units*
weights (artefacts) 1, **233-4**, **239-42**, **245-9**, **251-72**
auncel 233, 234
St. Bartholomew Fair 235
bread-weight 215, 217
Carysfort Committee 247-8
Celtic 18-19, 23
clove 129-30, 136-7, *Fig 29*
comparative tests *see* standards
early 3, *Fig 5*
Founders' standards 330-31
France 244
glass 3
grains 244, 245, 251
Imperial (Avoirdupois) **267-70**, 296, 307, *Fig 54*
inaccuracy 291-2
Imperial (Troy) **256-62**, 267-8
kilogram 296, 307
museum collections **272**
oxidation 268
'PC'/'PS'/'RS'/'Sp'/'T' 268-70, 307
pennyweights 244, 245
Plumbers' standards 329-30
pound *and* multiples/fractions 108-13, 233-4, 270, *Figs 48-51*
fifty-six pound 129, 233, 248, *Fig 28*
ninety-one pound 129, 142, 233-4, *Fig 28*
regulation/enforcement 313-22, **323-40**
Richard I standard 150
Roman 21, 132*n*, *Fig 12*
Royal Society 244, 268
shield-weights 128, 129, 136-7, **141-2**, 233
steelyard 22, 134, *Fig 16*
stone 3, *Fig 5*
Troy 120, 239, 241, 244-5, 251, **256**, *Fig 27*
sixteen-ounce pound 128, 241
weights lying/*couchants* 139
Winchester 129, 233, *Fig 28*
wool-weights 128, 129, 141-2, *Figs 31-3*
weights and measures
administration/control **323-40**
privilege 323-4

statutes *etc see* Parliament; *and individual monarchs*
Welsh foot 28
wergeld 100*n*, **101-2**, 104, 105, 106, 106*n*
wey/weigh (weight unit) 130, 132, 139, **170**, 320, 324
and chaldron 180*n*
wheat (as dry goods) 150, 157
for bread 194-220
corn laws 214
loaf-yield **185**, 194, 202, **204**, **210-13**, 216*n*
Middle Ages 194-205
for brewing 220-21
and gallon 151-4
and pint 154
weight of 151-60
wheat grains (as weight units) **2-4**, 108, 123, **124-5**, 127
and penny 151, 197
whisky quart 187
Whitehill (Oxon) 48-9, 60-61, *Fig 22*
wholemeal bread 194-6
'Widsith' (poem) 43-4, 102
Wihtred, king of Kent 103
William I the Conqueror
Battle Abbey 56
currency 120, 121, 125*n*
Domesday 60
knight's fee 62
Laws 105-6
William II Rufus 43
William III (and Mary)
ale 224
capacity 164-5, 179, 181, 183-4
US measures 188
corn laws 214
William IV 260
bread 215, 216
capacity 180, 182
inspectors 332
William de Turnemire 110*n*, 111
Winchester (city) **46-8**
acre's-breadths 38-9
capacity measures 160, 161, 165, *Fig 38*
currency bars 17, *Fig 10*
Edgar's standards 96
Godbegot (Goodbegot) 39, **41-3**, 61, 82, *Figs 19, 20*
hidage 39, **40-41**
High Street 41, 43
Lankhills 31
Old Minster 47-8, 81
population, medieval 31
rod 39, 46-7
Roman withdrawal 31
rulers, Roman 10
St. George Street 41
Saxon capital 39, 96
weights 129, 142, **233-4**, 239-40, *Fig 28*
Wolvesey Palace 81
yard **238**, 239, *Fig 46*
Winchester measure 159, **164-6, 260**
bushel 164-6, 181-4, 252, 272
gallon 160, 165
pint 178
quart 166
wine 157, **170-73**, 193
assize 316, 317
barrels 170-73, 177
bottle-sizes 187-8
gallon 152-5, **158-64**, *Fig 37*
Imperial measure 256-7
museum collections 272
reputed quart 187-8
Richard I standard 90, 150
sester 149
weight of 123, 151-5, 158-60
wineglassful 185
Winnipeg Free Press 308
Wirksworth Dish 177-8, *Fig 39*
wist 55
Wollaston, Dr W.H. 251, 253
Wood, Sir Charles, MP 281
wood (weighing of) 127
woodland perch **45-6**, 48
wool 128, 129-30, **130-32**, **135-41**, **270**
burial in 142-3
customs duty 131
trade 131-3, 142-3
weights for 128, 129, **141-2**, 320,

Figs 28, 31-3
Woolsack 131
Worcester (weights) 239-40
Worcestre, William 71, 72-3, 75, 84, *Fig 23*
Wordsworth, William 309
wormwood 127
Worshipful Company of Clockmakers 243, 244-5
Wrottesley, Sir John (*later* Lord), MP 280

Xenophanes of Colophon 7

Yale University 120, *Fig 27*
yard ('yd') **83-5**, 89, 96, **256**
'and a hand' 87
'and an inch' 87-8, 92-4
'and a nail' 93
cloth measure 87-9, 90-94
cubic 289*n*, 306
'gyrd' 42
king's *see* rules
as measuring instrument *see* rules
Merchant Taylors' *see* rules
metric equivalent 292, 294, **296**
and perch 69-70
post-metrication 306
as rood or virgate 37
square 306
as three feet 81
yard-land 55
yardsticks *see* rules
'yd' *see* yard
ynce (inch) 79
yoke *(jugum)* 56
Young, T. 253